Gastroenterology Dept
St Richards Hospital
Chichester
Aug '89

CAMPYLOBACTER PYLORI AND GASTRODUODENAL DISEASE

Campylobacter pylori and gastroduodenal disease

EDITED BY

B.J. RATHBONE MRCP
Senior Medical Registrar,
St James's University Hospital,
Leeds

R.V. HEATLEY MD, MRCP
Senior Lecturer in Medicine,
University of Leeds;
Consultant Physician,
St James's University Hospital,
Leeds

SUPPORTED BY
AN EDUCATIONAL GRANT FROM
GIST-BROCADES N.V.
PHARMACEUTICAL DIVISION
DELFT, THE NETHERLANDS

BLACKWELL SCIENTIFIC PUBLICATIONS
OXFORD LONDON
EDINBURGH BOSTON MELBOURNE

Editorial offices:
Osney Mead, Oxford OX2 0EL
(*Orders*: Tel: 0865 240201)
8 John Street, London WC1N 2ES
23 Ainslie Place, Edinburgh EH3 6AJ
3 Cambridge Center, Suite 208
Cambridge, Massachusetts 02142, USA
107 Barry Street, Carlton
Victoria 3053, Australia

First published 1989

Set by the Alden Press, London
Printed in Great Britain
at the Alden Press, Oxford,
and bound by the Green Street Bindery,
Oxford

DISTRIBUTORS
USA
Year Book Medical Publishers
200 North LaSalle Street
Chicago, Illinois 60601
(*Orders*: Tel: (312) 726-9733)

Canada
The C.V. Mosby Company
5240 Finch Avenue East
Scarborough, Ontario
(*Orders*: Tel: 416 298-1588)

Australia
Blackwell Scientific Publications
(Australia) Pty Ltd
107 Barry Street
Carlton, Victoria 3053
(*Orders*: Tel: (03) 347-0300)

British Library
Cataloguing in Publication Data

Campylobacter pylori and gastroduodenal disease.
1. Man. Intestines. Pathogens: Campylobacters
I. Rathbone, B.J. II. Heatley, R.V. (Richard Val)
616.3′4014

ISBN 0-632-01778-3

Contents

List of contributors

A.T.R. AXON, MD, FRCP, *Consultant Physician, Dept. of Gastroenterology, Leeds General Infirmary, Leeds LS1 3EX, UK.*

G.D. BELL, MD, MSc, FRCP, *Consultant Physician, Dept. of Medicine, The Ipswich Hospital, Heath Road, Ipswich, Suffolk, UK.*

M.J. BLASER, MD, *Chief, Infectious Disease Section, Veterans Administration Medical Center, University of Colorado School of Medicine, Denver, Colorado 80220, USA.*

T. BOHANE, MBBS, FRACP, *School of Microbiology, University of New South Wales and The Prince of Wales Children's Hospital, Randwick, Australia.*

M. BORROMEO, *Monash University Dept. of Medicine, Prince Henry's Hospital, Melbourne 3001, Australia.*

J.G. COGHLAN, *Meath/Adelaide Hospitals, Trinity College, Dublin 8, Ireland.*

P. CORREA, *Professor of Pathology, Louisiana State University Medical Center, 1901 Perdido Street, New Orleans, LA 70112-1393, USA.*

A. CURRY, BSc, PhD, *Principal Microbiologist, Public Health Laboratory, Withington Hospital, Manchester M20 8LR, UK.*

M.F. DIXON, MD, FRCPath, *Senior Lecturer and Consultant, Dept. of Pathology, University of Leeds, Leeds LS2 9JT, UK.*

B.S. DRASAR, PhD, DSc, FRCPath, *Reader in Bacteriology, London School of Hygiene and Tropical Medicine, Keppel Street, London WC1E 7HT, UK.*

B. DWYER, MB BS, FRACP, FRCPA, *Director, Clinical Pathology Laboratory, Fairfield Hospital, Melbourne, Australia.*

D.Y. GRAHAM, MD, *Professor of Medicine and Virology, Baylor College of Medicine, and Chief, GI Section, Digestive Disease, VA Medical Center, Houston, Texas 77030, USA.*

S.F. GRAY, BSc, FIMLS, *Senior MLSO, Dept. of Pathology, Leeds General Infirmary, Leeds LS1 3EX, UK.*

R.V. HEATLEY, MD, MRCP, *Senior Lecturer in Medicine, University of Leeds and Consultant Physician, St James's University Hospital, Leeds LS9 7TF, UK.*

A.M. HIRSCHL, MD, *Hygiene Institute, University of Vienna, Kinderspitalgasse 15, A-1095, Vienna, Austria.*

M.J. HUDSON, *Pathology Division, PHLS Centre for Applied Microbiology and Research, Porton Down, Salisbury, Wilts SP4 0JG, UK.*

R.H. HUNT, FRCP, FRCPC, FRCP(Edin), FACG, *Professor, Dept. of Medicine, and Head, Division of Gastroenterology, McMaster University, Hamilton, Ontario L8N 3Z5, Canada.*

D.M. JONES, MD, FRCPath, *Director, Public Health Laboratory, Withington Hospital, Manchester M20 8LR, UK.*

J. KALDOR, MAppSci, *Senior Research Scientist, Clinical Pathology Laboratory, Melbourne, Australia.*

P.D. KLEIN, PhD, *Professor of Pediatrics and Medicine, Baylor College of Medicine, and Director, Stable Isotope Laboratory, USDA/ARS Children's Nutrition Research Laboratory, Houston, Texas 77030, USA.*

A. LEE, PhD, *Associate Professor in Microbiology, University of New South Wales, Kensington, Australia.*

J.R. LAMBERT, MMed BS, PhD, *Senior Lecturer, Monash University Dept. of Medicine, Prince Henry's Hospital, Melbourne 3004, Australia.*

J.M. LITTLEWOOD, MD, FRCPE, DCH, *Consultant Paediatrician, St James's University Hospital, Beckett Street, Leeds LS9 7TF, UK.*

C.A.M. McNULTY, MRCPath, *Senior Registrar in Medical Microbiology, Gloucester Public Health Laboratory, Gloucestershire Royal Hospital, Great Western Road, Gloucester GL1 3NN, UK.*

M.J. MAHONY, MB, MRCPI, MRCP, DCH, *Senior Registrar in Paediatrics, St James's University Hospital, Beckett Street, Leeds LS9 7TF, UK.*

F. MEGRAUD, MD, *Bacteriology Laboratory, Children's Hospital and University of Bordeaux, Bordeaux, France.*

H.M. MITCHELL, BSc, *School of Microbiology, University of New South Wales, Kensington, Australia.*

A. MORRIS, BSc, MBChB, *Microbiology Registrar, Auckland Hospital, Auckland 1, New Zealand.*

D.G. NEWELL, *Pathology Division, PHLS Centre for Applied Microbiology and Research, Salisbury, Wilts SP4 0JG, UK.*

G. NICHOLSON, FRCP, FRACP, *Gastroenterologist, Auckland Hospital, Auckland 1, New Zealand.*

H.J. O'CONNOR, MD, MRCPI, *Senior Medical Registrar, Liver Unit, Queen Elizabeth Hospital, Birmingham B15 2TH, UK.*

C. O'MORAIN, *Consultant Gastroenterologist, Meath/Adelaide Hospitals, Trinity College, Dublin 8, Ireland.*

R.J. OWEN, MPhil, PhD, *Deputy Curator, National Collection of Type Cultures, Central Public Health Laboratories, Colindale Avenue, London NW9 5HT, UK.*

M. PLETSCHETTE, *Hygiene Institute, University of Vienna, Kinderspitalgasse 15, A-1095 Vienna, Austria.*

K. RAIOS, BSc, *Scientist, Clinical Pathology Laboratory, Fairfield Hospital, Melbourne, Australia.*

B.J. RATHBONE, MRCP, *Senior Medical Registrar, St James's University Hospital, Leeds LS9 7TF, UK.*

E.A.J. RAUWS, *Dept. of Gastroenterology and Hepatology, Academic Medical Center Amsterdam, Meibergdreef 9, 1105 AZ Amsterdam, The Netherlands.*

B. RUIZ, MD, *Dept. of Pathology, Louisiana State University Medical Center, 1901 Perdido Street, New Orleans, Louisiana 70112, USA.*

T.M. SHALLCROSS, *Dept. of Medicine, St James's Hospital, Leeds, LS9 7TF, UK.*

A.R. STACEY, *Pathology Division, PHLS Centre for Applied Microbiology and Research, Porton Down, Salisbury, UK.*

H.W. STEER, BSc, PhD, FRCS, *Dept. of General Surgery, Southampton General Hospital, Southampton, UK.*

C. STEWART GOODWIN, MD, FRCPath, FRCPA, *Head of Dept. of Microbiology, Royal Perth Hospital, Western Australia, and Associate Professor in Clinical Microbiology, University of Western Australia, Australia.*

WEE TEE, MSc, *Specialist Microbiologist, Campylobacter Laboratory, Fairfield Hospital, Melbourne, Australia.*

A. TOBIN, *Sir Patrick Dun's Research Unit, St James's Hospital, James's Street, Dublin 8, Ireland.*

D.S. TOMPKINS, MB, ChB, MRCPath, *Consultant Microbiologist, Bradford Royal Infirmary, Duckworth Lane, Bradford, BD9 6RJ, UK.*

G.N. TYTGAT, *Dept. of Gastroenterology and Hepatology, Academic Medical Center Amsterdam, Meibergdreef 9, 1105 AZ Amsterdam, The Netherlands.*

J. WEIL, MB, MRCP, *Registrar, Dept. of Medicine, The Ipswich Hospital, Heath Road, Ipswich, Suffolk, UK.*

J.I. WYATT, MB ChB, MRCPath, *Consultant Histopathologist, St James's University Hospital, Leeds LS9 7TF, UK.*

Preface

For many years interest has focused on biochemical and humoral factors in the maintenance of gastroduodenal mucosal defences. It is, therefore, somewhat ironic that we should now find our attention concentrated on a single microorganism as having a potential leading role in the pathogenesis of peptic ulcer disease and other possibly related conditions. It is also perhaps surprising that one agent should leap from apparent obscurity to such a position of prominence in as short a time as a few years. It is clear, however, from the literature, that microorganisms have attracted attention in the pathogenesis of ulcer disease periodically over the past hundred or so years. Although none has attracted so much attention as *Campylobacter pylori*, suspicions appear to have lingered that a microbiological component may be relevant in the aetiology of peptic ulcers.

Considerable research work has been carried out recently on *C. pylori* and discoveries are constantly forthcoming. In compiling this text we have been fortunate to enlist the support of this group of leading international experts in the field. We have confined contributions to authors' own and other published work and have deliberately excluded all unpublished references, except those relating to current work from authors' and collaborators' own studies, which is readily identifiable. Our aim has been to provide a comprehensive and up-to-date account of this field which will, without doubt, continue to develop in the future.

The discovery of *C. pylori* has stimulated work in many different directions. Not only is this organism of interest in relation to gastroenterology and microbiology, but also may provide a tool to understand more about host–bacteria interactions, the classification of this type of bacteria and the biochemical ecology of bacteria within the gut. Immunologists, biochemists, physicists, pharmacologists and many other specialties are now finding their work influenced by these discoveries and we hope this text will be of interest to, and stimulate research work within, a multi-disciplinary field.

B.J.R.
R.V.H.
Leeds, 1989

1 The historical associations between bacteria and peptic-ulcer disease

B.J. RATHBONE AND R.V. HEATLEY

Bizzozero [1] in 1893 has been credited with the first demonstration of bacterial colonization of mammalian stomachs with spiral organisms. Studying the microscopic appearance of the gastrointestinal epithelium of various animals he noted the presence of spirochaetes in the gastric glands and parietal cells of six dogs he examined. Salomon [2] in 1896 confirmed the presence of spiral organisms in the dog and reported similar organisms in the gastric mucosa of cats and rats. He also examined human tissue, apes, monkeys, cattle, pigs and mice without finding similar organisms. Balfour [3] demonstrated spirochaetes in gastric and intestinal ulcers of dogs and monkeys. Lucet [4] observed similar organisms in a dog suffering from haemorrhagic gastroenteritis. The transmission of spiral organism from dog or cat to mouse was demonstrated by Salomon [2] and Kasai & Kobayashi [5].

Human gastric spirochaetes were first reported in necrotic material at the surface of ulcerating carcinomas and in gastric secretions by Krienitz [6] and Luger [7]. Although little attention was paid to the gastric spirals at this time, considerable interest and research was directed to the possible role of non-spiral bacteria in peptic ulceration. In 1888, Letulle [8] reported that the oral or parenteral administration of *Staphylococcus pyogenes* to guinea-pigs resulted in gastric ulcers. Similar results were reported with other bacteria including dysentery organisms, pyogenic streptococci and lactobacilli. The implication from these reports was that infection may be an occasional or accessory cause of ulceration. The subsequent work mainly concentrated on two organisms, a colonic bacillus and a specific strain of streptococcus.

Whereas the above work reported acute lesions, Turck [9] claimed that chronic ulcers could be produced in 100% of dogs after they had been fed colonic bacilli daily for several months. The photographs illustrating these lesions, however, showed only non-specific gastric and duodenal changes [10]. Completely negative results were obtained when these experiments were repeated [11]. When rabbits were injected with the bacillus, gastric and duodenal lesions occurred within 24 hours but histological examination demonstrated them to have an embolic aetiology. The role of streptococci in ulceration was championed by Rosenow [12–15], who demonstrated that streptococci isolated from humans, especially those with ulceration, produced gastric ulcers when injected into laboratory animals. Rosenow maintained that the streptococcus initiating gastric ulceration had a selective affinity for gastric mucosa where it produced local destruction of the glandular tissue. This was then digested by the acid gastric juice. He postulated that an acute lesion so produced may be made chronic by the constant discharge into the blood of the specific organisms lurking in distant foci. That injection of streptococci produced ulcers and organisms isolated from ulcer patients appear to selectively localize in the stomach when injected, was confirmed by other workers [16–18]. In 1930 Saunders [19] isolated an α-streptococcus from human peptic ulcers and demonstrated specific agglutinins to this organism which were present in higher titres in ulcer patients compared to controls with other streptococcal infection. Attempts to produce ulcers in animals by injection of the specific organism, however, were unsuccessful. Saunders suggested that the lesions produced by Rosenow were due to an anaphylactic shock-type reaction, due to foreign protein present in the broths injected into the animals. Celler & Thalhimer [20] isolated a non-haemolytic streptococcus from human ulcers and like Saunders [19] failed to produce ulcers when they injected it into animals. Streptococci can thus be isolated from many ulcers in the stomach and duodenum and when injected into animals have often been associated with acute lesions of the gastric mucosa. These short-lived lesions, however, bear little relation to the ulcer crater seen in human peptic-ulcer disease.

No further work was published concerning human gastric spiral organisms until 1939 when Doenges [21] in a histological study of 242 autopsy stomachs, using haematoxylin & eosin staining of sections, found spirochaetes present in 43% of cases. The organisms occurred predominantly in the glandular lumen but were seen also in the parietal cells. Autolysis made detailed assessment of the gastric histology impossible. To avoid this problem Freedberg & Barron [22] studied gastric resection specimens using haematoxylin & eosin, and a silver stain, which they found best for identifying the bacteria. Overall, 37.1% of their 35 patients were positive for spirochaetes. However, in many cases bacteria were seen only after examining multiple sections. The organisms were mostly associated with gastric ulcers and gastric carcinoma. In 13 duodenal ulcer patients with associated chronic gastritis, only two were positive. In an attempt to confirm the presence of spirochaetes in gastric mucosa, Palmer [23] studied 1180 suction biopsies (mostly from the gastric body) from 1000 patients using haematoxylin & eosin stains. This author reported no spirochaetes nor any structure which could reasonably be of a spirochaetal nature.

An indirect clue to the possible presence of gastric epithelial bacteria came in 1924 when Luck & Seth [24] described the presence of considerable urease activity in the stomach. That the urease might be bacterial in origin was suggested in 1959 by Lieber & Lefevre [25] who demonstrated that the hypoacidity found in many patients with uraemia could be reversed with antibiotic therapy. This view was opposed by Mossberg *et al.* [26]. However, studies using conventional and germ-free animals confirmed that the gastric urease activity in animals was bacterial in origin [27]. No link was made between this urease activity and the presence of gastric spiral bacteria until 1984 [28].

Following the observations by Palmer in 1954 [23], no further reports of gastric spiral bacteria occurred until 1975, when bacteria were discribed on the lumenal surface of epithelial cells of gastric-ulcer patients [29, 30]. The bacteria were seen to have at least one filum and to be deep to the mucus layer, with the gastric epithelial cells having a diminished mucus content compared to those from a normal stomach. Electron micrographs illustrated phagocytosed bacteria within polymorphonuclear leukocytes and the absence of the bacteria from areas of intestinal metaplasia was noted. Attempted culture of gastric biopsies resulted in the growth of *Pseudomonas aeruginosa* which was almost certainly a contaminant from the endoscope.

Fung *et al.* [31] in 1979, correlating endoscopic, histological and ultrastructural appearances of chronic gastritis, described and illustrated bacteria in the crevices between surface mucus cells and in gastric pits. It was noted that many of these bacteria abutted directly onto the plasmalemma of the epithelial cell, but were never seen within the cell and were thus assumed to be of little significance.

Working in the same institution, Warren noted that the majority of endoscopic biopsies from patients with chronic gastritis and peptic ulceration were colonized with curved campylobacter-like organisms [32]. These organisms were poorly stained by haematoxylin & eosin, but seen easily with the Warthin–Starry silver stain. A prospective study was set up and attempts made to culture the organisms from gastric biopsies using non-selective and standard campylobacter media for 48 hours. No growth was observed until the 35th biopsy was incubated during an Easter holiday and was thus examined after 5 days' incubation. A heavy growth of campylobacter-like organisms was found on the non-selective media. When other biopsies were also incubated for 3–4 days, a similar growth of Gram-negative bacteria were seen [33–35].

Independently, Rollason *et al.* [36], in a retrospective study of endoscopic gastric biopsies, described spiral bacteria in association with chronic superficial and chronic atrophic gastritis. Steer, continuing his morphological studies of gastroduodenal mucosa in peptic ulceration, described the scanning electron microscopy appearance of bacteria in patients with ulcer-associated chronic gastritis [37]. The three groups of workers thus independently recognized and reported the association of the gastric spiral bacteria with chronic gastritis.

Thus, although the recent interest in *Campylobacter pylori* (*C. pylori*) has focused attention on bacteria and the gastroduodenal mucosa, research into these associations has been carried out over much of the past 100 years [38]. However, the role of bacteria in gastroduodenal disease has never been noted to be as close as that between *C. pylori* and active chronic gastritis.

References

1 Bizzozero, G. Über die schlauchformigen Drüsen des Magendarmkanals und die Bezienhungren ihres Epithels zu dem Oberflächenepithel der Schleimhaut. *Arch. f. mikr. Anat.* 1893, **42**, 82–152.

2 Salomon, H. Über das spirillum des Saugetiermagens und sein Verhalten zu den Belegzellen. *Centralbl. f. Bakt.* 1896, **19**, 433–42.

3 Balfour, A. A haemogregarine of mammals and some notes on trypanosomiasis in the Anglo-Egyptian Sudan. *J. Trop. Med.* 1906, **9**, 81–92.

4 Lucet, M. Sur la présence de spirochetes dans un case de gastro-enterite haémorrhagique chez le chien. *Bull. Soc. centre de med. vet. Par.* 1910, **64**, 376–9.

5. Kasai, K. & Kobayashi, R. The stomach spirochete occurring in mammals *J. Parasitol.* 1919, **6**, 1–10.

6 Krienitz, W. Über das Auftreten von Mageninhalt bei Carcinoma Ventriculi. *Dtsch. med. Wochenschr.* 1906, **22**, 872.

7 Luger, A. Über Spirochäten und fusiforme Bazillen im darm, mit einem Beitrag zur Frage der lamblienenteritis. *Wein. klin. Wochnschr.* 1917, **52**, 1643–7.

8 Letulle, M. Origine infectieuse de certains ulceres simples de l'estomac ou du duodenum. *Soc. Med. D. Hôp. de Paris* 1888, **5**, 360–87.

9 Turck, F.B. Ulcer of the stomach: pathogenesis and pathology. Experiments in producing artificial gastric ulcer and genuine induced peptic ulcer. *J.A.M.A.* 1906, **46**, 1753–63.

10 Ivy, A.C., Grossman, M.I. & Bachrach, W.H. 1950. *Peptic Ulcer*, 1st edn. The Blakiston Company, Philadelphia, p. 267.

11 Gibelli, C. Contributo critico sperimentale alleziologia dellulcera gastrica in rapporto coi traumi. *Arch. Internat. de Chir.* 1908–10, **4**, 127–71.

12 Rosenow, E.C. The production of ulcer of the stomach by injection of streptococci. *J.A.M.A.* 1913, **61**, 1947–50.

13 Rosenow, E.C. The causation of gastric and duodenal ulcer by streptococci. *J. Infect. Dis.* 1916, **19**, 333–63.

14 Rosenow, E.C. Etiology of spontaneous ulcer of stomach in domestic animals. *J. Infect. Dis.* 1923, **32**, 384–99.

15 Rosenow, E.C. The specificity of the *Streptococcus* of gastroduodenal ulcer and certain factors determining its localization. *J. Infect. Dis.* 1923, **33**, 248–68.

16 Hardt, L.L.J. Contributions to the physiology of the stomach: XXXIII. The secretion of gastric juice in cases of gastric and duodenal ulcers. *Am. J. Physiol.* 1916, **40**, 314–31.

17 Haden, R.L. The elective localization of bacteria in peptic ulcer. *Arch. Int. Med.* 1925, **35**, 457–71.

18 Nickel, A.C. & Hufford, A.R. Elective localization of streptococci isolated from cases of peptic ulcer. *Arch. Int. Med.* 1928, **41**, 210–30.

19 Saunders, E.W. The serologic and etiologic specificity of the alpha-*streptococcus* of gastric ulcer. *Arch. Int. Med.* 1930, **45**, 347–82.

20 Celler, H.L. & Thalhimer, W. Experimental studies of the etiology of gastric and duodenal ulcer. *Med. Record* 1916, **90**, 389.

21 Doenges, J.L. Spirochaetes in the gastric glands of *Macacus rhesus* and of man without related disease. *Arch. Path.* 1939, **27**, 469–77.
22 Freedberg, A.S. & Barron, L.E. The presence of spirochaetes in human gastric mucosa. *Am. J. Dig. Dis.* 1940, **7**, 443–5.
23 Palmer, E.D. Investigation of the gastric mucosa spirochetes of the human. *Gastroenterology* 1954, **27**, 218–20.
24 Luck, J.M. & Seth, T.N. Gastric urease. *Biochem. J.* 1924, **18**, 1227–31.
25 Lieber, C.S. & Lefevre, A. Ammonia as a source of gastric hypoacidity in patients with uremia. *J. Clin. Invest.* 1959, **38**, 1271–7.
26 Mossberg, S.M., Thayer, W.R. & Spiro, H.M. Azotemia and gastric acidity: the effect of intravenous urea on gastric acid and gastric ammonium production in man. *J. Lab. Clin. Med.* 1963, **61**, 469–75.
27 Delluva, A.M., Markley, K. & Davies, R.E. The absence of gastric urease in germ-free animals. *Biochim. Biophys. Acta* 1968, **151**, 646–50.
28 Langenberg, M-L., Tytgat, G.N.J., Schipper, M.E.I., Rietra, P.J.G.M. & Zanen, H.C. *Campylobacter*-like organisms in the stomach of patients and healthy individuals. *Lancet* 1984, **i**, 1348.
29 Steer, H.W. & Colin-Jones, D.G. Mucosal changes in gastric ulceration and their response to carbenoxolone sodium. *Gut* 1975, **16**, 590–7.
30 Steer, H.W. Ultrastructure of cell migration through the gastric epithelium and its relationship to bacteria. *J. Clin. Pathol.* 1975, **28**, 639–46.
31 Fung, W.P., Papadimitriou, J.M. & Matz, L.R. Endoscopic, histological and ultrastructural correlations in chronic gastritis. *Am. J. Gastroenterol.* 1979, **71**, 269–79.
32 Warren, J.R. Unidentified curved bacilli on gastric epithelium in active chronic gastritis. *Lancet* 1983, **i**, 1273.
33 Marshall, B. Unidentified curved bacilli on gastric epithelium in active chronic gastritis. *Lancet* 1983, **i**, 1273–5.
34 Marshall, B.J. & Warren, J.R. Unidentified curved bacilli in the stomach of patients with gastritis and peptic ulceration. *Lancet* 1984, **i**, 1311–15.
35 Marshall, B.J., Royce, H., Annear, D.I., Goodwin, C.S., Pearman, J.W., Warren, J.R. & Armstrong, J.A. Original isolation of *Campylobacter pyloridis* from human gastric mucosa. *Microbios Lett.* 1984, **25**, 83–8.
36 Rollason, T.P., Stone, J. & Rhodes, J.M. Spiral organisms in endoscopic biopsies of the human stomach. *J. Clin. Pathol.* 1984, **37**, 23–6.
37 Steer, H.W. Surface morphology of the gastroduodenal mucosa in duodenal ulceration. *Gut* 1984, **25**, 1203–10.
38 Rathbone, B.J., Wyatt, J.I. & Heatley, R.V. 1988 Bacteria and gastroduodenal inflammation. In Rees, W.D.W. *Advances in Peptic Ulcer Pathogenesis.* MTP Press, Lancaster, pp. 101–19.

2 The bacterial flora of the stomach

B.S. DRASAR

The bacteria present in the stomach and their role in disease has excited interest since the earliest days of microbiology. Among the first bacteria to be described was an organism present in the stomach contents [1]. This organism, *Sarcina ventriculi*, is still recognized as a valid species. Indeed many of the earlier and neglected studies foreshadowed many recent discoveries [2, 3].

Investigations of the bacteria found in the gastro-intestinal tract have focused on the examination of their metabolic impact. Major themes include studies of the role of the small intestinal flora in the blind or stagnant loop syndrome [4] and of the large bowel flora in large bowel cancer. Much of the information about the gut bacteria is reviewed and the development of understanding traced in a series of books [5–9].

The assumption has been made, since the successful investigation of the stagnant loop syndrome, that the bacterial flora of the stomach and small intestine are well understood. However, problems do remain. Investigations have focused on the possible risks of gastric cancer resulting from the effects of anti-secretagogues on bacterial colonization of the stomach [10, 11]. The discovery of *Campylobacter pylori* has awakened fresh interest in the bacteriology of the stomach and there are good reasons to believe that the subject is ripe for reassessment.

Methodological considerations

The stomach is not readily accessible, particularly in normal people. Furthermore, while gastric contents may be available from volunteers, gastric mucosa is less readily obtained. Most studies on the gastric microflora of normal subjects have concentrated on gastric contents. Even if the problems related to obtaining specimens are overcome or ignored, the bacteriological difficulties remain.

It has been estimated that the human intestine may contain some 400–500 distinct species of bacteria [12]. There is no way to be sure that the methods used will detect all the bacteria present. As the emergence of *Campylobacter* as a significant pathogen illustrates, until an adequate detection system is available the importance of individual observations cannot be delineated.

There is at present no simple means of isolating and identifying all the bacteria that may be present in a sample. To attempt this task would be beyond the means of most microbiology departments. Some compromise must be made either by restricting the study to the search for particular bacterial species or by using a system that will enable the nature of colonization to be categorized. For ecological studies an attempt to isolate aerobic, facultative, microaerophilic and anaerobic bacteria would be expected at the present state of knowledge. The tables in this chapter list those bacterial groups that would be expected to be detected by adequate bacteriology. In the clinical context, the categorization of samples on the basis of the type of flora present seems to be of value. The basic categories used are (i) the salivary/nasal or oral/respiratory type-I colonization and (ii) the faecal type-II colonization. Each of these broad groups may be further subdivided [13, 14].

Type I is characterized by the presence of oral streptococci, lactobacilli, fusobacteria, neisseria, staphylococci and other oral and nasal organisms. In practical terms such a flora is most readily detected by the use of sucrose agar to identify *Streptococcus salivarius*. Very few if any of the bacteria in type-I colonization will grow in the presence of bile salts.

Type II is characterized by the presence of coliforms (*Enterobacteriaceae*), faecal streptococci and bile-salt-resistant anaerobes including bacteroides and clostridia. The separation of the two types of colonization is not absolute. Type-I colonization may result from the pooling of saliva in the stomach, while a mixture of salivary and faecal organisms is often isolated.

As a response to these problems attempts have been made to use indirect measures of bacterial activity in the stomach and small intestine to detect bacterial activity. The levels of hydrogen in the end-expired air following a carbohydrate-rich meal have been used to assess bacterial activity in the stomach of normal people, patients who have undergone Billroth II partial gastrectomy and patients with pentagastrin-fast achlorhydria [15]. This technique has also been used to examine the effect of medication on gastric colonization [16].

The gastric flora

Normal people

Bacteria can be isolated from the stomach contents of fasting people (Table 2.1). The stomach functions as a reservoir that controls the entry of materials into the small intestine. The bacterial content of material entering the small intestine is controlled by antibacterial mechanisms in the stomach. Similarly, these same mechanisms alter the bacteriology of stomach contents. Acid stomach contents are usually sterile and the germicidal action of gastric juice is considered to be due mainly to free hydrochloric acid. The stomach sampled at operation has been found to contain num-

Table 2.1 Bacterial flora of gastric contents from normal people, based on country. Log_{10} bacteria per gram wet weight

	UK		Sweden	USA	Japan	India
Number of patients	25	5	10	7	N	12
Reference	17	18	19	20	21	22
Bacterial group						
Bacteroides	–	–	0.3	–	–	6.0
Eubacteria	–	–	–	–	–	–
Peptococcaceae	–	–	–	–	–	–
Bifidobacteria	–	–	–	–	–	–
Lactobacilli	–	3*	+	4	3.5	
Clostridia	–	–	–	–	–	–
Fusobacteria	–	–	–	–	–	6.5
Veillonella	–	2.0*	0.6	–	2.5	6.0
Enterobacteriaceae	–	–	–	–	–	4.5
Streptococci	–	3.3*	0.7	+	5	7.0
Staphylococci	–	3.2*	–	+	–	–
Micrococci	–	2.6*	1.3	–	–	–
Neisseria	–	–	0.3	–	–	–
Total count	<2	<3	1	1	N	8.0

*Not present in all patients.
+ = Present.
N = No data.

Table 2.2 Bacterial flora of gastric contents from achlorhydric patients. Log_{10} bacteria per gram wet weight

	Achlorhydria		Pernicious anaemia	Hypogammaglobulinaemia
Number of patients	5	16	7	6
Reference	17	15	18	18
Bacterial group				
Bacteroides	2.6*	–	–	4.7*
Eubacteria	–	–	–	–
Peptococcaceae	–	–	6.6*	4.7*
Bifidobacteria	+	–	5.9*	7.3*
Lactobacilli	+	–	4.6*	3.8*
Clostridia	–	–	3.0*	–
Fusobacteria	2*	–	–	–
Veillonella	+	–	4.7*	6.1*
Enterobacteriaceae	3.8	+	4.2*	4.1*
Streptococci	3.8	–	7.1	6.6*
Staphylococci	–	–	5.8*	4.3*
Micrococci	–	–	5.0*	6.3*
Neisseria	+	–	–	5.0
Total count	6.0	6.2	7.1	6.0

*Not present in all patients.
+ = Present.

bers of salivary organisms probably because anaesthesia and premedication depresses gastric secretion. Immediately after a test meal, a count of around 10^5 bacteria per ml of gastric juice can be recorded. The bacteria include streptococci, enterobacteria, bacteroides and bifidobacteria derived from the mouth and from the meal. As the pH falls the bacterial count declines, and relatively few bacteria are grown after a pH of about 3 has been attained.

The results from the various studies undertaken in developed countries are remarkably uniform; only in normal people in south India has there been evidence of permanent colonization of the stomach [22].

Achlorhydric patients

Achlorhydria of whatever aetiology seems to result in colonization of the stomach with a range of bacteria. This colonization does not conform to a strict pattern and the type and numbers of bacteria present have been shown to be different in the various series (Table 2.2). However, total bacterial contents of 10^6 per gram of stomach contents are common and higher numbers have been reported. The existence of this flora in the stomach has been known for many years [23] and its significance is subject to reassessment in the light of advances in our understanding of various disease states [24–27]. These studies emphasize the importance of gastric pH in control of colonization.

Gastric surgery

Surgery for the treatment of peptic ulcers is designed to reduce or eliminate gastric acid secretion. Gastric surgery for other reasons also reduces acid secretion. Thus, colonization of the stomach is to be expected (Table 2.3). Polya and Billroth II gastrectomies are more likely to result in type-II (faecal) colonization than Billroth I gastrectomy or vagotomy and pyloroplasty but in patients with a gastro-

Table 2.3 Bacterial flora of gastric contents from patients who have undergone gastric surgery. Log_{10} bacteria per gram wet weight

	Polya gastrectomy	Billroth II gastrectomy	Gastric resection
Number of patients	4	10	10
Reference	28	15	19
Bacterial group			
Bacteroides	–	–	+*
Eubacteria	–	–	–
Peptococcaceae	–	–	–
Bifidobacteria	6.0	–	–
Lactobacilli	4.0	N	+*
Clostridia	3.0*	–	–
Fusobacteria	–	–	+*
Veillonella	–	–	+*
Enterobacteriaceae	6.0	+	+*
Streptococci	6.0	N	+*
Staphylococci	2.0	N	+*
Micrococci	–	N	+*
Neisseria	–	N	+*
Total count	6.0	7.2	N

*Not present in all patients.
+ = Present.
N = No data.

jejunostomy, type-II colonization occurs probably because of the afferent loop [13, 26, 28, 29]. The extent to which colonization of the small bowel occurs is different in various series. In one series [30] about one-third of 43 patients with a Polya partial gastrectomy had gross contamination of the upper small bowel with a faecal type flora.

The total bacterial contents present in the stomach of patients with type-II colonization are high, like those of patients with non-surgical achlorhydria (Table 2.2). Patients without some form of afferent loop are likely to develop type-I colonization. This may reflect the comparative health of the mucosa when compared to patients with pernicious anaemia or hypogammaglobulinaemia.

Gastric diseases

There have been many studies of the bacteria present in the stomach of patients with gastric ulcers, gastritis and gastric cancer. Other than those designed to look for *C. pylori*, most have been concerned with the gastric flora as a source of infection. The data do not give a clear idea of the extent of colonization but it seems that when gastric acid is produced, very few bacteria are present. Gastric colonization is most likely to be found in patients with gastric cancer (Table 2.4).

Ecological consideration

The germicidal activity of gastric acid has been realized for many years. The paucity of bacteria in the stomach of fasting subjects and the normal decline in the bacterial content of the stomach after a meal are related to the acidity. The importance of gastric acidity is reflected in the concern as to the effect of anti-secretagogues on the gastric flora [10,

Table 2.4 Bacterial flora of gastric contents from patients with diseases of the stomach

	Gastric ulcer		Gastritis	Gastric cancer	
Number of patients	10	13	10	10	33
Reference	19	31	19	19	31
Bacterial group					
Bacteroides	–	–	+	+	+
Eubacteria	–	–	–	–	–
Peptococcaceae	–	+	+	+	+
Bifidobacteria	–	–	+	+	+
Lactobacilli	+	+	+	+	+
Clostridia	–	–	+	+	+
Fusobacteria	–	–	–	–	+
Veillonella	–	+	+	+	+
Enterobacteriaceae	–	–	+	+	+
Streptococci	+	+	+	+	+
Staphylococci	–	+	+	+	+
Micrococci	–	+	+	+	+
Neisseria	–	+	+	+	–
Total count	N	4.8	N	N	7.3

+ = Present.
N = No data.

16] and indeed the impact of nutrition on acid secretion may account for geographical differences in the normal gastric flora (Table 2.1) [32]. Differences in acid secretion alone, however, are not adequate to explain the difference in the types of bacteria isolated from gastric contents of different patient groups.

Adhesion to the gastric mucosa is a major determinant of gastric colonization in animals and indeed bacteria of various morphological types have been reported as adherent to the gastric mucosa of humans [33, 34].

It may be that, like other primates, man has a permanent gastric flora [35]. Should this prove to be the case, the mechanisms by which the adherent bacteria are protected from the gastric acid will be of major interest. The mucus layer will be of major importance. It may be best to regard such a flora as a special type of biofilm ecosystem. A number of biofilm ecologies have been described including mucus and tissue-associated systems in the intestine and vagina [36]. In the special case of the putative gastric mucus biofilm ecology, pH control would be of major importance. It is worthwhile remembering, however, that the toxin of *Vibrio cholerae* may have evolved to enable this organism to control its ionic environment. Thus there is no reason why the gastric bacteria should not have the ability to stimulate bicarbonate buffer accumulation in the mucus layer.

It may be that the role of adhesion is not concerned with the putative gastric flora of the acid stomach but rather with the selection of those organisms that colonize when acid secretion is impaired. Receptors on the gastric mucosa may determine which bacteria can colonize, only those with suitable cell-wall ligands being able to gain a permanent foothold.

At present these conjectures are purely speculative. The importance of the gastric acid as a barrier to pathogens [37] and in the control of other bacteria remains the sole well-understood ecological determinant.

References

1 Goodsir, J. History of a case in which a fluid periodically ejected from the stomach contained vegetable organisms of an undescribed form. With a chemical analysis of the fluid by George Wilson. *Edinburgh Med. Surg. Journal* 1842, **57**, 430–43.

2 Henning, N. Die Bakterienbesiedelung des gesunden und kranken Magens. *Archiv f. Verdauungskrankheiten* 1930, **48**, 1–59.

3 McNulty, C.A.M. & Wise, R. Gastric microflora. *Br. Med. J.* 1985, **291**, 367–8.

4 Tabaqchali, S. The pathophysiological role of the small intestinal bacteria flora. *Scand. J. Gastroenterol.* 1970, **5**(suppl. 6), 139–63.

5 Clarke, R.T.J. & Bauchop, T. 1977. *Microbial Ecology of the Gut*. Academic Press, London, New York.

6 Drasar, B.S. & Barrow, P.A. 1985. *Intestinal Microecology*. Van Nostrand Reinhold, Wokingham.

7 Drasar, B.S. & Hill, M.J. 1974. *Human Intestinal Flora*. Academic Press, London.

8 Hentges, D.J. 1983. *Human Intestinal Microflora in Health and Disease*. Academic Press, New York.

9 Rowland, I.R. 1988. *Role of the Gut Flora in Toxicity and Cancer*. Academic Press, London.

10 Stockbrugger, R. Cotton, P.B., Eugenides, N. Bartholomew, B.A., Hill, M.J. & Walters, C.L. Intragastric nitrites, nitrosamines and bacterial overgrowth during cimetidine treatment. *Gut* 1982, **23**, 1048–54.

11 Leach, S.A., Thompson, M. & Hill, M. Bacterially catalysed N-nitrosation reactions and their relative importance in the human stomach. *Carcinogenesis* 1987, **8**, 1907–12.

12 Moore, W.E.C. & Holdeman, L.V. Human fecal flora: The normal flora of 20 Japanese-Hawaiians. *Appl. Microbiol.* 1974, **27**, 961–79.

13 Browning, G.G., Buchan, K.A. & MacKay, C. The effect of vagotomy and drainage on the small bowel flora. *Gut* 1974, **15**, 139–42.

14 Hill, M.J. Normal and pathological microbial flora of the upper gastrointestinal tract. *Scand. J. Gastroenterol.* 1985, **20**(Suppl. III), 1–5.

15 Ambrecht, U., Boseus, I., Gillberg, R., Seeberg, S. & Stockbrugger, R. Hydrogen (H_2) breath test and gastric bacteria in acid-secreting subjects and in achlorhydric and postgastrectomy patients before and after antimicrobial treatment. *Scand. J. Gastroenterol.* 1985, **20**, 805–13.

16 Ambrecht, U., Seeberg, S. & Stockbrugger, R.W. Hydrogen (H_2) breath excretion in peptic disease before and during treatment with ranitidine. *Scand. J. Gastroenterol.* 1987, **22**, 1211–16.

17 Drasar, B.S., Shiner, M. & McLeod, G.M. The bacterial flora of the gastrointestinal tract in healthy and achlorhydric persons. *Gastroenterology* 1969, 71–9.

18 Borriello, S.P., Reed, P.J., Dolby, J.M., Barclay, F.E. & Webster, A.D.B. Microbial and metabolic profile of achlorhydric stomach: comparison of pernicious anaemia and hypogammaglobulinaemia. *J. Clin. Pathol.* 1985, **38**, 946–53.

19 Sjostedt, S., Heimdahl, A., Kager, L. & Nord, C.E. Microbial colonisation of the oropharynx, esophagus and stomach in patients with gastric disease. *Eur. J. Microbiol.* 1985, **4**, 49–51.

20 Gorbach, S.L., Plant, A.G., Nahas, L., Weinstein, L., Spanknebel, G. & Levitan, R. Studies of intestinal microflora. II. Microorganisms of the small intestine and their relations to oral and fecal flora. *Gastroenterology* 1967, **53**, 856–67.

21 Mitsuoka, T. Recent trends in research on intestinal flora. *Bifidobacteria microflora*. 1982, **3**, 3–24.

22 Bhat, P., Shantakumari, S., Rajan, D., Mathan, V.I., Kapadia, C.R., Swamabi, C. & Baker, S.J. Bacterial flora of the gastrointestinal tract in South Indian control subjects and patients with tropical sprue. *Gastroenterology* 1972, **62**, 11–21.

23 Knott, F.A. Addison's (pernicious) anaemia and subacute combined degeneration of the cord: the role of achlorhydria and intestinal infection. *Guy's Hosp. Reports* 1927, **77**, 1–12.

24 Dellipiani, A.W. & Girdwood, R.H. Bacterial changes in the small intestine in malabsorptive states and pernicious anaemia. *Clin. Sci.* 1964, **26**, 359–74.

25 Ruddell, W.S.J., Bone, E.S., Hill, M.J. & Walters, C.L. Pathogenesis of gastric cancer in pernicious anaemia. *Lancet* 1978, **i**, 521–3.

26 Gray, J.D.A. & Shiner, M. Influence of gastric pH on gastric and jejunal flora. *Gut* 1967, **8**, 574–81.

27 Parkin, D.M., McClelland, D.B.L., O'Moore, R.R., Percy-Robb, I.W., Grant, I.W.B. & Shearman, J.C. Intestinal bacterial flora and bile salt studies in hypogammaglobulinaemia. *Gut* 1972, **13**, 182–8.

28 Gorbach, S. & Tabaqchali, S. Bacteria, bile and the small bowel. *Gut* 1969, **10**, 963–72.

29 Muscroft, T.J., Deane, S.A., Youngs, D., Burdon, D.W. & Keighley, M.R.B. The microflora of the post-operative stomach. *Br. J. Surg.* 1981, **68**, 560–4.

30 Drasar, B.S. & Shiner, M. Bacterial flora of the small intestine in patients with gastro-intestinal disorders. *Gut* 1969, **10**, 812–19.

31 Gatehouse, D., Dunock, F., Burdon, D.W., Alexander Williams, J. & Keighley, M.R.B. Prediction of wound sepsis following gastric operations. *Br. J. Surg.* 1978, **65**, 551–4.

32 Gilman, R.H., Partanen, R., Brown, K.H., Spira, W.M., Khanam, S., Greenberg, B., Bloom, S.R. & Ali, A. Decreased gastric acid secretion and bacterial colonisation of the stomach in severely malnourished Bangladeshi children. *Gastroenterology* 1988, **94**, 1308–14.

33 Savage, D.C. Associations of indigenous microorganisms with gastro-intestinal mucosal epithelia. *Am. J. Clin. Nutr.* 1970, **23**, 1495–1501.

34 Savage, D.C. Microbial ecology of the gastrointestinal tract. *Ann. Rev. Microbiol.* 1977, **31**, 107–33.
35 Bauchop, T. Stomach microbiology of primates. *Ann. Rev. Microbiol.* 1971, **25**, 429–36.
36 Costerton, W.J., Cheng, K.J., Geesey, G.G., Ladd, T.I., Nickel, J.C., Dasgupta, M. & Marrie, T.J. Bacterial biofilms in nature and disease. *Ann. Rev. Microbiol.* 1987, **41**, 435–64.
37 Giannella, R.A., Broitman, S.A. & Zamcheck, N. Gastric acid barrier to ingested microorganisms in man: studies *in vivo* and *in vitro*. *Gut* 1972, **13**, 251–6.

3 Taxonomy of Campylobacter pylori

R.J. OWEN

The genus *Campylobacter* has provided a convenient home for the unidentified gastric spiral bacteria observed and first cultured in 1982 by Warren [1] and Marshall [2] and subsequently named *C. pylori* [3]. However, doubts have existed about the correct classification of these bacteria because ultrastructure details pointed to greater affinities to *Spirillum* rather than to *Campylobacter*. The difficulty of classifying Marshall and Warren's bacteria is mainly attributable to the fact that *Campylobacter* and other allied genera of Gram-negative spiral bacteria such as *Spirillum* and *Wolinella* are defined on a relatively limited number of positive traditional test characteristics, and their taxonomic affinities are not clearly established.

The aim of this chapter is to review the taxonomic data now available on *C. pylori* and to discuss the evidence either for its continued classification in *Campylobacter* or for reclassification elsewhere. Only the key taxonomically important characters of *C. pylori* will be considered.

Development of a classification for gastric spiral campylobacters

Spiral bacteria or 'spirochaetes' have been observed from time to time over the past 75 years in gastric specimens of the dog [4] and humans [5–8]. But it was not until Marshall [2] successfully cultured the small curved S-shaped bacilli observed by Warren [1] in the biopsy material from patients with gastritis, that any attempt at identification was possible. Warren [1] commented that the bacteria resembled *C. jejuni* by light microscopy but more detailed electron microscopy revealed they were quite different. *Campylobacter* strains characteristically have one polar unsheathed flagellum whereas the gastric isolates have 4–5 polar sheathed flagella, which according to Marshall [2] suggested affinities to the genus *Spirillum*. Further studies by Marshall & Warren [9] revealed closer similarities to campylobacters as the gastric isolates were microaerophilic, non-saccharolytic and had a DNA base composition of 36 mol% G + C. Skirrow [10] argued that multiple polar flagella were not an exclusive feature of *Spirillum* because some strains of *C. fetus* subsp. *venerealis* had such an arrangement, and he concluded '. . . their specific location and association makes the provisional name of "pyloric campylobacter" particularly apt; *pylorus* is Greek for gatekeeper — one who looks both ways. Should these bacteria prove to be campylobacters then *Campylobacter pyloridis* would be an appropriate name'. Marshall & Warren [11] cautiously concluded that their bacilli appeared to be a new species closely resembling the campylobacters although the flagella morphology was not that of the genus *Campylobacter*. They suggested that it was premature to talk of '*C. pyloridis*' and better to use the term 'pyloric campylobacter'.

Nevertheless, the name '*Campylobacter pyloridis*' was proposed by Marshall *et al.* [12] and the culture Royal Perth Hospital 13487 (= NCTC 11637) was designated as the type strain. It should be noted that they described the new species as negative in the ability to hydrolyse urea. However, this reported characteristic was incorrect because the ability to produce an active urease was later shown to be a distinctive diagnostic feature of

C. pyloridis [13, 14]. The unusual flagellar arrangement of *C. pyloridis* was the principal feature that first conflicted with its inclusion in *Campylobacter*. However, Marshall *et al.* [12] argued that other multiflagellate forms of *Campylobacter* had been observed [10, 15]. Furthermore, there was evidence that a specific antiserum against the *Campylobacter* group antigen fluoresced with *C. pyloridis* [16]. The new name '*C. pyloridis*' was subsequently validated by the IUMS [17] in 1985 as required by the international code of nomenclature of bacteria [18]. However, two years later Hartmann & von Graevenitz [19] pointed out that the specific epithet *pyloridis* was grammatically incorrect and should be *pylori* (the genitive of the noun *pylorus*). The species name was subsequently revised to *C. pylori* [3].

Various vernacular names or acronyms of *C. pylori* have appeared in the literature. For example Owen *et al.* [14] used the designation GCLO-1 (gastric campylobacter-like organisms) to distinguish them from a different type of gastric campylobacter (GCLO-2) described by Kasper & Dickgiesser [20] and now identified as a new subspecies of *C. jejuni* [21]. The name '*C. pyloridi*' has also been used [22] although the reason for this variation of the name was not explained.

The results of an ultrastructure study of *C. pylori* by Jones *et al.* [23] led them to conclude that the species had greater affinities to *Spirillum*, in particular *Aquaspirillum*, than to the genus *Campylobacter*. However, they observed that there was insufficient data to place these spiral bacteria in their exact taxonomic position and it was suggested that they and other spiral organisms noted earlier this century 'might represent a previously unrecognised taxonomic group of organisms'.

Relationships to Campylobacter

The classification of the genus *Campylobacter* has always been problematic. Member species are relatively inert in most traditional biochemical tests of which only a small number are suitable for identification and classification purposes. Until recently, the relationships between species within the genus were uncertain and relationships to other genera of allied spiral bacteria were highly speculative. Consequently the assignment of a new species to the genus has proved to be a difficult task, particularly if the prospective species had any aberrant characteristics.

Development of the genus

Microaerophilic vibrios were first recognized about 75 years ago as a cause of abortion in pregnant ewes [24]. Apparently similar organisms were also recovered from aborted calves [25] and were subsequently named *Vibrio fetus* [26]. This classification was generally unsatisfactory as *V. fetus* differed phenotypically from *V. cholerae*, the type species of *Vibrio*. No attempt was made to improve the classification for almost 40 years until Sebald & Véron [27] proposed a new genus, *Campylobacter*, with *C. fetus* as the type species. The genus was defined as comprising Gram-negative, slender and curved bacteria which were motile by means of a single, polar flagellum, microaerophilic with a strictly respiratory metabolism, produced no acid in media with carbohydrates, and had DNA with G + C contents between 29 and 36 mol%.

The taxonomic concept of the genus was later broadened by Véron & Chatelain [28] who proposed that the species described under the names *V. coli* Doyle, *V. jejuni* Jones, *V. sputorum* Prevot and *V. bubulus* Florent should be transferred to *Campylobacter*. These workers also suggested that *C. fetus*, the type of the species, should be divided into two subspecies (subsp. *fetus* and subsp. *venerealis*), and they designated a neotype strain (NCTC 10842) for *C. fetus* subsp. *fetus*.

These studies formed the basis of the genus *Campylobacter* as it is now recognized and provided the official nomenclature used in the Approved Lists of Bacterial Names [29]. There

Table 3.1 Phylogenetic groups within *Campylobacter*, *Wolinella* and allied taxa based on the analysis of partial base sequences of 16S rRNA. From the analysis of Thompson *et al.* [51]

Phylogenetic homology group		Proposed genus	Member species	G + C content (mol%)
I*	(subgroup 1)	*Campylobacter*	*C. coli*	31–33
	(92.0%)		*C. jejuni* (both subspecies)	29–32
			C. laridis	31–33
			'*C. upsaliensis*' (CNW)	35–36
I	(subgroup 2)		*C. fetus* (both subspecies)	33–36
	(88.0%)		*C. hyointestinalis*	35–36
			C. concisus	38–39
			C. mucosalis	38–39
			C. sputorum (3 biovars)	31–33
II	(80.1–88.8%)	*Wolinella*	*W. succinogenes*	47
			C. pylori	36–37
			'*C. cinaedi*'	37–38
			'*C. fennelliae*'	37–38
III	(86.9%)	Not named	*C. cryaerophila*	29–30
			C. nitrofigilis	28–29

*Homology Group I would also include *W. recta* (42–47 mol%), *W. curva* (43–47 mol%), *B. gracilis* (42–44 mol%) and *B. ureolyticus* (28–31 mol%) according to the results of Paster & Dewhirst [49].

are several excellent reviews of the historical development of the genus [28,30] and of the significance and features of the present species [31]. Prior to 1980, considerable confusion was caused by the use of a nomenclatural scheme based on the work of Smibert [32], and adopted in the 8th edition of *Bergey's Manual of Determinative Bacteriology*. Smibert [32] speculated, incorrectly as it transpired, that the original strains of *V. fetus* were more likely to have been *V. fetus* var. *venerealis* (Florent). He designated the latter variety as the type species and renamed it *C. fetus* subsp. *fetus*. He also gave the name *C. fetus* subsp. *jejuni* to strains that Véron & Chatelain [28] called *C. jejuni* and *C. coli*, respectively.

The definition of *Campylobacter* given subsequently by Smibert [32] was based on the nomenclature of Véron & Chatelain [28] and it is now widely adopted. The genus includes species with essentially the same primary characteristics as those in the original description of Sebald & Véron [27] except it also includes strains that are motile by means of a flagellum at *one or both ends* of the cell. The number of species in the genus has increased and 11 names are now validly published (Table 3.1). Until recently, little was known about the taxonomic diversity or interrelationships between these species or between *Campylobacter* and other bacterial genera. Primary interest has concentrated on species differentiation and strain identification because of the medical importance of *C. jejuni*. However, the use of RNA sequence analysis to examine the phylogenetic position and structure of the genus has heralded a new era in the taxonomy of *Campylobacter* and the significance of this development and others based on chemical analysis of key cellular macromolecules will be discussed below in relation to *C. pylori*.

Relationships between C. pylori and other Campylobacter species

Conventional bacteriological tests. The general morphological and primary characteristics of *C. pylori* conform with the description of the genus *Campylobacter*. The species comprises curved rod-shaped bacteria that are Gram-negative, motile, microaerophilic, and grow at 37°C, on conventional culture media with moist chocolate or blood agar as the preferred medium. They form coccoid bodies, produce catalase and oxidase but are non-saccharolytic. The principal conventional test differences between *C. pylori* and other campylobacters are in flagella arrangement and cell morphology (see below), although strains also differ from the description of *Campylobacter* as they hydrolyse urea and do not reduce nitrate. However, it is now evident that these are not useful defining tests for *Campylobacter* as proven members of *C. jejuni* and *C. laridis* exhibit such features [21,33].

Ultrastructure. *Campylobacter pylori* possess 4–6 unipolar, sheathed flagella whereas other campylobacters have a single polar unsheathed flagellum at one or both ends of the cell [11,23,34]. These detailed ultrastructure studies on *C. pylori* also revealed a number of other unique features: (i) the ends of the bacteria do not taper; (ii) there is no terminal concavity at the location of the flagella; (iii) cells possess flagella discs without radial structures; (iv) terminal structures internal to the plasma membrane are present; (v) the cell wall is smooth; (vi) flagellar sheaths are in continuity with the unit membrane of the cell wall.

Fatty acid content. The cellular fatty acid (FA) contents of bacteria are very useful taxonomic markers as they are reproducible and relatively unaffected by the growth conditions or medium [35]. Species of *Campylobacter* have been thoroughly investigated by the analysis of FA-methyl esters (FAME) and FA-picolinyl esters [36–38]. Quantitative as well as qualititive differences were found between species and up to ten profile groups have been assigned to date [39]. The FA profiles of *C. pylori* have been particularly well documented [34, 37–41]. All investigators found this species had a distinctive FAME profile characterized by the presence of relatively large amounts (about 25% of total FA content) of 14:0 and cyc 19:0 and the absence of 3-OH-14:0 and 16:1, which are found in other *Campylobacter* species. The profile of *C. pylori* was also quite distinct from those of other *Campylobacter* species as the latter contained significant amounts of 16:0 and 18:1 acids. Additional, but minor, FA differences between *C. pylori* and other campylobacters were detected, e.g. *C. pylori* contained moderate amounts (8–12%) of 18:0 and 3-OH-18:0 acids, which were absent or present in small amounts (< 2%) in other species.

Respiratory quinones. Menaquinones (MK) are constituents of bacterial plasma membranes and play important roles in electron transport and active transport. Their value in bacterial systematics as taxonomic markers lies in the variations in the length and degree of saturation of the C-3 prenyl side chain [42]. Collins *et al.* [43] found that respiratory quinones may be useful as generic markers for *Campylobacter* as the nine species they examined had similar patterns consisting of an unsaturated menaquinone-6 (MK-6) and a novel unsaturated methyl-substituted MK-6 compound, which had not been reported previously in bacteria. Similar quinones were found by other investigators [44, 45]. However, the novel compound, which was recently identified as thermoplasmaquinone-6 (TPQ-6) [39], was not found in *C. pylori* [34,39].

16S rRNA sequence comparisons. The comparison of 16S ribosomal RNA sequences for phylogenetic analysis has proved to be a powerful tool for accurate classification of microorganisms above the level of species [46]. The application of the approach in bacterial classification has been facilitated by the development of a technique for the rapid generation of partial 16S rRNA sequences [47]. Its use in the past few years in the study of species of

Campylobacter has provided a novel insight into the classification of the genus.

Romaniuk *et al.* [48] were the first to report that the 'true' *Campylobacter* species formed a previously undescribed eubacterial phylum, which was related only by very deep branching to other major phyla. With the exception of *C. pylori*, the representative *Campylobacter* strains examined formed a well-defined phylogenetic group with high interspecies rRNA sequence homology (92.4–98.1%). *Campylobacter pylori* was clearly distinct from all the other *Campylobacter* species tested with a reduced level of interspecies homology (84.6–85.5%) and was more closely related (90%) to *Wolinella succinogenes*. Several other investigators using similar methods of phylogenetic analysis also demonstrated that campylobacters with the exception of *C. pylori* formed a tight cluster and that *C. pylori* was more similar to *W. succinogenes* [49–51].

The phylogenetic relationships of all the species in the genus *Campylobacter* were included in the study of Thompson *et al.* [51] who found that they fell into three separate ribosomal RNA sequence homology groups (Table 3.1). These groups were quite distinct from representatives of the three major branches of the purple bacteria. Thompson *et al.* [51] concluded that members of these different homology groups should not continue to be included within a single genus because of their low intergroup rRNA sequence homology. Group I contained the type species, *C. fetus*, so only this group could represent the genus *Campylobacter*. However, *C. pylori* and the other species in Group II were all related to *W. succinogenes* which is the type species of the genus *Wolinella*. Thus the genus represented by Group II should in their view bear the name '*Wolinella*'. A name was not suggested for the Group III strains. Results on hybridizations between labelled 16S rRNA and chromosomal DNA provided further evidence of distinct genetic subgroups within *Campylobacter* although data on *C. pylori* were not given [52].

DNA characteristics. Campylobacter pylori has been characterized with respect to DNA base composition and DNA–DNA sequence homology. The base compositions of two strains (NCTC 11639 = RSB6; and NCTC 11637) were reported to be 36–37 mol%, which fell within the 29–37% G + C range for the genus *Campylobacter*. It should be emphasized, however, that similar base compositions are not a conclusive indicator of relatedness because taxonomically-distinct organisms can still have identical G + C values [53]. DNA–DNA hybridizations using whole-cell DNA showed that *C. pylori* had no or low (< 10%) homology with *C. jejuni*, *C. coli*, *C. fetus*, *C. hyointestinalis*, *C. laridis* and '*C. upsaliensis*' [54, 55].

Total protein electrophoretic profiles. Campylobacter pylori strains have a distinctive electrophoretic (SDS-PAGE) protein profile comprising seven major bands with sizes of 22, 27, 46, 57, 60, 65 and 93 kD. This particular pattern appears to be unique to *C. pylori* and has not been observed in any other *Campylobacter* species [33, 54, 56–59].

Intraspecific variation of C. pylori

Campylobacter pylori is a remarkably homogeneous species as strains isolated worldwide from human gastric mucosa exhibit little variation in traditional test characteristics except some variants have been reported in production of catalase and urease [54]. The *C. pylori* strains isolated from various animal sources appear to be identical to the human isolates with the exception of the ferret strains which exhibit enough differences to indicate they probably represent a separate subspecies or even species (see below). However, there is evidence of strain variation in plasmid profiles [22], chromosomal DNA restriction digest patterns [60, 61] and electrophoretic protein profiles [59].

Some *C. pylori* strains are deposited in culture collections and Table 3.2 lists a selection of strains from different animal hosts that are available for reference purposes.

Table 3.2 Reference strains of *C. pylori*

Species	Collection No*	Strain	Host
C. pylori	NCTC 11637^{T}(ATCC 43504)	RPM 13487	Human
	NCTC 11638	RPM 13491	Human
	NCTC 11639	RSB6	Human
	NCTC 11961	–	Porcine
	NCTC 12199	87008	Rhesus monkey
	NCTC 12031	F2/86	Ferret
	NCTC 12032	F7/86	Ferret
	NCTC 12198** (ATCC 43772)	–	Ferret
Wolinella succinogenes	NCTC 11488^{T} (ATCC 29543)	FDC602	Bovine

* *Abbreviations used*: T, Type strain; NCTC, National Collection of Type Cultures, London; ATCC, American Type Culture Collection, Rockville, MD.
** Reference strain for *C. pylori* subsp. "*mustelae*".

Campylobacter pylori is not a Campylobacter sensu stricto

Table 3.3 summarizes the various taxonomic characteristics which provide the basis for concluding that *C. pylori* is not a *Campylobacter sensu stricto*. The 16S rRNA homology data show without question that *C. pylori* is the most divergent of all the species of *Campylobacter* and is sufficiently distinct to warrant exclusion from the genus. Other proven chemotaxonomic generic markers such as cellular fatty acid profiles and menaquinone contents provide further evidence that *C. pylori* is not a *Campylobacter*. These features correlate with the unusual cell and flagella morphology of the species which are more convenient features for identification purposes. The value of total protein patterns as generic markers for *Campylobacter* are unclear as there is considerable variation between species and there are no particular bands which are consistently different. However, the protein profile of *C. pylori* is quite distinct from those of all other *Campylobacter* species. The differences at the species level are also reflected in low DNA–DNA sequence homologies between *C. pylori* and other species. Another feature of *C. pylori*, which possibly reflects basic structural and physiological differences from other *Campylobacter* and *Wolinella*, is its marked susceptibility to damage during the freeze-drying process [59].

Table 3.3 Key taxonomic features distinguishing *C. pylori* from other *Campylobacter* species

Divergent 16S rRNA sequence
4–5 unsheathed flagella
Lacks the respiratory quinone TPQ-6
Major fatty acids present are 14:0 and cyc19:0, but lacks 3-OH-14:0 and 16:1
Distinctive SDS-PAGE protein profile with 7 major bands

Towards a phylogenetic definition of Campylobacter

The 16S rRNA sequence data now available for *Campylobacter* provides the basis of a provisional phylogenetic definition for the genus. The species comprising homology group I of Thompson *et al.* [51] (see Table 3.1) were linked at an average similarity level of 89% in the RNA dendrogram. Similar results (87–93%) were obtained in other phylogenetic analyses of *Campylobacter* and allied species [48–50] but a hierarchical definition of *Campylobacter* needs to be validated by reference to other chemotaxonomic markers and conventional phenotypic characteristics. It is important to note that the genus *Campylobacter* based on homology Group I would have a G + C content range of 29–39 mol% G + C, which is typical of the heterogeneity level found in many bacterial genera. However, the inclusion of species such as *W. curva* and *W. recta* in a revised and broadened concept of the genus, a possibility suggested by Paster & Dewhirst [49], would result in a wider range of base compositions (29–47 mol% G + C), indicating considerable genetic diversity.

Taxonomic relationships between C. pylori and Wolinella

The 16S rRNA phylogenetic analyses showed that *C. pylori* had the closest affinities to *Wolinella succinogenes*, which is the type species of the genus *Wolinella*, but the two other species in the genus, *Wolinella recta* and *W. curva*, were clustered with *Campylobacter sensu stricto*. This was a surprising discovery because the name '*Wolinella*' proposed for a group of curved motile bacteria from the bovine rumen and human buccal cavity clearly distinguished them from *Campylobacter* in having G + C contents of 42–48 mol% and in growing anaerobically [62, 63].

Similarities to W. succinogenes

Conventional test characters. *Wolinella succinogenes*, like *C. pylori*, is a Gram-negative, curved rod-shaped bacterium. It is rapidly motile, grows on blood and is non-saccharolytic. The 14 or so principal differences between the two species in conventional phenotypic and chemotaxonomic characteristics are shown in Table 3.4. Tanner & Socransky [62] described *Wolinella* as anaerobic but noted that strains would grow on the surface of blood agar plates in an atmosphere including 5% oxygen when a heavy cell suspension was used. Another metabolic feature of *W. succinogenes* was an ability to grow in broth stimulated by a combination of formate and fumarate. Thus, Thompson *et al.* [51] argued that the species was not an anaerobe but an H_2-requiring microaerophile.

Chemotaxonomic characters. *Wolinella succinogenes* has a DNA base composition of 47 mol% G + C, which is significantly different from the 36–37% G + C value of *C. pylori*. A difference of 10% indicates a significant amount of genomic diversity. The G + C differences were also reflected in marked differences between the two species in their major and minor fatty acids (Table 3.4), which led Hudson *et al.* [39] to conclude that *C. pylori* did not have a close affinity to *W. succinogenes*. Both species contained MK-6 as the main respiratory quinone but *W. succinogenes* also contained TPQ-6 [64]. The 16S rRNA homology data do not, however, provide a clear-cut answer to the relationships between *C. pylori* and *W. succinogenes*. Although Thompson *et al.* [51] proposed that the two species should be included in the genus *Wolinella* with '*C. fennelliae*' and '*C. cinaedi*', the results obtained in other analyses [48–50] were less conclusive. The details of these analyses are summarized in Table 3.5. Romaniuk *et al.* [48] suggested, even though *C. pylori* was more closely related to *W. succinogenes* than to other campylobacters in the RNA dendrogram, that the differences between the two species was probably sufficient to warrant separate genera. Likewise, Lau *et al.* [50] suggested on a more limited set of

Table 3.4 Differential characteristics of *C. pylori* and *Wolinella succinogenes**

Characteristics	*C. pylori*	*W. succinogenes*
Flagella number	6	1
Flagella sheathed	+	−
Anaerobic growth**	−	+
Urea hydrolysis	+	−
Nitrate reduction	−	+
Catalase production	+	−
H_2S production	−	+
G + C content (mol%)	36–37	47
Respiratory quinone: TPQ-6	−	+
Major fatty acid components:		
14:0	+	±
16:0	−	+
16:1	−	+
18:1	−	+
cyc-19:0	+	−

* Symbols: +, positive/present; ±, present in small quantities (~ 1%); −, negative/not present.
** Atmosphere consisting of 80% N_2, 10% CO_2, 10% H_2.

Table 3.5 Summary of 16S rRNA sequence analyses on *Campylobacter* including *C. pylori* and *W. succinogenes*

Study reference	No. species		No. base positions compared	% sequence homology*			*Campylobacter* cluster inter-species homologies
	Campylobacter	*Wolinella*		*C.f* × *C.j*	*C.p* × *W.s*	*C.p* × *C.f*	
Romaniuk *et al.* [48]	6	1	745	96	90	85	93–98
Lau *et al.* [50]	6	1	460	–**	86	87	–**
Paster & Dewhirst [49]	7	3	902–998	95	90	84	93–98
Thompson *et al.* [51]	14	1	750	96	92	87	93–98

*Uncorrected data. C.f = *C. fetus*; C.j = *C. jejuni*; C.p. = *C. pylori*; W.s. = *Wolinella succinogenes*.
**Not available because data matrix was not published.

sequences that *C. pylori* should be assigned to a new genus, whereas Paster & Dewhirst [49] concluded that the taxonomic relationships between the two species needed further examination. Even though differences in the interpretation of the phylogenetic dendrograms may occur, the four groups of workers agreed remarkably well in the levels of sequence homology calculated between *C. pylori* and *W. succinogenes* and between *C. fetus*, *C. pylori* and *C. jejuni* (Table 3.5).

Phylogenetic position of C. pylori and Wolinella

About ten major taxonomic groups have been recognized within the eubacteria by analysis of 16S rRNA catalogue data [46]. It is not clear, however, from the phylogenetic studies on *C. pylori* and other campylobacters and wolinellas where they belong among the eubacteria although it has been established that they belong on the same phylogenetic branch [48, 49, 51]. The campylobacters and wolinellas are apparently only distantly related to representatives of the alpha, beta and gamma branches of the purple photosynthetic bacteria recently named the *Proteobacteria* [65] and do not belong to any previously defined branch of that class. Studies of 5S rRNA sequences and semiconserved secondary structural features (signature sequences), which also reflect phylogenetic history, have failed to clarify the position of these bacteria [50] but it is possible they may eventually be accommodated in an expanded version of the *Proteobacteria*.

Relationships to other gastric spiral bacteria

Ferret gastric CLO

Campylobacter-like organisms (CLO) resembling *C. pylori* have been isolated from ferret gastric mucosa in the USA [66] and UK [67]. These CLO, unlike human *C. pylori*, were able to reduce nitrate, produced little or no leucine arylamidase, were sensitive to nalidixic acid and were resistant to cephalothin [68, 69]. The human and ferret strains also differed in their SDS-protein patterns [68], isoelectric points of ureases [69] and cellular fatty acid content, notably in the relative amounts of 14:0 and 16:0 acids [39, 69]. DNA–DNA hybridization data of Fox *et al.* [55] showed that the ferret and human strains were 72% related at the optimum reassociation temperature but only 55% related at the stringent temperature. Phylogenetic studies based on 16S rRNA sequences confirmed that the ferret CLO were members of the *W. succinogenes–C. pylori* cluster [55] and it appears likely from the available data that they are either a subspecies of *C. pylori* or a closely-related new species.

Baboon and pig gastric CLO

Curry *et al.* [70] described spiral organisms isolated from the gastric mucosa of baboon and pig stomachs. Their similarities to *C. pylori* were subsequently confirmed in conventional tests and SDS-protein patterns [68] although the FAME profile of the baboon strain apparently resembled those of the *C. jejuni* group and the ferret gastric CLO [39].

Rhesus monkey gastric CLO

In a study of experimental and natural *C. pylori* infections in the Rhesus monkey, Newell & Baskerville [71] isolated campylobacter-like organisms from the gastric epithelium of 4/7 monkeys. These bacteria were found to be identical to *C. pylori* in flagella and cell morphology, in SDS-PAGE protein profiles, urease activity and fatty acid content [39] as well as in their immunological characteristics. Apparently similar bacteria were recently observed in the gastric mucosa of rhesus monkeys in the USA [72].

'Flexispira rappini'

Bryner [73] proposed the name '*Flexispira rappini*' for a group of microaerophilic, Gram-negative, motile, urease-producing bacteria isolated from the gastrointestinal mucosa of man and lower mammals (dog, pig and sheep). Members of the species had a G + C content of 33 mol% and possessed seven sheathed polar flagella. They were distinguished from *C. pylori* by the unique coilin fibres found in the periplasmic space, by the flexibility of the cell wall and the ability to grow at 43°C. The phylogenetic position of this new taxonomic group was not established.

Relationships to Thiovulum and spirilla

An interesting but unexpected result of the rRNA sequence analysis was that *C. pylori* and *Wolinella succinogenes* were shown to be as closely related to a strain of the genus *Thiovulum* as they were to *Campylobacter* [48, 49]. Little is known about *Thiovulum*, however, except that it is a sulphide-dependent marine bacterium. Although it was suggested that *C. pylori* might have affinities to *Spirillum* and *Aquaspirillum* [2,23], there is at present no chemotaxonomic data supporting such relationships. Likewise the relationships of *C. pylori* and *Wolinella succinogenes* to other genera of aerobic/microaerophilic/anaerobic motile, helical, Gram-negative bacteria, which include genera such as *Oceanospirillum*, *Azospirillum* and *Anaerobiospirillum*, remain uncertain and require further phylogenetic studies.

Comments and conclusions

Phylogenetic analyses based on 16S rRNA sequences and various other chemotaxonomic data provide incontrovertible evidence that *C. pylori* is not a member of the genus *Campylobacter*. However, its precise taxonomic position is still debatable as the phylogenetic analyses revealed unexpected affinities to *Wolinella succinogenes*. It has been suggested [51] that *C. pylori* should be reclassified in *Wolinella* with the organisms currently referred to as '*C. fennelliae*' and '*C. cinadei*' (these names have not yet been validated). Such an approach might prove problematic as *Wolinella* would then include organisms that were morphologically, metabolically and genetically diverse (e.g. their G + C contents would range from 36 to 47 mol%). An alternative would be to reclassify *C. pylori* in a new genus. It has been recommended that the complete DNA sequence should be the reference standard to determine phylogeny and that phylogeny should determine taxonomy, although at present there is no satisfactory phylogenetic definition of a genus [74]. The points at which genera are distinguished along a given phylogenetic branch must therefore be determined by phenotypic consistency. In the case of *C. pylori*, this would strongly indicate the need for a new genus, because of the significant number of differences between it and *W. succinogenes* in both traditional and chemotaxonomic characters. The new genus would probably comprise two species, one for the human, pig, monkey and baboon strains, and a second for the ferret strains.

References

1 Warren, J.R. Unidentified curved bacilli on gastric epithelium in active chronic gastritis. *Lancet* 1983 **i**, 1273.

2 Marshall, B. Unidentified curved bacilli on gastric epithelium in active chronic gastritis. *Lancet* 1983 **i**, 1273–5.

3 Marshall, B.J. & Goodwin, C.S. Revised nomenclature of *Campylobacter pyloridis. Int. J. Systematic Bacteriol.* 1987, **37**, 68.

4 Bizzozero, G. Über die Schlauchformigen drüsen des Magendarmkanals und die Beziehungen ihres Epithels zu dem Oberflächenepithel der Schleimhaut. *Archiv f. Mikrobiologie u. Anatomie.* 1893, **42**, 82–94.

5 Krienitz, W. Über das Auftreten von Spirochäten verschiedener Form im Mageninhalt bei Carcinoma ventriculi. *Dtsch. med. Wochenschr.* 1906, **28**, 872–90.

6 Doenges, J.L. Spirochaetes in the gastric glands of *Macacus rhesus* and humans without definite history of related disease. *Arch. Path.* 1938, **27**, 469–77.

7 Freedburg, A.S. & Barron, L.E. The presence of spirochaetes in human gastric mucosa. *Am. J. Digest. Dis.* 1940, **7**, 443–5.

8 Steer, H.W. & Colin-Jones, D.G. Mucosal changes in gastric ulceration and their response to carbenoxolone sodium. *Gut* 1975, **16**, 590–7.

9 Marshall, B.M. & Warren, J.R. 1983. Spiral bacteria in the human stomach: a common finding in patients with gastritis and duodenal ulcer. In Pearson, A.D., Skirrow, M.B., Rowe, B., Davies, J. & Jones, D.M. (eds),*Campylobacter II. Proceedings of the Second International Workshop on Campylobacter Infections.* Public Health Laboratory Service, London, pp. 11–12.

10 Skirrow, M. 1983. Report on the session: taxonomy and biotyping. In Pearson, A.D., Skirrow, M.B., Rowe, B., Davies, J. & Jones, D.M. (eds), *Campylobacter II. Proceedings of the Second International Workshop on Campylobacter Infections,* Public Health Laboratory Service, London, pp. 33–8.

11 Marshall, B.J. & Warren, J.R. Unidentified curved bacilli in the stomach of patients with gastritis and peptic ulceration. *Lancet* 1984, **i**, 1311–14.

12 Marshall, B.J., Royce, H., Anwar, D.I., Goodwin, C.S., Pearmans, J.W., Warren, J.R. & Armstrong J.A. Original isolation of *Campylobacter pyloridis* from human gastric mucosa. *Microbios Lett.* 1984, **25**, 83–8.

13 Langenberg, M.L., Tytgat, G.N.J., Schipper, M.E.I., Rietra, P.J.G.M. & Zanen, H.C. Campylobacter-like organisms in the stomachs of patients and healthy individuals. *Lancet* 1984, **i**, 1348.

14 Owen, R.J., Martin, S.R. & Borman, P. Rapid urea hydrolysis by gastric campylobacters. *Lancet* 1985, **i**, 111.

15 Ritchie, A.E., Keller, R.F. & Bryner, J.H. Anatomical features of *Vibrio fetus*: electron microscopic survey. *J. Gen. Microbiol.* 1966, **43**, 427–38.

16 Price, A.B., Dolby, J.M., Dunscombe, P.R. & Stirling, J. Detection of *Campylobacter* by immunofluorescence in stools and rectal biopsies of patients with diarrhoea. *J. Clin. Pathol.* 1984, **37**, 1007–13.

17 International Union of Microbiological Societies. Validation of the publication of new names and new combinations previously effectively published outside the IJSB. List No. 17. *Int. J. Systematic Bacteriol.* 1985, **35**, 223–5.

18 Lapage, S.P., Sneath, P.H.A., Lessel E.F., Skerman V.B.D., Seeliger H.P.R., & Clark, W.A. (ed.) 1975 *International Code of Nomenclature of Bacteria. 1976 Revision.* American Society for Microbiology, Washington, DC.

19 Hartmann, D. & von Graevenitz, A. A note on name, viability and urease tests of *Campylobacter pylori. Eur. J. Clin. Microbiol.* 1987, **6**, 82–3.

20 Kasper, G. & Dickgiesser, N. Isolation from gastric epithelium of campylobacter-like bacteria that are distinct from *'Campylobacter pyloridis'. Lancet* 1985, **i**, 111–12.

21 Steele, T.W. & Owen, R.J. *Campylobacter jejuni* subsp. *doylei* subsp. nov., a subspecies of nitrate-negative campylobacters isolated from human clinical specimens. *Int. J. Systematic Bacteriol.* 1988, **38**, 316–18.

22 Tjia, T.N., Hooper, W.E.S., Goodwin, C.S. & Grubb, W.B. Plasmids in *Campylobacter pyloridi. Microbios Lett.* 1987, **36**, 7–11.

23 Jones, D.M., Curry, A. & Fox, A.J. An ultrastructure study of the gastric Campylobacter-like organism *'Campylobacter pyloridis'. J. Gen. Microbiol.* 1985, **131**, 2335–41.

24 McFadyean, J. & Stockman, S. 1913 Great Britain Board of Agriculture and Fisheries, Dept. Comm. Epiz. Abortion, London. p. 22 and Appendix part III.

25 Smith, T. Spirilla associated with disease of the fetal membranes in cattle (injections abortion). *J. Exptl Med.* 1918, **28**, 701–19.

26 Smith, T. & Taylor, M.S. Some morphological and biological characters of the spirilla (*Vibrio fetus* n.sp.) associated with disease of the fetal membranes of cattle. *J. Exptl Med.* 1919, **30**, 299–311.

27 Sebald, M. & Véron, M. Teneur en bases de l'ADN et classification des vibrions. *Annals de l'Institut Pasteur* 1963, **105**, 897–910.

28 Véron M. & Chatelain, R. Taxonomic study of the genus *Campylobacter* Sebald and Véron and designation of the neotype strain for the type species *Campylobacter fetus* (Smith and Taylor) Sebald and Véron. *Int. J. Systematic Bacteriol.* 1973, **23**, 122–34.

29 Skerman, V.B.D., McGowan, V. & Sneath, P.H.A. (eds), Approved lists of bacterial names. *Int. J. Systematic Bacteriol.* 1980, **30**, 225–420.

30 Karmali, M.A. & Skirrow, M.B. Taxonomy of the

genus *Campylobacter*. In Butzler, J.P. (ed.) *Campylobacter Infection in Man and Animals*. CRC Press, Boca Raton, pp. 1–20.
31 Penner, J.L. The genus *Campylobacter*: a decade of progress. *Clin. Microbiol. Rev.* 1988, **1**, 157–72.
32 Smibert, R.M. 1974. Genus *Campylobacter*. In Krieg, N.R. & Holt, H.G.(eds), *Bergey's Manual of Systematic Bacteriology, Vol. 1* Williams & Wilkins, Baltimore, pp. 111–18.
33 Owen, R.J., Costas, M, Sloss, L. & Bolton, F.J. Numerical analysis of electrophoretic protein patterns of *Campylobacter laridis* and allied thermophilic campylobacters from the natural environment. *J. Appl. Bacteriol.* 1988, **65**, 69–78.
34 Goodwin, C.S., McCulloch, R.K., Armstrong, J.A. & Wee, S.H. Unusual cellular fatty acids, and distinctive ultrastructure in a new spiral bacterium (*Campylobacter pyloridis*) from the human gastric mucosa. *J. Med. Microbiol.* 1985, **19**, 257–67.
35 Jantzen, E. & Bryn, K. 1985. Whole cell and lipopolysaccharide fatty acids and sugars of Gram-negative bacteria. In Goodfellow, M. & Minnikin, D.E. (eds), *Chemical Methods in Bacterial Systematics*. Academic Press, London, pp. 245–71.
36 Leaper, S. & Owen, R.J. Identification of catalase producing *Campylobacter* species based on biochemical characteristics and on cellular fatty acid composition. *Current Microbiology* 1981, **31**, 31–5.
37 Wait, R. & Hudson, M.J. The use of picolinyl esters for the characterization of microbial lipids: application to the unsaturated and cyclopropene fatty acids of *Campylobacter* species. *Lett. Appl. Microbiol.* 1985, **1**, 95–9.
38 Lambert, M.A., Patton, C.M., Barrett, T.J. & Moss, C.W. Differentiation of *Campylobacter* and Campylobacter-like organisms by cellular fatty acid composition. *J. Clin. Microbiol.* 1987, **25**, 706–13.
39 Hudson, M.J., Bhavsor, P. & Wait, R. 1988. Chemotaxonomy of campylobacters. In Kaijser, B. & Falsen, E. (eds), *Campylobacter IV*. Kungälv, Goterna, pp. 34–5.
40 Goodwin, C.S., Blincow, E., Armstrong, J., McCulloch, R.K. & Collins, D. *Campylobacter pyloridis* is unique: GCLO-2 is an ordinary campylobacter. *Lancet* 1985, **i**, 38–9.
41 Itoh, T., Yassagawa, Y., Shingaki, M., Takohashi, M., Kai, A., Ohashi, M. & Hamana, G. Isolation of *Campylobacter pyloridis* from human gastric mucosa and characterization of the isolates. *Microbiol. Immunol.* 1987, **31**, 603–14.
42 Collins, M.D. & Jones, D. Distribution of isoprenoid quinone structural types in bacteria and their taxonomic implications. *Microbiol. Rev.* 1981, **45**, 314–54.
43 Collins, M.D., Costas, M. & Owen, R.J. Isoprenoid quinone composition of representatives of the genus *Campylobacter*. *Arch. Microbiol.* 1984, **137**, 168–70.
44 Carlone, G.M. & Anet, F.A.L. Detection of menaquinone-6 and a novel methyl substituted menaquinone-6 in *Campylobacter jejuni* and *Campylobacter fetus* subsp. *fetus*. *J. Gen. Microbiol.* 1983, **129**, 3385–93.
45 Moss, C.W., Kai, A., Lambert, M.A. & Patton, C. Isoprenoid quinone content and cellular fatty acid composition of *Campylobacter* species. *J. Clin. Microbiol.* 1984, **19**, 772–6.
46 Woese, C.R. Bacterial evolution. *Microbiol. Rev.* 1987, **51**, 221–7.
47 Lane, D.J., Pace, B., Olsen, G.J., Stahl, D.A., Sogin, M.L. & Pace, N.R. Rapid determination of 16S ribosomal RNA sequences for phylogenetic analysis. *Proc. Natl Acad. Sci. USA* 1985, **82**, 6955–9.
48 Romaniuk, P.J., Zoltouska, B., Trust, T.J., Lane, D.J., Olsen, G.J., Pace, N.R. & Stahl, D.A. *Campylobacter pylori*, a spiral bacterium associated with human gastritis, is not a true *Campylobacter* sp. *J. Bacteriol.* 1987, **169**, 2137–41.
49 Paster, B.J. & Dewhirst, F.E. Phylogeny of campylobacters, Wolinellas, *Bacteroides gracilis*, and *Bacteroides ureolyticus* by 16S ribosomal ribonucleic acid sequencing. *Int. J. Systematic Bacteriol.* 1988, **38**, 56–62.
50 Lau, P.P., DeBrunner-Vossbrinck, B., Dunn, B., Miotto, K., MacDonell, M.T., Rollins, D.M., Pilledge, C.J., Hespell, R.B., Colwell, R.R., Segin, M.L. & Fox, G.E. Phylogenetic diversity and position of the genus *Campylobacter*. *Systematic Appl. Microbiol*, 1987, **9**, 231–8.
51 Thompson III, L.M., Smibert, R.M., Johnson, J.L. & Krieg, N.R. Phylogenetic study of the genus *Campylobacter*. *Int. J. Systematic Bacteriol.* 1988, **38**, 190–200.
52 Vandamme, P. & Deley, J. 1988. Phylogenetic relationships in and of the genus *Campylobacter*. In Kaijser, B. & Falsen, E. (eds), *Campylobacter IV* Kungälv, Goterna, pp. 47–8.
53 Owen, R.J. & Pitcher, D. 1985. Current methods for estimating DNA base composition and levels of DNA–DNA hybridization. In Goodfellow, M. & Minnikin, D.E. (eds), *Chemical Methods in Bacterial Systematics*. Academic Press, London, pp. 67–93.
54 Von Wulffen, H. Low degree of relatedness between *Campylobacter pyloridis* and enteropathogenic *Campylobacter* species as revealed by DNA–DNA blot hybridization and immunoblot studies. *FEMS Microbiol. Lett.* 1987, **42**, 129–33.
55 Fox, J.G., Edwards, P., Taylor, N.S., Paster, B. & Dewhirst, F. 1988. Comparison of the phylogenetic, biochemical, phenotypic, and morphological characteristics of gastric campylobacters from humans and ferrets. In Kaijser, B. & Falsen, E. (eds), *Campylobacter IV*. Kungälv, Goterna, pp. 50–4.
56 Pearson, A.D., Bamforth, J., Booth, L., Holdstock, G., Ireland, A., Walker, C., Hawtin, P. & Millward-Sadler, H. Polyacrylamide gel electrophoresis of spiral

bacteria from the gastric antrum. *Lancet* 1984, **i**, 1349–50.

57 Megraud, F., Bonnet, F., Garnier, M. & Lamouliatte, H. Characterization of '*Campylobacter pyloridis*' by culture, enzymatic profile, and protein content. *J. Clin. Microbiol.* 1985, **22**, 1007–10.

58 Costas, M., Owen, R.J. & Jackman, P.J. Classification of *Campylobacter sputorum* and allied campylobacters based on numerical analysis of electrophoretic protein patterns. *Systematic Appl. Microbiol.* 1987, **9**, 125–31.

59 Owen, R.J., Costas, M., Morgan, D.D., On, S.L.W., Hill, L.R., Pearson, A.D. & Morgan, D.R. Strain variation in *Campylobacter pylori* detected by numerical analysis of one-dimensional electrophoretic protein patterns. *Antonie van Leeuwenhoek J. Microbiol.* 1989, in press.

60 Langenberg, W., Rauws, E.A.J., Widjojokusomo, A., Tytgat, G.N.J. & Zanen, H.C. Identification of *Campylobacter pyloridis* isolates by restriction endonuclease DNA analysis. *J. Clin. Microbiol.* 1986, **24**, 414–17.

61 Majewski, S.I.H. & Goodwin, C.S. Restriction endonuclease analysis of the genome of *Campylobacter pylori* with a rapid extraction method: evidence for considerable genomic variation. *J. Infect. Dis.* 1988, **157**, 465–71.

62 Tanner, A.C.R. & Socransky, S.S. 1981. Genus *Wollinella*. In Krieg, N.R. & Holt, J.G. (eds), *Bergey's Manual of Systematic Bacteriology*, Vol. 1. Williams & Wilkins, Baltimore, pp. 646–50.

63 Tanner, A.C.R., Badger, S. Lai., C.-H., Listgarten, M.A., Visconti, R.A. & Socransky S.S. *Wolinella* gen. nov., *Wolinella succinogenes (Vibrio succinogenes* Wolin *et al.*) comb. nov., and descriptions of *Bacteroides gracilis* sp. nov., *Wolinella recta* sp. nov., and *Eikenella corrodens* all from humans with periodontal disease. *Int. J. Systematic Bacteriol.* 1981, **31**, 432–45.

64 Collins, M.D. & Fernandez, F. Menaquinone-6 and thermoplasmaquinone-6 in *Wolinella succinogenes*. *FEMS Microbiol. Lett.* 1984, **22**, 273–6.

65 Stackebrandt, E., Murray, R.G.E. & Trüper, H.G. *Proteobacteria* classis nov., a name for the phylogenetic taxon that includes the 'purple bacteria and their relatives'. *Int. J. Systematic Bacteriol.* 1988, **38**, 321–5.

66 Fox, J.G., Edrise, B.M., Cabot, E.B., Beaucage, C., Murphy, J.C. & Prostak, K.S. *Campylobacter*-like organisms isolated from gastric mucosa of ferrets. *Am. J. Vet. Res.* 1986, **47**, 236–9.

67 Rathbone, B.J., West, A.P., Wyatt, J.I., Johnson, A.W., Tomkins, D.S. & Heatley, R.V. *Campylobacter pyloridis*, urease, and gastric ulcers. *Lancet* 1986, **ii**, 400.

68 Jones, D.M. & Eldridge, J. 1988. Gastric *Campylobacter*-like organisms (GCLO) from man ('*C. pyloridis*') compared with GCLO strains from the pig, baboon and ferret. In Kaijser, B. & Falsen, E. (eds), *Campylobacter IV*. Kungälv, Goterna, p. 44.

69 Tomkins, D.S., West, A.P., Goodwin, P.G.R., Wyatt, J.I. & Rathbone, B.J. 1988. Characterization of ferret gastric *Campylobacter*-like organisms. In Kaijser, B. & Falsen, E. (eds), *Campylobacter IV*. Kungälv, Goterna, p. 46.

70 Curry, A., Jones, D.M. & Eldridge, J. Spiral organisms in the baboon stomach. *Lancet* 1987, **ii**, 634–5.

71 Newell, D.G. & Baskerville, A. 1988. Experimental and natural *Campylobacter pylori* infections in the Rhesus monkey. In Kaijser, B. & Falsen, E. (eds), *Campylobacter IV*. Kungälv, Goterna, pp. 436–7.

72 Reed, K.D. & Berridge, B.R. Campylobacter-like organisms in the gastric mucosa of Rhesus monkeys. *Lab. Animal Sc.* 1988, **38**, 329–31.

73 Bryner, J.H. 1988. *Flexispira rappini* gen. nov., sp. nov. A motile, urease producing rod similar to *Campylobacter pyloridis*. In Kaijser, B. & Falsen, E. (eds), *Campylobacter IV*. Kungälv, Goterna, pp. 440–2.

74 Wayne, L.G., Brenner, D.J., Colwell, R.R., Grimont, P.A.D., Kandler, O., Krichevsky, M.I., Moore, L.H., Moore, W.E.C., Murray, R.G.E., Stackebrandt, E., Starr, M.P. & Trüper, H.G. Report of the *ad hoc* Committee on reconciliation of approaches to bacterial systematics. *Int. J. Systematic Bacteriol.* 1988, **37**, 463–4.

4 Survival and growth of Campylobacter pylori

D.S. TOMPKINS

The spiral bacteria which had been observed by histologists in gastric biopsies and gastrectomy specimens were first isolated when culture plates were left for prolonged incubation of six days over the Easter holiday period in 1982 [1]. In this original study gastric antral biopsies were obtained from patients at routine elective gastroscopy. The biopsies were transported to the laboratory in a chilled anaerobic broth transport medium, and cultured within 1 hour of collection. The biopsy specimens were minced and inoculated onto blood agar, 'chocolate' agar and Skirrow's campylobacter medium and incubated microaerobically at 37°C. After the initial isolation was made on 'chocolate' agar, it was found that colonies of this new organism were visible after 3–4 days' incubation. The organism was eventually named as *Campylobacter pylori*.

Since the preliminary report of this original isolation in 1983 [2], many workers all over the world have successfully isolated these organisms. The bacteria are present in a site previously thought to be colonized only by transient oral flora, except in patients with achlorhydria, because gastric acid rapidly destroys most bacteria [3]. They live closely attached to gastric epithelial cells beneath a protective layer of mucus, in a situation where the pH is virtually neutral [4]. *Campylobacter pylori* has been visualized histologically in association with metastatic gastric-type epithelium in the duodenum, oesophagus and in one case, in the rectum [5–7]. In the stomach, *C. pylori* does not colonize areas where the normal epithelium has been replaced by intestinal metaplasia [5]. There are no reports of isolation of *C. pylori* from sites other than the mucosa of the stomach and adjacent oesophagus and duodenum. There are no reports of invasion and spread to other sites by these bacteria.

Survival in gastric juice

Most bacteria are rapidly killed in an acid environment of pH < 4.0 [3]. It has been stated that *C. pylori* is rapidly destroyed by a pH of < 5, few cells remaining after 2 hours' exposure to these acidic conditions [8]. However, in one study [9] using pH adjusted brucella broth, counts of *C. pylori* and *C. jejuni* were not affected by exposure to pH 4.0 over a period of 2 hours. At pH 3.1, counts of one strain of *C. pylori* were reduced by a factor of 10^{-2} after 2 hours and *C. jejuni* did not survive; *C. pylori* survived for > 1 hour at pH 2.5 whereas *C. jejuni* was killed after 10 min in this more acidic solution. Two other groups have suggested that the acid susceptibilities of *C. pylori* and *C. jejuni* are equal. In one study [10], experiments were carried out at room temperature using 0.1 M citrate–phosphate buffer and in the second study [11], suspensions of bacteria were incubated in phosphate buffered saline at 37°C. Both groups found that survival of *C. pylori* and *C. jejuni* decreased after 30 min exposure to pH < 4 but addition of physiological concentrations of urea protected *C. pylori* but not *C. jejuni* at pH as low as 1.5.

Goodwin *et al.* [12] suggested that the production of substantial urease activity would enable *C. pylori* to survive in a cloud of ammonia, protected from the potentially lethal gastric acid. Urease is a characteristic of *C. pylori* isolates from gastric tissues of monkeys, baboons and pigs, and is also produced by similar organisms isolated from ferrets [13, 14]. A different spiral-shaped organism, which colonizes the stomachs of cats, also produces abundant urease activity, and this organism has also been visualized in human gastric biopsy specimens [15]. However, some authors consider that the production of large amounts of this enzyme by bacteria attached to the gastric mucosa is unlikely to be primarily an acid-resistance mechanism [15].

Campylobacter pylori may be isolated from acidic gastric juice but isolation rates are lower than those from biopsy specimens taken from the same patient. In one Japanese study [16], culture of gastric juice was only one-quarter as sensitive as biopsy culture but in the author's laboratory, juice culture was positive in 17 (38%) of 45 patients with gastritis compared with 37 (82%) of biopsy cultures. Presence of bile may inhibit the recovery of *C. pylori*; the organism is seen less often in patients with reflux gastritis and following gastric surgical procedures resulting in the reflux of bile, than in other patients with chronic gastritis [17]. *In vitro* work has demonstrated that *C. pylori* is inhibited by 1% bile salts [1] and is less tolerant of bile than other campylobacters or other bacteria commonly found in the human intestinal tract [11].

Microaerophilism

Oxygen is potentially toxic to all living cells and tolerance of different bacterial species varies [18]. Campylobacters are microaerophilic organisms; they cannot grow anaerobically in the absence of oxygen but die when exposed to the atmosphere with 21% oxygen [19]. In aerobic metabolism molecular oxygen is used as a terminal electron acceptor and as a substrate in metabolic oxygenase-catalysed reactions in which oxygen atoms are incorporated into organic compounds [20]. Campylobacters are probably unable to generate enough energy for growth in an anaerobic atmosphere although some campylobacters will grow anaerobically when supplied with an appropriate respiratory electron acceptor such as fumarate [18, 20]. The reasons for the toxicity of oxygen for microaerophilic organisms are not clear. Oxygen may be converted into superoxide anion radical, hydrogen peroxide, hydroxyl radical and singlet molecular oxygen [21, 22]. These toxic products may be produced either by metabolism within the bacterial cells or by auto-oxidation and photochemical oxidation processes in growth media [21–23]. Bacteria produce enzymes such as catalases, superoxide dismutase (SOD) and peroxidases to protect themselves from these toxic products but the relationship of oxygen toxicity to production of these enzymes is not direct [18, 20]. Catalase and SOD activity is present in *C. jejuni* and *C. pylori* [24, 25]; SOD activity has been shown to be related directly to the aero-tolerance of certain strains of *C. jejuni* [24].

Campylobacter pylori grows best in a microaerobic atmosphere of 5% O_2 with 5–10% CO_2 [26]; it will not grow in the complete absence of oxygen (anaerobic culture). A suitable atmosphere can be achieved in anaerobe jars by the use of commercial gas-producing sachets which are activated by adding water. Envelopes generating a gas mixture suitable for isolation of anaerobes may be used in jars without a catalyst, so that a microaerobic atmosphere results [27]. Envelopes are available, however, that, in the presence of a catalyst, produce the correct gaseous environment for the isolation of campylobacters [28, 29]. A moist atmosphere is generated in jars using sachets, which is useful as *C. pylori* requires a high (approximately 98%) relative humidity for growth [1, 26]. If the microaerobic atmosphere is produced in the jar by evacuation and replacement of air by a suitable gas mixture [26] the plates should not be dried and the humidity in the jar can be increased by including some blotting paper soaked in water [30]. It is possible to grow *C.*

pylori on sub-culture in a humidified carbon dioxide incubator with 8–10% CO_2 [1], but this is not recommended for primary isolation [26, 28], particularly if the incubator is used for other purposes so that the oxygen concentration is frequently increased when the door of the incubator is opened. Other campylobacters will grow in atmospheric oxygen with 8–10% CO_2 on media supplemented with blood or other additives which reduce oxygen toxicity [22, 23]. Carbon dioxide is assimilated in many key carboxylation reactions; capneic bacteria require an elevated concentration of CO_2 (5–10%) having a lower affinity rather than a special use for carbon-dioxide [20].

Specimens taken for isolation of *C. pylori* should be transported to the laboratory as quickly as possible. Gastric biopsies should be maintained in either broth or buffer solution to prevent the organisms from drying out and also to protect them from the effects of atmospheric oxygen. Nutrient broth will contain very little dissolved oxygen [20]. Nutrient broth [31], thioglycollate broth [32], phosphate buffered saline [29] and normal saline [27] have all been used as transport media; biopsies have been stored for up to 5 hours at 4°C without adverse effects on the recovery of *C. pylori* [26]. A small volume of transport medium should be used as the biopsy may then be ground in this fluid before inoculation of solid media and any bacteria detached from the biopsy in transit will be recovered. An alternative approach is to use a small amount of hypertonic fluid, e.g. 20% glucose, as the transport medium [26]. This will maintain the structure of the biopsy so that bacteria and the mucus layer will not become detached.

Solid media

A variety of basal media with agar and added blood have been used for the isolation of *C. pylori* from clinical specimens. On these solid media, incubated in a microaerobic atmosphere at 37°C, small colonies (1 mm diameter) are seen usually after incubation for 3–4 days. The small translucent colonies produce slight haemolysis on blood agar and are greyish in colour but take up a tan coloration on heated blood agar. Strains should be sub-cultured every 2–3 days to maintain viability and should be exposed to atmospheric oxygen for as short a time as possible. Basal media used include Blood Agar Base No. 2 (Oxoid) [29, 31], Columbia agar base [29], brucella agar [27, 28] and brain–heart infusion agar [26, 32]. These have been supplemented with 5–10% sheep or horse blood; one group reported that whole blood gave better results than lysed blood [26]. Others have, however, demonstrated an increased recovery on primary isolation using a selective medium with lysed blood compared with a non-selective medium using whole blood [33]. Skirrow's medium, a selective medium widely used for the isolation of thermophilic campylobacters from faeces, has been used by a number of workers [9, 33, 34]. This medium incorporates lysed blood, as do the selective media used for the isolation of *Neisseria gonorrhoeae*, modified Thayer Martin medium [28] and New York City medium (used by the author) which have both been used successfully for the primary isolation of *C. pylori*. Many studies quote the use of 'chocolate' agar [1, 27, 30]. 'Chocolate' agar may be heated blood agar or made with GC agar base supplemented with yeast extract, haemoglobin or haemin and various other additives.

A wide variety of solid media have been used for sub-culture and sensitivity testing; these have usually incorporated blood or serum. In addition to the media described for primary isolation the basal media include tryptic soy agar [27, 30], Mueller–Hinton agar [27, 30], Wilkins–Chalgren agar [27] and Isosensitest agar [35]. Goodwin *et al.* found that their brain-heart infusion agar with 7% horse blood and 1% Isovitalex could be stored at room temperature in an air tight container for over two weeks without deterioration [26]. Plates should not be dried before use as high humidity is required for optimal isolation rates.

Triphenyl-tetrazolium chloride (TTC) (0.4%) has been used in a clear solid medium (brain–heart infusion agar with 10% horse serum and 0.25% yeast extract) to demonstrate colonial growth, as growing bacteria reduce TTC and produce red colonies [26]. A selective blood agar incorporating TTC has also been used for primary isolation [32]. *Campylobacter pylori* grew as tiny golden yellow colonies and no other bacterial species tested produced this particular sort of pigmentation.

Although blood agar media are satisfactory for the primary isolation and subculture of *C. pylori* a number of studies have investigated the use of other supplements and blood-free media in an attempt to determine the nutritional requirements for growth, to produce more rapid growth and better aero-tolerance. *C. pylori* will not grow on nutrient agar (Blood Agar Base No. 2, Oxoid) incubated microaerobically at 37°C but growth has been described on this medium satelliting around an X-Factor (haemin) disc [34]. However, blood, serum and haemin are not absolute requirements for growth [30, 36]. Slight growth is obtained on GC agar base and Mueller–Hinton agar which both contain starch. GC agar base supplemented with an extra 1% starch or 0.2% activated charcoal supports growth as well as 'chocolate' agar (GC agar base with 1% haemin) and brucella agar with 5% horse serum is almost as good as 'chocolate' agar for the growth of *C. pylori* [36]. The role of charcoal, serum, starch and haemin may be to inactivate or absorb toxic factors; the production of toxic radicals by bacteria and in media has been described above. The growth and aero-tolerance of *C. jejuni* is improved by addition of various media components, which may neutralize the effects of toxic radicals produced in the media rather than providing extra nutritional factors [19, 21–23]. Blood contains catalase, peroxidase and SOD; 1% blood and haematin, also probably acting as a detoxifying agent, will support the growth of *C. jejuni* on basal media incubated microaerobically [37]. Catalase and SOD have been shown to reduce the deleterious effects produced in media by storage in UV light [21, 22, 24]. Ferrous sulphate, sodium pyruvate and sodium metabisulphite have also been used singly and in combination, as FBP supplement, to reduce the effects of oxygen toxicity on the survival and growth of *C. jejuni* [19, 23, 37]. Sodium metabisulphite is however inhibitory for some isolates of *C. pylori* [26], so commercial FBP supplement should not be used for isolation of this species. A number of other scavengers of toxic radicals such as dithionite and histidine have been used to improve the isolation and aero-tolerance of *C. jejuni* [22] as has noradrenaline, which increases iron uptake by *C. jejuni*, and stimulates growth [38]. These substances are under investigation in the author's laboratory for the improvement of culture media for *C. pylori*.

'Isovitalex' (BBL), identical in composition to 'Vitox' (Oxoid), is a chemically-defined mixture of growth factors, and has been used to supplement blood agar media for the culture of *C. pylori* [26]. The individual components that may be of value have not been investigated, and 'Vitox' was found to have no effect on growth in broth in another study [39].

Liquid media

Culture in broth is dependent on an adequate dispersion of a microaerobic gas mixture throughout the medium and this may be achieved in static culture by incubation of a small volume of medium with a large area of gas–liquid interface, e.g. using tissue culture flasks [12, 40], or by shaking the broth in a microaerobic atmosphere [39, 41]. Liquid media described include brain–heart infusion broth with 10% horse serum and 0.25% yeast extract [40], brucella broth with 1–10% fetal calf serum [39] or 5–10% horse serum [36, 42], nutrient broth with 0.25% yeast extract and 5% horse serum [12] and Isosensitest, Columbia and Mueller–Hinton broths all with 10% fetal calf serum [41]. In the author's laboratory better growth has been achieved with broths containing fetal calf rather than horse serum. As stated above, 1% 'Vitox' supplement did not stimulate growth in serum enriched broth [39] but 1% soluble starch may replace serum in liquid media, as well as in solid media, with equivalent growth [36]. All bacteria colonizing intestinal mucus tend to be helical or spiral in shape and this may facilitate rapid movement in this environment [15], but hog gastric mucin has little stimulatory effect on growth of *C. pylori* in broth [36].

Campylobacter pylori has never been recovered from blood cultures or sites distant from the stomach and adjacent oesophagus and duodenum, but, in one study, *C. pylori* failed to grow in three commercial blood culture systems inoculated with the organism and fresh human blood [42]. This blood was not serologically tested and could have included anti-*C. pylori* antibodies but the organism also failed to grow when the broths were inoculated with sheep blood which did support growth on solid media.

Optimum pH and temperature for growth

The survival at different pHs has been described above. A number of studies have investigated the pH range for growth of *C. pylori*. This organism will grow over a wide range of pH from 5.5 to 8.5 with good growth between pH 6.9 and 8.0 [9, 39, 43]. The studies quoted demonstrated positive growth by measuring turbidity in serum broth or viable counts. There has been one report of growth of *C. pylori* over an extremely wide pH range of 2.9 to 9.2 [44]. However, in this study, reduction of triphenyl-tetrazolium chloride in a brucella medium with agar was used to demonstrate active growth.

The temperature range at which growth of *C. pylori* occurs is described as 30–40°C [9, 43]. There are no reports of growth at temperatures below 30°C but some authors have found certain strains capable of growth at 42°C [27, 29, 30]. However, many strains will not grow in this temperature and it is not recommended that the incubation conditions used for the isolation of thermophilic campylobacters from faeces are used for the attempted isolation of *C. pylori*.

Storage of C. pylori

These bacteria cannot be freeze-dried to reliably produce a good recovery on sub-culture [45]. One group of workers has maintained cultures by monthly sub-culture in 0.17% brain–heart infusion agar containing 10% horse serum and 0.25% yeast extract [40]. This medium was incubated at 37°C in a carbon dioxide (8%) incubator with 98% humidity. The organisms may be stored frozen at −70°C or in liquid nitrogen and a variety of cryopreservative media have been employed: 1% proteose peptone with 15% glycerol [29], brucella broth [32], tryptone soy broth with 15% glycerol [46] and the 'Protect' Bacterial Preserver System [46]. There has been one report, disputed by others, that combinations of brain–heart infusion broth, tryptic soy broth, heat-inactivated serum and glycerol all failed to sustain viability of *C. pylori* for more than a few weeks at −70°C [45]. Skimmed milk was better than other media for cryopreservation but the best medium for storage of frozen *C. pylori* was whole defibrinated horse blood. In blood, the organisms were maintained even after repeated freeze–thawing for over 6 months. Other groups have also used whole rabbit blood [39] and sheep blood [36]. Defibrinated horse blood with 10% glycerol is used in the author's laboratory for storage in liquid nitrogen or at −70°C and organisms have remained viable for over 2 years stored under these conditions.

The optimum conditions for growth of *C.*

pylori have probably not yet been defined. Coccoid forms occur readily in cultures, and with *C. jejuni* the production of coccoid forms occurs when conditions for growth are unfavourable, particularly when cells are exposed to atmospheric oxygen [24]. Culture techniques are more sensitive than microscopic methods for the detection of most other bacterial species but presently histology is more sensitive than gastric biopsy culture for the detection of *C. pylori* in patients with gastritis. Improvement in the methodology for culture should lead to a greater sensitivity which may reveal more details of the sources and transmission of *C. pylori* and the pathogenesis of *C. pylori*-associated disease.

References

1 Marshall, B.J., Royce, H., Annear, D.I., Goodwin, C.S., Pearman, J.W., Warren, J.R., & Armstrong, J.A. Original isolation of *Campylobacter pyloridis* from human gastric mucosa. *Microbios Lett.* 1984, **25**, 83–8.

2 Warren, J.R. & Marshall, B.J. Unidentified curved bacilli on gastric epithelium in active chronic gastritis. *Lancet* 1983, **i**, 1273–5.

3 Giannella, R.A., Broitman, S.A. & Zamcheck, N. Gastric acid barrier to ingested micro-organisms in man: studies *in vivo* and *in vitro*. *Gut* 1972, **13**, 251–6.

4 Quigley, E.M.M. & Turnberg, L.A. pH of the microclimate lining human gastric and duodenal mucosa *in vivo*. *Gastroenterology* 1987, **92**, 1876–84.

5 Rathbone, B.J., Wyatt, J.I. & Heatley, R.V. *Campylobacter pyloridis*—a new factor in peptic ulcer disease? *Gut* 1986, **27**, 635–41.

6 Paull, G., Yardley, J.H., Rawles, J.W. & Walters, C.C. Campylobacter-like organisms in patients with Barrett's oesophagus. *Gastroenterology* 1986, **90**, 1580.

7 Pambianco, D.J., Dye, K.R., Marshall, B.J., Frierson, H.F., MacMillan, R.H, Franquemont, D. & McCallum, R.W. Gastritis in the rectum: campylobacter-like organisms in heterotopic inflamed gastric mucosa. *Gastroenterology* 1988, **94**, A340.

8 Hazell, S.L. & Lee, A. 1985. The adaptation of motile strains of *Campylobacter pyloridis* to gastric mucus and their association with gastric epithelial intercellular spaces. In Pearson, A.D., Skirrow, M.B., Lior, H. & Rowe, B. (eds) *Campylobacter III: Proceedings of the Third International Workshop on Campylobacter Infection, Ottawa, 1985*. Public Health Laboratory Service, London, pp. 189–91.

9 Itoh, T., Yanagawa, Y., Shingaki, M., Takahashi, M., Kai, A., Ohashi, M. & Hamana, G. Isolation of *Campylobacter pyloridis* from human gastric mucosa and characterization of the isolates. *Microbiol. Immunol.* 1987, **31**, 603–14.

10 Marshall, B.J., Barrett, L.J., Prakesh, C., McCallum, R.W. & Guerrant, R.L. 1987. Protection of *Campylobacter pyloridis*, but not *Campylobacter jejuni* against acid susceptibility by urea. In Kaijser, B. & Falsen, E. (eds) *Campylobacter IV: Proceedings of the Fourth International Workshop on Campylobacter Infection, Göteborg, 1987*. University of Göteborg, Göteborg, pp. 402–3.

11 Tompkins, D.S. & West, A.P. *Campylobacter pylori*, acid and bile. *J. Clin. Pathol.* 1987, **40**, 1387.

12 Goodwin, C.S., Armstrong, J.A. & Marshall, B.J. *Campylobacter pyloridis*, gastritis, and peptic ulceration. *J. Clin. Pathol.* 1986, **39**, 353–65.

13 Jones, D.M. & Eldridge, J. 1987. Gastric campylobacter-like organisms (GCLO) from man (*C. pyloridis*) compared with GCLO strains from the pig, baboon and ferret. In Kaijser, B. & Falsen, E. (eds) *Campylobacter IV: Proceedings of the Fourth International Workshop on Campylobacter Infection. Göteborg, 1987*. University of Göteborg, Göteborg, p. 44.

14 Bronsdon, M.A. & Schoenknecht, F.D. *Campylobacter pylori* isolated from the stomach of the monkey, *Macaca nemestrina*. *J. Clin. Microbiol.* 1988, **26**, 1725–8.

15 Lee, A. & Hazell, S.L. *Campylobacter pylori* in health and disease: An ecological perspective. *Microbiol. Ecol. Health Dis.* 1988, **1**, 1–16.

16 Ishii, E., Kishi, T., Inoue, H., Shimoyama, T., Tanaka, T., Tsuyuguchi, T., Tamura, T. & Shoji, K. Isolation of *Campylobacter pyloridis* from gastric mucosa of patients with gastric complaints and susceptibility of the isolates to antibiotics. *J. Jap. Ass. Inf. Dis.* 1987, **61**, 668–75.

17 O'Connor, H.J., Wyatt, J.I., Dixon, M.F., Axon, A.T.R. Campylobacter-like organisms and reflux gastritis. *J. Clin. Pathol.* 1986, **39**, 531–4.

18 Stouthamer, A.H., De Vries, W. & Niekus, H.G.D. Microaerophily. *Antonie van Leeuwenhoek J. Microbiol* 1976, **45**, 5–12.

19 George, H.A., Hoffman, P.S., Smibert, R.M. & Krieg, N.R. Improved media for growth and aerotolerance of *Campylobacter fetus*. *J. Clin. Microbiol.* 1978, **8**, 36–41.

20 Morris, J.G. 1983. The metabolism, growth and death of bacteria. In Wilson, G.S., Miles, A.A. & Parker, M.T. (eds) *Topley and Wilson's Principles of Bacteriology, Virology and Immunity*, Vol. 1, 7th edn. Edward Arnold, London, pp. 39–69.

21 Hoffman, P.S. George, H.A., Krieg, N.R. & Smibert,

R.M. Studies of the microaerophilic nature of *Campylobacter fetus* subsp. *jejuni*. II. Role of exogenous superoxide anions and hydrogen peroxide. *Can. J. Microbiol.* 1979, **25**, 8–16.

22 Juven, B.J. & Rosenthal, I. Effect of free-radical and oxygen scavengers on photochemically generated oxygen toxicity and on the aerotolerance of *Campylobacter jejuni*. *J. Appl Bacteriol.* 1985, **59**, 413–19.

23 Bolton, F.J., Coates, D. & Hutchinson, D.N. The ability of campylobacter media supplements to neutralize photochemically induced toxicity and hydrogen peroxide. *J. Appl. Bacteriol.* 1984, **56**, 151–7.

24 Moran, A.P. & Upton, M.E. Effect of medium supplements, illumination and superoxide dismutase on the production of coccoid forms of *Campylobacter jejuni* ATCC 29428. *J. Appl. Bacteriol.* 1987, **62**, 43–51.

25 Lior, H. & Johnson, W.M. 1985. Catalase, peroxidase and superoxide dismutase activities in *Campylobacter* spp. In Pearson, A.D., Skirrow, M.B., Lior, H. & Rowe, B. (eds) *Campylobacter III: Proceedings of the Third, International Workshop on Campylobacter Infections* Public Health Laboratory Service, London, pp. 266–7.

26 Goodwin, C.S., Blincow, E.D., Warren, J.R., Waters, T.E., Sanderson, C.R. & Easton, L. Evaluation of cultural techniques for isolating *Campylobacter pyloridis* from endoscopic biopsies of gastric mucosa. *J. Clin. Pathol.* 1985, **38**, 1127–31.

27 Megraud, F., Bonnet, F., Garnier, M. & Lamouliatte, H. Characterization of *'Campylobacter pyloridis'* by culture, enzymatic profile and protein content. *J. Clin. Microbiol.* 1985, **22**, 1007–10.

28 Parsonnet, J., Welch, K., Compton, C., Strauss, R., Wang, T., Kelsey, P. & Ferraro, M.J. Simple microbiologic detection of *Campylobacter pylori*. *J. Clin. Microbiol.* 1988, **26**, 948–9.

29 Langenberg, W., Rauws, E.A.J., Widjojokusumo, A., Tytgat, G.N.J. & Zanen, H.C. Identification of *Campylobacter pyloridis* isolates by restriction endonuclease DNA analysis. *J. Clin. Microbiol.* 1986, **24**, 414–17.

30 Buck, G.E., Gourley, W.K., Lee, W.K., Subramanyam, K., Latimer, J.M. & Dinuzzo, A.R. Relation of *Campylobacter pyloridis* to gastritis and peptic ulcer. *J. Inf. Dis.* 1986, **153**, 664–9.

31 Jones, D.M., Lessells, A.M. & Eldridge, J. Campylobacter-like organisms on the gastric mucosa: culture, histological and serological studies. *J. Clin. Pathol.* 1984, **37**, 1002–6.

32 Queiroz, D.M.M., Mendes, E.N. & Rocha, G.A. Indicator medium for isolation of *Campylobacter pylori*. *J. Clin. Microbiol.* 1987, **25**, 2378–9.

33 Krajden, J., Bohnen, J., Anderson, J., Kempston, J., Fuksa, M., Matlow, A., Marcon, N., Haber, G., Kortan, P., Karmali, M., Corey, P., Petrea, C., Babida, C. & Hayman, S. Comparison of selective and nonselective media for recovery of *Campylobacter pylori* from antral biopsies. *J. Clin. Microbiol.* 1987, **25**, 1117–18.

34 Hazell, S.L., Lee, A., Brady, L. & Hennessy, W. *Campylobacter pyloridis* and gastritis: Association with intercellular spaces and adaptation to an environment of mucus as important factors in colonization of the gastric epithelium. *J. Inf. Dis.* 1986, **153**, 658–63.

35 McNulty, C.A.M., Dent, J. & Wise, R. Susceptibility of clinical isolates of *Campylobacter pyloridis* to 11 antimicrobial agents. *Antimicrob. Agents Chemother.* 1985, **28**, 837–8.

36 Buck, G.E. & Smith, J.S. Medium supplementation for growth of *Campylobacter pyloridis*. *J. Clin. Microbiol.* 1987, **25**, 597–9.

37 Bolton, F.J. & Coates, D. Development of a blood-free *Campylobacter* medium: screening tests on basal media and supplements, and the ability of selected supplements to facilitate aerotolerance. *J. Appl. Bacteriol.* 1983, **54**, 115–25.

38 Hoffman, P.S., Krieg, N.R. & Smibert, R.M. Studies of the microaerophilic nature of *Campylobacter fetus* subsp. *jejuni*. I. Physiological aspects of enhanced aerotolerance. *Can. J. Microbiol.* 1979, **25**, 1–7.

39 Morgan, D.R., Freedman, R., Depew, C.E. & Kraft, W.G. Growth of *Campylobacter pylori* in liquid media. *J. Clin. Microbiol.* 1987, **25**, 2123–5.

40 Goodwin, C.S., Blake, P. & Blincow, E. The minimum inhibitory and bactericidal concentrations of antibiotics and anti-ulcer agents against *Campylobacter pyloridis*. *J. Antimicrob. Chemother.* 1986, **17**, 309–14.

41 Leunk, R.D., Johnson, P.T., David, B.C., Kraft, W.G. & Morgan, D.R. Cytotoxic activity in broth-culture filtrates of *Campylobacter pylori*. *J. Med. Microbiol.* 1988, **26**, 93–9.

42 Westblom, T.U., Madan, E., Midkiff, B.R., Adkins, V.W. & Subik, M.A. Failure of *Campylobacter pylori* to grow in commercial blood culture systems. *J. Clin. Microbiol.* 1988, **26**, 1029–30

43 Goodwin, C.S., Collins, M.D. & Blincow, E. The absence of thermoplasmaquinones in *Campylobacter pyloridis*, and its temperature and pH growth range. *Microbios. Lett.* 1986, **32**, 137–40.

44 Goldie, J., Jalali, S., Hunt, R.H. & Richardson, H. Study of media and pH requirements for the growth of *Campylobacter pylori*. *Gastroenterology* 1988, **94**, A150.

45 Westblom, T.U., Barthel, J.S., Havey, A.D., Gonzalez, F.J., Tarka, E.F. & Everett, E.D. Long term freeze storage of *Campylobacter pyloridis*. *J. Clin. Pathol.* 1987, **40**, 353.

46 Ribeiro, C.D. & Gray, S.J. Long term freeze storage of *Campylobacter pyloridis*. *J. Clin. Pathol.* 1987, **40**, 1265.

5 Campylobacter pylori: biochemical characteristics

M.J. HUDSON

Campylobacter pylori (previously *C. pyloridis*) is a microaerophilic, motile, Gram-negative curved or spiral bacterium. Its principal habitat appears to be within and beneath the mucus blanket overlying human gastric-type epithelium in the upper gastro-intestinal tract; its presence is often associated with inflammatory and ulcerative changes [1–4]. The morphology, growth requirements and basic phenotypic characteristics of *C. pylori*, taken together with a DNA base ratio of 36–8 moles G + C%, are all consistent with its inclusion within the genus *Campylobacter* [5]; the multiple polar flagella arrangement seen in *C. pylori* has not been described for other campylobacters.

Campylobacter-like organisms (CLO) and spirals distinct from *C. pylori* have been seen and sometimes cultured from the gastric mucosae of man, non-human primates and other mammals [6–13]. Whether these other gastric organisms are more or less closely related to *C. pylori* awaits isolation and comparative studies, but clearly they comprise a unique group of bacteria sharing a very particular micro-ecological niche. Other non-spiral microbes are often present and commonly dismissed as contaminants from the oro-pharynx; that is clearly not the case where there is local or general atrophic changes and hypochlorhydria. Understanding the relationships of these other gastric microbes to gastroduodenal disease will greatly facilitate our understanding of the role of *C. pylori* in the aetiology and pathogenesis of active and atrophic gastritis, peptic ulcer disease and intestinal metaplasia. A clearer understanding of the biochemistry and physiology of growth *in vitro* may contribute to our understanding of the factors controlling gastric colonization in man and identify possible pathogenic mechanisms.

Basic morphological and biochemical characteristics

Primary culture of *C. pylori* from clinical material is a relatively simple but slow process, taking 3–7 days for colonies 0.5–1 mm in diameter to appear; apart from being slow-growing, the organism is fastidious, requiring both a carefully-controlled atmosphere and blood- or serum-supplemented media. It is oxygen-sensitive but has an absolute requirement for oxygen and grows best in a humid microaerobic atmosphere of 5–7% O_2 with 5–10% CO_2 [14, 15]. The apparent paradox of both an obligate requirement for and marked sensitivity to oxygen is common to many campylobacters and of course is also seen in other microaerophilic obligate aerobes such as the gonococcus. Subsequent passage *in vitro* often results in laboratory strains which grow in 1–3 days and are markedly more tolerant of oxygen.

C. pylori exhibits both catalase and cytochrome oxidase activity but does not ferment sugars. Neither nitrate nor nitrite are normally reduced; hydrogen sulphide is not detectable in FBP medium but lead acetate paper is blackened. Hippurate is not hydrolysed and triphenyl-tetrazolium chloride (TTC), although tolerated, is not reduced. *Campylobacter pylori* can grow in the presence of 1% glycine but not 1% bile salts or 3.5% NaCl [14, 15].

Urea is rapidly hydrolysed [15–17] in contrast to the original description of the organism [14]. *Campylobacter pylori* urease is produced constitutively and with exceptionally high enzyme activity. The large amounts of urease produced *in vivo* has been utilized for the rapid diagnosis of *C. pylori* colonization by direct demonstration of urease activity of biopsy tissue [18]. Similarly, gastric mucosal urease activity may be detected indirectly by a breath test after oral administration of isotopically-labelled urea [19, 20]. Hazell and Lee [21] have suggested that in gastric epithelia colonized with *C. pylori*, local rapid hydrolysis of urea to ammonia may interfere with the normal transfer of protons through the mucus and result in back diffusion; a consequence of the latter would be locally-reduced gastric acid secretion, loss of mucus integrity and ultimately ulceration. This hypothesis could, and should, be tested using specific urease inhibitors or urease-negative *C. pylori* mutants in an animal model of *C. pylori*-associated gastritis.

The optimal growth temperature of *C. pylori* is 37°C; only a minority of strains can grow at 30°C or 42°C and where growth does occur it is poor; no growth occurs at 25°C and 45°C. Batch culture in static liquid media is improved by the addition of serum, starch, mucin, charcoal or methylcellulose, apparently any of several substrates that 'detoxify' media [22]. Batch culture in liquid media is much more successful if small volumes of broth with a large surface area are shaken under a microaerobic atmosphere [23, 24]. Under such conditions, only small concentrations (1%) of serum are required for good and reproducible growth [23].

A weak haemolysin is commonly produced on horse and sheep blood agars [14, 15] and a mucin protease has been found by one group [25]. The significance of these potential virulence factors remains largely unknown; it has been suggested that the haemolysin may satisfy a requirement for haemin [4] although others have failed to show an obligate requirement for large amounts of haemin for growth *in vitro* [18, 19]. Mucin degradation might serve to facilitate motility in this viscous environment, although *C. pylori* appears to be adapted to such environments and is highly motile in viscous non-mucin solutions [4]; amino acids liberated by mucin degradation might serve some nutritional requirement although mucin does not greatly stimulate growth *in vitro* [22]. Damage to the integrity of the mucus barrier by the action of a mucin protease would clearly be a powerful mechanism by which *C. pylori* initiated inflammatory and ulcerative changes in the gastroduodenal mucosa; confirmation of, and further studies on the mucin protease activity of *C. pylori* and related organisms are eagerly awaited. Mannose-resistant haemagglutinin activity against erythrocytes from a number of animal species has been demonstrated [26], and may mediate adhesion of *C. pylori* to specific mucin or epithelial glycoproteins.

Little is known regarding the amino acids and carbon sources required and utilized by *C. pylori* for growth *in vitro* or *in vivo*. In preliminary continuous culture studies with *C. pylori*, we have shown fumarate to be reduced to succinate in the presence of molar excess formate and several amino acids, including glutamate, aspartate and serine disappeared from the medium in direct proportion to cell density. Under conditions of oxygen excess, nitrate was not readily reduced to nitrite although some nitrate reduction was observed under partial oxygen debt. Fumarate reductase activity is common among the campylobacters, wolinellas and bacteroides; the physiological role of an inducible nitrate reductase requires further study but may in part explain the variable result obtained with this test [14, 15].

Chemosystematic aspects

The isolation and characterization of *C. pylori* has coincided with a new era in microbiology in which various macromolecular and structural chemical markers are commonly used as adjuncts to the rapid identification of isolates rather than purely for taxonomic comparisons. SDS-polyacrylamide gel electrophoresis of proteins from whole cells or cell fractions has been shown to be a simple method for the comparison of *C. pylori* isolates and differentiation from other campylobacters [9, 15, 27]. Similarly, we and others have found analysis of structural fatty acids by gas–liquid chromatography (GLC) to be a simple and rapid technique for identifying *C. pylori* and differentiating this organism from other campylobacters [28, 29]. Both of these useful and rapid techniques for the identification of *C. pylori* isolates show this organism to be a distinct and largely homogeneous species.

Campylobacter pylori as a distinct species

The bacterial species is the only taxonomic unit in bacterial systematics where the phylogenetic definition usually fits species assigned by classical phenotyping [30]. DNA–DNA hybridization is used to measure the similarity in sequence of genomic DNA between strains. Von Wulffen [31] has shown high (> 70%) DNA–DNA homology between *C. pylori* strains and low (< 30%) homology with other campylobacters. These data are consistent with *C. pylori* as a distinct species and this is further supported by the uniform cellular fatty acid composition [27, 28, 32], enzyme activities and protein profiles between strains [15].

There is little phenotypic heterogeneity within *C. pylori* [15], but it is possible to biotype isolates beyond species level by gel electrophoresis of restriction endonuclease digests of genomic DNA [24, 33]. This technique results in an enormous variety of digest pat-

Table 5.1 FAME profile groups of *Campylobacter* species. (Data are presented semi-quantitatively for simplicity)*

FAME group		12.0	14.0	OH-14.0	16.1	16.0	OH-16.0	18.2	18.1	18.0	cyc-19.0
I	*C. jejuni*	−	+	(+)	(+)	+++	−	(+)	+++	(+)	++
	C. coli										
	C. jejuni s.s. *doylei*										
	C. upsaliensis										
II	*C. fetus* s.s. *fetus*	(+)	(+)	(+)	+	+++	−	(+)	+++	(+)	−
	C. laridis										
	UPTC										
	C. hyointestinalis										
	C. concisus										
III	*C. sputorum* b.v. *sputorum*	−	++	+	(+)	+++	−	(+)	+++	(+)	−
	C. sputorum b.v. *bubulus*										
	C. sputorum b.v. *fecalis*										
	C. upsaliensis										
IV	*C. fetus* s.s. *venerealis*	(+)	+	(+)	++	+++	−	−	++	−	−
V	*C. cinaedi*	(+)	+	−	−	+++	(+)	−	++	(+)	−
VI	*C. fennelliae*	(+)	+	(+)	−	+++	−	−	+++	(+)	−
VII	*C. cryaerophila*	(+)	(+)	(+)	+++	++	(+)	−	++	(+)	−
VIII	*C.* sp. DSM 806	(+)	+	(+)	+++	+++	−	−	++	−	−
	spirillum 5175										
IX	*C. pylori*	(+)	+++	−	−	+	(+)	(+)	++	++	+++
X	*C. mucosalis*	+	+	+	++	+++	−	(+)	+++	(+)	−
XI	*C. nitrofigilis*	−	(+)	−	+++	+++	−	−	+++	(+)	(+)

*(+) = 1–5%; + = 6–10%; ++ = 11–25%; +++ = >26%.

terns and suggests that there is considerable genetic diversity within the species. Immunoblot analysis [34] has been used to differentiate isolates and suggests that serotyping may be possible although no scheme exists at present.

Does Campylobacter pylori belong in the genus Campylobacter?

Ultrastructure

Campylobacter pylori differs from other campylobacters in several ultrastructural features, most notably in surface structure and the presence of a polar tuft of 2–6 sheathed flagella with terminal bulbs or blebs [14, 28, 35]. A gastric organism from the ferret possessed similar features except that the sheathed flagella arose apparently randomly around the cell [8, 10]. Both organisms are obviously well adapted to life in the specialized micro-environment of the gastric mucosa and these distinctive ultrastructural features could reflect evolutionary divergence over any time scale and their taxonomic significance remains unclear.

Fatty acid composition

Goodwin and colleagues reported [29] that the cellular fatty acid profile of *C. pylori* differed from that of *C. jejuni*. The major acids in *C. pylori* were 14.0 and cyclopropane-19.0 together with 18.0, 18.1 and hydroxy-16.0; in *C. jejuni*, the major acids were 16.0, 18.1 and cyclopropane-19.0. However, in this study the cyclopropane-19.0 acid was misidentified as the *cis*-9,10-isomer. Mass-spectrometric analysis [27] showed it to be the *cis*-11,12-isomer, identical to that present in *C. jejuni* or *C. coli* and the appropriate cyclopropane fatty acid produced from 11-octadecenoic acid, the precursor 18.1 acid present [27].

Comparison of bacterial cellular fatty acid composition is a relatively simple and reproducible procedure. Fatty acid methyl esters (FAME) are prepared by mild acid methanolysis of a few loopfuls of cells followed by extraction and analysis by GLC. Using such techniques, the campylobacters can be divided into at least 11 FAME profile groups [36] as shown in Table 5.1, and finer differentiation within groups to species level is sometimes possible. The fatty acid profile of *C. pylori* is distinctive and reproducible. The major acids are 14.0 and cyc-19.0 with moderate amounts of 18.0 and comparatively little 16.0; of the campylobacters studied, only in *C. pylori* is the ratio of 18.0 to 16.0 greater than unity. In these studies, no fatty acid isomers unique to *C. pylori* were identified. Differences between *C. pylori* and other campylobacters were quantitative rather than qualitative. Lambert and colleagues [32] have recently reported the presence of OH-18.0 in *C. pylori*; this is probably due to differences in FAME preparation and analysis. We have not found this hydroxy-acid to comprise a significant proportion of the fatty acids and the significance of this hydroxy acid as a taxonomic marker for *C. pylori* requires further study.

Table 5.2 shows comparative FAME data for a number of gastric CLO together with reference *Campylobacter* species and *Wolinella succinogenes* (*vide infra*). The cellular fatty acid profiles of gastric CLO isolated from rhesus monkeys were indistinguishable from *C. pylori* [9]. Single isolates from pig and baboon [10] also resembled *C. pylori* whereas two other, non-ureolytic gastric isolates from baboon resembled *C. jejuni*. *Campylobacter hyointestinalis*-like CLO may also be isolated from the pig stomach. Ferret gastric CLO from two separate colonies fall into the same general FAME group as *C. jejuni* although these can be distinguished by several minor differences.

Campylobacter pylori and ferret gastric CLO are strongly ureolytic. This is not a unique feature among the campylobacters; urease activity is also found in the UPTC (urease-positive thermophilic campylobacters) and C.

Table 5.2 FAME profiles of *Campylobacter pylori* and other gastric CLO compared to *C. jejuni* and *Wolinella succinogenes*. (Data are presented semi-quantitatively for simplicity)*

	12.0	14.0	OH-14.0	16.1	16.0	OH-16.0	18.2	18.1	18.0	cyc-19.0
C. pylori	(+)	+++	–	–	+	(+)	(+)	++	++	+++
Rhesus GCLO	(+)	+++	–	–	+	(+)	(+)	++	++	+++
Pig GCLO	(+)	+++	–	–	+	(+)	(+)	++	++	+++
Baboon GCLO (1)	(+)	+++	–	–	+	(+)	(+)	++	++	+++
W. succinogenes	–	(+)	(+)	+++	++	(+)	(+)	+++	(+)	–
Baboon GCLO (2)	(+)	++	(+)	(+)	+++	–	–	++	–	++
Ferret GCLO	–	++	(+)	(+)	+++	(+)	–	++	(+)	++
C. jejuni s.s. *doylei*†	–	++	(+)	+	+++	–	–	+++	–	++
C. jejuni	–	++	(+)	(+)	+++	–	–	++	–	++

*(+) = 1–5%; + = 6–10%; ++ = 11–25%; +++ = >26%.
†GCLO-2.

nitrofigilis. Comparison of FAME profiles shows that this characteristic is represented in four separate FAME groups: I, II, IX and XI (Table 5.1).

From a diagnostic standpoint, FAME profiling is an excellent method for the rapid and accurate identification of *C. pylori*. From a taxonomic standpoint, these cellular fatty acid data show campylobacters comprise a broadly-related group of organisms. There are few qualitative differences in fatty acids between species, but rather the distinctions are quantitative. *Campylobacter pylori*, together with the ferret gastric CLO, belong to this broad family of organisms.

Respiratory quinone composition

Two strains of *Campylobacter pylori* have been examined for respiratory quinones [37]. Menaquinone-6 (MK-6) was the sole isoprenoid quinone detected; no thermoplasmaquinone-6 (TPQ-6, methyl-substituted MK-6) was detected. All campylobacters previously studied contained both of these isoprenoid quinones [38, 39], although the proportions found for individual species in the two studies was sometimes quite different. Whether this is attributable to methodological differences or variation of these quinones under different growth conditions remains to be determined. Further strains of *C. pylori* need to be examined, together with other *Campylobacter* species and CLO. Interestingly, this combination of menaquinones is also found in *Wolinella succinogenes* [40], *Bacteroides gracilis* [41] and in spirillum 5175 [42]. The taxonomic significance of the quinone profile of *C. pylori* is uncertain; it may lack part of the quinone profile characteristic of the campylobacters and near relatives but equally it does not possess isoprenologues of MK or other respiratory quinones not found in the campylobacters.

Ribosomal RNA sequence comparison

Several groups have used 16S ribosomal RNA partial sequence analysis to demonstrate phylogenetic relationships between the campylobacters and with other prokaryotes. Romaniuk and colleagues [43] compared the rRNA sequences of six species of *Campylobacter*, including *C. pylori*, with a database of comparable sequences for a variety of other microbes. They found that the campylobacters and wolinellas are not related to any major bacterial group, i.e. they are probably an ancient branch of eubacteria. The five 'true' campylobacters clustered together while *C. pylori* was found to be an outlier to the campylobacter group and more closely related to *Wolinella succinogenes*. Paster and Dewhirst [44] extended the phylogeny with data for additional *Campylobacter* species, wolinellas and bacteroides. Lau and colleagues [45] independently confirmed that *C. pylori* was more closely related to *W. succinogenes* than other campylobacters. Recently, Thompson and colleagues [46] have compared rRNA sequence data for all of the *Campylobacter* species. In addition to confirming the phylogenetic relationship between *C. pylori* and *W. succinogenes*, they also found *C. cinaedi* and *C. fennelliae* to fall within this wolinella cluster.

Wolinella succinogenes is an anaerobic, Gram-negative curved rod and differs from *C. pylori* in several respects. Not only does it not share the distinctive ultrastructure of *C. pylori* but it has a higher G + C content of 42–48% and different fatty acid profile (Table 5.2). The FAME profile resembles that of two anaerobic CLO, an aspartate-fermenting *Campylobacter* sp. (DSM 806) and a sulphur-reducing CLO (spirillum 5175). *Wolinella succinogenes* and spirillum 5175, like the campylobacters, have both been shown to possess MK-6 and TPQ-6 [40, 42].

Conclusions

It is often difficult to reconcile bacterial systematics with conventional phenotyping. The compelling data of rRNA sequence comparisons indicate *C. pylori* is only distantly related to the true campylobacters. Classification based on rRNA sequence divergence assumes that the rate of genetic drift is constant between and within branches and is not greatly affected by adaptation to specialized environments. It has been recommended that some allowance be made for this in deciding the depth at which particular hierarchical divisions, such as genera, are assigned along rRNA dendrograms [30]. Thompson and colleagues recommend that *C. pylori*, *C. cinaedi* and *C. fennelliae* be excluded from the genus *Campylobacter* and the genus *Wolinella* be redefined to include these species [46]. Given that several anaerobic species of the genera *Wolinella* and *Bacteroides* fall within the main campylobacter cluster [44] it might be prudent to re-examine the interrelationships of these species using other molecular and chemosystematic methods before formal reclassification.

Should *C. pylori* and close relatives such as the ferret gastric CLO [47] be assigned to a new genus? The transfer of existing species to new or revised genera should not only reflect bacterial systematics but also be consistent with basic phenotypic characteristics and thus help (and not hinder) routine laboratory studies. The onus is on the proposer of taxonomic changes to demonstrate to the scientific community at large, and in particular to the microbiologist working with conventional phenotyping systems at the bench, the usefulness of such changes. *Campylobacter pylori* grows on the same blood-enriched media (and selective media) under the same microaerobic atmosphere as other campylobacters and shares with them the morphology and basic phenotypic characteristics of the genus. In bacterial systematics there are 'lumpers' of species and 'splitters' of genera. In the absence of a compelling reason for haste in assigning *C. pylori* to some new genus, common sense favours the 'lumper', and the appropriate genus for this most interesting organism should continue to be *Campylobacter*.

Taxonomy apart, there is an urgent need for more information concerning the basic nutritional requirements and physiology of *C. pylori*; a great deal of emphasis has been placed on the properties of the urease. Other biochemical aspects that require clarification include the mechanism of oxygen utilization and toxicity, the role of the mucin protease, potential cytotoxins such as the haemolysin and other macromolecules such as lipopolysaccharides and haemagglutinins. A great deal has been learned about this organism in a relatively short period of time but many fundamental aspects of the physiology and biochemistry of *C. pylori* remain relatively unexplored.

References

1 Warren, J.R. Unidentified curved bacilli on gastric epithelium in active chronic gastritis. *Lancet* 1983, **i**, 1273.

2 Marshall, B. Unidentified curved bacilli on gastric epithelium in active chronic gastritis. *Lancet* 1983, **i**, 1273–5.

3 Marshall, B. & Warren, R. Unidentified curved bacilli in the stomach of patients with gastritis and peptic ulceration. *Lancet* 1984, **i**, 1311–15.

4 Hazell, S.L., Lee, A., Brady, L. & Hennessy, W. *Campylobacter pyloridis* and gastritis: Association with intercellular spaces and adaptation to an environment of mucus as important factors in colonization of the gastric epithelium. *J. Infect. Dis.* 1986, **153**, 658–63.

5 Smibert, R.M. 1984. Genus *Campylobacter* Sebald and Veron 1963,907. In Krieg, N.R. & Holt, J.G. (eds) *Bergey's Manual of Systematic Bacteriology*, Vol. 1 Williams & Wilkins, Baltimore, pp. 111–18.

6 Lee, A. & Hazell, S.L. *Campylobacter pylori* in health and disease: An ecological perspective. *Microb. Ecol. Health Dis.* 1988, **1**, 1–16.

7 Lee, A., Hazell, S.L., O'Rourke, J. & Kouprach, S. Isolation of a spiral-shaped bacterium from the cat stomach. *Infect. Immun.* 1988, **56**, 2843–50.

8 Fox, J.G., Edrise, B.M., Cabot, E.B., Beaucage, C., Murphy, J.C. & Prostak, K.S. Campylobacter-like organisms isolated from gastric mucosa of ferrets. *Am. J. Vet. Res.* 1986, **47**, 236–9.

9 Newell, D.G., Hudson, M.J. & Baskerville, A. Isolation of a gastric campylobacter-like organism from the stomach of four rhesus monkeys, and identification as *Campylobacter pylori*. *J. Med. Microbiol.* 1988, **27**, 41–4.

10 Jones, D.M. & Elridge, J. 1987. Gastric campylobacter-like organisms (GCLO) from man ('*C. pyloridis*') compared with GCLO strains from pig, baboon and ferret. In Kaijser, B. & Falsen, E. (eds) *Campylobacter IV*. University of Göteborg, Göteborg, p. 44.

11 Bronson, M.A. & Schoenknecht, F.D. *Campylobacter pylori* isolated from the stomach of the monkey, *Macaca nemestrina*. *J. Clin. Microbiol.* 1988, **26**, 1725–8.

12 Curry, A., Jones, D.M. & Elridge, J. Spiral organisms in the baboon stomach. *Lancet* 1987, **ii**, 634–5.

13 Dent, J., McNulty, C.A.M., Uff, J.C., Wilkinson, S.P. & Gear, M.W. Spiral organisms in the gastric antrum. *Lancet* 1987, **ii**, 96.

14 Marshall, B., Royce, H., Annear, D.I. *et al.* Original isolation of *Campylobacter pyloridis* from human gastric mucosa. *Microbios Lett.* 1984, **25**, 83–8.

15 Megraud, F., Bonnet, F., Garnier, M. & Lamouliatte, H. Characterization of '*Campylobacter pyloridis*' by culture, enzymatic profile, and protein content. *J. Clin. Microbiol.* 1985, **22**, 1007–10.

16 Langenberg, M.-L., Tytgat, G.N.J., Schipper, M.E.I., Rietra, P.J.G.M. & Zanen, H.C. Campylobacter-like organisms in the stomach of patients and healthy individuals. *Lancet* 1984, **i**, 1348.

17 Owen, R.J., Martin, S.R. & Borman, P. Rapid urea hydrolysis by gastric campylobacters. *Lancet* 1985, **i**, 111.

18 McNulty, C.A.M. & Wise, R. Rapid diagnosis of *Campylobacter*-associated gastritis. *Lancet* 1985, **i**, 1443–4.

19 Graham, D.Y., Klein, P.D., Evans, D.J. Jr. *et al.* *Campylobacter pylori* detected non-invasively by the ^{13}C-urea breath test. *Lancet* 1987, **i**, 1174–7.

20 Bell, G.D., Weil, J., Harrison, G. *et al.* ^{14}C-urea breath analysis, a non-invasive test for *Campylobacter pylori* in the stomach. *Lancet* 1987, **i**, 1367–8.

21 Hazell, S.L. & Lee, A. *Campylobacter pyloridis*, urease, hydrogen ion back diffusion, and gastric ulcers. *Lancet* 1986, **ii**, 15–7.

22 Buck, G.E. & Smith, J.S. Medium supplementation for growth of *Campylobacter pyloridis*. *J. Clin. Microbiol.* 1987, **25**, 597–9.

23 Morgan, D.R., Freedman, R., Depew, C.E. & Kraft, W.G. Growth of *Campylobacter pylori* in liquid media. *J. Clin. Microbiol.* 1987, **25**, 2123–5.

24 Majewski, S.I.H. & Goodwin, C.S. Restriction endonuclease analysis of the genome of *Campylobacter pylori* with a rapid extraction method: evidence for considerable genomic variation. *J. Infect. Dis.* 1988, **157**, 465–71.

25 Slomiany, B.L., Bilski, J., Sarosiek, J. *et al.* *Campylobacter pyloridis* degrades mucin and undermines mucosal integrity. *Biochem. Biophys. Res. Commun.* 1987, **144**, 307–14.

26 Emody, L., Carlsson, A., Ljungh, A. & Wadstrom, T. Mannose-resistant haemagglutination by *Campylobacter pylori*. *Scand. J. Infect. Dis.* 1988, **20**, 353–4.

27 Pearson, A.D., Bamforth, J. & Booth, L. *et al.* Polyacrylamide gel electrophoresis of spiral bacteria from the gastric antrum. *Lancet* 1984, **i**, 1349–50.

28 Wait, R. & Hudson, M.J. The use of picolinyl esters for the characterization of microbial lipids: application to the unsaturated and cyclopropane fatty acids of *Campylobacter* species. *Lett. Appl. Bacteriol.* 1985, **1**, 95–9.

29 Goodwin, C.S., McCulloch, R.K., Armstrong, J.A. & Wee, S.H. Unusual cellular fatty acids and distinctive ultrastructure in a new spiral bacterium (*Campylobacter pyloridis*) from the human gastric mucosa. *J. Med. Microbiol.* 1985, **19**, 257–67.

30 Wayne, L.G., Brenner, D.J., Colwell, R.R. *et al.* Report of the *ad hoc* committee on reconciliation of approaches to bacterial systematics. *Int. J. Systematic Bacteriol.* 1987, **37**, 463–4.

31 Von Wulffen, H. Low degree of relatedness between *Campylobacter pyloridis* and enteropathogenic *Campylobacter* species as revealed by DNA–DNA blot hybridization and immunoblot studies. *FEMS Microbiol.*

Lett. 1987, **42**, 129–33.

32 Lambert, M.A., Patton, C.M., Barrett, T.J. & Moss, C.W. Differentiation of *Campylobacter* and *Campylobacter*-like organisms by cellular fatty acid composition. *J. Clin. Microbiol.* 1987, **25**, 706–13.

33 Langenberg, W., Rauws, E.A.J., Widjojokusumo, A., Tytgat, G.N.J. & Zanen, H.C. Identification of *Campylobacter pyloridis* isolates by restriction endonuclease DNA analysis. *J. Clin. Microbiol.* 1986, **24**, 414–17.

34 Burnie, J.P., Lee, W., Dent, J. & McNulty, C.A.M. Immunoblot fingerprinting of *Campylobacter pylori*. *J. Med. Microbiol.* 1988, **27**, 153–9.

35 Jones, D.M., Curry, A. & Fox, A.J. An ultrastructural study of the gastric campylobacter-like organism '*Campylobacter pyloridis*'. *J. Gen. Microbiol.* 1985, **131**, 2335–41.

36 Hudson, M.J., Bhavsar, P. & Wait, R. 1987. Chemotaxonomy of the campylobacters. In Kaijser, B. & Falsen, E. (eds) *Campylobacter IV*. University of Göteborg, Göteborg, pp. 34–6.

37 Goodwin, C.S., Collins, M.D. & Blincow, E. The absence of thermoplasmaquinones in *Campylobacter pyloridis*, and its temperature and pH growth range. *Microbios Lett.* 1986, **32**, 137–40.

38 Collins, D.M., Costas, M. & Owen, R.J. The isoprenoid quinone composition of representation of the genus *Campylobacter*. *Arch. Microbiol.* 1984, **137**, 168–70.

39 Moss, C.W., Kai, A., Lambert, M.A. & Patton, C.M. Isoprenoid quinone content and cellular fatty acid composition of *Campylobacter* species. *J. Clin. Microbiol.* 1984, **19**, 772–6.

40 Collins, D.M. & Fernandez, F. Menaquinone-6 and thermoplasmaquinone-6 in *Wolinella succinogenes*. *FEMS Microbiol. Lett.* 1984, **22**, 273–6.

41 Collins, D.M. & Fernandez, F. Co-occurrence of menaquinone-6 and thermoplasmaquinone-6 in *Bacteroides gracilis*. *FEMS Microbiol. Lett.* 1985, **26**, 181–6.

42 Collins, D.M. & Widdel, F. Respiratory quinones of sulphate-reducing and sulphur-reducing bacteria: a systematic investigation. *System. Appl. Microbiol.* 1986, **8**, 8–18.

43 Romaniuk, P.J., Zoltowska, B., Trust, T.J., *et al*. *Campylobacter pylori*, the spiral bacterium associated with human gastritis, is not a true *Campylobacter* sp. *J. Bacteriol.* 1987, **169**, 2137–41.

44 Paster, B.J. & Dewhirst, F.E. Phylogeny of campylobacters, wolinellas, *Bacteroides gracilis* and *Bacteroides ureolyticus* by 16S ribosomal ribonucleic acid sequencing. *Int. J. Systematic Bacteriol.* 1988, **38**, 56–62.

45 Lau, P.P., DeBrunner-Vossbrinck, B., Dunn, B. *et al*. Phylogenetic diversity and position of the genus *Campylobacter*. *System. Appl. Microbiol.* 1987, **9**, 231–8.

46 Thompson III, L.M. Smibert, R.M., Johnson, J.L. & Krieg, N.R. Phylogenetic study of the genus *Campylobacter*. *Int. J. Systematic Bacteriol.* 1988, **38**, 190–200.

47 Fox, G.E., Edmonds, P., Taylor, N.S., Paster, B. & Dewhirst, F. 1987. Comparison of the phylogenetic, biochemical, phenotypic and morphological characteristics of gastric campylobacters from humans and ferrets. In Kaijser, B. & Falsen, E. (eds) *Campylobacter IV*. University of Göteborg, Göteborg, pp. 50–3.

6 Campylobacter pylori: enzymes

F. MEGRAUD

Microbial enzymes have always been studied in the past for the purpose of identification. However, the studies often concerned complex functions, for example, fermentation of Enterobacteriaceae. More recently, the availability of adapted coloured substrates have made possible the exploration of enzymes *per se*. For instance, β-D-galactosidase of the *Enterobacteriaceae* can be studied with O-nitrophenol-β-galactoside (colourless). The enzyme hydrolyses this substrate and O-nitrophenol, a yellow compound, is released.

The importance of enzymes in the pathogenicity of bacteria is also a matter of renewed interest for many microorganisms.

Concerning the genus *Campylobacter*, the low reactivity of the species in the traditional tests as well as the difficulty of determining pathogenetic pathways make the study of enzymes attractive. The importance of *C. pylori* urease activity is the key factor in this respect.

Enzymes for diagnosis

Urease

HISTORICAL BACKGROUND

In 1890, Miquel [1] showed that the culture supernatant of certain bacilli isolated from urine could break down urea. This was probably the first example of such activity. The most important industrial source of urease was discovered in 1916 by Mateer & Marshall [2] in the jack bean (*Canavalia ensiformis*). It contains a greater amount of the enzyme than any other source. Urease was the first enzyme to be successfully crystallized and, subsequently, characterized. Urease is widely distributed in the vegetable world. Numerous bacteria contain this enyzme: the genus *Proteus*, *Yersinia*, *Ureaplasma*, certain species or strains of *Klebsiella* (60%), *Enterobacter*, *Serratia*, *Pseudomonas aeruginosa* (35%), *Staphylococcus* and *Streptococcus*. Urease is also present in the genus *Bordetella* (*B. parapertussis* and *B. bronchiseptica*), *Moraxella* (*M. phenylpyruvica*), *Pasteurella* (*P. pneumotropica*, *P. urae*), *Mycobacterium* and *Brucella*. It is used in the biotyping of *Haemophilus influenzae* and *H. parainfluenzae*. In the animal world, molluscs possess urease but mammals do not.

Descriptions of gastric tissue of dogs and other animals which contain urease activity date back to 1924 [3]. Fitzgerald [4] was the first to document such activity in human stomach.

In 1968, experimental results led one to believe that urease was, in fact, of bacterial origin [5]. No particular organism, however, was mentioned.

In 1984, Langenberg *et al.* [6] pointed out that the newly discovered bacterium, *Campylobacter pylori*, known to have a restricted ecological niche — the human stomach — had a strong urease activity.

Since this original work, *C. pylori* urease has been the subject of many studies. It is now accepted that urease is the most important biochemical feature of *C. pylori*. All strains possess the enzyme at the time of diagnosis, and only a few strains have lost it in subcultures [7].

CHARACTERIZATION OF UREASE

Purification. This first step in the process of

characterization was performed by precipitation in acetone, pH 6, at 4°C. Subsequently, gel filtration on Sephadex was used, followed by an elution with 0.05 M Tris HCl. Currently, affinity chromatography or gel filtration is used.

pH_i and molecular weight. The jack bean urease has been studied extensively and is considered the reference material. Its molecular weight (MW) is 483,000 daltons as determined by ultracentrifugation and it is a polymer of a 30,000-dalton polypeptide. The main form is a hexadecapolymer [8]. Isoenzymes of jack bean urease exist. Its pH_i determined by isoelectrofocalization is 4.8. In comparison to the jack bean, *Proteus mirabilis* urease has a MW and a pH_i which are very similar, if not identical [9].

We have studied *C. pylori* urease and found that its MW was also 480,000 daltons but metabolically-active subunits of 240,000 and 120,000 daltons were also found (Babin & Megraud unpublished data). It was found to be 300,000, by Ferrero *et al.* [10]. Mobley *et al.* [11] obtained a value of 510,000 for the MW and 5.9 for the pH_i. Tompkins *et al.* [12] compared the pH_i of *C. pylori* which was also 5.9 to that of the ferret GCLO (*C. pylori* s.s. *mustelae*) which was slightly lower at 5.4.

ENZYMATIC PROPERTIES

Urease has a narrow substrate specificity. It hydrolyses urea according to the following reaction [13,14].

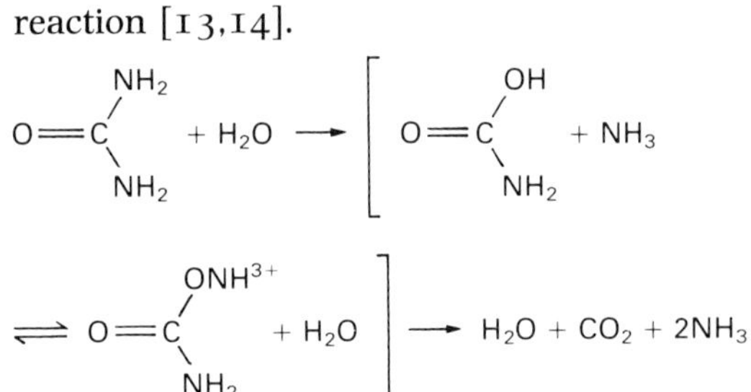

The activity is detected by the amount of NH_3 or CO_2 produced. Mobley *et al.* [11] found *C. pylori* urease to have the highest activity of all the bacterial urease tested. The amount of NH_3 produced was 36 ± 28 μmol/min/mg protein versus 14.6 ± 11.8 μmol/min/mg protein for *Proteus mirabilis*, 5.5 ± 4.9 for *P. vulgaris*, 5.2 ± 2.6 for *Morganella morganii*, and 4.6 ± 1.4 μmol/min/mg protein for *Providencia rettgeri*. In clinical bacteriology, only an increase in pH is used to detect the enzyme. Media used include Christensen medium with and without agar [15] and urea–indole medium [16]. Phenol red is the pH indicator in both cases. Urea–indole medium contains neither peptone nor glucose and does not allow growth of bacteria.

The strong activity of *C. pylori* urease is used in non-invasive tests as well as for a rapid diagnosis of *C. pylori* in biopsy specimens. Commercial preparations are now available (CLO test, CU test, Quick Campyl Test).

The main purpose of urease for the bacteria is most likely to make a source of nitrogen available. Ferrero *et al.* [10] have shown that *C. pylori* possesses a glutamate dehydrogenase which allows ammonia assimilation. In some species, urease is constitutive and in others it is induced. We found that the amount of urease produced was 70 times greater if the medium contained urea (Babin & Mégraud, unpublished data) which is in favour of a constitutive but also inducible enzyme. Ferrero *et al.* [10] noted that nitrogen starvation was not needed to detect *C. pylori* urease and that an elevated concentration of ammonia did not affect urease synthesis. Therefore, they concluded that the enzyme was constitutive.

Inhibitors of urease have been tested by Mobley *et al.* [11]. Acetohydroxamic acid inhibits *C. pylori* urease activity at a concentration twice that which inhibits *Proteus mirabilis* urease.

ANTIGENIC PROPERTIES

The urease molecule is a high MW protein and is highly antigenic. Dent *et al.* [17] were the first to propose an ELISA test using urease as antigen.

UREASE IN THE GENUS CAMPYLOBACTER AND IN OTHER SPIRAL-SHAPED BACTERIA

Urease is not widely distributed in the genus *Campylobacter*. Certain strains of *C. nitrofigilis*, a bacterium which is a plant parasite, can produce urease [18]. More recently, a variant of *C. laridis*, formerly known as a urease-positive thermophilic *Campylobacter* (UPTC), was found to be a strong urease producer. This bacterium, in contrast to *C. nitrofigilis*, was found in humans but only in faecal specimens and not in the stomach [19]. GCLO which is found in the ferret [20] also produces urease. Organisms similar if not identical to *C. pylori*, and therefore urease positive, have also been isolated from the stomach of the monkeys *Macaca mulatta* [21] and *Macaca nemestrina* [22], and baboons (*Papio* spp.) [23].

Corkscrew bacteria different from *C. pylori* and spiral GCLO have been observed in the stomach of humans, monkeys, cats and dogs. One such bacteria has been cultured from the cat stomach [24] and exhibited a strong urease activity.

Other important enzymes

OXIDASE

The respiratory chain of bacteria contains several cytochromes, which are iron-porphyrin proteins. The terminal cytochrome which varies from one species to another will transfer the electron to the oxygen molecule and is named cytochrome oxidase.

The detection of the cytochrome oxidase is known in clinical bacteriology as the oxidase test. It consists of the oxidation of molecules of N-tetramethyl *p*-phenylenediamine (TMPD) [25] or other reagents, and a violet colour is produced. Only the bacteria with a cytochrome C whose redox potential is close to that of TMPD will give a positive oxidase test.

All *Campylobacter* species, including *C. pylori*, are oxidase positive and this particular feature is part of the genus definition [26].

CATALASE

Catalase is an iron-porphyrin enzyme that catalyses the reaction:

$$H_2O_2 + H_2O_2 \longrightarrow 2H_2O + O_2$$

Catalase allows the detoxification of H_2O_2. This enzyme is present in most cytochrome-containing bacteria. For a long time, it has been a criterion in differentiating two categories of *Campylobacter* (catalase positive and negative). This division has limited value since it was found that *C. faecalis* was, in fact, close to *C. sputorum* s.s. *bubulus* with the exception of its catalase activity, and that '*C. upsaliensis*' is related to thermophilic *Campylobacter* but gives a negative or weak catalase test. *Campylobacter pylori* is catalase positive. Lior and Johnson studied catalase activity in culture filtrates as well as in sonicated cell lysates. *Campylobacter pylori* catalase was found to be mainly extracellular and to have a high specific activity [27].

γ-GLUTAMYL TRANSPEPTIDASE

γ-Glutamyl transferase (γGT), also called γ-glutamyl transpeptidase, is a glycoprotein associated with membranes widely distributed in epithelial cells. γGT from the kidney of various mammalians has mainly been studied. Its functions include mediation of amino acid transport, uptake of γ-glutamyl compounds and translocation of oligo-peptides. It catalyses the transfer of γ-glutamyl residue of glutathione (GSH) and other γ-glutamyl peptides to various amino acids [28].

$$\text{GSH} + \begin{matrix}\text{amino}\\ \text{acid}\end{matrix} \xrightarrow{\gamma\text{GT}} \begin{matrix}\gamma\text{-glutamyl}\\ \text{amino acid}\end{matrix} + \begin{matrix}\text{cysteinyl}\\ \text{glycine}\end{matrix}$$

γGT is present in certain bacteria and can be of interest in differentiating between species, for instance, in the genus *Moraxella* (*M. phenylpyruvica* and *M. urethralis* are γGT positive while the others are negative) and *Neisseria* (*N. gonorrohoeae* and *N. lactamica* are γGT negative while the others are positive). In the genus *Campylobacter*, few species harbour γGT. *Campylobacter pylori* has been found to be the only species with a consistently strong γGT activity. γGT is also present in 30% of *C. jejuni* strains [29] and in some strains of *C. laridis* and '*C. upsaliensis*'. This enzyme has not yet been characterized.

ALKALINE PHOSPHATASE

The phosphatase enzyme is a phosphomonoesterase which acts on phosphate monoester substrates. By hydrolysis, inorganic phosphate is produced:

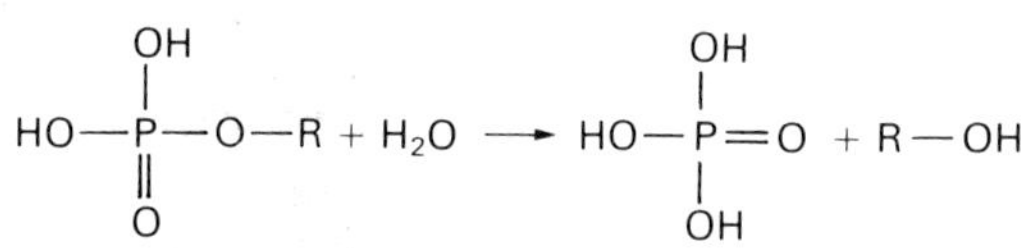

The enzyme has its optimal catalytic activity at an alkaline pH, hence the name alkaline phosphatase [30].

Alkaline phosphatase has been detected in some *Campylobacter* species such as *C. jejuni*, but the results have varied between different studies. *Campylobacter pylori* harbour a constantly strong alkaline phosphatase activity [7,31].

DNAse

Deoxyribonuclease (DNAse) is an endonuclease which is specific for the hydrolysis of DNA. Extracellular DNAses have been reported in a few bacterial species. Lior & Patel [32] described a medium to detect DNAse in *C. jejuni* and *C. coli* which allows a biotyping of these strains. It has been named Toluidine Blue DNA agar. This medium is derived from a medium designed to detect DNAse in staphylococci. A positive reaction is observed 1–2 days after inoculation. On this medium, *C. pylori* gives a consistently positive reaction.

HIPPURICASE

Hippuricase (hippurate hydrolase) hydrolyses hippuric acid (usually sodium salt) to benzoic acid and glycine. This enzyme is mainly intracellular.

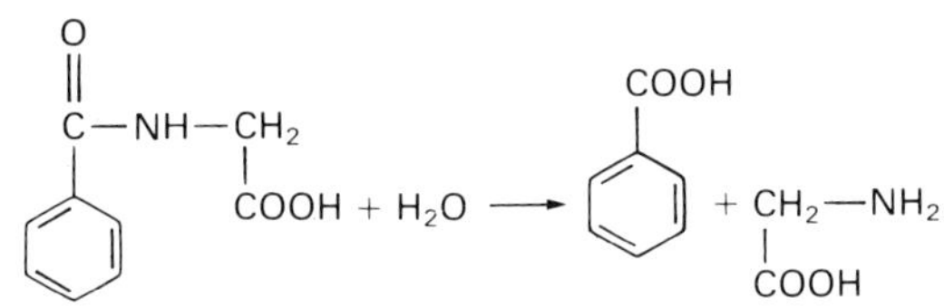

Hippurate hydrolysis may be determined by detection of either end product. The most common is to detect glycine by the addition of ninhydrin. The procedure described by Hwang & Ederer [33] for group B streptococci is usually employed for the genus *Campylobacter*. Hippuricase is only present in *C. jejuni* and not in the other species including *C. pylori*.

A group of *Campylobacter* strains isolated by Kasper & Dickgiesser [34], formerly known as gastric campylobacter-like organisms No. 2 (GCLO2), are sometimes present in the stomach. These strains are hippurate-positive. They have been classified as *C. jejuni* s.s. *doylei* [35].

NITRATE REDUCTASE

Nitrate reductase transforms nitrate to nitrite and allows growth in the absence of oxygen, with nitrate being the electron acceptor:

$$NO_3^- + 2e^- + 2H^+ \longrightarrow NO_2^- + H_2O$$

The medium used must support the growth of the microorganism and contain 0.1% potassium salt. The only observation made at the end is the presence or absence of nitrites in the medium after 24–48 hours. Sulphanilic

Table 6.1 Main enzymes used for the differentiation between *C. pylori*, related species and other *Campylobacter* species possibly found in humans*

Species	Urease	γ-Glutamyl transpeptidase	Alkaline phosphatase	Oxidase	Catalase	Nitrate reductase	Hippuricase
C. pylori	+	+	+	+	+	−	−
C. pylori s.s. *mustelae*	+	+	+	+	+	+	−
C. jejuni	−	v	+	+	+	+	+
C. jejuni s.s. *doylei*	−	+	+	+	V	−	+
C. coli	−	−	+	+	+	+	−
C. laridis	−	v,w	−	+	+	+	−
C. laridis UPTC variant	+	−	−	+	+	+	−
C. fetus s.s. *fetus*	−	−	−	+	+	+	−
C. hyointestinalis	−	−	+	+	+	+	−
C. sputorum s.s. *sputorum*	−	−	−	+	−	+	−
C. concisus	−	−	+	+	−	+	−
'*C. upsaliensis*'	−	v,w	+	+	−	+	−

* + = positive result; − = negative result; v = variable result; w = weak result.

acid and α-naphthylamide are added for this detection.

All *Campylobacter* species have a nitrate reductase except the two isolated in the human stomach (*C. pylori* and *C. jejuni* s.s. *doylei*). In contrast, *C. pylori* s.s. *mustelae*, the ferret GCLO, is nitrate-reductase positive.

The results obtained in the detection of these various enzymes in different *Campylobacter* species are presented in Table 6.1.

Additional enzymes

AMINOPEPTIDASES

Fifty-six aminopeptidases have been tested by us with API strips. The results are presented in Table 6.2 [7]. The substrates consisted of β-naphthylamide derivatives of tetra-, tri- and di-peptides and amino acids. Fast blue BB, a diazonium salt, was added at the end of the reaction and combined with the released naphthol to give a coloured compound [36]. McNulty & Dent [31] have tested nine aminopeptidases by using Rosco tablets which contain the same types of substrate. Agreement exists for most of the enzymes tested in both studies. A few discrepancies (alanine aminopeptidase, pyrrolidonyl aminopeptidase), positive in some strains with API strips and negative with Rosco tablets, could be due to minor substrate differences.

Table 6.2 Aminopeptidases found in 20 *C. pylori* strains

Aminopeptidase of:	No. positive strains	Aminopeptidase of:	No. positive strains
L-Tyrosine	5	L-Glutamine	13
L-Pyrrodidonyl	6	α-L-Glutamate	2
L-Lysyl	14	L-Tryptophane	10
L-Arginyl	14	L-Alanyl-L-arginine	17
L-Alanyl	2	L-Alanyl-L-phenylalanyl-L-proline*	19
N-Benzoyl-leucine*	20	L-Argininyl-L-arginine	3
S-Benzyl-cysteine*	20	L-Phenylalanyl-L-arginine	4
D,L-Methionine	17	L-Phenylalanine-L-proline*	20
Glycyl-phenylalanine	2	L-Prolyl-L-arginine	3
Glycyl-proline	14	L-Seryl-L-methionine	8

*Strong activity.

Table 6.3 Esterases studied in 20 *C. pylori* strains

Esterase of:	No. positive strains
Butyrate (C_4)	20
Valerate (C_5)*	20
Caproate (C_6)*	20
Caprylate (C_8)*	20
Nonanoate (C_9)*	20
Caprate (C_{10})*	20
Laurate (C_{12})*	19
Myristate (C_{14})	0
Palmitate (C_{16})	0
Stearate (C_{18})	0

* Strong activity

Leucine aminopeptidase is an enzyme present in all *C. pylori* strains.

OSIDASES

None of the osidases is present in *C. pylori* strains, as in the entire genus *Campylobacter* [29,31,37].

ESTERASES

Ten esterases have been studied by us [7] and two by McNulty & Dent [31]. We found esterases from fatty acids C_4 to C_{12} to be positive (Table 6.3), while laurate esterase was negative in the other study. Again, a substrate difference could be the cause.

Phosphonate esterase as well as three thymidine esterases were found in all strains of *C. pylori* but the reaction was weak for the former. A phosphorylcholine esterase was present in some strains but its activity was always weak (unpublished observations). Indoxylacetate esterase which has been helpful in the identification of *Campylobacter* species, is negative in *C. pylori* [38].

OTHERS

Enzymes such as arginine dihydrolase, lysine and ornithine decarboxylases, tetrathionate reductase and sulphatase have not been found in *C. pylori*.

In conclusion, *C. pylori* appears to be a homogeneous species with respect to the enzymatic profile. Some differences between strains have been observed among aminopeptidases but the positive reactions were too weak to be used in differentiating the strains. A micromethod including tests for 10 enzymes has been proposed to differentiate campylobacters

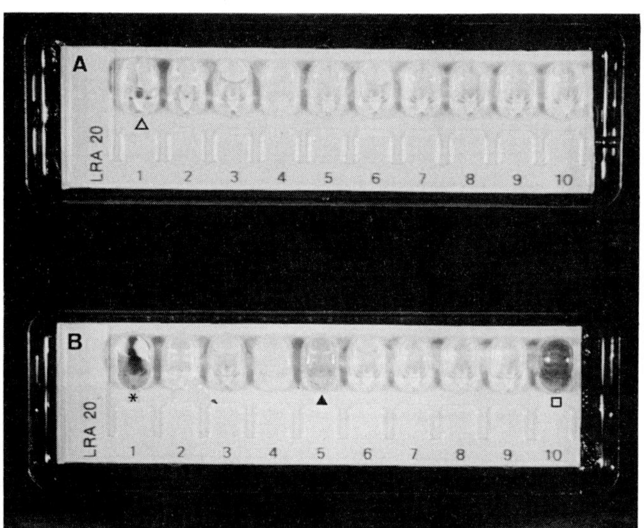

Fig. 6.1 API 20 Campi strip. A: control, B: *C. pylori* after a 4-hour incubation period. 1 urea hydrolysis, 2 nitrate reduction, 3 indoxyl acetate esterase, 4 hippurate hydrolysis, 5 γ-glutamyl transpeptidase, 6 triphenyl tetrazolium chloride reduction, 7 pyrrolidonyl arylamidase, 8 arginyl arylamidase, 9 aspartate arylamidase, 10 alkaline phosphatase. △ = yellow; * = magenta; ▲ = orange; □ = red.

[39]. The last prototype (Fig. 6.1.) will be marketed by API System in 1989.

Enzymes possibly involved in pathogenicity

The pathogenetic pathways of *C. pylori* are not presently well understood. It can be hypothesized that different mechanisms probably have cumulative effects such as adherence, cytotoxins and immunologic factors [40] with none of them being dominant. In this context, some enzymes, and particularly urease, are important.

Urease

Urease has been considered as a pathogenic factor of *C. pylori*. Marshall *et al.* [41] have shown that by urea hydrolysis, *C. pylori* can create a micro-environment with an increased pH around the bacterial cells. This allows *C. pylori* to survive at a pH of 1.5 while *C. jejuni* cannot and, therefore, to cross the first barrier against microorganisms in the gastrointestinal tract. Once in the mucus, *C. pylori* is mainly found near the intercellular junctions. Hazell *et al.* [42] explained this observation by the fact that the preferred metabolite for *C. pylori*, e.g. urea as well as a growth factor (hemin), would be present there at higher concentrations diffusing across the mucosa via the paracellular route. Confirmation of this hypothesis has not been provided. However, NH_3, a product of urea hydrolysis, is known to be toxic to cells and it probably contributes to the cell alteration observed in *C. pylori* infection. Barer *et al.* [43] claimed that the cytotoxic factor described in *C. pylori* was, in fact, NH_3 released from urea.

Hazell & Lee [44] developed a theory to explain gastric ulcer formation. They speculate that the rapid hydrolysis of urea would alter the milieu of the gastric epithelium preventing the normal passage of hydrogen ions from the gastric glands through the mucus to the lumen and would permit back diffusion. This hypothesis has not found enthusiastic support among specialists in the field.

Mucinase

The integrity of the gastric mucus layer is successfully maintained by the factors controlling the synthesis, secretion and breakdown of its constituents. Impairment of the protective function of the mucus can be the consequence of a modification in these steps.

Slomiany *et al.* [37] have shown that *C. pylori* produces a protease which is able to hydrolyse gastric mucin. Dialysed *C. pylori* filtrates were tested on pig gastric mucus as well as on purified mucus glycoprotein. The release of α-amino residues was followed by the trinitrophenyl methylation method. This enzyme of endopeptidase type can disrupt the gastric mucus in the same way as pepsin. The consequences include a modification of the mucus properties and eventually the possibility of H^+ back-diffusion. Modifications of mucus have been shown by scanning electron microscopy [45].

Lipase and phospholipase A

Campylobacter pylori filtrate is also able to break down glycerol trioleate and phosphatidylcholine [46]. These enzymes could contribute to the elevated concentrations of non-amidated bile acids and cytotoxic lysolecithin in the gastric juice of patients infected by *C. pylori*.

Catalase

The presence of a strong catalase [27] can be considered as a way of resisting the killing mechanisms of polymorphs. However, it is so rare to observe polymorphs with intracellular *C. pylori* that it is unlikely that this enzyme plays an important role in pathogenesis. The capacity of *C. pylori* to escape ingestion by polymorphs seems to be a more realistic mechanism.

Conclusion

In conclusion, we have, in *C. pylori*, an organism with a unique feature, a powerful urease activity. This is a key point for its routine identification. However, because of the possibility of finding another urease-positive *Campylobacter* in humans (e.g. *C. laridis* UPTC variant) and because of the corkscrew-shaped bacteria in the stomach which are known to be urease positive, *C. pylori* diagnosis cannot be reduced to a single urease test. The other important enzymes for diagnosis are γ-glutamyl transpeptidase and alkaline phosphatase which are consistently strongly positive, and nitrate reductase and hippuricase which are negative. This species appears to be very homogeneous and, unfortunately, it has not been possible to find enzymes allowing a biotyping scheme.

More work is needed to understand the importance of enzymes in the pathogenesis of *C. pylori* infection.

References

1 Miquel, P. Sur le ferment soluble de l'urée. *Comptes Rendus de l'Académie des Sciences (Paris)* 1890, **III**, 397–9.

2 Mateer, J.G. & Marshall, E.K. The urease content of certain beans, with special reference to the jack bean. *J. Biol. Chem.* 1916, **25**, 297–305.

3 Luck, J.M. & Seth, T.N. Gastric urease. *Biochem. J.* 1924, **18**, 1227–31.

4 Fitzgerald, O. Urease in the gastric mucosa and its increase after a meat diet, soy bean flour diet or urogastrone injections. *Nature* 1946, **158**, 305.

5 Delluva, A.M., Markley, K. & Davies R.E. The absence of gastric urease in germ-free animals. *Biochim. Biophys. Acta* 1968, **151**, 648–50.

6 Langenberg, M.L., Tytgat, G.N.J., Schipper, M.E., Rietra, P.J.G. & Zanen, H.C. Campylobacter-like organisms in the stomach of patients and healthy individuals. *Lancet* 1984, **i**, 1348.

7 Mégraud, F., Bonnet, F., Garnier, M. & Lamouliatte H. Characterization of '*Campylobacter pyloridis*' by culture, enzymatic profile, and protein content. *J. Clin. Microbiol.* 1985, **22**, 1007–10.

8 Boyer P.D. 1971. *The Enzymes*, Vol. IV. Academic Press, New York.

9 Babin, L. Identification Biochimique de l'Uréase de *Proteus mirabilis* dans l'Urine. Application au Diagnostic des Infections Urinaires à Germes Uréasiques. Thesis of Medicine No. 59, 1985, University of Bordeaux 2, Bordeaux, France.

10 Ferrero, R.L., Hazell, S.L. & Lee A. The urease enzymes of *Campylobacter pylori* and a related bacterium. *J. Med. Microbiol.* 1988, **27**, 33–40.

11 Mobley, H.L.T., Cortesia, M.J., Rosenthal, L.E. & Jones, B.D. Characterization of urease from *Campylobacter pylori*. *J. Clin. Microbiol.* 1988, **26**, 831–6

12 Tompkins, D.S., Millar, M.R. & West A.P. 1988. Isoelectric focusing of ureases from campylobacters. In Kaijser, B. & Falsen E. (eds) *Campylobacter IV*. Kungälv, Goterna, p. 427.

13 Blakeley, R.L., Hinds, J.A., Kunze, H.E., Webb, E.C. & Zerner B. Jack bean urease. Demonstration of a carbamoyl-transfer reaction and inhibition by hydroxamic acids. *Biochemistry* 1969, **8**, 1991–2000.

14 Gorin, G. On the mechanism of urease action. *Biochim. Biophys. Acta* 1959, **34**, 268–9.

15 Christensen, W.B. Urea decomposition as a means of differentiating *Proteus* and paracolon cultures from each other and from *Salmonella* and *Shigella* types. *J. Bacteriol.* 1946, **52**, 461.

16 Roland, F., Bourbon, D. & Sztrum, S. Différenciation rapide des entérobactéries sans action sur le lactose. *Annales de l'Institut Pasteur (Paris)* 1947, **73**, 914–16.

17 Dent, J., McNulty, C.A.M., Uff, J.S., Gear, M.W.L. & Wilkinson S.P. *Campylobacter pylori* urease: a new serological test. *Lancet* 1988, **i**, 1002.

18 McClung, C.R., Patriquin, D.G. & Davis, R.E. *Campylobacter nitrofigilis* sp. nov., a nitrogen-fixing bacterium associated with roots of *Spartina alterniflora* Loisel. *Int. J. Systematic Bacteriol.* 1983, **33**, 605–12.

19 Mégraud, F., Chevrier, D., Desplaces, N., Sedallian, A. Guesdon, J.L. Urease positive thermophilic *Campylobacter* (*Campylobacter laridis* variant) isolated from appendix and human feces. *J. Clin. Microbiol.* 1988, **26**, 1050–1.

20 Fox, J.G., Edrise, B.M., Cabot, E.B., Beaucoge, C., Murphy, J.C. & Prostack, K.S. Campylobacter-like organisms isolated from gastric mocase of ferrets. *Am. J. Vet. Res.* 1986 **47**, 236–9.

21 Newell, D.G., Hudson, M.J. & Baskerville, A. Isolation of a gastric campylobacter-like organism from the stomach of four Rhesus monkeys and identification as *Campylobacter pylori*. *J. Med. Microbiol.* 1988, **27**, 41–4.

22 Bronsdon, M.A. & Schoenknecht, F.D. *Campylobacter pylori* isolated from the stomach of the monkey, *Macaca nemestrina*. *J. Clin. Microbiol.* 1988, **26**, 1725–8.

23 Jones, D.M. & Eldridge, J. 1988. Gastric campylobacter-like organisms (GCLO) from man ('*C. pyloridis*') compared with GCLO strain from the pig, baboon and ferret. In Kaijser, B. & Falsen, E. (eds), *Campylobacter IV*. Kungälv, Goterna, p. 44.

24 Lee, A., Hazell, S.L., O'Rourke, J. & Kouprach, S. Isolation of a spiral-shaped bacterium from the cat stomach. *Infect. Immun.* 1988, **56**, 2843–50.

25 Kovacs, N. Identification of *Pseudomonas pyocyanea* by the oxidase reaction. *Nature* 1956, **178**, 703.

26 Smibert, R.M. 1984. Genus *Campylobacter* Sebald and Véron 1963. In Krieg, N.R. & Holt, J.H. (eds), *Bergey's manual of systematic bacteriology*, Vol. 1. Williams & Wilkins, Baltimore, pp. 111–18.

27 Lior, H. & Johnson, W.M. 1985. Catalase, peroxidase and superoxide dismutase activites in *Campylobacter* spp. In Pearson, A.D., Skirrow, M.B., Lior, H. & Rowe B. (eds), *Campylobacter III*. Health Laboratory Service, London, pp. 226–7.

28 Siest, G. & Heusghem, C. 1982. *Gamma-glutamyl-transferases*, Masson, Paris.

29 Elharrif, Z. & Mégraud, F. Characterization of thermophilic *Campylobacter*. II. Enzymatic profiles. *Current Microbiology* 1986, **13**, 317–22.

30 MacFaddin, J.F. 1980. *Biochemical Tests for Identification of Medical Bacteria*, 2nd edn. Williams & Wilkins, Baltimore.

31 McNulty, C.A.M. & Dent, J.C. Rapid identification of *Campylobacter pylori* (*C. pyloridis*) by preformed enzymes. *J. Clin. Microbiol.* 1987, **25**, 1683–6.

32 Lior, H. & Patel, A. Improved Toluidine Blue–DNA agar for detection of DNA hydrolysis by campylobacters. *J. Clin. Microbiol.* 1987, **25**, 2030–1.

33 Hwang, M. & Ederer, G.M. Rapid hippurate hydrolysis method for presumptive identification of group B streptococci. *J. Clin. Microbiol.* 1975, **1**, 114.

34 Kasper, G. & Dickgiesser, N. Isolation from gastric epithelium of campylobacter-like baceria that are distinct from '*Campylobacter pyloridis*'. *Lancet* 1985, **i**, 111–12.

35 Steele, T.W. & Owen, R.J. *Campylobacter jejuni* subsp. *doylei* subsp. nov., a subspecies of nitrate-negative campylobacters isolated from human clinical specimens. *Int. J. Systematic Bacteriol.* 1988, **38**, 316–18.

36 Monget, D. 1978. Mise au Point d'une Microméthode de Détection et de Mesure d'Activités Enzymatiques (API ZYM). Résultats Obtenus dans Différents Domaines d'Application. Thesis No. 297, University Claude Bernard, Lyon, France.

37 Slomiany, B.L., Bilski, J., Sarosiek, J., Murty, V.L.N., Dworkin, B., Van Horn, K., Zielenski, J. & Slomiany, A. *Campylobacter pyloridis* degrades mucin and undermines gastric mucosal integrity, *Biochem. Biophys. Res. Commun.* 1987, **144**, 307–14.

38 Mills, C.K. & Gherna, R.L. Hydrolysis of indoxyl-acetate by *Campylobacter* species. *J. Clin. Microbiol.* 1987, **25**, 1560–1.

39 Mégraud, F., Belbouri, A., Monget, D. & Gayral, J.P. 1988. A micromethod to identify *Campylobacter* species. Preliminary results. In Kaijser, B. & Falsen, E. (eds), *Campylobacter IV*. Kungälv, Goterna, p. 72.

40 Rathbone, B.J. Wyatt, J.I. & Heatley, R.V. Possible pathogenetic pathways of *Campylobacter pylori* in gastroduodenal disease. *Scand. J. Gastroenterol.* 1988, **23** (Suppl. 142), 40–3.

41 Marshall, B.J., Barret, L.J., Prakesh, C., McCallum, R.W. & Guerrant, R.L. 1988. Protection of *Campylobacter pyloridis* but not *Campylobacter jejuni* against acid susceptibility by urea. In Kaijser B. & Falsen, E. (eds), *Campylobacter IV*. Kungälv, Goterna, pp. 402–3.

42 Hazell, S.L., Lee, A., Brady, L. & Hennessy W. *Campylobacter pyloridis* and gastritis: association with intercellular spaces and adaptation to an environment of mucus as important factors in colonization of the gastric epithelium. *J. Infect. Dis.* 1986, **153**, 658–63.

43 Barer, M.R., Elliott, T.S.J., Berkeley, D., Thomas, J.E. & Eastham E.J. Cytopathic effects of *Campylobacter pylori* urease. *J. Clin. Pathol.* 1988, **41**, 597.

44 Hazell, S.L. & Lee, A. *Campylobacter pyloridis*, urease hydrogen ion back diffusion, and gastric ulcers. *Lancet* 1986, **ii**, 15–17.

45 Droy-Lefaix, M.T., Mégraud, F., Lamouliatte, H., Geraud, G. & Jutteau, J. 1988. *Campylobacter pylori* and gastric mucosal lesions: a scanning electron microscopy study. In Kaijser, B. & Falsen, E. (eds), *Campylobacter IV*. Kungälv, Goterna, pp. 392–3.

46 Sarosiek, J., Slomiany, A., Van Horn, K, Zalesna, G. Slomiany B. Lipolytic activity of *Campylobacter pylori*: effect of sofalcone. *Gastroenterology* 1988, **94**, A399 (Abstr).

7 The ultrastructure of Campylobacter pylori

D.M. JONES AND A. CURRY

Early literature on spiral organisms seen in the stomach often refers to them loosely as 'spirochaetes' [1–5]. Spirochaetes have a characteristic morphology distinct from *Campylobacter pylori*, campylobacters and the recently reported tightly spiralled organism [6, 7]. For example, spirochaetes do not possess free flagellar filaments; these structures are present but are located between the plasma membrane and body wall. True spirochaetes undoubtedly occur in the gut [8], but reports of such organisms in the stomach may represent misclassification in the assumption that all spiral organisms are spirochaetes.

External appearance and general description of Campylobacter pylori

The 'usual' appearance of *Campylobacter pylori* is that of a robust non-fimbriate, gently-sinusoidal organism, ranging from 2 to 6.5 μm long by 0.5 to 0.6 μm wide which undergoes one to three turns over this length. It has hemispherical ends and from one pole originate several sheathed flagellar filaments (Fig. 7.1). Other forms of *C. pylori* are seen in cultures and include spherical, V-shaped, U-shaped ('ox bow') and straightened forms. The organism has the cell-wall structures characteristic of Gram-negative organisms; in some preparations the outer membrane-like wall component can be seen blebbing from the body of the bacterium, although this is not as constant as the loose 'envelope' seen in true campylobacters.

In favourable negatively-stained prepara-

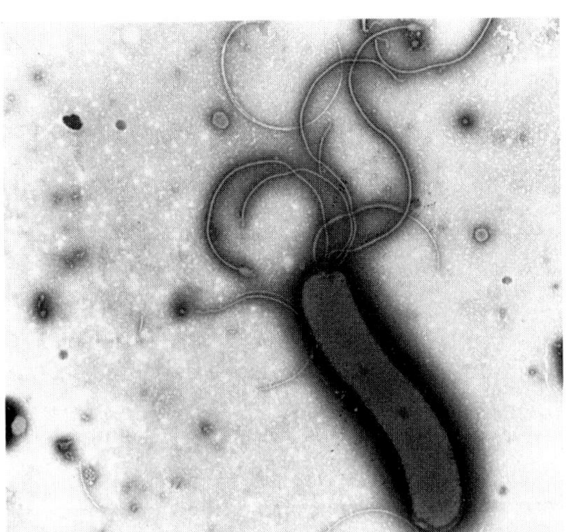

Fig. 7.1 Negatively-stained preparation of cultured human isolate of *C. pylori*. Note sheathed flagellar filaments originating from one of the dome-shaped ends of the organism. Magnification × 14 200.

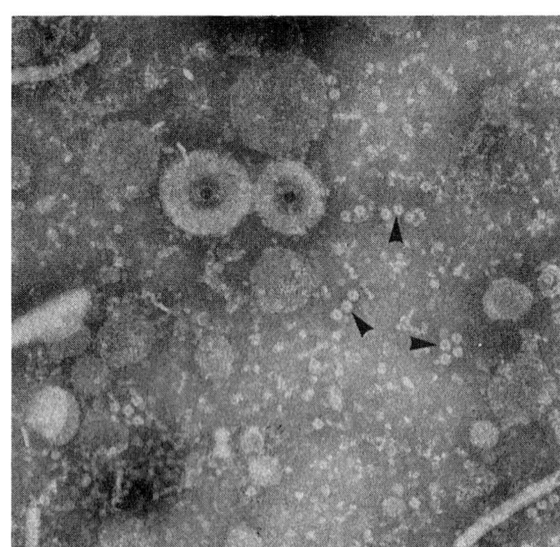

Fig. 7.2 Negatively-stained preparation of sarcosinate-treated culture of *C. pylori*. In addition to fragmented flagellar filaments and fragments of membrane, discs perforated by a central hole and surface proteins shaped like doughnuts are present (arrowheads). Magnification × 152 000.

tions, tightly-packed structures are seen covering the cell surface. These surface structures are roughly 12 nm in diameter with a central darkly-staining hole 4 nm in diameter. In sarcosinate preparations (Fig. 7.2) which disrupts the plasma membrane and membrane-like wall components covering the body of the organism, these 'doughnut' structures are readily dissociated from the surface. Proteolytic enzymes degrade these 'doughnuts' showing that they are wholly or partially composed of protein. Such protein structures are not readily seen in campylobacters but are not unusual in other bacteria; the function of those identified in *C. pylori* remains to be clarified but in other bacteria they may be structural and restrict access of substances through the wall before absorption [9].

Internal appearance

Thin sections of *C. pylori* show typical prokaryotic structures. The central region of the organism is filled with the filamentous nucleoid and is surrounded by a dense granular cytoplasm comprising large numbers of bacterial ribosomes. Occasional elecronlucent spherical (up to 150 nm) storage bodies are located in the protoplasmic matrix. The protoplasmic cylinder is covered by the true plasma membrane which, in turn, is covered by the membrane-like outer wall component. A small gap 7 nm wide (the periplasmic space) separates the plasma membrane and the outer envelope. In favourable sections, a thin (2-nm) electron-dense layer can be seen in the periplasmic space which most probably represents the location of peptidoglycan layer. The terminal region of the flagellated end of the organism shows specializations which are restricted to that region and probably represent terminal strengthening associated with the flagellar insertion (Fig. 7.3). The terminal specializations consist of an electron-lucent region immediately below the terminal region of the plasma membrane. Additional structures are found on the protoplasmic face of the plasma membrane but are restricted to a distance of about 400 nm from the tip of the organism although the extreme tip appears to be devoid of these additions. These comprise of additional decorations to this membrane, a 6-nm thick electron-dense layer is intimately associated with the membrane, with another electron-dense layer approximately 20-nm thick being found internally. In favourable sections these two layers (separated by 15 nm) appear to have links joining them together. Similar structures are to be seen in many polar-flagellated organisms including true campylobacters [10, 11].

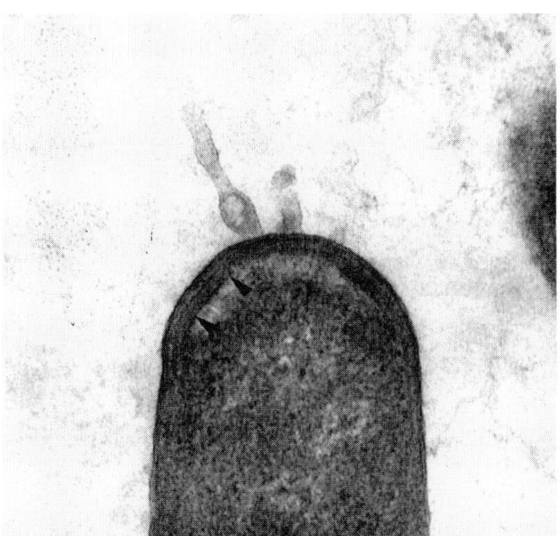

Fig. 7.3 Thin section through flagellated terminal dome of *C. pylori*. Note membrane-like nature of flagellar sheath, electronlucent zone and terminal plasma membrane decorations on inner (protoplasmic) surface (arrowheads). Magnification × 80 000.

Flagellar structure

The flagellar apparatus consists of the external filaments with associated membrane-like sheaths and basal rotatory structures embedded in the wall of the terminal region. Although in some cultures non-flagellated cells can be found, usually from one to eight filaments are inserted into the terminal dome. Two-day cultures more commonly show cells

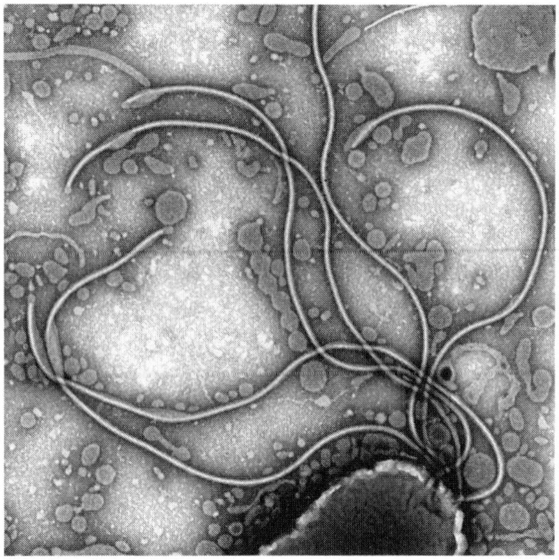

Fig. 7.4 Negatively-stained preparation of flagella terminations showing dilated sheath and occasional terminal 'paddles'. Magnification × 28 000.

with five or six filaments, four-day cultures often possess up to nine flagellar filaments. The average number of filaments present may increase with age of the culture.

Individual sheathed filaments are helical but assume a sinusoidal or curved form when dried down onto electron microscope specimen support films. They are up to 3 μm in length and have an overall diameter (including the sheath) of 30 nm. In favourable preparation the wavelength would appear to be 2–3 μm. The terminal regions or sub-terminal regions of the sheathed filaments often show a paddle-like or dilated form (Fig. 7.4). These dilatations are expressed in the membrane-like sheaths. Empty terminal sheaths can also be found devoid of their internal filament. Occasionally the internal filament, 15 nm in diameter, is seen to extend through and beyond the termination of the sheath. This form may be an artefact produced during preparation for examination under the electron microscope.

The sheathed nature of the filaments probably reflects an adaptation to living in and moving through mucus, and as protection against stomach acid which would de-polymerize naked filaments [12]. Such degradation of the filament would not allow the organism to move away from hostile acid conditions [13]. Other mucus-inhabiting bacteria show similar adaptations and a spiral body form and sheathed flagella may be optimal requirements for living in and moving through mucus [14]. The flagellar sheaths are continuous with the outer membrane-like component of the body wall. The internal flagellar filament, which probably rotates within the flexible sheath [15], penetrates a large disc associated with the outer wall membrane. The insertion discs can sometimes be seen in favourably-stained whole organisms (Fig. 7.5), but are best viewed in sarcosinate preparations where they separate from the membranous wall components (Fig. 7.2). The insertion discs are about 90 nm in diameter, have a central hole of 19-nm diameter which is surrounded by a more densely-staining zone about 40 nm in diameter. This disc is probably located on the inner surface of the outer membrane, the annular arrangement reflecting adherence to this membrane. The function of the disc is not fully understood but probably acts both as a bearing for the rotational movements of the flagellar filaments

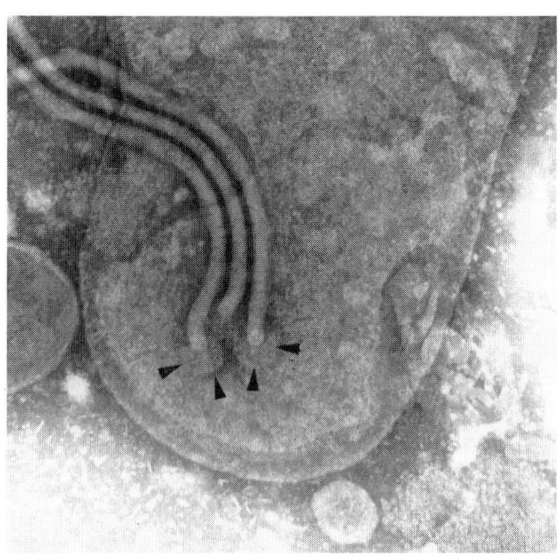

Fig. 7.5 Fortuitous negatively-stained preparation showing location of large discs of the flagellar insertions (arrowheads). Magnification × 92 000.

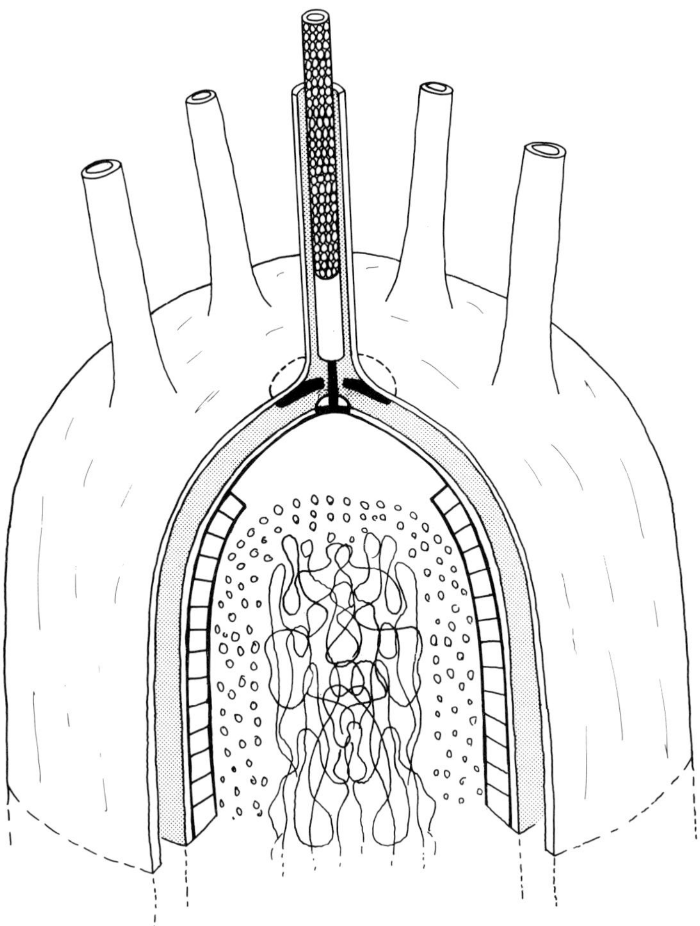

Fig. 7.6 Diagram of terminal region showing flagellar insertions and terminal specializations.

and as a strengthening structure, stabilizing the flagellar insertion [11, 16]. Two smaller discs, 22 nm in diameter can also be found associated with the proximal region of the flagellar apparatus where it originates from the plasma membrane. Such small discs have been identified in other motile bacteria, particularly *Escherichia coli*, *Salmonella typhimurium* and *Bacillus subtilis* as providing the motive force to drive the rotational movements of the flagellar filament [17]. Such discs are fixed to or rotate about a central rod, which in turn is connected to the flexible hook region at the cell surface and the rigid filament. The structure of this flagellar apparatus is summarized in Fig. 7.6.

Division

Division in *C. pylori* is by transverse fission (Fig. 7.7). The plasma membrane initially invaginates to divide the protoplasmic cylinder into halves. The membrane-like outer-wall component then completes separation by invagination between the plasma membranes of the two incipient daughter cells. During this division process one half of the cell retains the original flagellar apparatus. The daughter cell without the original flagellar apparatus produces these structures sequentially at the opposite pole from the original (Fig. 7.8), but will not have produced a full complement at the time of separation. Thus the division sep-

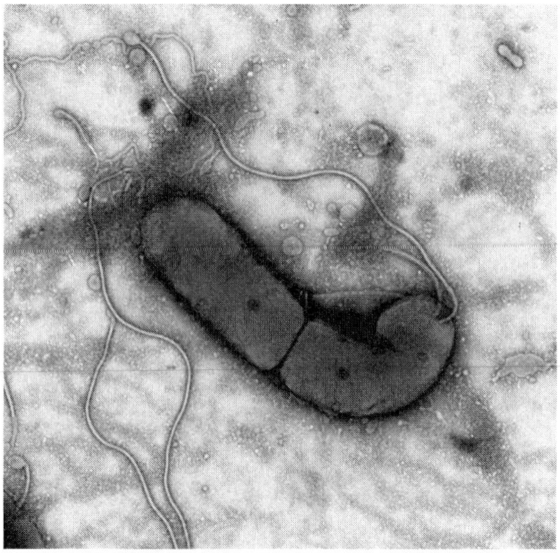

Fig. 7.7 Negatively-stained preparation of dividing *C. pylori* in culture. Note loose outer membrane where organism has bent which indicates that this organism may develop into an atypical form. Magnification × 16 800.

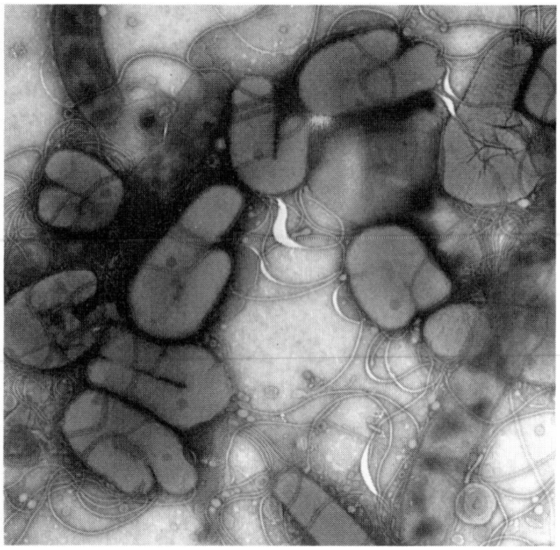

Fig. 7.9 Negatively-stained preparation of *C. pylori* culture showing abnormal, mainly U-shaped forms. Magnification × 11 200.

tum region does not produce flagellar filaments.

Spherical and other forms

Atypical forms of *C. pylori* are seen in some cultures of various age and passage. Such

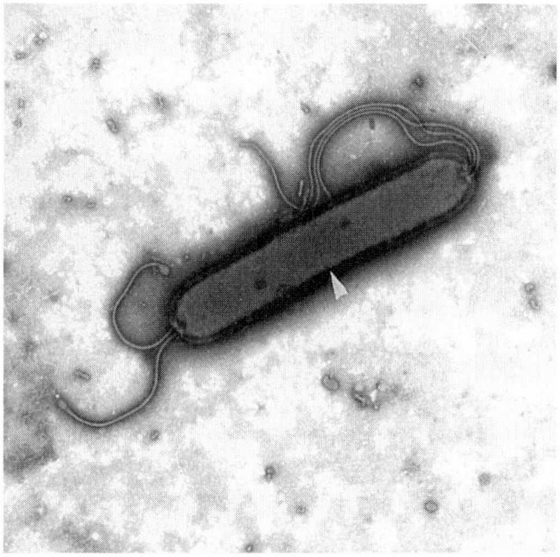

Fig. 7.8 Negatively-stained preparation of dividing organism. Note central division septum (arrowhead) and two flagella filaments from previously non-flagellated pole. Magnification × 16 000.

forms have not been observed *in vivo*. This phenomenon is also apparent with true campylobacters and other bacteria [18, 19]. Spherical forms have also been seen in campylobacters from aqueous environments (Pearson, A.D., personal communication). Thus 'atypical' forms may be adaptations to marginal or hostile environments or may reflect a response to defects in the cultural conditions, nutrients provided, build up of waste products, prolonged passage or antibiotics.

In culture U-shaped, V-shaped and spherical forms of *C. pylori* can be found (Fig. 7.9). The mechanism by which such bizarre forms are produced seems to involve protoplasmic growth within a restricted outer envelope which has failed to expand in synchrony with the other cell components. Some of these forms show features of early division septum formation. If the protoplasmic cylinder bounded by the plasma membrane continues to elongate within a non-expanding or more slowly-expanding outer envelope then bending of the cylinder occurs producing U-shaped forms. Equally, if the organism divides but the outer envelope fails to separate (Fig. 7.10) or elongate adequately, then growth of the

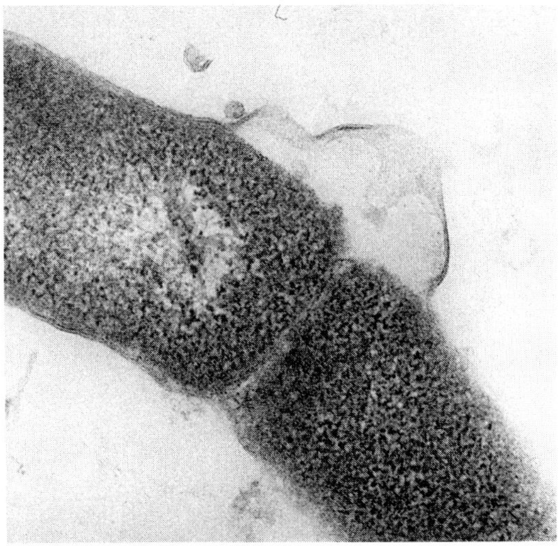

Fig. 7.10 Thin section of dividing organism showing septum and outer membrane-like wall component blebbing away from plasma membrane. Magnification × 90 400.

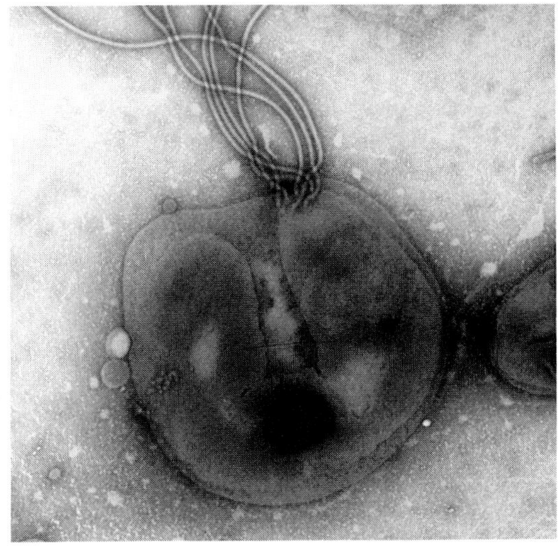

Fig. 7.12 Negatively-stained U-shaped organism showing single bunch of filaments and loose outer membrane enclosing both arms of the U form. Magnification × 296 000.

daughter cells tends to produce the V-shaped forms. Rupture of the plasma membrane would fill the space enclosed by the outer envelope with protoplasm although this seems to self-heal. Entirely spherical forms then result from the outer envelope losing the last vestiges of its rigidity or cylindrical form, presumably because of defects in this wall. Evidence for this mechanism can be found by examination of both negatively-stained and thin-sectioned cultured organisms (Fig 7.7, 7.10–7.15).

By negative staining, organisms bent double can be found enclosed within a loose

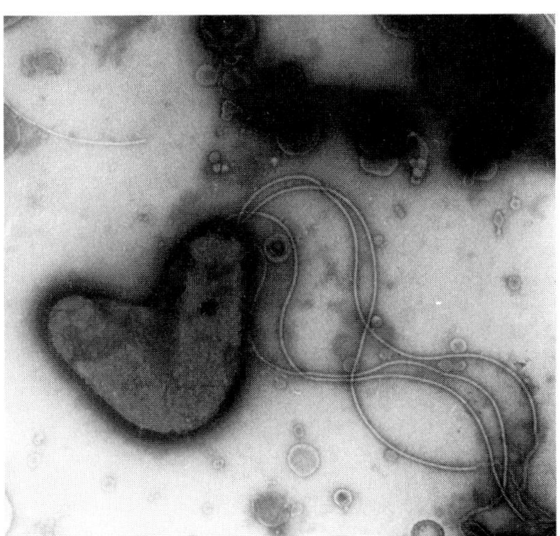

Fig. 7.11 Negatively-stained V-shaped organism. Note flagella from one pole only and loose outer wall membrane between arms of V. Magnification × 19 200.

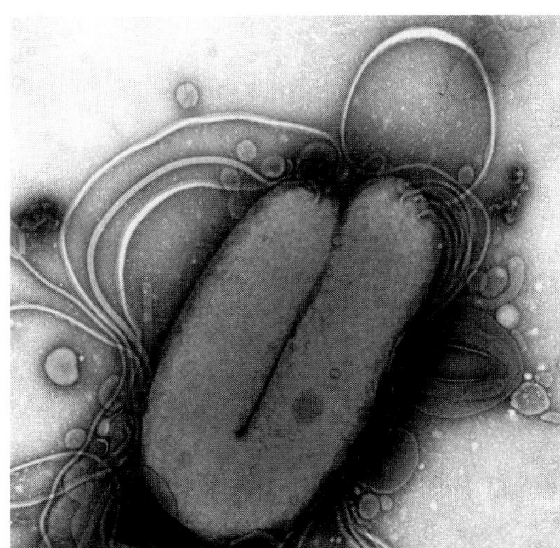

Fig. 7.13 U-shaped form showing flagellar filaments from both poles of both arms. A division septum is faintly visible. Magnification × 31 200.

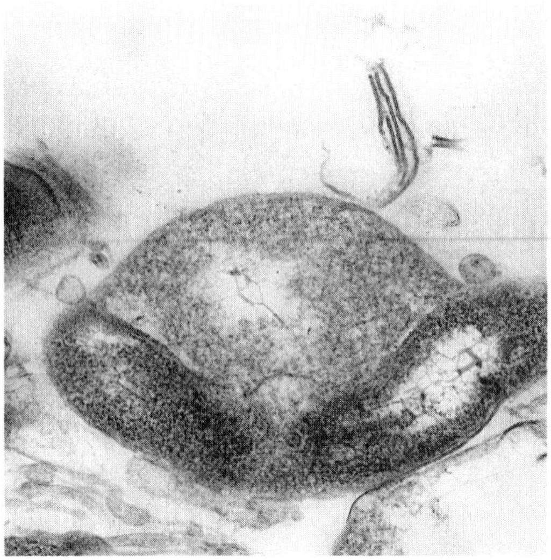

Fig. 7.14 Thin section of abnormal organism showing bend in protoplasmic cylinder with less electron-dense protoplasm filling the space between it and the outer membrane-like wall. Magnification × 51 200.

outer envelope (Figs 7.11–7.13). A division septum can be found in some of the aberrant forms. Occasionally an organism bent double within an envelope shows flagellation on one side of the bend but not the other, indicating that bending had occurred in an organism with previously normal morphology. Thin sections of these organisms show a membrane-like wall enclosing a U-shaped protoplasmic cylinder surrounded by the plasma membrane. Some less-dense protoplasm can be found in the space between the halves of the protoplasmic cylinder, indicative of protoplasmic leakage (Fig. 7.14). The resultant spherical forms have the appearance of viability and do not show degenerative changes (Fig. 7.15).

Appearance in vivo

In biopsy material, *C. pylori* is an extracellular, mucus-living organism which is found either in crypts or on the microvillus surface. Only the sinusoidal form has been observed *in vivo*. In crypts the organisms do not appear to be attached to the epithelial cells (Fig. 7.16) but on the mucosal surface a proportion of organisms is intimately associated with the cell surface. Such organisms are attached by short, approximately 50-nm strands to the epithelial cell membrane (via the glycocalyx) (Figs 7.17 and 7.18). Pedestals and surface pits with associated *C. pylori* are sometimes found. Occasionally *C. pylori* can be found directly above the junction (Fig. 7.19) between epithelial cells and there is

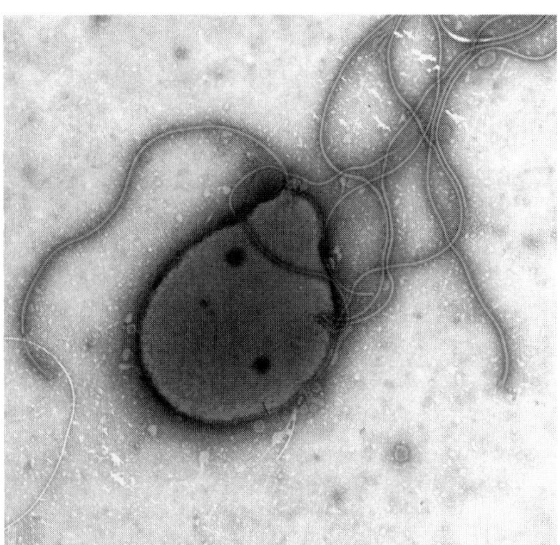

Fig. 7.15 Spherical form of *C. pylori*. Note non-degenerative look of this organism. Magnification × 18 400.

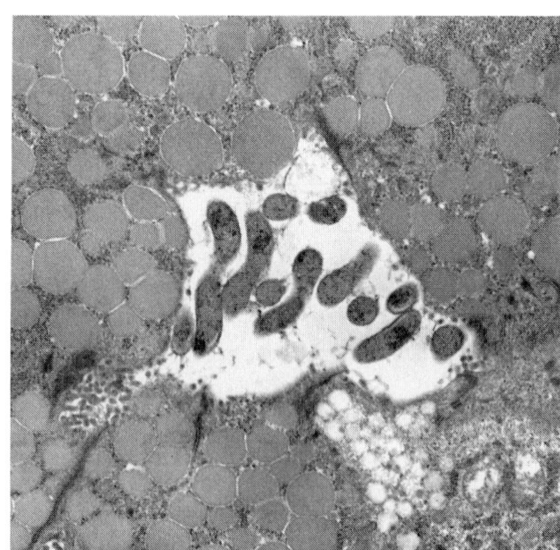

Fig. 7.16 *C. pylori* in crypt of baboon stomach. Magnification × 7360.

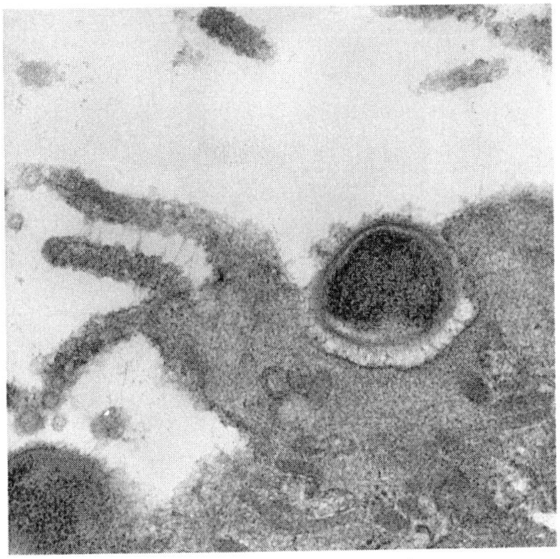

Fig. 7.17 Thin section of *C. pylori* from a human biopsy showing organism partially embedded in epithelial cell. Note filamentous material surrounding the organism and faint links anchoring it to cell surface. Magnification × 48 000.

some evidence that this is a preferred location because of the leakage of metabolites at these sites [20, 21]. The present authors, however, have not seen this commonly in their biopsy material. There are also reports [22–24] of organisms being seen in cross-sections of the canaliculi of oxyntic cells. The present auth-

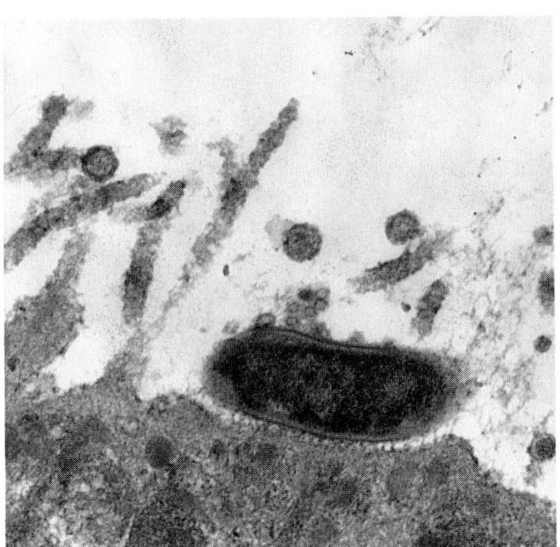

Fig. 7.18 Thin section of *C. pylori* on epithelial cell surface. Magnification × 34 000.

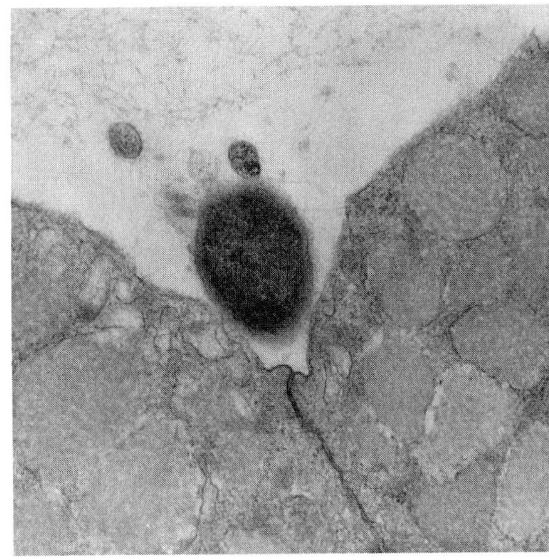

Fig. 7.19 Thin section of *C. pylori* above the junction between two epithelial cells. Magnification × 36 000.

ors have not found such an association with *C. pylori* in humans but have seen a different, tightly-spiralled organism, resembling the organism seen by Dent *et al.* [6], with a predilection for the canaliculi of HCl-secreting cells in baboons [7]. Further examination of biopsy material from different areas of the stomach may clarify the situation of any association of *C. pylori* with oxyntic cells.

Ultrastructure of colonies

Little work has been performed on the ultrastructure of *C. pylori* in colonies. Forty-eight-hour cultures show organisms to be flagellated and many organisms are orientated in parallel arrays. A few abnormal forms are found (see above). Four-day cultures show many more abnormally-shaped forms and also degraded organisms (Fig. 7.20).

Appearance in other animals

Organisms closely resembling *C. pylori* have been found in various animal species in addition to man [25]. Primates, pigs, and ferrets have been shown to harbour these bacteria [Figs 7.21–7.23]. From baboons, monkeys

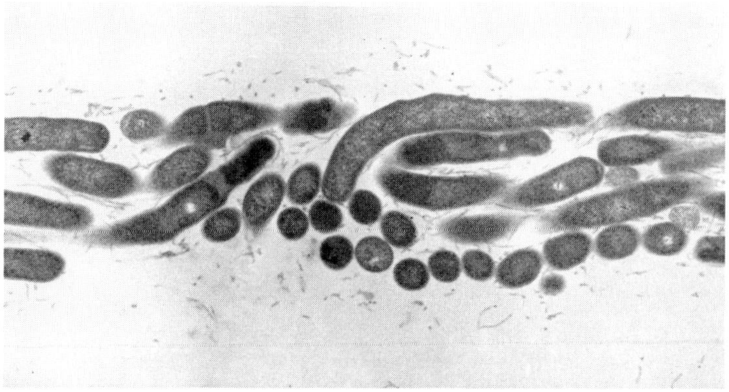

Fig. 7.20 Thin section through 2-day old colony of *C. pylori*. Note that some degree of alignment is apparent. Magnification × 9200.

and pigs, organisms that are structurally, biochemically and serologically identical to *C. pylori* have been isolated. The ferret organism has a number of differences including some important morphological features. It tends to be short, less sinusoidal and has sheathed flagellar filaments, but notably these originate from lateral or sub-terminal locations and are not arranged at one pole. In addition, the surface 'doughnuts' have a different appearance being slightly smaller in diameter (10 nm, with a 5-nm central hole).

All these organisms are easily distinguishable on morphological grounds from true campylobacters and the tightly-spiralled organism described in baboons and humans.

Taxonomic position of C. pylori

Since the first observations of Warren [26] and Marshall [27], the spiral bacterium isolated from gastric mucus has been included in the genus *Campylobacter*. This genus contains a number of species and these are morphologically identical and quite distinct from *C. pylori*. Indeed a true campylobacter has been isolated from human gastric epithelium [28, 29] which both sets of authors describe

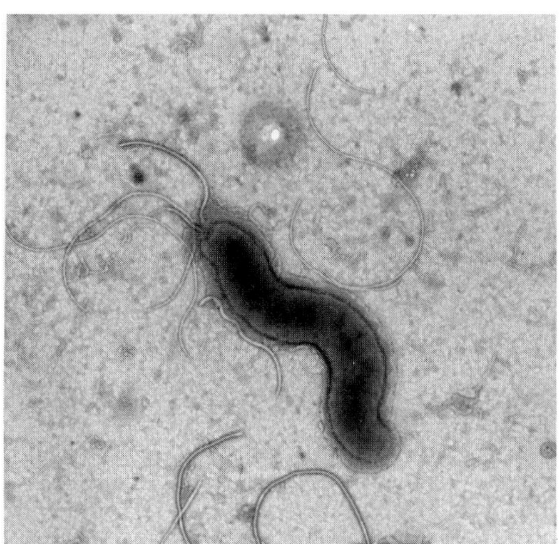

Fig. 7.21 *C. pylori* isolated from a baboon stomach. Magnification × 16 800.

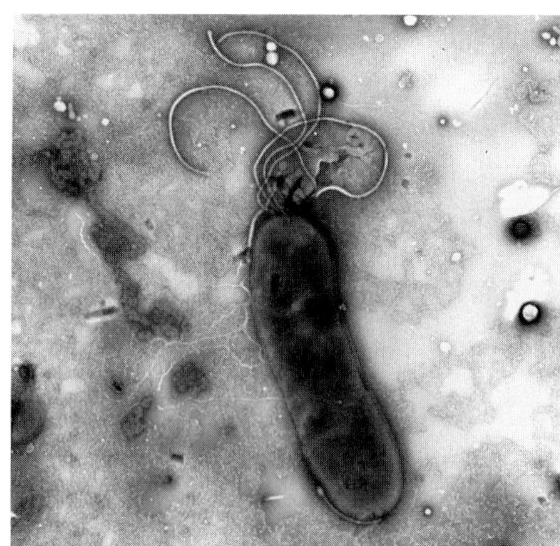

Fig. 7.22 *C. pylori* isolated from a pig stomach. Magnification × 16 000.

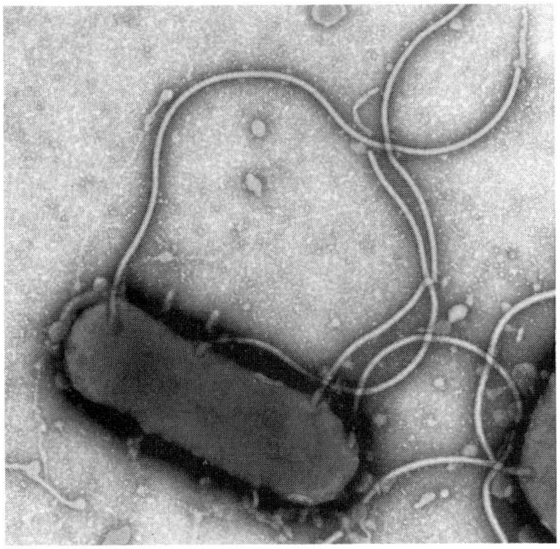

Fig. 7.23 Organism isolated from the ferret stomach. Note lateral insertions of flagellar filaments. Magnification × 41 600.

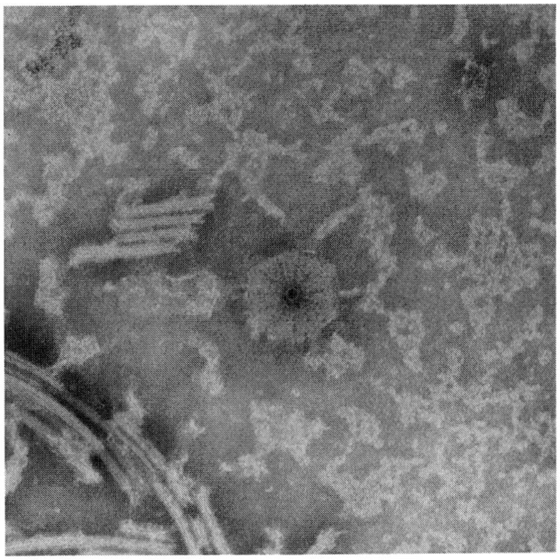

Fig. 7.25 Flagellar disc isolated from a sarcosinate-treated *C. jejuni* preparation. Note radial spokes originating from the central darkly staining hole. Magnification × 131 200.

as being distinct from *C. pylori*. It is morphologically quite different and is now designated *C. jejuni* s.s. *doylei*. In addition, partial 16S RNA sequencing from various campylobacter species have shown significant differences between *C. pylori* and other campylobacters [30, 31]. True campylobacters, such

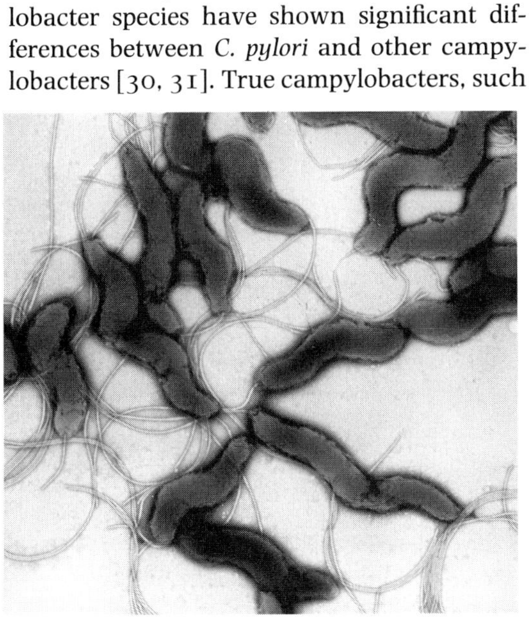

Fig. 7.24 Negatively-stained preparation of *C. jejuni*. Note different form of this organism compared to '*C. pylori*'. Magnification × 18 400.

as *C. jejuni* have a single, unsheathed flagellum at each end of the organism which originates in a shallow terminal pit (Fig. 7.24). In addition, the ends of the organism are truncated cones, rather than the hemispherical form of *C. pylori*. The flagellar discs associated with the flagellar apparatus are different, those of *C. jejuni* and *C. fetus* having radial spokes (Fig. 7.25), whereas those of *C. pylori* show an annular differentiation. There are therefore two easily separable, morphological groups of organisms — the true campylobacters and the isolates of *C. pylori* from man and various animal species. To classify such morphologically-diverse groups under the same generic name is taxonomic nonsense.

It has been suggested that *C. pylori* is related to species in the genus *Wolinella* [30–32]. Significant morphological differences are apparent between *C. pylori* on the one hand and *W. recta* and *W. succinogenes* on the other, particularly as these wolinellas appear to possess a single polar unsheathed flagellum. However, until the taxonomic situation is clarified, '*C. pylori*' will probably remain in common usage.

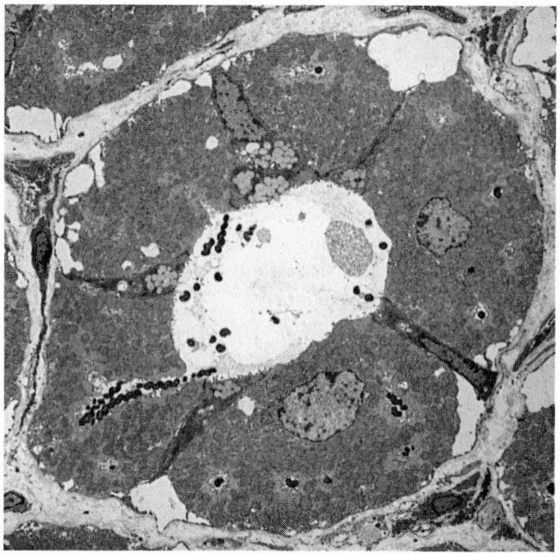

Fig. 7.26 Thin section of baboon stomach showing tightly-spiralled organisms in a crypt. Note that this organism lives in crypt lumen but is also commonly found in the canaliculi of the oxyntic cells. Magnification × 1840.

Comparison with other spiral organisms

Another spiral organism has recently been found in the human stomach which is distinct from both *C. pylori* and the true campylobacters [6]. A very similar tightly spiralled organism has been found in samples from baboon stomachs [7]. Both these new organisms show similar overall morphology, but have not been cultured so their true affinities remain to be elucidated. In baboons, this new tightly-spiralled organism which has multiple sheathed flagella at each end, is found in the canaliculi of the HCl-secreting oxyntic cells as well as in the mucus of the mucosal surface and crypts (Fig. 7.26).

Further and still morphologically-different spiral organisms can be seen in the stomachs of the dog and the cat. Therefore several types of spiral bacteria can exist in stomachs of different hosts and workers should be careful to distinguish these morphologically distinctive bacteria.

References

1 Freedberg, A.S. & Barron, L.E. The presence of spirochaetes in human gastric mucosa. *Am. J. Digest. Dis.* 1940, **7**, 443–5.

2 Palmer, E.D. Investigation of the gastric mucosa spirochaetes of the human. *Gastroenterology* 1954, **27**, 218–20.

3 Kasia, K. & Kobayashi, R. The stomach spirochete occurring in mammals. *J. Parasitol.* 1919, **6**, 1–10.

4 Doenges, J.L. Spirochetes in the gastric glands of *Macacus rhesus* and of man without related disease. *Arch. Pathol.* 1939, **27**, 469–77.

5 Lim, R.K.S. A parasitic spiral organism in the stomach of the cat. *Parasitology*, 1920, **12**, 108–12.

6 Dent, J.C., McNulty, C.A.M., Uff, J.C., Wilkinson, S.P. & Gear, M.W.L. Spiral organisms in the gastric antrum. *Lancet* 1987, **ii**, 96.

7 Curry, A. Jones, D.M. & Eldridge, J. Spiral organisms in the baboon stomach. *Lancet* 1987, **ii**, 634–5.

8 Tompkins, D.S., Foulkes, S.J., Godwin, P.G.R. & West, A.P. Isolation and characterisation of intestinal spirochaetes. *J. Clin. Pathol.* 1986, **39**, 535–41.

9 Dickson, M.R., Downing, K.H., Wu, W.H. & Glaeser, R.M. Three-dimensional structure of the surface layer protein of *Aquaspirillum serpens* VHA determined by electron crystallography. *J. Bacteriol.* 1986, **167**, 1025–34.

10 Rowles, C.R., Parton, R. & Jeynes, M.H. 1976. Some aspects of the cell walls of *Vibrio* spp. In Fuller, R. & Lovelock, D.W. (Eds), *Microbial Ultrastructure: The use of the Electron Microscope*. Academic Press, London, pp. 109–15.

11 Curry, A., Fox, A.J. & Jones, D.M. A new bacterial flagellar structure found in *Campylobacters*. *J. Gen. Microbiol.* 1984, **130**, 1307–10.

12 Ibrahim, G.F., Fleet, G.H., Lyons, M.J. & Walker, R.A. Method for the isolation of the highly purified *Salmonella* flagellins. *J. Clin. Microbiol.* 1985, **22**, 1040–4.

13 Hazell, S.L. & Lee, A. The adaption of motile strains of *Campylobacter pyloridis* to gastric mucus and their association with gastric epithelial intercellular spaces. In Pearson, A.D., Skirrow, M.B., Lior, H. & Rowe, B. (eds), *Campylobacter III Proceedings of the Third International Workshop on Campylobacter Infections, Ottawa, 7–15 July, 1985* Public Health Laboratory Service, London, pp. 189–91.

14 Ferrero, R.L. & Lee, A. Motility of *Campylobacter jejuni* in a viscous environment: Comparison with conventional rod-shaped bacteria. *J. Gen. Microbiol.* 1988, **134**, 53–9.

15 Thomashow, L.S. & Rittenberg, S.C. Isolation and composition of sheathed flagella from *Bdellovibrio bacteriovorus* 109J. *J. Bacteriol.* 1985, **163**, 1047–54.

16 Jones, D.M., Curry, A. & Fox, A.J. An ultrastructural study of the gastric campylobacter-like organism '*Campylobacter pyloridis*'. *J. Gen. Microbiol.* 1985, **131**, 2335–41.

17 De Pamphilis, M.L. & Adler, J. Fine structure and isolation of the hook–basal body complex of flagella from *Escherichia coli* and *Bacillus subtilis*. *J. Bacteriol.* 1971, **105**, 384–95.

18 Buck, G.E., Parshall, K.A. & Davis, C.P. Electron microscopy of the coccoid form of *Campylobacter jejuni*. *J. Clin. Microbiol.* 1983, **18**, 420–1.

19 Jai-King, N.G., Sherburne, R., Taylor, D.E. & Stiles, M.E. Morphological forms and viability of *Campylobacter* species studied by electron microscopy. *J. Bacteriol.* 1985, **164**, 338–43.

20 Hazell, S.L., Lee, A., Brady, L. & Hennessy, W. *Campylobacter pyloridis* and gastritis: association with intercellular spaces and adaption to an environment of mucus as important factors in colonization of the gastric epithelium. *J. Infect. Dis.* 1986, **153**, 658–63.

21 Lee, A. & Hazell, S.L. *Campylobacter pylori* in health and disease: An ecological perspective. *Microb. Ecol. Health Dis.* 1988, **1**, 1–16.

22 Rollason, T.P., Stone, J. & Rhodes, J.M. Spiral organisms in endoscopic biopsies of the human stomach. *J. Clin. Pathol.* 1984, **37**, 23–6.

23 Chen, X.G., Correa, P., Offerhaus, J., Rodriguez, E., Janney, F., Hoffmann, E., Fox, J., Hunter, F. & Diavolitsis, S. Ultrastructure of the gastric mucosa harboring campylobacter-like organisms. *Am. J. Clin. Path.* 1986, **86**, 575–82.

24 Marshall, B.J. *Campylobacter pyloridis* and gastritis. *J. Infect. Dis.* 1986, **153**, 650–7.

25 Curry, A., Jones, D.M., Eldridge, J. & Fox, A.J. Ultrastructure of *Campylobacter pyloridis*–not a *Campylobacter*? In Pearson, A.D., Skirrow, M.B., Lior, H. & Rowe, B. (eds), *Campylobacter III Proceedings of the third International Workshop on Campylobacter Infection, Ottawa, 7–10 July 1985* Public Health Laboratory Service, London, Abst. No. 117.

26 Warren, J.R. Unidentified curved bacilli on gastric epithelium in active chronic gastritis. *Lancet*, 1983, **i**, 1273.

27 Marshall, B. Unidentified curved bacilli on gastric epithelium in active chronic gastritis. *Lancet* 1983, **i**, 1273–5.

28 Kasper, G. & Dickgiesser, N. Isolation from gastric epithelium of campylobacter-like bacteria that are distinct from *'Campylobacter pyloridis'*. *Lancet* 1985, **i**, 111–12.

29 Goodwin, S., Blincow, E., Armstrong, J., McCulloch, R., & Collins, D. *Campylobacter pyloridis* is unique: GCLO-2 is an ordinary *Campylobacter*. *Lancet* 1985, **ii**, 38–9.

30 Romaniuk, P.J., Zoltowska, B., Trust, T.J., Lane, D.J., Olsen, G.J., Pace, N.R. & Stahl, D.A. *Campylobacter pylori*, the spiral bacterium associated with human gastritis, is not a true *Campylobacter* sp. *J. Bacteriol*, 1987, **169**, 2137–41.

31 Paster, B.J. & Dewhirst, F.E. Phylogeny of *Campylobacters, Wolinellas, Bacteroides gracilis* and *Bacteroides ureolyticus* by 16S ribosomal ribonucleic acid sequencing. *Int. J. Systematic Bacteriol.* 1988, **38**, 56–62.

32 Thompson, L.M., Smibert, R.M., Johnson, J.L. & Krieg, N.R. Phylogenetic study of the genus *Campylobacter*. *Int. J. Systematic Bacteriol.* 1988, **38**, 190–200.

8 Campylobacter pylori: detection and culture

C. STEWART GOODWIN

In late 1981 a young gastroenterologist, Dr Barry Marshall, asked the Microbiology Department of the Royal Perth Hospital to culture, from gastric biopsy specimens, spiral bacteria that had been seen in histological preparations by Dr Robin Warren. The head of the Microbiology Department, Dr Stewart Goodwin and Dr Marshall drew up a protocol: 100 patients with duodenal ulcers would be endoscoped, and biopsy specimens of the stomach cultured for the spiral bacteria. Dr John Pearman and Mr Frank Kosaras processed the specimens, but for the first 34 specimens no culture of these bacteria was obtained, in spite of changing the media and temperature of incubation. The plates were frequently overgrown with contaminating bacteria, and were discarded after 48 hours' incubation in a microaerophilic atmosphere. At this time the Easter holiday intervened, and agar plates from the 35th patient were left incubating for 5 days. On April 15th, 1982, instead of a mass of contaminating commensal bacteria, small translucent colonies were seen. We had discovered that *Campylobacter pylori* requires longer than 48 hours for primary isolation, and we were fortunate that commensals were absent in that patient. Gram stain of the colonies did not show spiral bacteria, but many were straight rods. Subsequently we have used culture conditions which allow the production of spiral bacteria *in vitro*. This first culture of *C. pylori* was reported by Marshall [1].

Transportation of biopsy specimens

Many different liquid transport media have been used, but few formal comparisons have been made. *Campylobacter pylori* rapidly loses viability at room temperature in 0.9% NaCl [2] so this should not be used. To prevent the mucus layer from detaching from the mucosa, with loss of bacteria, one study used 0.5 ml of 20% glucose solution as a transport medium [3]. Others have used thioglycollate, nutrient broth, Brucella broth [4] or Stuart's transport medium [5]. A biphasic transport medium with 2 ml Brucella broth, 10% horse serum and 1% IsoVitaleX plus a solid phase of brain–heart infusion agar with 10% bovine blood and 1% IsoVitaleX on a 20° slant in a 16-mm glass tube was compared to a liquid medium. All media contained vancomycin, nalidixic acid and amphotericin B. A higher recovery rate was found with the slant, but this did not reach statistical significance [6]. A biopsy specimen can be kept in transport medium for five hours at 4°C without any loss of viability of *C. pylori* [3]. If a biopsy specimen is put into −70°C in a medium of 1% peptone water containing 25% glycerol, *C. pylori* remains viable for months. Gastric brush biopsy specimens yield a lower rate of detection of *C. pylori* than biopsy specimens; occasionally gastric juice can be found to contain *C. pylori*.

Inoculum preparation

A corner of the biopsy specimen is cut off to prepare a smear on a slide for Gram stain to detect spiral bacteria. An optimal inoculum is prepared by grinding the specimen in a ground-glass grinder in 0.3 ml 20% glucose solution [3]. Alternatively, the specimen can be macerated in 2–3 drops of transport solution with the tip of an unsealed pipette.

Current best routine plate culture technique

Freshly poured media are ideal, but plates kept at room temperature in a closed plastic box for 6–19 days gave as good culture growth as freshly poured plates [3]. Whatever medium is used, the incorporation of tetrazolium chloride 40 mg/l will yield unique golden colonies which may be more easy to detect if there are also many commensals [7]. A blood-containing medium is probably required for the highest rate of primary isolation of *C. pylori*. However, it has been reported that brain–heart infusion agar or Brucella agar supplemented with 0.2% activated charcoal and 1% corn starch will support the growth of *C. pylori*. Whether these media are ideal for primary isolation has not been reported [8].

Blood agar, chocolate agar or lysed blood agar yield similar rates of isolation although we have found that brain–heart infusion agar with 7% horse blood lysed with saponin plus 1% IsoVitaleX yields a heavier growth of *C. pylori* more rapidly than other media. Selective media with antibiotics are essential to ensure detection of *C. pylori*. In one series, media without antibiotics yielded only 24/29 positive cultures [3]. Vancomycin 6 mg/l, amphotericin B 10 mg/l and cefsulodin 5 mg/l with or without trimethoprim 20 mg/l are the combinations of choice. The use of nalidixic acid may result in a few isolates of *C. pylori* being inhibited. Concentrations of blood used in media vary from 5% to 10% and may be horse blood or sheep blood. No definite comparisons have been reported. If Skirrow's medium is used, FBP should not be added because sodium metabisulphite is inhibitory to *C. pylori* [9].

For primary isolation at least one non-selective medium and one selective medium should be used, and incubation should be in a microaerophilic atmosphere with a concentration of 5–7% O_2. Although *C. pylori* will grow after sub-culture in a 10% CO_2 incubator in air, some primary isolates will not grow unless the oxygen concentration is reduced [3]. We prefer to evacuate an anaerobic jar to 220 mmHg and fill it with an anaerobic gas mixture (10% CO_2, and 10% H_2 and 80% N_2) giving an atmosphere of 5% O_2, 7% CO_2, 8% H_2 and 18% N_2. Such an atmosphere certainly gives as good and occasionally better growth than a standard campylobacter gas mixture ('Campy-Pak' Systems, BBL, Cockeysville, MD, USA). Plates should be examined after 3 days' incubation, then daily, and discarded after 6 days.

In most laboratories it is found that in a low percentage of biopsy specimens the Gram stain is positive for spiral bacteria but culture is negative. Thus, the ideal method for culturing *C. pylori* has probably not yet been determined. Because it is such a frequent organism in human beings, most laboratories are lulled into a false sense of efficiency because they can culture it from many patients. However, in our laboratory we are able to culture *C. pylori* in a few patients when the Gram stain is negative.

Culture as a diagnostic test — points for and against

The detection of *C. pylori* is undertaken either as an initial event or after antibacterial treatment. There is some benefit to the patient if the initial study includes endoscopy and biopsy, because an endoscopic specimen can be cultured and the isolate of *C. pylori* studied for its antibiotic susceptibilities. *Campylobacter pylori* can become resistant to metronidazole [10] and also to quinolones such as ciprofloxacin, when one of these drugs has been given alone. If the infecting strain is resistant to either of these drugs, then they should not be used. Up to now, all *C. pylori* isolates have proved to be sensitive to penicillin, amoxycillin and erythromycin. However, as more people with *C. pylori* are treated, wider resistance may be detected.

The other value of culture to diagnose *C. pylori* infection is that it provides unequivocal proof of the presence of *C. pylori*. Although studies of breath tests using ^{13}C-urea [11] or ^{14}C-urea [12] have indicated their reliability and specificity, we now know that there are other urease-producing bacteria which can colonize the stomach, and these could possibly give rise to a positive breath test, not due to *C. pylori*. However, after antibacterial treatment there is no need for culture to determine whether *C. pylori* has been eliminated, a breath test is adequate. Culture is expensive because of endoscopy, and involves much more discomfort to the patient than a breath test. Serology can also be useful.

Diagnosis of *C. pylori* by a urease test can be useful but requires biopsy, and therefore endoscopy, in which case culture can be undertaken anyway. If endoscopy is being performed for gastroenterological reasons, biopsy and culture should be undertaken. Histology can be used for diagnosis, but the only difference between these two is the fact that an additional site has been sampled.

References

1 Marshall, B. Unidentified curved bacilli on gastric epithelium in active chronic gastritis. *Lancet* 1983, **i**, 1273–5.

2 Hartmann, D. & von Graevenitz, A. A note on name, viability and urease tests of *Campylobacter pylori*. *Eur. J. Clin. Microbiol.* 1987, **6**, 82–3.

3 Goodwin, C.S., Blincow, E.D., Warren, J.R. *et al.* Evaluation of culture techniques for isolating *Campylobacter pyloridis* from endoscopic biopsies of gastric mucosa. *J. Clin. Pathol.* 1985, **38**, 1127–31.

4 Coghlan, J.G., Gilligan, D., Humphries, H. *et al.* *Campylobacter pylori* and recurrence of duodenal ulcers — a 12-month follow-up study. *Lancet*, 1987, **ii**, 1109–11.

5 Faoagali, J.L., Troughton, D., Aitken, J. & Gwynne, J.F. Campylobacter-like organisms in gastric biopsies: a Christchurch study. *N. Z. Med. J.* 1986, **99**, 50–2.

6 Westblom, T.U., Midkiff, B.R., Madan, E. *et al.* Improved isolation of *Campylobacter pylori* from gastric biopsies using a biphasic transport media. *Gastroenterology*, 1988, **94**, A494.

7 Queiroz, D.M.M., Mendes, E.N., Rocha, G.A. Indicator medium for isolation of *Campylobacter pylori*. *J. Clin. Microbiol.* 1987, **25**, 2378–9.

8 Buck, G.E. & Smith, J.S. Medium supplementation for growth of *Campylobacter pyloridis*. *J. Clin. Microbiol.* 1987, **25**, 597–9.

9 Goodwin, C.S., Blake, P. & Blincow, E. The minimum inhibitory and bactericidal concentrations of antibiotics and anti-ulcer agents against *Campylobacter pyloridis*. *J. Antimicrob. Chem.* 1986, **17**, 309–14.

10 Goodwin, C.S., Marshall, B.J., Blincow, E.D., Wison, D.H., Blackbourn, S. & Phillips, M. Prevention of nitroimidazole resistance in *Campylobacter pylori* by co-administration of colloidal bismuth subcitrate: clinical and *in vitro* studies. *J. Clin. Pathol.* 1988, **41**, 207–10.

11 Graham, D.Y., Klein, P.D., Evans, D.J., Evans, D.G., Alpert, L.C., Opekun, A.R. & Boutton, T.W. *Campylobacter pylori* detected noninvasively by the ^{13}C-urea breath test. *Lancet* 1987, **i**, 1174–7.

12 Marshall, B.J. & Surveyor, I. Carbon-14 urea breath test for the diagnosis of *Campylobacter pylori* associated gastritis. *J. Nucl. Med.* 1988, **29**, 11–16.

9 Detection of Campylobacter pylori by histology

J.I. WYATT AND S.F. GRAY

Campylobacter pylori is unique amongst bacteria associated with human disease in that it can be diagnosed accurately on the basis of its appearance in tissue sections. Histology provides a reliable method for detection of *C. pylori*.

The recognition by a histopathologist, J.R. Warren [1], that the morphology of the 'unidentified spiral bacteria' resembled members of the *Campylobacter* genus, led to its initial isolation using campylobacter culture techniques. Spiral bacteria had been reported sporadically by histologists for many decades before this. They attracted little attention however, until culture of the organism [2] made its characterization and systematic investigation possible. *Campylobacter pylori* is identifiable in tissue sections by virtue of its characteristic morphology, position, diffuse distribution and high population density (Fig. 9.1). The organisms appear curved or S-shaped, are usually numerous, closely applied to the epithelial cell surface and also present in overlying mucus. They colonize the surface and foveolar regions of the gastric mucosa of antral, body and cardiac types, but tend to be

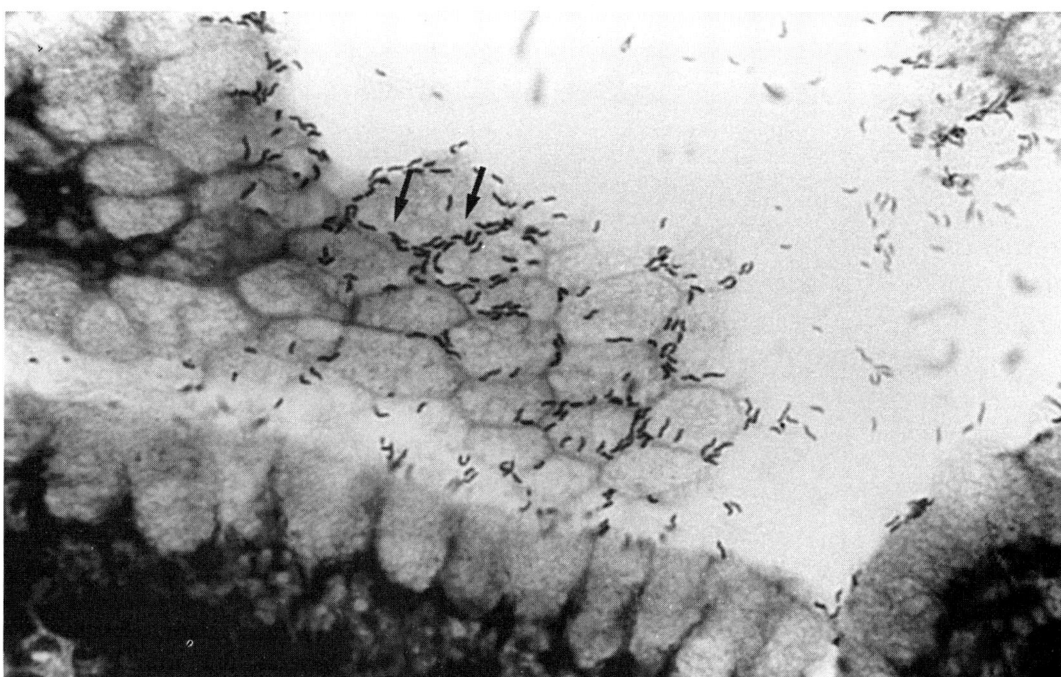

Fig. 9.1 The characteristic appearance of *C. pylori* in gastric biopsies. Numerous, uniform curved bacteria are present on and above the surface epithelium. Where the surface is seen *en face* due to tangential sectioning bacteria are localized over intercellular junctions (arrow). Modified Giemsa. × 256.

most numerous in the antrum. They do not colonize metaplastic intestinal epithelium in the stomach [3,4], but may colonize gastric epithelium present in the oesophagus (Barrett's oesophagus) [5] or duodenum (gastric metaplasia) [6,7].

Staining characteristics

Campylobacter pylori is often visible on routine haematoxylin & eosin (H&E) stained biopsies, although different preparations of haematoxylin differ in the degree to which they stain the organisms. Mayer's haematoxylin gives the strongest staining of *C. pylori*, probably because it does not require differentiation. In general, progressive rather than regressive histological staining techniques give the best demonstration of *C. pylori*. Many alternative histological stains have been recommended to demonstrate the morphology of the bacteria more clearly. In practice, *C. pylori* when numerous can be readily recognized on H&E, but if the organisms appear infrequent or absent, or if their morphology is not well seen on H&E, another stain should be used.

The Warthin–Starry silver stain (conventionally a technique for demonstrating spirochaetes in tissue sections) was used by Warren and Marshall in their original work [1,2]. This shows *C. pylori* particularly clearly, since deposition of silver onto the bacterial surface makes *C. pylori* appear larger in tissue sections (Fig.9.2). The disadvantages of this stain in routine use are that it is technically demanding, time consuming and prone to artefactual stain precipitate which may mimic bacteria. Because of this, other techniques have been explored.

Stains such as Giemsa [8], Gimenez [9], half-Gram [10], Brown-Hopps [11], cresyl fast violet [12], and acridine orange (with fluorescence microscopy) [13] are all suitable, and have advantages of speed and simplicity over

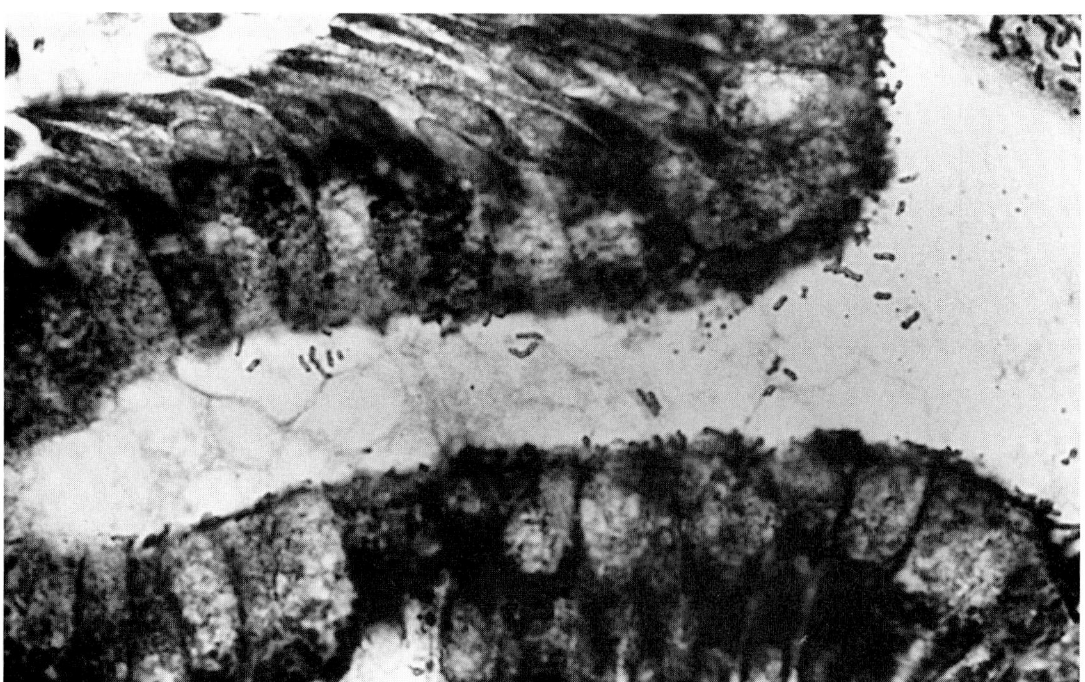

Fig. 9.2 *C. pylori* within the gastric pit (= foveola) of the gastric mucosa (Warthin–Starry stain). The magnification is the same as in Fig. 9.1, but the bacteria appear larger due to coating with the silver stain.

the Warthin–Starry silver stain. Excellent agreement was found when the Warthin–Starry silver stain was compared with H&E, acridine orange, Giemsa and Gimenez stains by Peterson *et al.* [14]. Which stain is chosen depends on local preference. We have found the Giemsa stain without differentiation [8] to be the most satisfactory, and Potters *et al.* have confirmed its correlation with the Warthin–Starry technique [15].

Specificity

None of the stains described above is specific for *C. pylori* but instead demonstrate the characteristic appearance of the organism which allows *C. pylori* to be recognized. Strictly speaking, bacteria recognized histologically (where microbiological culture has not been performed) should be described as 'campylobacter-like organisms' [16]. However, in practice these correspond almost invariably to *C. pylori*, the only occasional confusion being with 'GCLO-2' — a bacterium closely related to *C. jejuni*, which is occasionally found in the stomach [17]. The morphologically-distinct, long, spiral bacterium described by Dent *et al.* [18], can be readily distinguished from *C. pylori* in histological sections (Fig. 9.3).

Immunostaining using specific anti-*C. pylori* antibodies does allow positive identification of *C. pylori* in tissue sections. Polyclonal and monoclonal antibodies have been raised by different groups [19,20], but are not as yet commercially available. The antibodies can be used in conjunction with immunofluorescent or immunoperoxidase techniques (Fig. 9.4).

Sensitivity

The sensitivity of detecting *C. pylori* in tissue sections is dependent on the bacterium col-

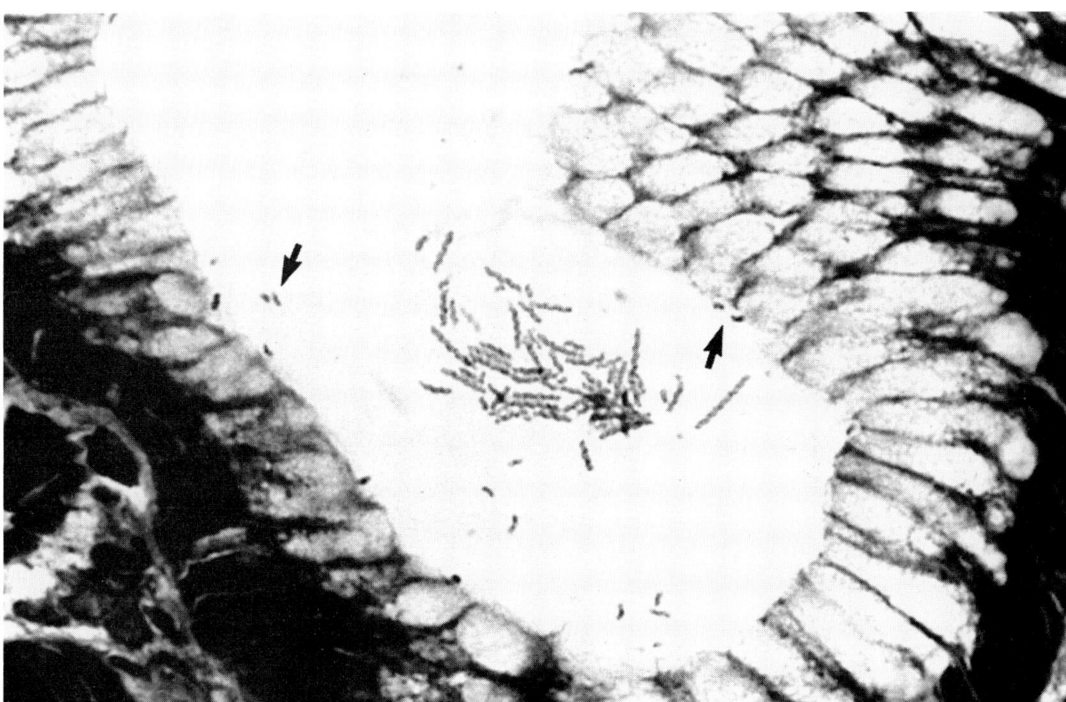

Fig. 9.3 Long spiral-shaped bacteria as described by Dent *et al.* [18] seen in a gastric biopsy from a patient with gastritis. These bacteria lie within the gastric pits and do not appear to attach to the epithelial surface. In this biopsy, scanty *C. pylori* were also present (arrows). Modified Giemsa. × 297.

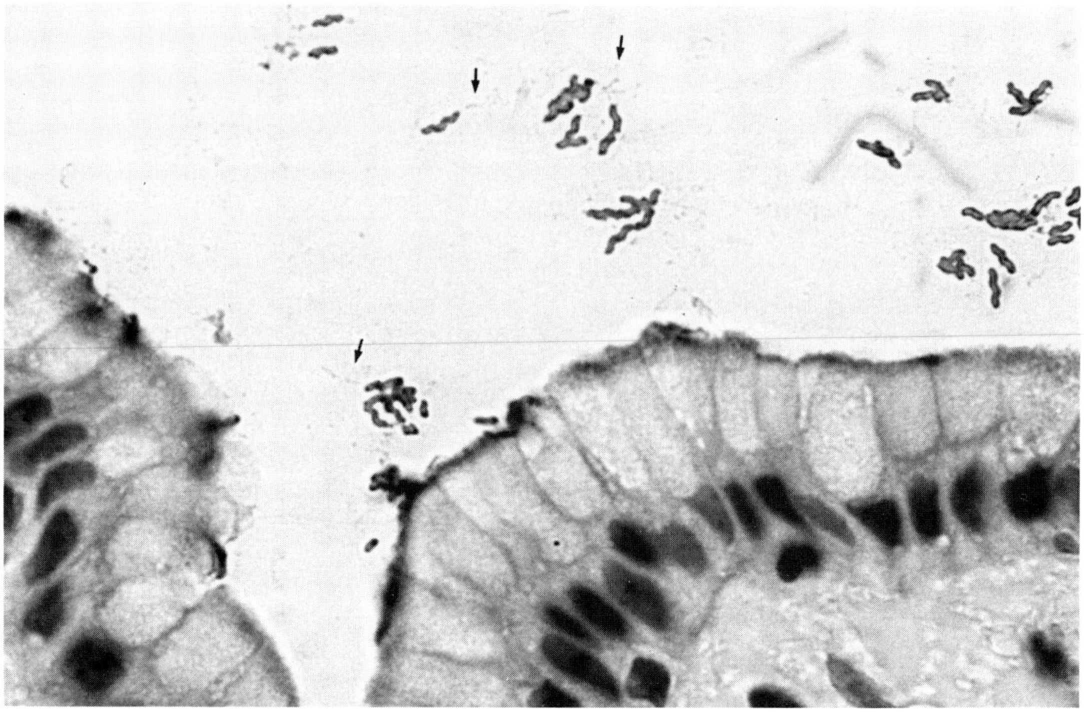

Fig. 9.4 *C. pylori* in a fixed gastric biopsy stained with a rabbit polyclonal anti-*C. pylori* antiserum using an indirect immunoperoxidase technique. Some flagellae are faintly visible (arrows).

onizing the gastric mucosa diffusely. False-negative results would be found if the distribution of *C. pylori* were patchy, and this problem would apply also to culture and biopsy urease tests on biopsies. However, in a study of multiple gastric biopsies from 50 patients, we found that both gastritis and *C. pylori* were diffuse in the antrum. Falsely-negative antral biopsies were only rarely encountered (3% of antral biopsies from *C. pylori* positive subjects appeared negative histologically). This was not the case in body type mucosa where both inflammation and *C. pylori* were quite often of patchy distribution [21]. On the basis of these results, two antral biopsies should be sufficient to avoid sampling error and establish *C. pylori* status.

Comparison with other techniques

The sensitivity and specificity of histological detection of *C. pylori* depends on the expertise of the pathologist, and in experienced hands is at least as good as microbiology [22,23]. Some comparisons of a range of techniques for identification of *C. pylori* have been published [24]. Which method is used in practice depends largely on local facilities, and many workers have used at least two techniques employed in parallel. We feel that the choice of method of detection depends on the nature of information sought from the study, and the degree of certainty with which the presence of *C. pylori* must be established. For example, most of the techniques using biopsies will be less sensitive in the minority of cases when the bacterial density is low [22]. The most appropriate test(s) can be selected from the choice of histology, culture, Gram smear, biopsy urease test, serology and urea breath test. Although we do not yet know whether the density of colonization is important clinically at the time of initial diagnosis, the detection of small numbers of persisting *C. pylori* following treatment will require a more thorough study using more than one test for the organism.

The advantages of histological detection are convenience — biopsies which are routinely taken for histology can be used — cheapness, and the permanence of the material. Additionally, from a research point of view, histology gives specific information on the relationship between *C. pylori* and the nature of the mucosal inflammation, and also allows retrospective diagnosis in archival material.

Histological diagnosis is accurate in conditions where no other bacteria are likely to be encountered in the stomach. In patients with achlorhydria, or with fungating tumours, there is a potential risk that other bacteria of similar morphology may be mistaken for *C. pylori* in tissue sections. Under these circumstances specific identification of *C. pylori* would be desirable. *Campylobacter pylori* has a very characteristic appearance in tissue sections from biopsy material. In contrast, it is our impression that histology is less reliable for detecting *C. pylori* in surgical resection specimens. Although mucosa-associated bacteria are often present, they may not have the curved shape of *C. pylori* and often appear coccoid. Poor or delayed fixation seems to aggravate the problem. These perhaps represent degenerative forms of *C. pylori*, but the correlation between histology and other diagnostic techniques has not as yet been established in surgically resected specimens. Histologically, a diagnosis of campylobacter-associated gastritis should be restricted to those cases showing bacteria of typical *C. pylori* morphology.

References

1 Warren, J.R. Unidentified curved bacilli on gastric epithelium in active chronic gastritis. *Lancet*, 1983, **i**, 1273.

2 Marshall, B.J. Unidentified curved bacilli on gastric epithelium in active chronic gastritis. *Lancet* 1983, **i**, 1273–5.

3 Meyrick-Thomas, J. Campylobacter-like organisms in gastritis. *Lancet* 1984, **i**, 1217.

4 Wyatt, J.I. Rathbone, B.J. & Heatley, R.V. Local immune response to gastric *Campylobacter* in non-ulcer dyspepsia. *J. Clin. Pathol.* 1986, **38**, 863–70.

5 Paull, G. & Yardley, J.H. Gastric and oesophageal *Campylobacter pylori* in patients with Barrett's oesophagus. *Gastroenterology* 1988, **95**, 216–18.

6 Wyatt, J.I. Rathbone, B.J. Dixon, M.F. & Heatley, R.V. *Campylobacter pyloridis* and acid-induced gastric metaplasia in the pathogenesis of duodenitis. *J. Clin. Pathol.* 1987, **40**, 841–8.

7 Caselli, M., Bovolenta, M.R., Aleotti, E. *et al.* Epithelial morphology of duodenal bulb and campylobacter-like organisms. *J. Submicrosc Cytol. Pathol.* 1988, **20**, 237–42.

8 Gray, S.F., Wyatt, J.I. & Rathbone, B.J. Simplified techniques for identifying *Campylobacter pyloridis*. *J. Clin. Pathol.* 1986, **39**, 1279–80.

9 McMullen, L., Walker, M.M., Bais, L.A., Karim, G.N. & Baron, J.H. Histological identification of *Campylobacter* using Gimenez technique in gastric antral mucosa. *J. Clin. Pathol.* 1987, **40**, 464–5.

10 Trowell, J.E., Yoong, A.K.H., Saul, K.J., Gant, P.W. & Bell, G.D. Simple half-Gram stain for showing presence of *Camplylobacter pyloridis* in sections. *J. Clin. Pathol.* 1987, **40**, 702.

11 Westblom, T.U., Madan, E., Kemp, J. *et al.* Improved visualisation of mucus penetration by *Campylobacter pylori* using a Brown–Hopps stain. *J. Clin. Pathol.* 1988, **41**, 232–6.

12 Burnett, R.A., Brown, I.L. & Findlay, J. Cresyl-fast violet staining method for campylobacter-like organisms. *J. Clin. Pathol.* 1987, **40**, 353.

13 Walters, L.L., Budin, R.E. & Paull, G. Acridine orange to identify *Campylobacter pyloridis* in formalin-fixed paraffin-embedded gastric biopsies. *Lancet* 1986, **i**, 42.

14 Peterson, W.L. Lee, E. & Feldman, M. Relationship between *Campylobacter pylori* and gastritis in healthy humans after administration of placebo or indomethacin. *Gastroenterology* 1988, **95**, 1185–97.

15 Potters, H.V.P.J., Loffeld, R.J.L.F., Stobberingh, E. *et al.* Rapid staining of *Campylobacter pylori*. *Histopathology* 1987, **11**, 1223.

16 Blaser, M. Gastric campylobacter-like organisms, gastritis, and peptic ulcer disease. *Gastroenterology* 1987, **93**, 371–83.

17 Kasper, G. & Diekgiesser, N. Isolation from gastric epithelium of campylobacter-like bacteria that are distinct from *Campylobacter pyloridis*. *Lancet* 1985, **i**, 111–12.

18 Dent, J.C., McNulty, C.A.M., Uff, J.C. *et al.* Spiral organisms in the gastric antrum. *Lancet* 1987, **ii**, 96.

19 Barbosa, J., Queiroz, D.M.M. Edilberto, N. *et al.* Immunocytochemical identification of *Campylobacter pylori* in gastritis and correlation with culture. *Arch. Pathol. Lab. Med.* 1988, **112**, 523–5.

20 Andersen, L.P. Holck, S. & Povlesen, C.O. 1987. Detection of *Campylobacter pylori* in tissue sections by an indirect immunohistochemical method. In Kaijser, B. & Falsen, E. (eds), *Campylobacter IV, Proceedings of the 4th International Workshop on Campylobacter infections, Göteborg, Sweden* University of Göteborg, Göteborg, pp. 413–14.
21 Wyatt, J.I., Primrose, J. & Dixon, M.F. Distribution of *Campylobacter pylori* in gastric biopsies. *J. Pathol.* 1988, **155**, 350A.
22 McNulty, C.A.M., Dent, J.C., Uff, J.G. *et al.* The prevalence of *Campylobacter pylori* in 1447 patients at endoscopy. *Am. J. Gastroenterol.* 1988, **83**, 1035.
23 Hazell, S.L., Hennessy, W.B., Borody, T.J. *et al. Campylobacter pylori* gastritis. II. Distribution of bacteria and associated inflammation in the gastroduodenal environment. *Am. J. Gastroenterol.* 1987, **82**, 297–301.
24 Megraud, F. Comparison of different tests for *C. pylori*. *Scand. J. Gastroenterol.* 1988, **23** (suppl. 142), 64–8.

10 Detection of Campylobacter pylori by the biopsy urease test

C.A.M. McNULTY

The presence of urease in the gastric mucosa of mammals was first described in 1924 by Luck *et al.* who found copious amounts in the cat and dog [1]. Fitzgerald & Murphy (1950) investigated the presence of the enzyme in the gastric mucosa of many species, including the human, dog, cat, pig, mouse, rat, rabbit and the frog [2]. They found that urease was widely distributed in all animal species, on the surface of the gastric mucous membrane, in the neighbourhood of the oxyntic cells and the floor of peptic ulcers. They suggested that the function of urease was to protect the gastric mucosa by neutralization of hydrochloric acid. Kornberg *et al.* later showed in the cat that this enzyme activity was of bacterial origin [3].

After Langenberg *et al.* first reported that *C. pylori* produced large amounts of urease [4], it became evident that human gastric urease originated from this organism. McNulty's initial observation that there was enough preformed urease produced by the organisms to detect *C. pylori* in biopsy specimens led to a small pilot study that confirmed the ease and value of this test in the rapid diagnosis of *C. pylori* gastritis [5]. Since this first report many workers worldwide have now modified the biopsy urease test and it is used routinely in many centres for the diagnosis of this infection from gastric biopsy specimens.

Principle of the biopsy urease test

A gastric biopsy specimen is placed in urea broth or agar. The test detects the presence of preformed urease produced by *C. pylori* in the biopsy specimen; growth of the organisms is not required. The urease present hydrolyses urea in the broth with the production of ammonium ions which raise the pH. The ingredients for Christensen's urea broth are: urea, 20 g/l; phenol red, 0.012 g/l; KH_2PO_4, 2 g/l; peptone, 1 g/l; NaCl, 5 g/l; glucose, 10 g/l. The urease reaction is:

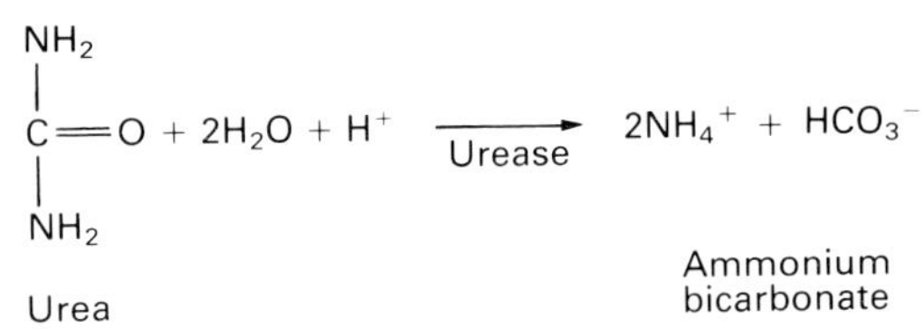

This pH change is detected by a pH indicator, phenol red, which changes colour from yellow/brown at pH 6.8 to pink at pH 8.4. A buffer is present in the broth to increase its stability and reduce false-positives. A colour change from yellow/brown to pink is considered positive. The time taken for the colour change to occur is usually noted.

The different biopsy urease tests (Table 10.1)

To increase the predictive value and speed of the biopsy urease test many centres have modified the original 2% Christensen's urea broth used by McNulty and Wise [5]. Nutrients in the Christensen's broth allow the growth of some aerobic urease-producing organisms that theoretically can lead to false-positive results. In most modified broths the nutrients are excluded and in several a bacteriostatic agent is present.

Most of the tests are presented as liquids but

Table 10.1 Characteristics of the different biopsy urease tests

Test name	Reference	Phenol red (mg/l)	Urea (%)	Volume (μl)	Temperature (°C)	Buffer (+/−)	Bacteriostatic agent (+/−)
Christensen	McNulty *et al.* [8]	12	2	500	Room	+	−
Microtitre BUT*	Hazell *et al.* [7]	*	2	50	Room	+	−
CLO test**	Marshall *et al.* [6]	*	*	−	30	+	+
Modified BUT*	McNulty *et al.* [8]	40	2	500	Room	+	−
6% RUT***	Vaira *et al.* [10]	*	6	*	Room or 50	−	*
4-hour RUT***	Vaira *et al.* [14]	*	2	500	37	+	*
1-minute test	Arvind *et al.* [11]	2 drops (1%)	10	1000	Room	−	−

*BUT = Biopsy urease test.
***RUT = Rapid urease test.
**CLO = Campylobacter-like organism.
* = Not known.

in the CLO test developed by Marshall *et al.* [6] the urea is incorporated in a small bleb of agar into which the biopsy specimen is placed. The volume of broth used in the tests is variable but most use about 0.5 ml. Hazell *et al.* [7] use a microtitre tray for the test. Biopsy specimens are dropped into only 50 μl of urea broth which because of the small volume improves the sensitivity and speed of the test. The wells of the microtitre tray in this method must be sealed or the ammonia produced may diffuse to neighbouring wells and cause false-positive results.

The concentration of the pH indicator has been increased in some of the tests [8] which may decrease the time taken for the colour change by making it easier to detect. So far, the pH of the urea broths has not been changed. Changing the pH of the broth so it is closer to the ideal pH for urease activity (pH 8.2) [9] might increase the sensitivity and speed of the test. This change will necessitate the use of an indicator with a higher pH range, for example *m*-cresol purple (pH range 7.4–9.0) or thymol blue (pH range 8.0–9.6), and a buffer with a higher pH range.

The concentration of the urea has been increased in the tests described by Vaira *et al.* [10] and Arvind *et al.* [11] to 6% and 10% urea, respectively, to increase the speed of the results, and attain a diagnosis before the patient leaves the endoscopy department. The ideal concentration of substrate in an enzyme reaction is usually 50–100 times the K_m value of the enzyme (a measure of affinity of an enzyme for its substrate), so that the substrate is in a large excess, and the reaction rate is dependent only on the enzyme activity. *Campylobacter pylori* urease has a K_m value of 0.32 g/l (8 mM) [9]; the ideal concentration of urea would therefore be about 16–32 g/l (1.6–3.2%). Too much substrate may result in inhibition of an enzyme. The 6% and 10% solutions therefore probably do not increase the sensitivity of the test and may even inhibit the urease activity. The rapid results produced with these tests may be due to the absence of buffer in the formula, so that the pH change occurs much more rapidly.

The method and timing of inoculation is also important. Deltenre *et al.* [12] have found that immediate inoculation in the endoscopy unit, crushing the specimen and avoiding the use of transport media increases sensitivity.

Value of the biopsy urease tests (Table 10.2)

When different urease tests are compared it is important to note the characteristics of the patient population tested and how *C. pylori* positivity is defined. If the test is compared with detection by both histology and/or mi-

Table 10.2 Value of the different biopsy urease tests

Test name	Reference	Number	Sensitivity (%)	Specificity (%)	Speed′
Christensen″	McNulty *et al.* [8]	597	92	99.4	61% at 1 h 71% at 2 h
Microtitre BUT*″	Hazell *et al.* [7]	376	91	100	75% at 1 h
CLO test**	Marshall *et al.* [6]	141	98		75% at 20 min 92% at 3 h
Modified BUT*″	McNulty *et al.* [8]	847	96	100	68% at 1 h 74% at 2 h
4-hour RUT***	Vaira *et al.* [14]	256	89	100	100% at 4 h
6% RUT***	Vaira *et al.* [10]	90	94	100	90% at 20 min
1-minute test	Arvind *et al.*[11]	40	90	100	100% at 1 min

*BUT = Biopsy urease test.
**CLO = Campylobacter-like organism.
***RUT = Rapid urease test.
′ Percentage positive at given time.
″ Compared with culture.

crobiology rather than microbiology alone, or the prevalence of *C. pylori* in the test population is low, the calculated predictive value of the tests will be lower. The biopsy urease tests described so far all detect more than 90% of *C. pylori*-positive specimens. The number of specimens tested in most studies has been small and although some tests are reported to be more sensitive than others these differences are not statistically significant. The campylobacter-like organism (CLO) test as described by Marshall *et al.* has the highest reported sensitivity [6]. This test is incubated at 30°C which is nearer the optimum temperature for urease (45°C) [9], so fewer organisms may be needed to produce a positive result. The high sensitivity of the CLO test has been confirmed by Morris *et al.* [13]; however Vaira *et al.* found that in their hands the CLO test had only a 58% sensitivity [10].

The modified 2% urea broth tests used by Hazell *et al.* [7], McNulty *et al.* [8], Vaira *et al.* [10], and Deltenre *et al.* [12], all have a sensitivity of between 94 and 98%. The increased sensitivity with the modified broths is either due to the smaller volume of broth used or the increased concentration of phenol red which makes the colour change easier to detect. Specimens yielding false-negatives with these tests probably contain insufficient organisms to produce the pH change required. McNulty *et al.* [8] found that 70% (9/13) of specimens yielding a false-negative modified urease test, grew < 10 colonies on culture.

Although false-positives are a potential problem with these tests the specificity is near 100% even in the nutrient containing Christensens's broth. There are several reasons for this high specificity: (i) contaminating bacteria are usually present in low numbers compared with *C. pylori*; (ii) no other organism produces urease with such high affinity or rate of activity [9]; (iii) specimens are left at room temperature or a bacteriostatic agent has been incorporated so preventing the rapid growth of other organisms.

Speed of results

As all the urease tests described thus far have similar sensitivity and specificity the choice of test will be determined by the cost, the time within which the result is required and the ease of preparing and performing it.

All the tests produce more rapid results than the originally described Christensen's broth; at least 70% of positives can be reported within 1 hour. The CLO test reported by

Marshall *et al.* [6] and Morris *et al.* [13] produce 75% of positives within 20 min. The two very rapid biopsy tests which are designed for use in the endoscopy unit can be read at 1 min or 20 min after inoculation [10,11]. The 1-minute test of Arvind *et al.* must be freshly prepared in the endoscopy unit to avoid false-positive results [11]. It contains 1 ml of 10% urea in deionized distilled water with 2 drops of 1% phenol red solution added. The rapid results produced with this test may outweigh the disadvantages of the extra nursing time required to prepare the test. As the 1-minute test has only been assessed in 40 patients thus far, further studies are required to confirm its high predictive value. Vaira *et al.* have described several rapid tests including the 6% rapid urease test in short communications [10,14]. The constituents and methods described in their reports are not complete and have been modified with each communication. Their latest test contains 6% urea with no buffer, plus another ingredient that increases the stability and shelf-life of the test so that it can be made in advance and stored [10]. The test has similar specificity and sensitivity to all the other tests but is read at 20 min post-incubation — although many tests are positive immediately. This test may become available commercially and if its specifity and sensitivity are confirmed, it would have significant advantages over the CLO test, which is the other most commonly-used commercial product.

With all the 2% urea broth tests so far described there is a correlation between the degree of infection with *C. pylori* and the speed of the test. This difference in speed is not seen with the CLO test, Vaira's 4-hour RUT [14], or the rapid tests. If in the future it became evident that the severity of infection was important it may be useful to have a test which will differentiate scanty from profuse colonization.

Cost

Cost may be an important factor in the choice of biopsy urease test. The CLO test is a commercial product and is therefore the most expensive test (currently, £2 per test from Bripharm Ltd, UK), but it can be performed with ease without the use of a microbiology department. In contrast the 'home-made' 2% urea broth tests can be made by the majority of microbiology departments for use in endoscopy and cost less than 5 pence per test.

Advantages and disadvantages of the biopsy urease tests

The major advantage of the biopsy urease tests for the diagnosis of *C. pylori* infection are that they can be performed very simply and results are obtainable much more rapidly than by culture or microscopy. The more rapid tests now available will allow the confident diagnosis of *C. pylori* infection before the endoscopy report is written by the clinician and the patient leaves the department. All the tests are extremely cheap when compared to the cost of culture, or microscopy of histology sections.

There are several disadvantages. Despite the high sensitivity of the biopsy urease test, some positives will be missed. In a study of 1445 patients attending endoscopy, *C. pylori* was detected in 628 patients. Gram stain was positive in 582 patients, culture positive in 583, histology positive in 575 and the biopsy urease test positive in 553 [8]. The small lack of concordance is probably due to sampling variation; *C. pylori* infection can be patchy and organisms may be present in one specimen and not another. The tests are not as sensitive when performed with specimens from patients who have received antimicrobial or bismuth treatment and Deltenre *et al.* would not recommend their use in these circumstances [12].

If biopsy specimens are not sent for culture and histology strains of *C. pylori* will not be available for serotyping or sensitivity testing and histology sections will not be available for histological assessment of the gastric mucosa

— a major disadvantage during treatment trials. The author considers the gold standard diagnostic method required for research studies should still include detection by culture and histology, with the addition of the biopsy urease test if a rapid diagnosis is required. However, the biopsy urease test can usefully and reliably replace the other diagnostic methods when these other parameters are not needed.

Conclusion

In summary, all the biopsy urease tests have a specificity of near 100% and sensitivity of at least 90%. Several very rapid tests have now been described that give almost immediate results. All the tests so far described can be made up quite easily for use by endoscopy units in most microbiology departments; alternatively several tests are now available commercially. The tests can replace, at negligible expense, the diagnosis by culture or histology. If sensitivities and serotyping, or histology of the gastric mucosa are required, the other diagnostic methods should still be used. Although non-invasive diagnostic tests are now available, clinicians will probably continue to perform endoscopy and biopsy for the investigation of upper gastrointestinal disease and the biopsy urease test will continue to be a very useful rapid diagnostic test for *C. pylori* infection for many years to come.

Acknowledgement

Caroline Gaston is thanked for her help in the preparation of this chapter.

References

1 Luck, J.M. & Seth, T.N. Gastric urease. *Biochem. J.* 1924, **18**, 1227–31.
2 Fitzgerald, O. & Murphy, P. Studies on the physiological chemistry and clinical significance of urease and urea with special reference to the stomach. *Ir. J. Med. Sci.* 1950, **292**, 97–159.
3 Kornberg, H.L., Davies, R.E. & Wood, D.R. The activity and function of gastric urease in the cat. *Biochem. J.* 1954, **56**, 363–72.
4 Langenberg, M.L., Tytgat, G.N., Schnipper, M.E.I., Rietra, P.J.G.M. & Zanen, H.C. Campylobacter-like organisms in the stomach of patients and healthy individuals. *Lancet* 1984, **i**, 1348.
5 McNulty, C.A.M. & Wise, R. Rapid diagnosis of *Campylobacter*-associated gastritis. *Lancet* 1985, **i**, 1443–4.
6 Marshall, B.J., Warren, J.R., Francis, G.J., Langton, S.R., Goodwin, C.S. & Blincow, E.D., Rapid urease test in the management of *Campylobacter pyloridis*-associated gastritis. *Am. J. Gastroenterol.* 1987, **82**, 200–10.
7 Hazell, S.L., Borody, T.J., Gal, A & Lee, A. *Campylobacter pyloridis* gastritis. I. Detection of urease as a marker of bacterial colonization and gastritis. *Am. J. Gastroenterol.* 1987, **82**, 292–6.
8 McNulty, C.A.M., Dent, J.C., Uff, J.C., Gear, M.W.L. & Wilkinson, S.P. Detection of *Campylobacter pylori* by the biopsy urease test: an assessment in 1445 patients. *Gut*, 1989, in press.
9 Mobley, H.L.T., Cortesia, M.J., Rosenthal, L.E. & Jones, B.D. Characterization of urease from *Campylobacter pylori. J. Clin. Microbiol.* 1988, **26**, 831–6.
10 Vaira, D., Holton, J., Cairns, S., Polydorou, A., Falzon, M., Dowsett, J. & Salmon, P.R. Urease tests for *Campylobacter pylori*: care in interpretation. *J. Clin. Pathol.* 1988, **41**, 812–13.
11 Arvind, A.S., Cook, R.S., Tabaqchali, S. & Farthing, M.J.G. One-minute endoscopy room test for *Campylobacter pylori. Lancet* 1988, **i**, 704.
12 Deltenre, M., Nyst, J.F., Glupczinski, Y., De Prez C., Dekoster, E. & Burette, A. Diagnosis of *Campylobacter pylori* infection. *Scand. J. Gastroenterol.* 1989 (Supplement), in press.
13 Morris., McIntyre, D., Rose, T. & Nicholson, G. Rapid diagnosis of *Campylobacter pyloridis* infection. *Lancet* 1986, **i**, 149.
14 Vaira, D., Holton, J., Cairns, S., Falzon, M. & Salmon, P. Four hour rapid urease test (RUT) for detecting *Campylobacter pylori*: is it reliable enough to start treatment? *J. Clin. Pathol.* 1988, **41**, 355–6 (Letter).

11 The serology of Campylobacter pylori infections

D.G. NEWELL AND A.R. STACEY

It is now widely accepted that there is an association between *Campylobacter pylori* colonization of the gastric antrum and non-autoimmune (type B) gastritis and peptic ulceration. Histologically-definable gastritis is undoubtedly a consequence of the local immune response of the host to this infection and involves the infiltration of lymphocytes, plasma cells, histiocytes and, frequently, polymorphonuclear cells, into the lamina propria.

The role of antibodies in C. pylori associated gastritis

A large proportion of the lymphoid cells infiltrating the gastric mucosa are immunoglobulin secreting B cells. These mature B cells in the gastric mucosa produce a local antibody response which is primarily of the IgA and IgG classes [1]. The role of these local antibodies in the disease process is debatable. The extent of the response does not correlate with either the degree of inflammation or the presence of ulceration [2–5]. Although organisms attached to the gastric epithelium are coated with human immunoglobulin [6], the antibodies neither eliminate the colonization of the gastric mucosa nor prevent the re-establishment of the infection after clearance [7]. As *C. pylori* is sensitive *in vitro* to both antibody-dependent and -independent complement-mediated bactericidal mechanisms [8, 9], this suggests that the organisms have a means of protecting themselves from the host response. Moreover, the localized immune response is so extensive that the possibility of a pathological role for host antibody in the disease process cannot be ignored.

The host circulating antibody response during infection

Most patients colonized with *C. pylori* elicit a measurable systemic antibody response, which may reflect the specificity of those antibodies produced at the gastric mucosa (Stacey, unpublished data).

The immunoglobulin classes and subclasses of these circulating anti-*C. pylori* antibodies are consistent with a prolonged chronic mucosal infection, with IgG and IgA predominating and IgM antibodies rarely seen [1, 4] even in children with acute infections [10] (Newell, unpublished data). Serum levels of the IgG_1 and IgG_4 antibodies are significantly raised, as is specific IgG_2, but not in all cases. IgG_3 antibodies, which are synonymous with acute infection, are not detected [4]. In the single, well-documented, volunteer study, where ingestion of *C. pylori* eventually led to chronic gastritis, IgG seroconversion occurred between 22 and 33 days post-infection [11]. Interestingly, no initial IgM response was observed and no specific seroconversion of IgA was noted during this study.

This specific IgG response remains constant throughout the infection [12, 13]. Several studies now suggest that both IgG and IgA levels decrease after treatment and concomitant clearance of the organism [7, 14, 15]. Nevertheless, the fall is generally slow with patients attaining normal levels only after 6 months or so. Re-infection is accompanied by a rapid rise in titre [12].

The antigenic specificity of anti-C. pylori antibodies

The response of the host to those bacterial antigens to which it is exposed during infection, and thereby the antibodies produced, is directed by a variety of both host and bacterial factors. Little is known of the immune response to infection at the gastric mucosal surface or of how that response spills over into the circulation. However, epithelial damage, the massive infiltration of immunocompetent cells, the efficiency of the antigen presenting cells and the chronic nature of the infection, will all affect the molecular basis of the host response. There are two main methods for analysing antibody specificity at the molecular level. The first is radioimmunoprecipitation (RIPA), which involves the separation and identification, by SDS-PAGE, of ^{125}I-labelled bacterial surface antigens bound to the antibodies and immunoprecipitated onto agarose beads. The second and easier method is Western blotting, which involves the electrophoretic transfer onto nitrocellulose of bacterial antigens separated by polyacrylamide gel electrophoresis (PAGE). Antibodies bound to these immobilized antigens are detected by an enzyme-linked antibody system. A sophisticated alternative to Western blotting is the electro-elution of separated polypeptides from the PAGE gel and the analysis of their antigenicity by ELISA [16]. Because of conformational changes potentially induced in the antigen by PAGE or the limitations of the radiolabelling techniques, these techniques may not provide identical results but should be considered as complementary.

Significant differences in the molecular weights of the major protein antigens have been reported. Such variations are largely due to the PAGE systems used. However, it seems that most *C. pylori*-positive patients produce antibodies that react against 61 000, 56 000 and 31 000 dalton polypeptides, which are specific for *C. pylori* and which are probably urease components (unpublished observations). Other, apparently specific, antigens frequently detected by positive sera include a 110 000–120 000 dalton protein [17–19] and several series of high (87 000–86 000, 63 000–57 000) and low (20 000–22 000, 14 000–16 000) molecular weight proteins [3, 5, 19]. Almost all sera react with a doublet at 54 000 and 56 000 daltons, which are putative flagella proteins and are not specific for *C. pylori* [20, 21]. At present, no single protein antigen has been detected which consistently reacts with all the positive sera investigated. Most reports, therefore, agree that the antibody response to *C. pylori* is both complex and highly variable. The observation that individual sera will detect the same major antigens in heterologous strains as in the homologous strain [20], supports the contention that this variation in antibody specificity is host-mediated rather than a reflection of antigenic differences between infecting strains.

Nevertheless, serological tests to detect circulating anti-*C. pylori* antibodies, could be useful, not only for epidemiological and research investigations but also as tools for the pre-endoscopic screening of dyspeptic patients and the monitoring of treatment.

Techniques for serodiagnosis

Serum anti-*C. pylori* antibodies can be detected by a variety of serological assays. Such tests include complement fixation (CFT), haemagglutination, bacterial agglutination, immunofluorescence and enzyme-linked immunosorbent assay (ELISA) [22–24]. ELISA is currently the technique of choice in most laboratories, because of its speed, low cost, simplicity and reproducibility. However, both Western blotting and RIPA may be ultrasensitive tools in serodiagnosis. Although the variation in human antibody specificity at the molecular level is great, sufficiently specific antigens are present to allow accurate diagnosis. Such techniques have been of particular value in diagnosing infection in primates (unpublished observations).

The ELISA technique involves the absorption of antigen onto a plastic support, usually the well of a 96-well microtitre plate. Excess antigen is removed by washing and diluted serum incubated with the immobilized antigen. After washing, human immunoglobulin bound to the antigen is detected using an anti-human immunoglobulin coupled to an enzyme, such as peroxidase. The peroxidase indirectly bound to the antigen can then be visualized using a chromogenic enzyme substrate. The assay is quantitative with the optical density of the coloured product being directly proportional to the amount of antibody bound. The antibody concentration in the serum can be measured by reference to standard curves or antibody titres obtained by serial dilution of the serum.

Antigens for ELISA serodiagnosis

The sensitivity and specificity of the ELISA are largely dependent on the nature of the antigenic material bound to the solid plastic support. The criteria essential to the selection of a suitable antigenic material for ELISA are given in Table 11.1.

Several antigens have now been investigated for ELISA serodiagnosis. Originally crude antigens, such as whole-cell sonicates, were used but these tended to give high backgrounds, a high proportion of false-positive sera and a poor sensitivity. It is now widely accepted that these crude antigens are generally unsatisfactory for the accurate diagnosis of *C. pylori*-related gastritis.

More recently, partly purified antigens have been developed. The most common of these antigens in use at present is an acid-extractable material eluted from the surface of whole organisms in 0.2 M glycine buffer (pH 2.2). This material is easily prepared and has reduced amounts of antigens which are cross-reactive with other *Campylobacter* species [20]. Both this antigen and an ultracentrifuged whole-cell sonicate antigen comprise a number of polypeptides as determined by SDS-PAGE (Fig. 11.1). ELISAs with these complex antigens detect approximately 95% of patients with *C. pylori* infection with a specificity of about 90% [25].

Without doubt the ability of the serodiagnostic test to accurately detect *C. pylori*-associated disease would be further improved with the development of even better antigens. Several major problems, discussed below, in-

Table 11.1 Criteria for the selection of suitable ELISA antigens

Contains a high proportion of immunogenic components
Immunogens should be common to all *C. pylori* strains
Absence of antigenic cross-reactions with other organisms
Absence of non-specific immunoglobulin binding
Easily prepared and purified
Bind effectively to the plastic support
Stable on storage

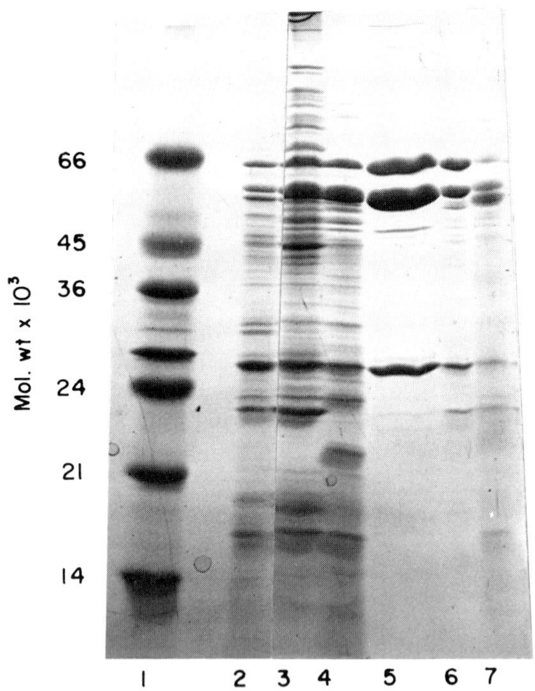

Fig. 11.1 Total protein profile of antigens used in serodiagnosis determined by SDS-PAGE using 10–25% gradient gel. Track 1 = standards; Track 2 = acid extract; Track 3 = whole-cell sonicate; Track 4 = French press urease preparation; Tracks 5 and 6 = FPLC purified urease; Track 7 = ultracentrifuged cell sonicate (prepared by B.J. Rathbone). (Reproduced from Newell & Rathbone [25].)

herent in the antigenic material have now been identified which influence the specificity and sensitivity of the ELISA test.

Antigenic variation between C. pylori strains

A proportion of sera from patients with proven *C. pylori* infection do not appear to have a detectable antibody response. As most antigen preparations are the product of a limited number of *C. pylori* strains (up to five strains), the presence of antigenic variation within the species could prevent the detection of antibody responses in those patients infected with the less-common variants. There is some evidence for the presence of serotype-specific antigens on *C. pylori*, detectable by hyperimmune rabbit sera in agglutination systems [26] and by Western blotting [27]. This serotype specificity is geographically distributed [23]. It is unlikely that antibody specificity observed with hyperimmune rabbit antisera will reflect the specificity of the human antibody responses induced by infection. In agglutination systems, the serotype differences probably result from the diversity in *C. pylori* lipopolysaccharides [28]. Lipopolysaccharides bind indifferently to micro-ELISA plate plastics and may only marginally influence ELISA results. However, the inclusion of more strains in the antigen preparation only marginally increases the sensitivity of the serodiagnostic test, if at all [29]. The serotypic differences, observable in agglutination techniques, therefore are probably not relevant when using the ELISA technique.

Non-specific immunoglobulin binding

Campylobacter pylori can non-specifically bind human immunoglobulin. Such processes could be advantageous to the host, enhancing opsonophagocytosis [8], but may also, by virtue of Fc receptor-like properties, form the basis of the organism's protection from host antibody responses. Nevertheless, this non-specific immunoglobulin binding can also cause unacceptable high backgrounds in the ELISA technique. The use of Fab_2 fragments conjugated to the enzyme reduces some of this non-specific binding (Stacey, unpublished data). However, the major components mediating this non-specificity appear to be confined to a high-molecular-weight fraction of sonicated *C. pylori*, which can be largely eliminated from the whole cell sonicate (unpublished observations).

Antigenic cross-reactivity with other bacteria

Some of the surface antigens of *C. pylori*, in particular the flagella proteins, have epitopes in common with other *Campylobacter* species [21, 30–32]. Unfortunately these flagella proteins are extremely immunogenic. The presence of such proteins in the antigen preparation would prevent the exclusion of those seropositive patients with, for example, a recent *Campylobacter jejuni* infection [20]. Some authors advise the screening of sera for anti-*C. jejuni* as well as anti-*C. pylori* antibodies [3]. Because flagellins would be difficult to selectively exclude from a crude antigen without significantly reducing sensitivity, the development of purified ELISA antigens has become necessary.

The choice and isolation of purified antigens for ELISA

Both the sensitivity and specificity of the serodiagnostic test may be significantly improved using species-specific antigens. Preferably these should be proteins or glycoproteins which bind most effectively to ELISA plates and are readily purified.

As mentioned previously, few proteins, if any, are detected by all *C. pylori*-positive sera using techniques like Western blotting or RIPA. One such antigen is a 120 000 molecular-weight protein [5, 19]. This antigen, which is probably equivalent to the antigen 9 described by Hirschl *et al.* [16] can be isolated by electro-elution after preparative SDS-PAGE with retention of antigenicity. However, there is evidence to suggest that this protein is not expressed by all strains of *C. pylori* [19], and electro-elution is not an easy technique for large-scale preparation.

Another protein apparently specific to *C. pylori* is the urease. Several methods of urease production and purification have been described. The supernatant retained following ultra-centrifugation of antigens disrupted by French Press [33] or by sonication (unpublished observation), has a high urease activity. As an ELISA antigen this crude urease preparation has a very good sensitivity, however, the specificity of antigens prepared by these methods is similar to the whole-cell sonicate [29], which they closely resemble as seen by SDS-PAGE analysis (Fig. 11.1).

A significantly purified urease preparation can also be obtained from disrupted cells by means of size exclusion gel filtration by fast protein liquid chromatography (FPLC). The material obtained by this method comprises a single protein with urease activity, by PAGE, under non-denaturing conditions and has a molecular weight of approximately 500 000 daltons. By SDS-PAGE analysis, under reduced conditions, this material comprises three major polypeptide bands with molecular weights of 61 000, 56 000 and 31 000. At least two of these polypeptides have been shown to be antigenic during human infections by Western blotting techniques (Stacey, unpublished data). In ELISA, this purified urease antigen has, once again, an excellent specificity but appears to have a slightly reduced sensitivity when compared with the acid extract antigen [29]. The majority of *C. pylori*-infected patients produced anti-urease antibodies. However, some patients, known to have anti-*C. pylori* antibodies by the acid extract-ELISA and to be ^{14}C-urea breath test-positive, did not have detectable circulating anti-urease antibodies by ELISA or Western blotting.

The results obtained so far with purified material confirm that the antibody response to *C. pylori* is, in general, unique to every patient, with individuals responding to different immunogenic proteins on the surface of the organism. Consequently, antigen preparations comprising single proteins, like the urease, cannot provide the sensitivity necessary for an ideal ELISA antigen. However, the purified urease could constitute a highly specific component of a synthetic antigen. To this end, the expression of urease genes in *E. coli* [34] and the use of monoclonal anti-urease antibodies for affinity purification will allow the large-scale production of this pure antigenic material.

The value of immunoglobulin class analysis in serodiagnosis

The detection of IgA, as well as IgG antibodies, can significantly increase sensitivity [15], though in some cases IgG antibodies are present in the absence of raised specific IgA [4]. It is possible that an assay which detects both classes may be the best method of enhancing the sensitivity. The value of IgG subclass detection has yet to be investigated. It seems likely that some subclasses of IgG, such as IgG_2, may fall rapidly on eradication of the organism and could therefore be used to monitor infection.

Circulating anti-C. pylori antibodies in the asymptomatic population and seroepidemiology

In studies of normal healthy blood donors the overall incidence of anti-*C. pylori* antibodies appears to be about 20–30% [3, 35, 36]. However, this incidence appears to be both age and geographically dependent. In Great Britain the frequency of anti-*C. pylori* antibodies in asymptomatic children under the age of 15 years is about 5% rising gradually to between 50 and 75% in the general population over the age of 50 years [3, 29]. This trend is similar in populations throughout the developed world [13, 37–39]. However, raised antibodies are common in children from Third

World areas. Geographical differences are also observed in the incidence of seropositivity in adulthood. The maximum incidence occurs at roughly the same age (the sixth decade of life) in all populations, but the percentage of individuals who are seropositive at that age may vary between 50 and 100%. Interestingly, the higher incidences are also found in populations of the underdeveloped areas of the world such as China, Africa and South America [38, 40] (Newell unpublished data).

It should be remembered that a high incidence of asymptomatic carriage of *C. jejuni* occurs in underdeveloped countries, which could influence the serological data. However, biopsy culture and the ^{13}C-urea breath test also confirm these geographical differences [40]. Moreover, regional differences within Great Britain, where asymptomatic *C. jejuni* carriage is uncommon, have recently been discovered by us. In patients presenting at endoscopy units, the incidence of *C. pylori* antibody positivity was significantly higher in Glasgow than in London, North Staffordshire or Slough. Similar observations have now been made in asymptomatic blood donors from these regions (Stacey & Newell, unpublished observations).

Seroepidemiological observations will be profitable in attempts to locate the source and mode of transmission of the organism. It would appear that close contact with infected people, for example within family units and in endoscopy units, is reflected in a significantly higher incidence of seropositivity [10, 41], though this is not always the case [42]. Most infections with *C. pylori*, at least in the developed world, occur early in adult life. If human-to-human transmission occurs so readily then the low infection rate in children is difficult to understand, given the high incidence in adults with which they will have close contact.

The differences between developed and underdeveloped parts of the world suggest a role for poorer hygiene standards in transmission. However, this would hardly account for regional differences within Great Britain. Socioeconomic status appears to have little effect on the incidence of *C. pylori* infection [43]. Moreover, some primitive populations have extremely low seropositivity, with a concomitantly low incidence of peptic ulcers [38]. Nevertheless, there is some evidence that in situations of poorer hygiene, for example institutions for mentally retarded patients [37], the incidence of raised-*C. pylori* antibodies increases significantly.

The question of whether *C. pylori* infection is a zoonosis has also been addressed using seroepidemiological techniques. *Campylobacter pylori* has only been isolated from primates [44, 45] and one pig [46] apart from man. However, antibodies to *C. pylori* are raised in abattoir workers [14] and these antibody titres fall after treatment to eradicate *C. pylori* infection. There are, of course, other *Campylobacter* species, which are commonly present in slaughtered animals and antigens common to this group of organisms would confuse such studies.

The role of serology in diagnosing histological gastritis and peptic ulceration

Dyspepsia is a considerable clinical problem in general practice [47]. Traditionally these patients are investigated by endoscopy or radiology — both expensive and time-consuming techniques. Up to 75% of patients attending gastroenterology clinics with upper abdominal pain are diagnosed as having non-ulcer dyspepsia [48]. A large proportion of these patients will have *C. pylori*-associated gastritis, identifiable by histology and bacterial culture of endoscopic biopsies. However, up to 50% of patients could be histologically normal. The prior selection of those patients who would benefit from further investigation could, therefore, release substantial UK National Health Service resources [25].

The incidence of anti-*C. pylori* antibodies in patients with histologically-defined gastritis is between 80 and 90% (specificity $>90\%$, sensitivity $> 80\%$ and positive predicative value 80–95%) [22]. These figures depend largely on the biopsy sampling techniques and the ability of pathologists to diagnose chronic gastritis and distinguish this from normal histology. Histological gastritis is now recognized as a spectrum of diseases. Some histological states, such as autoimmune gastritis (type A) and reflux gastritis, have a poor association with *C. pylori*, while patients with lymphocytic gastritis, although *C. pylori*-negative by histology, are generally antibody-positive [49].

Currently-available serological tests diagnose chronic gastritis with acceptable accuracy. Seropositive individuals would also include the majority of high-risk patients, especially those with peptic ulcer and gastric cancer. The serological test can, therefore, effectively screen patients prior to endoscopy. Such a test may be used either to exclude seronegative patients from further investigation by endoscopy or to treat seropositive patients. It would appear that the first option is preferable to avoid missing the high-risk patients such as those with treatable gastric cancer. The incorporation of such a screening procedure into the routine management of the patient will require evaluation in controlled trials. It would appear that the identification of *C. pylori* infection in those patients with peptic ulcers will not necessarily influence patient management, although the choice of those anti-ulcer drugs with antimicrobial activity may affect the ulcer relapse rate [22].

In those patients with non-ulcer dyspepsia the association between gastritis and *C. pylori* infection is debatable [50–52]. However, should eradication of the infection be shown to alleviate symptoms, then a serodiagnostic test would have therapeutic implications to the patient. In this case, a non-invasive technique such as ELISA could become an essential diagnostic procedure in patients under 40–45 years. In this group of patients the risk of malignancy is very low and any ulcers present would also be treatable with an anti-*C. pylori* regimen, such as colloidial bismuth subcitrate [25].

Conclusion

The majority of people colonized with *C. pylori* elicit a specific antibody response. Western blotting and RIPA techniques have shown that the host response to *C. pylori* appears to be not only complex but extremely variable. This, coupled with the fact that up to 20% of the normal population are infected with *C. pylori* but remain asymptomatic, indicates that the ability of the host to respond to the organism may be an important factor in the aetiology of *C. pylori*-associated gastritis. It seems likely, therefore, that research focusing on these areas of investigation should be profitable.

Serological tests have become an important tool in the detection of *C. pylori*-associated gastritis. Several major advances have been made in serodiagnosis. In comparative studies, the sensitivity and specificity of both ELISA and Western blotting correlate well with the invasive methods of detection, biopsy, bacterial culture and histology. Western blotting is costly, time-consuming and requires considerable expertise, so the ELISA system would be preferred since the technique is already routinely in use in many laboratories.

The antigens currently employed routinely in the serodiagnosis of *C. pylori* have been improved in terms of their sensitivity, specificity and reliability. Nevertheless, the development of better antigens is still necessary. The unicomponent antigens, such as the purified urease, whilst having good specificity, are insufficiently sensitive. This is probably due to the variation in the patient antibody response. Future research should concentrate on the identification and isolation of multiple specific antigens from the organism and the establishment of techniques to prepare these antigens in reasonable quantities.

References

1 Rathbone, B.J., Wyatt, J.I., Worsley, B.W., Shires, S.E., Trejdosiewicz, L.K., Heatley, R.V. & Losowsky, M.S. Systemic and local antibody responses to gastric *Campylobacter pyloridis* in non-ulcer dyspepsia. *Gut*, 1986, **27**, 642–647.

2 Booth, L., Holdstock, G., McBride, H., Hawtin, P., Gibson, J.R., Ireland, A., Bamforth, J., Duboulay, C.E., Lloyd, R.S. & Pearson, A.D. Clinical importance of *Campylobacter pyloridis* and associated serum IgG and IgA antibody responses in patients undergoing upper gastrointestinal endoscopy. *J. Clin. Pathol.* 1986, **39**, 215–219.

3 Jones, D.M., Eldrige, J., Fox, A.J., Sethi, P. & Whorwell, P.J. Antibody to the gastric campylobacter-like organism ('*Campylobacter pyloridis*') — clinical correlations and distribution in the normal population. *J. Med. Microbiol.* 1986, **22**, 57–62.

4 Steer, H.W., Hawtin, P.R. & Newell, D.G. An ELISA technique for the serodiagnosis of *Campylobacter pyloridis* infection in patients with gastritis and benign duodenal ulceration. *Serodiagnosis & Immunotherapy* 1987, **1**, 253–259.

5 Von Wulffen, H., Grote, H.J., Gaterman, S., Loning, T., Berger, B. & Buhl, C. Immunoblot analysis of immune response to *Campylobacter pylori* and its clinical associations. *J. Clin. Pathol.* 1988, **41**, 653–659.

6 Wyatt, J.I., Rathbone, B.J. & Heatley, R.V. Local immune response to gastric *Campylobacter* in non-ulcer dyspepsia. *J. Clin. Pathol.* 1986, **39**, 863–870.

7 Langenberg, W., Rauws, E.A.J., Widjojokusumo, A., Tytgat, G.N.J. & Zanen, H.C. Identification of *Campylobacter pyloridis* isolates by restriction endonuclease DNA analysis. *J. Clin. Microbiol.* 1986, **24**, 414–417.

8 Das, S.S., Karin, Q.N. & Easmon, C.S.F. Opsonophagocytosis of *Campylobacter pylori*. *J. Med. Microbiol.* 1988, **27**, 125–130.

9 Pruul, H., Lee, P.C., Goodwin, C.S. & McDonald, P.J. Interaction of *Campylobacter pyloridis* with human immune defence mechanisms. *J. Med. Microbiol.* 1987, **23**, 233–238.

10 Mitchell, H.M., Bohane, T.D., Berkowicz, J., Hazell, S.L. & Lee, A. Antibody to *Campylobacter pylori* in families of index children with gastrointestinal illness due to *Campylobacter pylori*. *Lancet* 1987, **ii**, 681–682.

11 Morris, A. & Nicholson, G. Ingestion of *Campylobacter pyloridis* causes gastritis and raised fasting pH. *Am. J. Gastroenterol.* 1987, **82**, 192–199.

12 Langenberg, W., Rauws, E.A.J., Houthoff, H.J., Oudbier, J.H., van Bohemen, C.G., Tytgat, G.N.J. & Rietra, P.J.G.M. Follow-up of individuals with untreated *Campylobacter pylori*-associated gastritis and of non-infected persons with non-ulcer dyspepsia. *J. Infect. Dis.* 1988, **157**, 1245–1248.

13 Perez-Perez, G.I., Dworkin, B.M., Chodos, J.E. & Blaser, M.J. *Campylobacter pylori* antibodies in humans. *Ann. Inter. Med.* 1988, **109**, 11–17.

14 Vaira, D., Holton, J., Londei, M., Beltrandi, E., Salmon, P.R., D'Anastasio, C., Dowsett, J.F., Bertoni, F. Grauenfels, P. & Gandolfi, L. *Campylobacter pylori* in abattoir workers: is it a zoonosis? *Lancet* 1988, **ii**, 725–726.

15 Von Wulffen, H. & Grote, H.J. Enzyme-linked immunosorbent assay for detection of immunoglobulin A and G antibodies to *Campylobacter pylori*. *Eur. J. Clin. Microbiol. Infect. Dis.* 1988, **7**, 559–565.

16 Hirschl, A.M., Pletschette, M., Hirschl, M.H., Berger, J., Stanek, G. & Rotter, M.L. Comparison of different antigen preparations in an evaluation of the immune response to *Campylobacter pylori*. *Eur. J. Clin. Microbiol. Infect. Dis.* 1988, **7**, 570–575.

17 Apel, I., Kist, M., & Jacobs, E. 1988. Characterization of *C. pylori* by PAGE-analysis and detection of a specific protein band with human immune serum. In Kasijer, B. & Falsen, E. (eds.) *Campylobacter IV*. pp. 183–184. University of Göteborg, Göteborg.

18 Kaldor, J., Tee, W., Nicolacopolous, C., Demirtzoglou, K., Noonan, D. & Dywer, B. Immunoblot confirmation of immune response to *Campylobacter pyloridis* in patients with duodenal ulcers. *Med. J. Australia* 1986, **145**, 133–135.

19 Von Wulffen, H. & Grote, H-J. 1988. Correlation of *Campylobacter pyloridis* serology by immunoblot studies with chronic antrum gastritis. In Kasijer, B. & Falsen, E. (eds.) *Campylobacter IV*. pp. 180–181. University of Göteborg, Göteborg.

20 Newell, D.G. Human antibody responses to the surface protein antigens of *Campylobacter pyloridis*. *Serodiagnosis & Immunotherapy* 1987, **1**, 209–217.

21 Newell, D.G. Identification of the outer membrane proteins of *Campylobacter pyloridis* and antigenic cross-reactivity between *C. pyloridis* and *C. jejuni*. *J. Gen. Microbiol.* 1987, **133**, 163–170.

22 Jones, D.M., Lessells, A.M. & Eldridge, J. Campylobacter-like organisms on the gastric mucosa: culture, histology, and serological studies. *J. Clin. Pathol.* 1984, **37**, 1002–1006.

23 Kaldor, J., Tee, W., McCarthy, P. & Dwyer, B. Immune response to *Campylobacter pyloridis* in patients with peptic ulceration. *Lancet* 1985, **i**, 921.

24 Marshall, B.J., McGechie, D.B., Francis, G.J. & Utley, P.J. Pyloric *Campylobacter* serology. *Lancet* 1984, **ii**, 281.

25 Newell, D.G. & Rathbone, B.J. The serodiagnosis of *Campylobacter pylori* infection — a review. *Serodiagnosis & Immunotherapy* 1989, **3**, 1.

26 Danielsson, D., Blomberg, B., Jarnerot, G. & Kosunen, T.U. Heterogeneity of *Campylobacter pylori* as demonstrated by co-agglutination testing with rabbit antibodies. *Scand. J. Gastroenterol.* 1988, **23**, (Suppl. 142), 58–63.

27 Burnie, J.P., Lee, W., Dent, J.C. & McNulty, C.A.M. Immunoblot fingerprinting of *Campylobacter pylori*. *J. Med. Microbiol.* 1988, **27**, 153–159.

28 Perez-Perez, G.I. & Blaser, M.J. Conservation and diversity of *Campylobacter pyloridis* major antigens. *Infection & Immunity* 1987, **55**, 1256–1263.

29 Newell, D.G. & Stacey, A. 1989. Antigens for the serodiagnosis of *Campylobacter pylori* infections. *Gastroentérologie Clinique et Biologique* 1989, **13**, 376–418.

30 Dunn, B.E., Altmann, M., Perez-Perez, G.I. & Blaser, M.J. 1988. Characterization of the outer membrane proteins of *Campylobacter pyloridis* by two-dimensional gel electrophoresis and immunoblotting. In Kasijer, B. & Falsen, E. (eds.) *Campylobacter IV*. pp. 98–99. University of Göteborg, Göteborg.

31 Lee, A., Logan, S.M. & Trust, T.J. Demonstration of a flagellar antigen shared by a diverse group of spiral shaped bacteria that colonize intestinal mucus. *Infection & Immunity* 1987, **55**, 828–831.

32 Newell, D.G. Monoclonal antibodies directed against the flagella of *Campylobacter jejuni*: cross-reacting and serotypic specificity and potential use in diagnosis. *J. Hygiene* 1986, **96**, 377–384.

33 Dent, J.C., McNulty, C.A.M., Uff, J.S., Gear, M.W.L. & Wilkinson, S.P. *Campylobacter* urease: A new serological test. *Lancet*, 1988, **i**, 1002.

34 Clayton, C.L., Topping, A.J., Wren, B.W., Mullany, P. & Tabaqchali, S. 1989. Molecular cloning and expression of species specific immunogenic antigens of *Campylobacter pylori* in *Escherichia coli* K12. In Megraud, F. & Lamouliatte, H. (eds.) *Proceedings of the Workshop on Gastroduodenal Pathology and Campylobacter pylori, Bordeaux*. Elsevier, Amsterdam, in press.

35 Goodwin, C.S., Blincow, E., Peterson, G., Sanderson, C., Cheng, W. Marshall, B., Warren, J.R. & McCullouch, R. Enzyme linked immunosorbent assay for *Campylobacter pyloridis*: correlation with presence of *C. pyloridis* in the gastric mucosa. *J. Infect. Dis.* 1987, **155**, 488–494.

36 Newell, D.G., Johnston, B.J., Ali, M.H. & Reed, P.I. An enzyme-linked immunosorbent assay for the serodiagnosis of *Campylobacter pylori*-associated gastritis. *Scand. J. Gastroenterol.* 1988, **23** (Suppl. 142), 53–57.

37 Bercowicz, J. & Lee, A. Person-to-person transmission of *Campylobacter pylori*. *Lancet* 1987, **ii**, 680-681.

38 Dwyer, B., Nanxiong, S., Kaldor, J., Tee, W., Lambert, J. Luppino, M. & Flannery, G. Antibody response to *Campylobacter pylori* in an ethnic group lacking peptic ulceration. *Scand. J. Infect. Dis.* 1988, **20**, 63–68.

39 Morris, A., Nicholson, G., Llyod, G., Haines, D., Rogers, A. & Taylor, D. Seroepidemiology of *Campylobacter pyloridis*. *N. Z. Med. J.* 1986, **99**, 657–659.

40 Graham, D.Y., Klein, P.D., Opekun, A.R., Boutton, T.W., Evans, D.J., Evans, D.G., Alpert, L.C., Michaletz, P.A., Yoshimura, H.H. & Adam, E. Epidemiology of *Campylobacter pylori* infection: ethnic considerations. *Scand. J. Gastroenterol.* 1988, **23** (Suppl. 42), 9–13.

41 Morris, A., Llyod, G. & Nicholson, G. *Campylobacter pyloridis* serology among gastroendoscopy clinic staff. *N. Z. Med. J.* 1986, **99**, 819–820.

42 Jones, D.M., Eldridge, J. & Whorwell, P.J. Antibodies to *Campylobacter pyloridis* in household contacts of infected patients. *Br. Med. J.* 1987, **294**, 615.

43 Ramirez-Ramos, A., Pearson, A.D., Gilman, R.H., Leon-Barua, R., Spira, W.M., Recavarren, S., Watanabe, P., Rodriguez, C., Bonilla, J., Montebette, M. & Lior, H. 1988. Epidemiology of *Campylobacter pylori* in Lima, Peru. In Kasijer, B. & Falsen, E. (eds.) *Campylobacter IV*. pp. 386–38. University of Göteborg, Göteborg.

44 Baskerville, A. & Newell, D.G. Naturally occurring chronic gastritis and *C. pylori* infection in the Rhesus monkey: a potential model for gastritis in man. *Gut* 1988, **29**, 465–472.

45 Bronston, M.A. & Schoenknecht, F.D. *Campylobacter pylori* isolated from the stomach of the monkey, *Macaca nemestrina*. *J. Clin. Microbiol.* 1988, **26**, 1725–1728.

46 Jones, D.M. & Curry, A. 1988. Ultrastructural study of gastric campylobacter-like organisms (GCLO) from man, pig and ferret. In Kasijer, B. & Falsen, E. (eds.) *Campylobacter IV*. p. 109. University of Göteborg, Göteborg.

47 Heatley, R.V. & Rathbone, B.J. Dyspepsia — a dilemma for doctors? *Lancet* 1987, **ii**, 779–782.

48 Gear, M.W.L. & Barnes, R.J. Endoscopic studies of dyspepsia in a general practice. *Br. Med. J.* 1980, **280**, 1136–1137.

49 Dixon, M.F., Wyatt, J.I., Burke, D.A. & Rathbone, B.J. Lymphocytic gastritis — relationship to *Campylobacter pylori* infection. *J. Pathol.* 1988, **154**, 125–132.

50 Borody, T.J., Carrick, J. & Hazell, S.L. Symptoms improve after the eradication of gastric *Campylobacter pylori*, *Med. J. Australia* 1987, **146**, 450–451.

51 Lambert, J.R., Borromeo, M., Korman, M. & Hansky, J. Role of *Campylobacter pyloridis* in non-ucler dyspepsia — a randomized controlled trial. *Gastroenterology* 1987, **92**, 1488.

52 Rokkas, T., Pursey, C., Simmons, N.A., Filipe, M.I. & Sladen, G.E. Colloidal bismuth subcitrate (De-nol) in non-ulcer dyspepsia. Placebo controlled trial with particular reference to the role of *Campylobacter pyloridis*. *Gut* 1987, **28**, A1337.

12 Detection of Campylobacter pylori by the ^{14}C-breath test

J. WEIL AND G.D. BELL

The organism *Campylobacter pylori* has very high urease activity [1], which has recently been shown to be between 20 and 70 times as active as that of, for instance, proteus [1]. This important observation has become the basis of the various different 'rapid urease tests' now available. Endoscopy is required to obtain gastric biopsies for the detection of the *C. pylori* organism, whether this be done by culture, the rapid urease test or examination of histological specimens by one of the various stains used.

The serological detection of *C. pylori* is obviously of considerable value for epidemiological studies as to whether particular groups of patients and healthy controls have or have not at some time in their life been infected by *C. pylori*. A positive test, however, does not give great information as to whether the patient's stomach is currently colonized by the organism. Furthermore, the demonstrated falls in antibody titres to the organism which have been reported following bismuth/antibiotic therapy, are unlikely to be sufficiently sensitive to accurately predict in which patient the organism has been successfully eradicated from the stomach, rather than, as so frequently happens, merely temporarily suppressed.

It is our belief that the ^{14}C-urea breath test is a reasonably sensitive, inexpensive, non-invasive and reproducible test for *C. pylori* detection and may have a particular role to play in serially monitoring the efficacy (or otherwise) of different forms of treatment designed to eradicate the organism. The amount of ^{14}C-urea required is very small (it has been calculated that a patient receiving 10 μCi ^{14}C-urea receives about one-tenth of the radiation dose of a patient having a single plain abdominal X-ray). The ^{13}C-urea breath test has the advantage of using a stable non-radioactive isotope [2], but has the major disadvantage that the $^{13}CO_2$ expired in the patient's breath requires the use of a mass spectrometer which costs over ten times as much as a simple scintillation beta counter and is thus unlikely to be widely available outside teaching hospital/university centres.

In this chapter, we will concentrate on discussing the technique involved for the ^{14}C-urea breath test, giving examples of its use before discussing the points for and against its use as a diagnostic test. In preparing this chapter, we have relied heavily on our own experience at the Ipswich Hospital. Since our preliminary report in *Lancet* [3] we have greatly expanded our numbers now to well over 300 patients studied. Quite independently of our own group, those of Marshall in Australia [4] and Tytgat and co-workers [5,6] in Amsterdam have also recently published results. Although several other groups have experience with the test, most of their results appear at present to be in abstract form in relatively small numbers of patients, and since their findings broadly agree with the findings from Australia [4], Holland [5,6] and Great Britain [3], they will not be discussed further in the present chapter.

The ^{14}C-urea breath test

Technique

Our patients are fasted overnight. At 9 a.m. each patient is given 350 ml of a liquid meal, followed immediately by 0.4 Mbq ^{14}C-urea

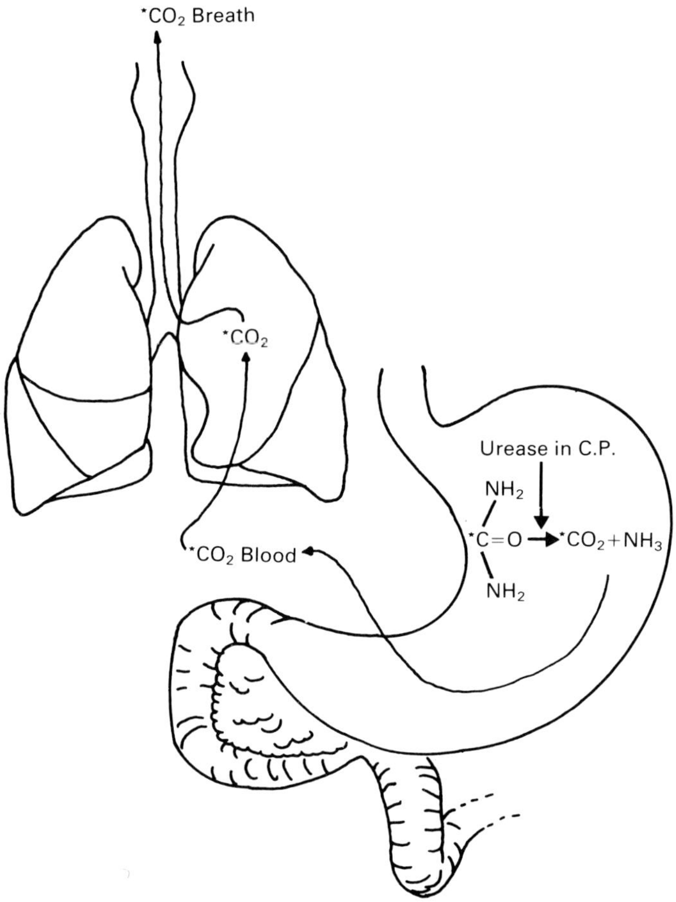

Fig. 12.1 Schematic drawing of the principle of the ¹⁴C-urea breath test. When ¹⁴C-labelled urea is administered orally, urea-derived labelled CO_2 appears in the breath of *C. pylori*-infected individuals.

(Amersham International) in 20 ml of water. The patients are asked to lie on their right side for a few seconds and then on the left for a few more seconds in an attempt to get the labelled urea evenly distributed over the lining of the stomach. Breath samples are collected at 10, 20, 30, 40, 60, 80, 100 and 120 min after ingestion of the ^{14}C-urea. The patients exhale through a tube of anhydrous calcium chloride into a vial containing 2 mmol of hyamine (a trapping agent for CO_2) in 2 ml of ethanol with phenolphthalein as indicator; decolorization of this solution indicates the collection of 2 mmol of exhaled CO_2. Then 10 ml of toluene-based scintillate is added, and the ^{14}C activity measured by liquid scintillation counting. A 1% standard is also counted and the activity expressed as the percentage of the administered dose per millimole of expired CO_2. This is multiplied by body weight (kg) to allow for endogenous CO_2 production. The area under the curve (AUC) is then computed, using the trapezoid rule for each individual 2-hour breath test. The single figure derived is then used as an expression of the cumulative excretion of $^{14}CO_2$.

The principle of the test is illustrated in Fig. 12.1, and an example of the breath test of a patient who was *C. pylori*-positive (by culture and histological criteria) before and again at the end of a one-month course of bismuth subcitrate (De-Nol, Brocades) is shown in Fig. 12.2. It can be seen that ^{14}C-labelled CO_2 rapidly appears in the breath of the patient.

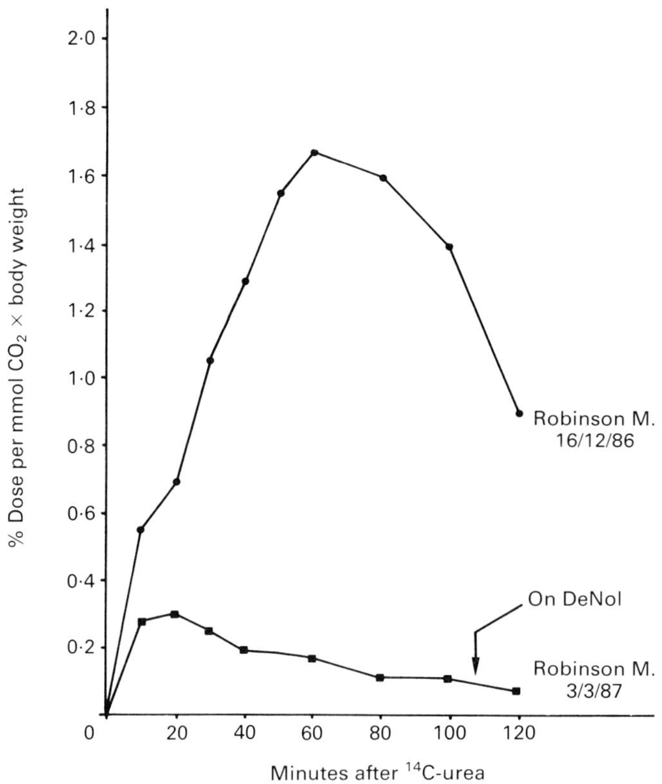

Fig. 12.2 Typical ^{14}C-urea breath test result in a *C. pylori*-positive individual before and again while taking the bismuth preparation De-Nol.

(The peak, in our experience, may be anywhere between 20 min and 60 min with a mean of 40 min.)

The patient may need to exhale into the hyamine for between 1 and 3 min to achieve decolorization of the solution (thus indicating the collection of 2 mmol of exhaled CO_2). Marshall's group [4] employ only 1 mmol of hyamine which halves the time the patient needs to blow into the solution, since only 1 mmol of CO_2 is trapped. Since the radioactive counts obtained are high (see below) and there is reasonably clear separation between *C. pylori*-positive and -negative groups, this appears to us to be a perfectly reasonable 'shortcut' and one we are considering adopting ourselves.

The dose of ^{14}C-urea used by both Marshall's group [4] and ourselves [3] of 0.4 MBq (10 μCi) is almost certainly much higher than is really necessary, since the Amsterdam group [5,6] have experience with over 130 cases in which very satisfactory detection of *C. pylori* colonization was achieved by giving the patients only 0.11 MBq (3 μCi).

Our own discussions with Amersham International lead us to suspect that it is probably not necessary to give the ^{14}C-urea mixed with 'cold' ^{12}C-urea. We have never included any ^{12}C-urea in our protocol, although we are aware that others are doing so for both the ^{14}C and ^{13}C-urea breath tests.

We do not agree with the Australian workers [4] that it is necessary to either avoid giving H_2-blockers for 48 hours prior to the ^{14}C-urea breath test, or to wash out the mouth of the patient prior to examination because of the theoretical risk of urease-splitting mouth organisms affecting the test.

Although we have always given the ^{14}C-urea with a test meal, in the hope that this might delay gastric emptying, and hence enhance the time for which the radioactive urea is in contact with the organism, we have no evidence to suggest that the test would not work equally well without giving a test meal. The Amsterdam group [6] have, however, done a number of paired studies and concluded that the addition of a test meal increased the resolution of the curves and suggested that possibly lower doses of ^{14}C-urea could be used to reliably detect *C. pylori* if a meal was given, rather than the isotope being given alone. As stated above, it has been our own custom [3] in Ipswich to collect a number of samples over a 2-hour period, and compute an AUC.

We have argued that this is likely to give a more accurate reflection of total bacterial urease activity in the stomach than simply taking one sample, for example 20–40 min after ingestion of the radioactive urea. In a busy district general hospital, however, it would appear to us to be perfectly acceptable simply to take one sample approximately 40 min after receiving the isotope. Marshall's group [4] only collect samples for 30 min and their results, like our own (see below), show good separation between patients whose stomachs are colonized with *C. pylori* and those that are not. Furthermore, the Amsterdam group [5,6] state that they collected samples at 10-min intervals for 90 min, but found that the most favourable cut-off level between the *C. pylori*-culture-positive and -negative individuals occurred at 40 min. Although, as stated above, our own group prefer to express our results as a 2-hour AUC, we have also looked at our data in terms of the percentage of the ^{14}C-urea excreted/mmol CO_2 × body weight (kg) at 30, 40 and 60 min and all seemed to give perfectly satisfactory separation between groups.

Presumably, anyone wishing to set up the ^{14}C-urea breath test in their own hospital would initially study their own culture-positive and -negative subjects to define their own normal range and upper limit of normal.

The Amsterdam group [6], for instance, reported that an arbitrary cut-off level of 0.07% of the dose of ^{14}C-urea/mmol CO_2 × body weight (kg) at 40 min conveniently separated culture-positive and -negative subjects. In their series, sensitivity of the test was 95% and specificity 98%.

We have presented our own data quite deliberately as distribution tables, rather than taking one single 'cut-off point' and giving sensitivity and specificity figures for that arbitrary value. For instance, one group setting up the test might wish to use the test for an epidemiological study. Another might wish to use it to monitor the success (or failure) of bismuth/antibiotic treatment to eradicate the organism from the stomach of a series of patients already known to be *C. pylori*-positive on endoscopic biopsy. The first group might thus opt for a higher 'cut-off point' where it was important not to get too many false-positives, while in the latter case, a much lower arbitrary cut-off figure would be more appropriate.

Results and examples of usage

Table 12.1 shows the Ipswich results for a series of 27 *C. pylori*-negative patients expressed both as percentage of administered dose per millimole of CO_2 × body weight (kg) for 10–40 min and also as the AUC over 2 hours. It will be seen that we have divided the *C. pylori*-negative patients arbitrarily into two groups: one in which our pathologist felt the mucosa was entirely normal, and a second group of 13 in which minor histological abnormalities were seen (though not active chronic gastritis — which is believed by many to be a histological 'hallmark' of *C. pylori* colonization). It will be seen that in both groups relatively low counts were obtained, but, interestingly, those with a normal mucosa were significantly lower. We did not perform serological tests on the *C. pylori*-negative controls without normal histology. It may be that some of these may have had the organism present in small numbers, which we failed to detect by simply taking two biopsies on the

Table 12.1 ^{14}C-urea breath test in *C. pylori*-negative patients (mean ± S.D.) expressed as a percentage of administered dose per millimole CO_2 × body weight (kg) for times 10–40 min and also as AUC over 2 hours

Group	*n*	Time (min)				AUC in 2 hours
		10	20	30	40	
C. pylori-negative controls with normal histology	14	0.21* (0.13)	0.16* (0.11)	0.16* (0.11)	0.18* (0.16)	18.4* (14.7)
C. pylori-negative controls with abnormal histology	13	0.47 (0.40)	0.46 (0.44)	0.50 (0.47)	0.52 (0.50)	54.6 (46.5)

* Significantly different from the *C. pylori*- negative controls with abnormal histology ($P < 0.05$).

antrum (one for histology and stain and the other for bacterial culture).

Table 12.2 shows the results we obtained in 75 patients who were known to be *C. pylori*-positive (on culture of antral biopsies). It can be seen that in this group, the $^{14}CO_2$ appears in the breath in much higher quantities than in the controls (see Table 12.1). The differences in the cumulative excretion expressed as AUC in 2 hours was not significantly different in the patients with duodenal ulcers, gastric ulcers, or those with an active chronic gastritis in whom no active ulcer was demonstrated.

The distribution and mean of estimated AUC of breath $^{14}CO_2$ excretion of ^{14}C-labelled urea by both endoscopic diagnosis and the presence of *C. pylori* of culture is shown in Table 12.3, and a similar exercise, taking the 40-min $^{14}CO_2$ excretion data, is shown in Table 12.4.

In order to assess the reproducibility of the ^{14}C-urea breath test, we studied 16 individuals on two occasions within one month of each other, while the subjects were on no treatment likely to affect the *C. pylori* status. As can be seen from Fig. 12.3, the test was fairly reproducible. The mean AUC on the first occasion of 147.3 ± 80.2 was not significantly different from that on the second occasion of 145.2 ± 87.2 (mean ± S.D.).

We have also studied the ^{14}C-urea breath test in *C. pylori*-positive individuals, both before and after being on cimetidine in a dose of 400 mg/b.d. There was the theoretical possibility that by suppressing acid secretion with an H_2-antagonist drug this would alter the environment of the *C. pylori* organism and either increase or decrease its number/enzyme activity. In our experience see Table 12.5), this was not the case, with the AUC on the first occasion of 184 ± 100 being similar

Table 12.2 ^{14}C-urea breath test in *C. pylori*-positive patients (on culture of antral biopsies) expressed as a percentage of administered dose per millimole CO_2 × body weight (kg) for times 10–40 min and also as AUC over 2 hours (mean ± S.D.)

Group	*n*	Time (min)				AUC in 2 hours
		10	20	30	40	
Active chronic gastritis	24	1.58 (1.07)	1.69 (1.10)	1.88 (1.02)	1.90 (0.98)	178.4 (76.1)
Duodenal ulcer	43	1.56 (0.85)	1.72 (0.76)	1.84 (0.77)	1.88 (0.71)	174.9 (66.0)
Gastric ulcer	8	1.74 (0.89)	2.21 (1.31)	2.35 (1.53)	2.35 (1.5)	209.2 (118.0)
Total	75	1.58 (0.92)	1.77 (0.94)	1.91 (0.95)	1.95 (0.90)	179.7 (75.5)

Table 12.3 Distribution and mean of estimated areas under curve (AUC) of breath $^{14}CO_2$ excretion of ^{14}C-labelled urea by endoscopic diagnosis and presence of *Campylobacter pylori* (CLO) on culture

AUC	Normal controls all CLO −ve (n = 19)	Abnormal controls all CLO −ve (n = 13)	Gastritis only		Duodenal ulcer		Gastric ulcer		Total	
			CLO −ve (n = 4)	CLO +ve (n = 24)	CLO −ve (n = 3)	CLO +ve (n = 43)	CLO −ve (n = 1)	CLO +ve (n = 8)	CLO −ve (n = 40)	CLO +ve (n = 75)
0–24	16	6	1*	1	1	1			24	2
25–49	2	1			1			1	4	1
50–74	1	3						1	4	1
75–99			1*			1			1*	1
100–124		1		2		9			1	11
125–149		2	1*	5	1*	6			4	11
150–174			1*	3		10		1	1*	14
175–199				6		2		1		9
> 200				7		14	1*	4	1*	25
Mean AUC	18.4	54.6	99.8	178.4	54.0	174.9	325.7	209.2	48.6	179.7
S.D.	14.7	46.5	68.6	76.1	67.7	66.0	–	118.0	63.7	75.5

* These subjects had *C. pylori* found in gastric biopsies on histology only.

Table 12.4 Distribution and mean breath $^{14}CO_2$ excretion at 40 min after ingestion of ^{14}C-labelled urea by endoscopic diagnosis and presence of *Campylobacter pylori* (CLO) on culture

Excretion of $^{14}CO_2$ at 40 min	Normal controls all CLO −ve (n = 19)	Abnormal controls all CLO −ve (n = 13)	Gastritis only		Duodenal ulcer		Gastric ulcer		Total	
			CLO −ve (n = 4)	CLO +ve (n = 24)	CLO −ve (n = 3)	CLO +ve (n = 43)	CLO −ve (n = 1)	CLO +ve (n = 8)	CLO −ve (n = 40)	CLO +ve (n = 75)
0	18	8	1*	1	2	1		1	29	3
0.5	1	3	1*	3	–	1		1	5	5
1.0		1	1*	5	–	13		1	2	19
1.5		1	–	6	1*	11		–	2	17
2.0			1*	2		8		1	1*	11
2.5				4		6		1	–	11
3.0				2		2		1	–	5
3.5				–		1		1	–	2
4.0				–			1*	1	1*	1
4.5				1						1
> 5										
Mean excretion	0.18	0.52	0.99	1.90	0.71	1.88	4.24	2.35	0.51	1.94
S.D.	0.16	0.50	0.86	0.98	0.88	0.71	–	1.50	0.79	0.90

*These subjects had *C. pylori* found in gastric biopsies on histology only.

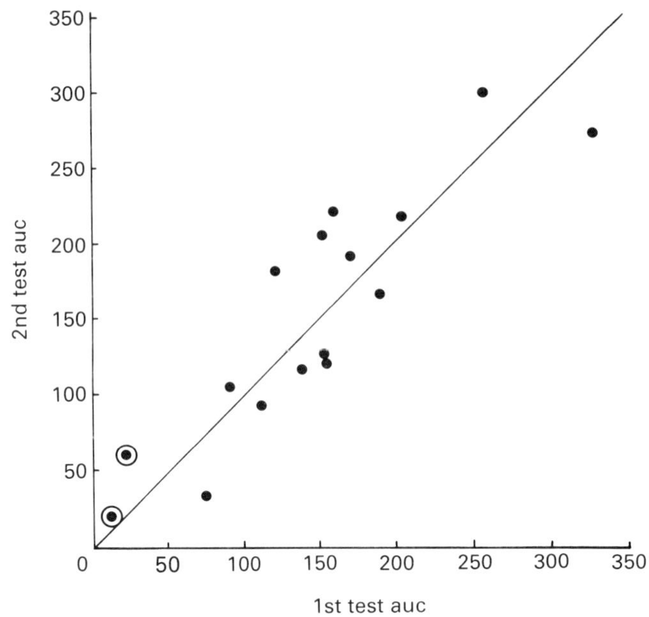

Fig. 12.3 Reproducibility results of ^{14}C-urea breath test on 16 individuals having tests repeated within 1 month on no treatment for *C. pylori*. The two ringed dots are biopsy-proven *C. pylori*-negative patients while the rest are all *C. pylori*-positive.

to that after treatment of 167 ± 64.6 (mean ± S.D.).

We have elected to take an AUC of 70 as our arbitrary cut-off point between *C. pylori*-positive and -negative individuals. In those with intact stomachs and who were *C. pylori*-positive (on either histological and/or bacterial criteria), 76/81 (93.8%) had AUCs > 70. No *C. pylori*-negative patient with normal antral histology had an AUC > 61.9, but 4/13 of those who were *C. pylori*-negative with abnormal histology were above this arbitrary figure.

As can be seen from Table 12.6, the urea breath test was extremely disappointing in patients who were *C. pylori*-positive and had undergone previous gastric surgery. Only 3/8 (37.8%) of these post-operative stomach patients had an AUC > 70. One must thus conclude that the ^{14}C-urea breath test is unreliable following gastric surgery. It could, however, be regarded as being a predictable

Table 12.5 ^{14}C-urea breath test in 10 *C. pylori*-positive individuals repeated after one month's treatment with cimetidine (400 mg b.d.) (mean ± S.D.)

Subject No.	Age	Sex	AUC_1	AUC_2
1	65	M	119.2	253.10
2	61	F	143.40	175.60
3	53	M	194.40	122.80
4	49	M	259.30	229.60
5	54	M	116.10	155.50
6	53	F	146.40	171.40
7	50	F	422.60	251.70
8	59	M	88.60	46.50
9	72	F	233.60	135.20
10	46	M	122.50	130.90
Mean ± S.D.	56.2 ± 8.0		184.6 ± 100.0	167.2 ± 64.6

Table 12.6 Individual details of 8 *C. pylori*-positive patients who had undergone previous gastric surgery

Subject No.	Age	Sex	Operation	*C. pylori* status		^{14}C-Urea breath test AUC
				Bacterial Culture	Histology	
1	43	M	Polya	+	+	32.0
2	80	F	Billroth I	+	+	109.6
3	77	M	Gastrojejunostomy	+	+	197.7
4	54	M	Polya	+	+	53.4
5	72	F	Polya	+	+	22.1
6	57	M	Polya	+	+	36.4
7	57	M	V & P	+	+	63.8
8	68	M	V & P	+	+	86.8
Mean ± S.D.	63.4 ± 12.9					72.3 ± 57.5

outcome if retention of the ^{14}C-urea in the stomach is a critical factor.

We feel that the ^{14}C-urea breath test is of particular value in monitoring the effect of various different treatments designed to suppress or hopefully eradicate the organism from the lining of the stomach. The mean AUC, for instance, of 40 of our *C. pylori*-positive patients re-tested while taking standard-dose bismuth subcitrate (De-Nol, Brocades) for periods of at least one month was 28.6 ± 29.4. The $^{14}CO_2$ excretion fell dramatically on active treatment in every case and in only 4 out of the 40 did the AUC fail to go below our arbitrary cut-off point of 70. Similarly, in a series of 12 further *C. pylori*-positive patients treated with a combination of De-Nol and Amoxil, the AUCs fell in every case, the mean being 19.7 ± 24.3 with 11 out of 12 falling below 70.

The fact that the ^{14}C-urea breath test is negative when re-studied while the patient is on bismuth subcitrate, either alone or in combination with an antibiotic, does not, of course, mean that the organism has been totally eradicated. In many cases, the number of organisms has simply been reduced to a level that it is no longer detectable, and once active treatment is withdrawn, there is a recrudescence of the *same* organism [7]. Our own experience [8], and that of Tytgat's group in Amsterdam [9], is that it is probably necessary to wait at least one month after discontinuing active treatment before either re-biopsing a patient, or doing a repeat urea breath test in order to obtain a realistic idea of true eradication rates. When tested one month after conventional 1–2 month courses of De-Nol, eradication rates of only about 20% are achieved. A combination of De-Nol and Amoxil increases this figure to approximately 40%. The value of the ^{14}C-urea breath test in serially monitoring patients is illustrated in Fig. 12.4. It can be seen that 15 of the 16 patients treated with De-Nol alone achieved an AUC < 70, while on treatment. Many had relapsed when retested a month later. Although not illustrated in Fig. 12.4, when some of the apparent 'successes' were followed up for 3 months, further relapses were seen.

It is our experience, using the ^{14}C-urea breath test to monitor patients serially, and also that of Tytgat's group [9] (using serial endoscopic biopsy criteria), that the majority, however, of patients whose test is negative at one month remain negative.

Although the dose of radioactivity received by a patient, using the ^{14}C-urea breath test, is extremely small, a single test done about 1–3 months after completing bismuth/antibiotic treatment may be all that is required to confirm successful elimination of the organism. Similarly, if a peptic ulcer has been found radiologically and thus no endoscopic biopsy is available for analysis, one could then per-

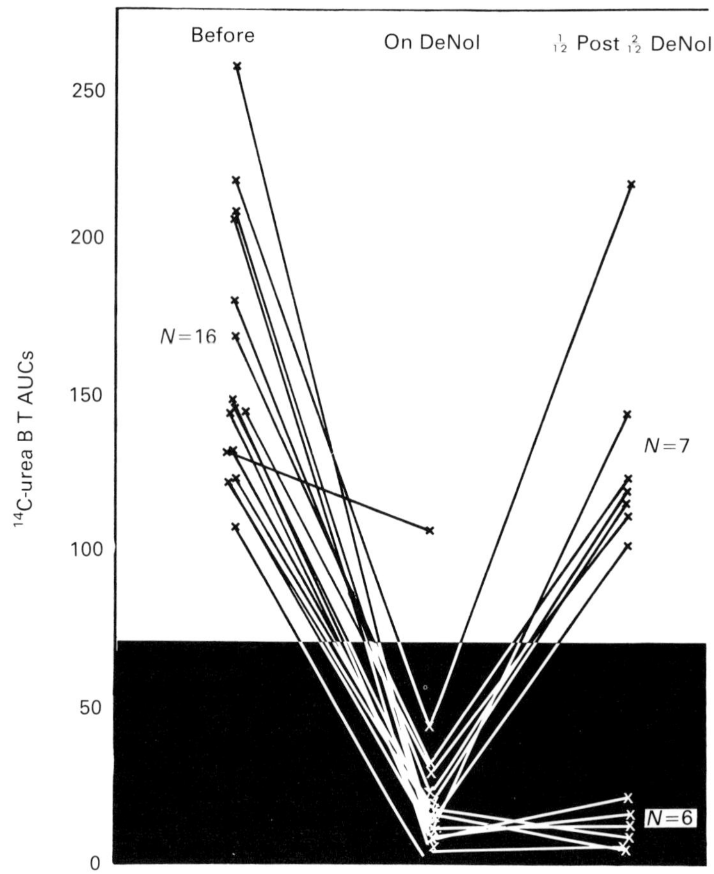

Fig. 12.4 ^{14}C-urea breath test in 16 *C. pylori*-positive patients before and again after 56 days on bismuth subcitrate: 15/16 subjects had become 'negative' (i.e. AUC <70); 7/15 *C. pylori*-'negative' subjects had relapsed when re-tested 1 month after discontinuing treatment.

form the ^{14}C-urea breath test twice: (i) prior to starting bismuth/antibiotic treatment, and (ii) 1–3 months after completing the course. As can be seen from the results above, patients can be treated throughout with an H_2-blocker without affecting the interpretation of the urea breath test.

Use as a diagnostic tool — points for and against

Clearly, if a patient is undergoing endoscopy, the easiest way to confirm or refute the presence of *C. pylori* in the stomach would be to take a series of biopsies. Depending on local interest and expertise, either the biopsies can be looked at by the histopathologist using an appropriate stain (in Ipswich we find the half-Gram stain simple and inexpensive [10]), or have the nearest co-operative microbiologist carry out the appropriate cultures. For a very rapid answer, then one of the many forms of rapid urease tests can be used. Additionally, if one cannot persuade one's local histopathologist or microbiologist to show an interest, one can rely almost entirely on the rapid urease test in most cases. Only if there is some contraindication to biopsy (e.g. a clotting disorder or the fact the patient was on oral anticoagulants) would one need the ^{14}C-urea breath test at this stage.

For instance, in a patient with a gastric ulcer, one would probably wish to do a further endoscopy 1–2 months after starting active anti-ulcer therapy to confirm healing of the ulcer and to make quite sure that a malignant process has not been missed. Again, in this case, since one would be repeating the endoscopy anyway, it would be reasonable to see whether a particular form of treatment has eradicated the bacteria, using biopsy criteria. In the case of duodenal ulcers which are, to all intents and purposes, never malignant, it is not normally necessary (outside clinical trials at least) to repeat the endoscopy to confirm healing. In this group and also in patients who have *C. pylori*-positive non-ulcer dyspepsia (endoscopy/biopsy proven), it might be more appropriate (and certainly much less invasive) to monitor the success or failure of a bismuth/antibiotic combination, using the ^{14}C-urea breath test, rather than subjecting the patient to an unnecessary second endoscopy.

The main disadvantage of the test appears to be the fact that in the post-operative situation it is unreliable, with a high percentage of false-negatives in *C. pylori* patients.

Since the test involves ^{14}C-urea, it might be considered unsuitable for certain groups such as children and pre-menopausal women. Certainly, in children, the ^{13}C-urea breath test [2] would have the advantages of not using a radioactive isotope, but the lack of general access to a mass spectrometer is always likely to limit its general usage.

Conclusion

The ^{14}C-urea breath test is a non-invasive accurate and non-expensive screening test which might be particularly useful for long-term follow-up studies, and particularly when looking at therapeutic efficacy of different drug treatments designed to eradicate the organism. It is unreliable in the post-operative stomach. A single breath sample taken 40 min after ingestion of the ^{14}C-urea is probably adequate to distinguish *C. pylori*-positive and -negative groups, but the 2-hour single AUC figure arguably gives a better assessment of total bacterial counts in the patient's stomach.

References

1 Ferrero, R.L., Hazell, S.L. & Lee, A. The urease enzymes of *Campylobacter pylori* and a related bacterium. *J. Med. Microbiol.* 1988, **27**, 33–40.

2 Graham, D.Y., Evans, D.J., Alpert, L.C. *et al.* *Campylobacter pylori* detected non-invasively by the ^{13}C-urea breath test. *Lancet* 1987, **i**, 1174–117.

3 Bell, G.D., Weil, J., Harrison, G. *et al.* ^{14}C-urea breath analysis, a non-invasive test for *Campylobacter pylori* in the stomach. *Lancet* 1987, **i**, 1367–1368.

4 Marshall, B.J. & Surveyor, I. Carbon-14 urea breath test for the diagnosis of *Campylobacter pylori* associated gastritis. *J. Nucl. Med.* 1988, **29**, 11–16.

5 Rauws, E.A.J., Tytgat, G.N.J., Langenberg, W. & Royen, E.V. 1988. Experience with ^{14}C-urea breath test in detecting *Campylobacter pylori*. In Menge, H., Gregor, M., Tytgat, G.N.J. & Marshall, B.J. (eds.) *Campylobacter pylori*. pp. 151–156. Springer-Verlag, Berlin.

6 Tytgat, G.N.J. & Rauws, E.A.J. Significance of *Campylobacter pylori*. *Aliment Pharmacol Therap.* 1987, **i**, 527S–539S.

7 Langenberg, W., Rauws, E.A.J., Widjojokusumo, A., Tytgat, G.N.J., & Zanen, H.C. Identification of *Campylobacter pyloridis* isolates by restriction endonuclease DNA analysis. *J. Clin. Microbiol.* 1986, **24**, 414–417.

8 Weil, J., Bell, G.D., Harrison, G., Trowell, J.E., Gant, P.W. & Jones P.H. *Campylobacter pylori* survives high dose bismuth subcitrate (De-Nol) therapy. *Gut* 1988, **29**, A1438.

9 Rauws, E.A.J., Langenberg, W., Houthoff, H.J., Zanen, H.C. & Tytgat, G.N.J. *Campylobacter pyloridis*-associated chronic active antral gastritis. *Gastroenterology* 1988, **94**, 33–40.

10 Trowell, J.E., Yoong, A.K.H., Saul, K.J., Grant, P.W. & Bell, G.D. A simple half-Gram stain for demonstrating *Campylobacter pyloridis* in sections *J. Clin. Pathol.* 1987, **40**, 702.

13 Detection of Campylobacter pylori by the ^{13}C-urea breath test

P.D. KLEIN AND D.Y. GRAHAM

The principle of carbon breath tests is based on isotopic carbon tracers which exploit the concept of a target bond that links a low-molecular-weight, isotopically-labelled group to the remainder of the substrate molecule. Cleavage of the target bond releases the labelled moiety that may be, or may undergo subsequent conversion to, labelled carbon dioxide which is exhaled in breath. The presence of the labelled carbon in breath signals the presence of an enzyme which acts on the target bond, and the rate and extent of isotopic CO_2 production can be used to estimate the quantity of enzyme present.

It is useful to divide breath tests into two categories: endolytic and xenolytic. The former category encompasses all hydrolytic and oxidative enzymes of endogenous origin normally found in the digestive tract, including lipases, disaccharidases, amino acid decarboxylases, glycolytic enzymes, and mixed-function oxidase systems [1–5]. Xenolytic breath tests document the presence of foreign enzymes, usually of bacterial origin. For example, the bile acid conjugate cholylglycine, in which the glycine is isotopically labelled, may be used to detect the presence of bacterial overgrowth in the small intestine [6]. Under normal circumstances, conjugated bile acids are absorbed intact in the terminal ileum. If the bile acid conjugate is exposed to bacterial peptidases, the glycine is cleaved off and metabolized to CO_2. For more details the reader is referred to a recent overview of ^{13}C breath tests [7].

Differences between ^{14}C and ^{13}C breath tests

Breath test substrates have been labelled with either ^{14}C or ^{13}C depending on the availability of the appropriate analytical instrumentation. Determinations of $^{14}CO_2$ depend upon liquid scintillation counting, and determinations of $^{13}CO_2$ (properly, $^{13}CO_2/^{12}CO_2$) require use of a gas isotope ratio mass spectrometer. Although the natural abundance of $^{14}CO_2$ is very low, that of ^{13}C is relatively high: for every million $^{12}CO_2$ molecules in breath, there are approximately 11 238 molecules of $^{13}CO_2$. This means that a person producing 150 μmol/kg/min of CO_2 has a baseline production of 1.685 μmol/kg/min of $^{13}CO_2$. This baseline $^{13}CO_2$ abundance can be measured with an analytical precision of 3 parts per million of total CO_2, or 0.03% of $^{13}CO_2$ concentration. This represents an uncertainty in baseline production values of 0.0005 μmol/kg/min. During the day, the baseline abundance of $^{13}CO_2$ normally fluctuates over a narrow range in response to dietary consumption of plant materials enriched in ^{13}C, notably carbohydrates derived from corn or cane [8, 9]. Stabilization of the baseline abundance can be achieved by fasting the subject or, as in the urea breath test, by use of a test meal.

Urea breath tests for Campylobacter pylori infection

Initially, detection of *C. pylori* infection of the gastric or duodenal mucosa required endoscopy for biopsy and subsequent culture or histological analysis. Such procedures are appropriate for symptomatic subjects but are sufficiently expensive and invasive to preclude their use in the general population. Moreover, the need to undergo serial endoscopic examinations to chart progression or regression of disease discourages participation in studies designed to investigate the natural history of *C. pylori* infection and in therapeutic trials designed to eliminate *C. pylori*. It became clear that we needed a non-invasive, sensitive, accurate test that could provide information about changes in the level of *C. pylori* infestation and that could be repeated as frequently as required by a patient's clinical condition or by a study protocol. The high level of endogenous urease activity and the ready availability of ^{13}C-urea provided the opportunity to develop a ^{13}C-breath test that would identify gastric infection caused by urease-producing bacteria.

Antecedents to the urea breath test for C. pylori

Although our urea breath test was developed independently, we subsequently discovered that localization of urease in the gastric mucosa by the use of labelled urea had been accomplished as long ago as the late 1940s [10]. Kornberg and his coworkers (1954) studied the activity and function of gastric urease by infusing ^{14}C-urea intravenously into cats that had been eviscerated from pylorus to anus [11]. The investigators observed a constant excretion of $^{14}CO_2$ in breath that was unchanged by the instillation of 0.1 N HCl into the cats' stomachs. Stimulation of gastric juice secretion by Priscol (2-benzylglyoxaline hydrochloride) infusion, however, resulted in a sharp rise in the specific activity of expired CO_2 (Fig. 13.1). These results demonstrated that urea was hydrolysed during passage from the blood to the lumen and was not dependent on the presence of acid in the stomach. In the same study, Kornberg and coworkers [11] showed that the breakdown of urea was also associated with the passage of the urea solution from the lumen of the stomach to the blood (Fig. 13.2), and concluded that 'the activity of gastric urease was dependent upon the flow of urea-containing liquid past the enzyme, which is located in the stomach at sites protected from inactivation by acid'. In a forerunner to modern clinical trials of *C. pylori* eradication, Kornberg *et al.* [11] also demonstrated that treatment of the cats with antibiotics abolished the urease activity which was therefore clearly of bacterial origin.

Performance of the ^{13}C-urea breath test

We designed our breath test so that sufficient substrate would be provided to saturate any urease present and so that the urea would be kept in the stomach (and out of the colon) for

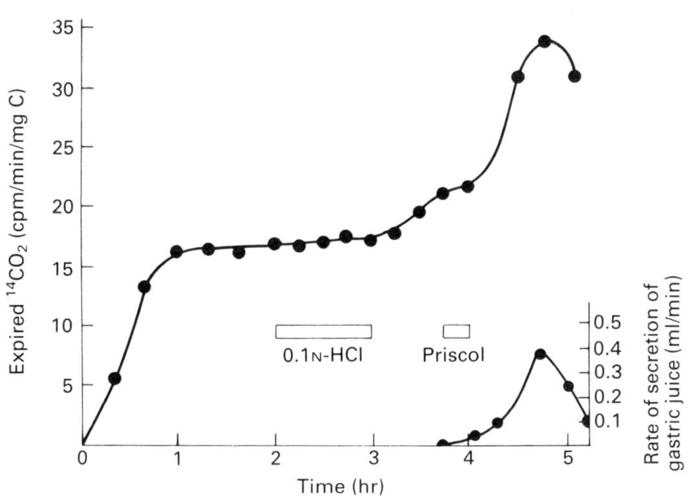

Fig. 13.1 The effect of the presence of acid in the stomach and of an infusion of Priscol (2-benzylglyoxaline hydrochloride) on the specific activity of expired CO_2 and on the rate of secretion of gastric juice in a cat whose kidneys, pancreas, and intestine from pylorus to anus were removed before (intravenous) injection of ^{14}C urea at 0 hr. Upper curve: specific activity of CO_2. Lower curve: rate of secretion of gastric juice. (Drawn from Kornberg *et al.* [11].)

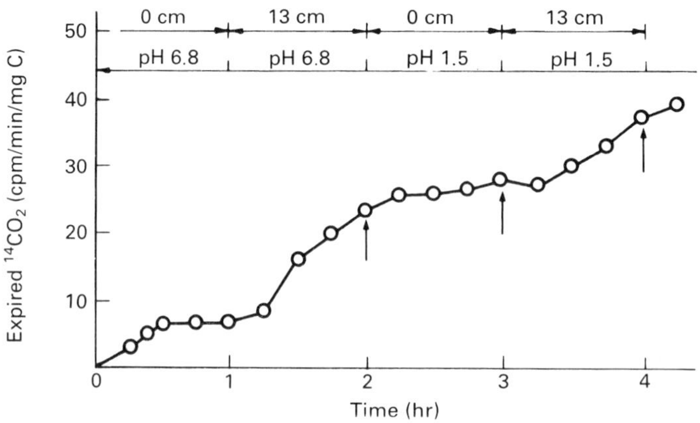

Fig. 13.2 The effect of changes in the pressure of ^{14}C urea solution in the stomach of the cat on the specific activity of expired CO_2. The pH and the pressures (in centimetres of solution above the top surface of the stomach) are given. (Drawn from Kornberg *et al.* [11].)

as long a period as possible. We performed a number of experiments to make the test as effective as possible. These investigations included the determination of the optimal amount of substrate, the appropriate timing of sample collection, and the need for a test meal. The components of our current breath-test kit are shown in Fig. 13.3. They include a nutrient-dense meal (pudding), the ^{13}C-urea substrate, a plastic bag with valve and mouthpiece, a needle for transferring the breath samples to each of five 20-ml Vacutainers®

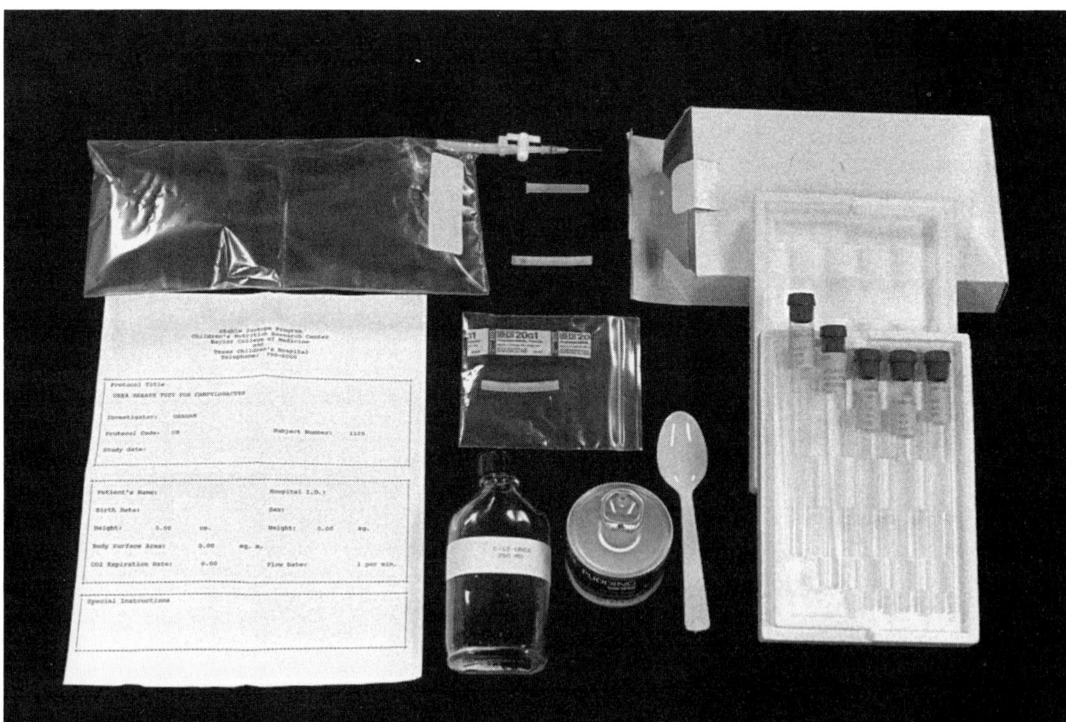

Fig. 13.3 Components of the ^{13}C urea breath test kit. Clockwise from bottom centre: Spoon, can of Sustacal® or Ensure® pudding, ^{13}C-urea, protocol data sheet, inflated plastic bag with needle attached, needle guard, mouth piece (centre, bag with needle and mouthpiece as supplied), carboard mailer carton, styrofoam mailer container for five Vacutainers® labelled with sample times.

(Becton-Dickinson, East Rutherford, NJ) and a shipping container for the samples.

The fasting subject reports to the clinic where his/her weight and height are measured and recorded. A baseline breath sample is collected, and the subject consumes a nutrient-dense meal to inhibit gastric emptying. We use a can of Sustacal® pudding (Mead Johnson, Evansville, IN) or 120 ml of 25% glucose polymer (Polycose®, Ross Laboratories, Columbus, OH). In paediatric studies, a scoop of ice-cream has proved both acceptable and effective. Ten minutes later the subject drinks a solution of distilled water containing 250 mg of ^{13}C-urea (99% ^{13}C, MSD Laboratories, Montreal, Canada), and breath samples are collected at 20, 30, 40 and 50 min. The samples are analysed for excess $^{13}CO_2$ by gas isotope ratio mass spectrometry in the automated system previously described by Schoeller and Klein [12]. Total CO_2 production is calculated from body surface area [13, 14], and these values are used to compute percentage dose recovery at each time point. Urease activity is computed and expressed as μmoles of urea hydrolysed per minute. No correction is made for incomplete ^{13}C recovery as determined by bicarbonate infusion studies [15].

Factors in scoring the ^{13}C-urea breath test

Figure 13.4 illustrates the outcome of the ^{13}C-urea breath test applied to uninfected and infected individuals. A negative breath test results in minimal changes in $^{13}CO_2$ abundance, e.g. less than 3 per ml, or 0.3% of the ^{13}C abundance over the four time points covering 30 min. A positive breath test is reflected by a rapid and sustained increase of 6 to 150 per ml (0.6% to 15.0% ^{13}C). An occasional test (approximately 1%) will show a transient increase in $^{13}CO_2$ in the first post-dose sample that is not maintained in later samples. This increase may be attributable to urease activity of oral flora that becomes inactivated in the gastric medium.

Expression of changes in $^{13}CO_2$ abundance as enzymatic activity allows comparison of bacterial infection levels within the same individual or between different members of a cohort. Translation of these changes is possible only if the excretion of label is multiplied by the CO_2 production rate and integrated over time. The CO_2 production rate can be calculated from body surface area, assuming a rate of 5 mmol/m^2 in resting subjects as reported by Shreeve and coworkers [14]. CO_2 production values must also be taken into account when establishing the minimum levels of urease that will be considered positive test results. Children younger than 2 years produce substantially fewer millimoles of CO_2 per minute than adults. As a result, less hydrolysis of the substrate is necessary in children of that age to produce a significant change in $^{13}CO_2$ enrichment. Positive results in young children, therefore, have lower values (as low as $>0.2\ \mu$mol/min in 1-year-olds) than the adult positive values ($>0.50\ \mu$mol/min).

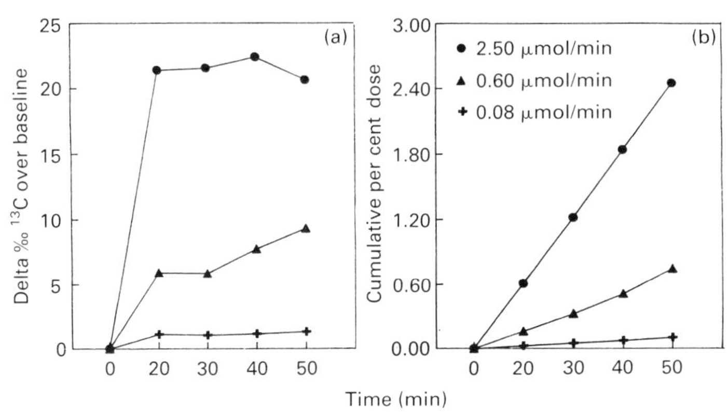

Fig. 13.4 Scoring the ^{13}C urea breath test. Left, isotopic enrichment of respiratory CO_2 over baseline values after ingestion of 250 mg ^{13}C urea by three individuals. Right, cumulative percentage of administered dose recovered, showing urease values of 0.08 μmol/min (negative), 0.60 μmol/min (low level of *C. pylori* activity) and 2.50 μmol/min (strongly positive for *C. pylori* activity).

Table 13.1 indicates the importance of providing a dose of ^{13}C-urea with sufficient substrate mass to saturate any level of enzyme that may be present. Note that the assayed level of urease does not decrease as the period of sampling increases from 60 to 120 min. Another important element of the breath test is the inclusion of a nutrient-containing meal. This meal ensures prolonged contact of the substrate with the gastric mucosa by preventing the substrate from emptying into the small intestine and travelling rapidly to the colon. Two benefits are derived from the test meal: (i) increased contact with gastric mucosa enhances detection of low levels of *C. pylori* and (ii) reduction of substrate escape into the colon prevents spurious $^{13}CO_2$ formation by urease of colonic flora.

The outcome of the test does not appear to be affected by subjects' consumption of food before the test. We retested six positive individuals on a subsequent day after each had consumed a hamburger (Big Mac®) and a chocolate milk shake. Thirty minutes later, each consumed a can of Sustacal®, and the ^{13}C-urea breath test was administered. The outcomes of the tests did not differ from those of the earlier tests given to the same subjects [16].

Table 13.1 The effect of measurement interval on urease activity obtained by the ^{13}C-urea breath test

C. pylori status	Measurement interval after substrate administration (mean ± S.D., μmol/min)		
	20–60 min	20–90 min	20–120 min
Negative	0.00 (0.10)	0.02 (0.12)	0.03 (0.14)
	0.05 (0.15)	0.00 (0.15)	0.01 (0.21)
	0.09 (0.16)	0.00 (0.20)	0.05 (0.24)
	0.33 (0.28)	0.38 (0.24)	0.33 (0.22)
Positive	1.21 (0.59)	1.66 (0.76)	1.93 (0.78)
	2.16 (0.31)	2.44 (0.48)	2.59 (0.48)
	2.20 (0.26)	2.22 (0.24)	2.05 (0.36)
	5.34 (1.21)	5.71 (1.11)	5.40 (1.12)

Validation of the ^{13}C-urea breath test

In our initial validation studies, we compared results of the ^{13}C-urea breath test with the outcomes of endoscopic examinations of normal subjects and patients with gastric or duodenal ulcers. The mucosa of the oesophagus, stomach, and duodenum of each subject was inspected, and four biopsy specimens were taken from the gastric antrum within 3 cm of the pylorus. For each subject, two biopsy specimens were submitted for culture and the remaining two were submitted for histological staining and examination. Fourteen of the 21 biopsied subjects had positive breath tests, and of these 14, 13 had positive cultures for *C. pylori*, and 13 had organisms visible in silver-stained sections. The one breath-test positive patient with a negative culture was not the same person as the one breath-test positive, histologically-negative patient. None of the seven subjects with a negative breath test had any evidence of *C. pylori* infection. We have continued to investigate *C. pylori*-infected and uninfected patients, and the urea breath test has proved to be an excellent standard.

Comparison of the ^{13}C-urea breath test with an ELISA test

An ELISA test that measures the presence of IgG antibody against purified *C. pylori* high-molecular-weight proteins (urease) has been developed recently by Evans and coworkers [17] in our laboratory. Results from the ELISA and ^{13}C-urea breath tests were compared in 300 individuals: 153 were *C. pylori* infected and 147 were negative. The ELISA test and urea breath test gave identical results on all but four individuals (false-negatives) indicating a positive predictivity of 100% for each test. Two subjects had false-negative ELISA tests and two had false-negative (below cut-off value) urea breath tests. The two false-negative ELISA tests were in subjects whose *C. pylori* infection was determined by breath tests and confirmed by gastric biopsy and culture of *C. pylori*; neither had developed detectable IgG response to their *C. pylori* infection. Thus, the sensitivity and negative predictive value of both the urea breath test and the ELISA test was 98.7%.

Relationship between the ^{13}C-urea and ^{14}C-urea breath tests

Soon after initial development of the ^{13}C-urea breath test, we shared the detail of our protocol with Dr Guido Tytgat in The Netherlands and Dr Richard Hunt in Canada. Each was rapidly able to adapt our test for use with ^{14}C-urea, and each used the important elements of our test, namely, sufficient substrate in the form of 'cold' urea and a nutrient-dense meal to delay gastric emptying. In contrast, Marshall & Surveyor [18] independently developed a different version of the ^{14}C-urea breath test. Marshall's test employed only a tracer dose of ^{14}C-urea and provided no nutrient-dense meal to delay gastric emptying, thereby allowing the substrate rapidly to be exhausted or emptied from the stomach. There are few current data concerning the accuracy of this test, but our experience indicates that Marshall's test would be only reasonably accurate for the diagnosis of *C. pylori* infections.

The ^{14}C-urea breath tests do not currently use CO_2 production values to enable the calculation of urease activity. When using ^{14}C-urea, it is the current practice to collect 2 mmol $^{14}CO_2$ in NaOH according to the method described by Fromm & Hofmann [19], then to divide the percentage dose recovered by the weight of the subject. This quantity has the dimensions of μBq/mmol CO_2/kg body weight and is not related in a consistent manner to *in vivo* bacterial enzyme levels.

Advantages of ^{13}C breath tests

Although the determination of $^{13}CO_2$ by gas isotope ratio mass spectrometry is more difficult for most laboratories than the determination of $^{14}CO_2$ by liquid scintillation counting, use of the stable isotope is safer and has operational advantages that warrant the extra effort. The absence of radioactivity enhances the acceptability of the breath test in infants and children, as well as in women of child-bearing age. The safety of ^{13}C enables repetitive longitudinal tests in the same individual as well as epidemiological studies of the general population. Finally, in an era of increasing waste management costs, the absence of sample disposal requirements is also a welcome benefit.

Advantages compared with alternative methods

The pre-eminent advantageous aspect of this test is its safety. It requires no invasive procedure, no blood samples, and it does not expose the subject to radioactivity. The ^{13}C-urea breath test has been used to demonstrate the presence of *C. pylori* infection in 4-month-old infants, and it has been performed on infants as young as 10 days and on adults in their late 80s. It is this level of safety that is responsible for the wide acceptability of the test and for the ease of its application to all types of populations.

In addition to non-invasiveness, the ^{13}C-urea breath test has other advantages when compared with endoscopic examination and biopsy. *Campylobacter pylori* infection is often a patchy disease, and the successful location by biopsies in the early stages of infection may be difficult. In addition, culture of *C. pylori* is both expensive and difficult and requires a dedicated and expert microbiologist. Because the breath test effectively integrates the response of the entire gastric mucosa, even low levels of the infection can be detected. Moreover, all the organisms present in the stomach can be expressed as a scalar term for comparisons within and between individuals. Such scalar comparisons are difficult, if not impossible to accomplish by histological examination or bacterial culture of biopsies.

The excellent selectivity and specificity of

our ELISA test will give it wide utility in epidemiological studies in which the general incidence or prevalence of *C. pylori* infection is of interest. None the less, the ^{13}C-urea breath test possesses an important advantage over the ELISA test: the breath test reflects current infection, not previous exposure. The breath test therefore is very useful in monitoring the outcome of therapeutic trials and thus provides information that cannot be duplicated by the ELISA test.

Portability is one of the more useful features of the ^{13}C-urea breath test because it allows the test to be performed in an outpatient setting or under field conditions. The kits shown in Fig. 13.3 can be used in any environment, and the samples can be returned to a central facility for analysis. These kits have been used in locations as diverse as Bahrain, Japan, Peru and the United States. The absence of radioactivity, in addition to providing safety, also enables investigators to clear the kits through customs inspections without difficulty.

Access to the use of the ^{13}C-urea breath test

Reaping the benefits of the ^{13}C-urea breath test requires access to $^{13}CO_2/^{12}CO_2$ isotopic ratio measurements, and many investigators believe that this requirement precludes the use of ^{13}C. It is true that a fully automated gas isotope ratio mass spectrometer system, capable of analysing 50–75 samples per day, requires an outlay of \$150 000 to \$200 000 and that such instruments are usually found only in tertiary medical centres. It is also true that most clinical investigators are unlikely to generate the numbers of urea breath tests required to justify exclusive use of such equipment. How then can they participate in the benefits of stable tracer methodology? The answer, at least for scientists in the United States, is to use commercial analytical laboratories [20]. A number of established, reliable, commercial facilities will accept breath samples in 20-ml Vacutainers® and carry out isolation, purification, and isotopic analysis of

Table 13.2 Commercial sources in the USA for respiratory CO_2 isotope analyses

Coastal Science Laboratories 5321 Industrial Oaks Austin, TX 78735 (512) 892-2097 Dr. J. Kenneth Winters	*Global Geochemical Corporation* 6919 Eton Avenue Camoda Park, CA 91303 (213) 992-4104, 873-1489 Dr. Paul Makiewicz
Geochron Laboratories Division Kreuger Enterprises, Inc. 24 Blackstone Street Cambridge, MA 02139 (617) 876-3691 Dr. Harold W. Kreuger	*Tracer Technologies, Inc* 20 Assembly Square Drive Somerville MA 02145 (617) 776-6410 Dr. Alfred M. Ajami

respiratory CO_2. A partial list of these facilities is shown in Table 13.2. Although specific prices cannot be guaranteed, a number of laboratories charge from \$10 to \$25 per sample, depending upon the total number of analyses. In a study with a large number of subjects, it is possible to pay as little as \$50 for analysis of a baseline and four time points.

A low-cost alternative to gas isotope ratio mass spectrometry, which would cost perhaps one-fifth as much, depends on a recently described [21] infrared heterodyne ratiometer concept. Such a system would greatly reduce the capital instrument costs and would also simplify the measurement process by enabling direct analysis of breath samples without prior purification. Commercial development of this system is under way at the present time (Amnion, Inc, 2408 Timberloch, Suite B11, The Woodlands, TX 77380).

The market cost of ^{13}C-urea is also responsive to the quantity of substrate purchased. With a standard dose of 250 mg per test and an annual rate of 200 tests, substrate costs would contribute approximately \$30 to the total cost per subject. For an annual rate of 2000 tests, the cost drops to \$20 per test, and at 4000 tests, the urea price would be only \$15 per subject. Further economies might be achieved by reducing the dose to 125 mg per subject. Changing the dose, however, would probably raise the lower limit at which the breath test can detect infection.

Applications of the ^{13}C-urea breath test

The most intriguing questions about *C. pylori* include those concerned with its incidence in the general population, the age-specific changes in its distribution, the variation between ethnic groups, and the familial association of infected individuals. All of these questions are readily addressed with the ^{13}C-urea breath test. We have been able to demonstrate, for example, that the frequency of *C. pylori* infection in asymptomatic Caucasian individuals below age 40 is approximately 20%. The number of infected individuals increases with age, reaching 70% by age 80 [22]. Chinese and Indian nationals living in Houston (Texas) were shown to have a higher proportion of infected individuals in the group aged 20–40 years [23]. Current studies are exploring the age-specific incidence in blacks compared with whites and in smokers compared with non-smokers.

The ability to quantify *C. pylori* infection has also demonstrated that the severity of infection, i.e. the number of organisms reflected by the urease levels, increases with age. Figure 13.5 illustrates the range of urease activity by age cohort and shows that values approaching 15 μmol/min are found in older individuals.

The ^{13}C-urea breath test also contributes to the study of disease in symptomatic patients. Graham *et al.* [24] have shown that patients whose ulcers are cured by surgical means, such as highly selective vagotomy, none the less continue to have active *C. pylori* infections. The continuing presence of *C. pylori* in these patients may indicate that the infection is necessary but not sufficient for ulcer formation. The persistence of *C. pylori*-positive breath tests in patients who are taking various medications is prima facie evidence that the organism is not eradicated by these treatments [25].

Although *in vitro* studies have shown that *C. pylori* is susceptible to most commonly used antibiotics, predictions from these studies have not led to a safe and universally effective therapy: indeed, *C. pylori* has proven to be very difficult to eradicate *in vivo*. The urea breath test provides a simple *in vivo* method to assess the response of *C. pylori* to various drugs. *Campylobacter pylori* infection is one of the few infectious diseases where such *in vivo* susceptibility testing is possible. This is extremely important because the outcome of *in vitro* susceptibility testing is often critically dependent upon the pH at which the test was

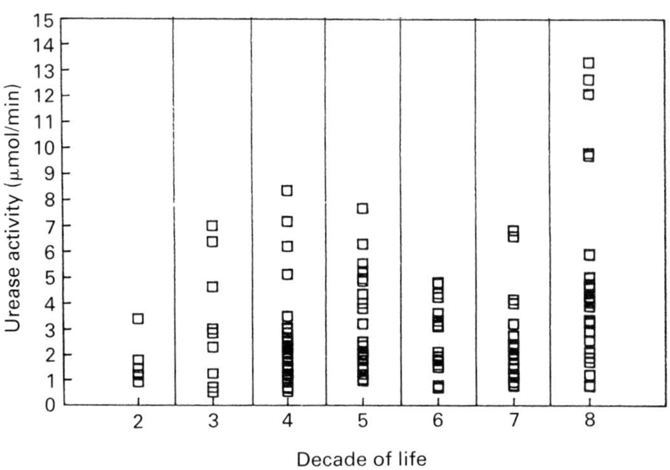

Fig. 13.5 Values for urease activity in *C. pylori*-positive individuals as a function of their decade of life.

Table 13.3 Effect of various drugs on *C. pylori* assessed by the urea breath test (urea breath test was positive in all cases before therapy)

Drug*	Dosage (mg/day)	Duration (days)	Urease result after therapy
Ferrous sulphate[1]	1300	7	Positive
Ferrous sulphate[1]	1300	7	Positive
Ferrous sulphate[1]	1300	7	Positive
Ferrous sulphate[1]	1300	4	Negative
Ferrous sulphate[2]	1300	7	Positive
Ferrous sulphate[2]	700	3	Positive
Ferrous sulphate[2]	700	3	Positive
Ferrous gluconate[3]	1200	3	Positive
Zinc sulphate[2]	700	3	Positive
Bismuth subsalicylate[1]	2096	10	Negative
Bismuth subsalicylate[1]	2096	10	Negative
Bismuth subsalicylate[1]	2096	10	Negative
Bismuth subnitrate	600	3	Negative
Misoprostol[3]	0.80	4	Positive
Misoprostol[3]	0.80	4	Positive
Misoprostol[3]	0.80	4	Positive
Misoprostol[3]	0.80	4	Positive
Misoprostol[3]	0.80	4	Positive
Misoprostol[3]	0.80	4	Positive
Sucralfate	4000	1	Positive
Maalox[3]	120 ml	1	Positive
Ranitidine	300	1	Positive
Cimetidine	1200	1	Positive
Furazolidone[3,4]	49	4	Negative
Furazolidone[3,4]	49	4	Borderline
Furazolidone[3,4]	49	4	Negative
Gold[5]	6	2	Positive
Gold[5]	6	2	Positive
Ursodeoxycholic acid	4000	2	Positive
Ursodeoxycholic acid	4000	2	Positive

*Key: [1]Tablet. [2]Capsule. [3]Liquid. [4]Famotidine 40 mg b.i.d. also given orally to reduce gastric acidity. [5]Ridaura.

conducted. Remarkably different results from *in vitro* susceptibility tests have been demonstrated with minor pH changes [26]. Table 13.3 shows the results of an *in vivo* susceptibility test of drugs for their ability to kill *C. pylori*.

Figure 13.6 shows that 24 hours of treatment with bismuth subsalicylate (524 mg q.i.d.) resulted in virtual abolishment of *C. pylori* urease activity, and treatment for an additional 9 days had no further effect. Two weeks after discontinuation of therapy, the breath test returned to positive values, indicating recrudescence of the infection to detectable levels. We have also seen another instance in which short-term therapy appeared to eradicate the infection (Fig. 13.7). Administration of unlabelled urea (5 g or 1 teaspoon, t.i.d.) in fruit juice resulted in a marked increase in urease activity, confirm-

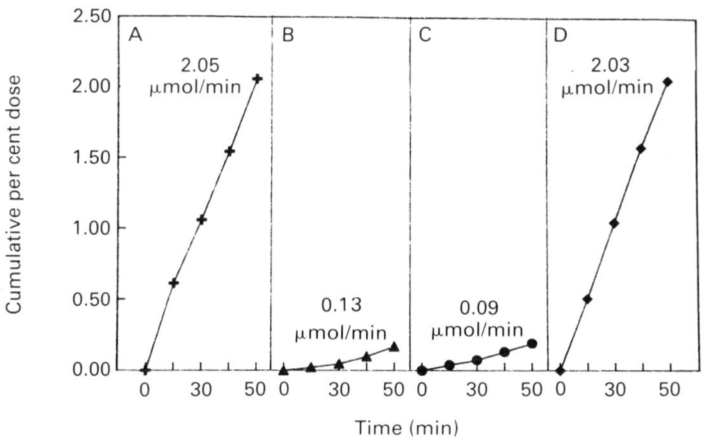

Fig. 13.6 Effect of bismuth therapy on *C. pylori* urease activity as measured by the ^{13}C urea breath test. (A) Pretreatment activity; (B) after 1 day of therapy with 4 × 524 mg bismuth subsalicylate; (C) after 10 days of therapy; (D) 14 days after discontinuation of therapy.

ing the presence of uneradicated *C. pylori*. Application of a urea challenge appears to be very useful to confirm that the infection has been eradicated or in instances where the breath test yields borderline values.

Clinical use of the breath test for diagnosis and management of patients

The most efficient and cost-effective method of evaluating patients with dyspepsia is still not established. As more clinicians become aware that *C. pylori* infection is strongly associated with many of the clinical diseases in which dyspepsia is present, a test to establish the presence of *C. pylori* infection will be one of the first diagnostic tests performed. The simplest, least expensive, and most rapid test would be serologic, i.e. a latex agglutination test for anti-*C. pylori* urease IgG. When a safe and effective therapy for the eradication of *C. pylori* has been developed, and physicians are convinced that this eradication will result in a positive benefit for the patient, such an approach will become the preferred one. Confirmation of the eradication of *C. pylori* infection will continue best to be assessed by the use of the urea breath test. The decision to begin treatment for *C. pylori* infection implies that a programme of surveillance is available to identify reinfection when it occurs. The ^{13}C-urea breath test will be invaluable for this purpose.

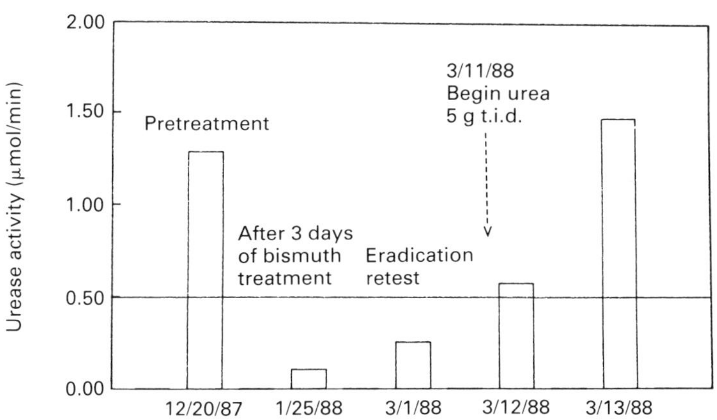

Fig. 13.7 Effect on urease levels of feeding urea to *C. pylori*-positive subject after treatment with bismuth subsalicylate has reduced the level of infection.

Acknowledgements

This work has been supported in part by the USDA/ARS Children's Nutrition Research Center, by the Veterans Administration, by grants DK 28129, DK 35369 and DK 39919 from the National Institute of Diabetes and Digestive and Kidney Diseases, by the Baylor College of Medicine General Clinical Research Center at The Methodist Hospital under grant RR-00350 from the Division of Research Services, National Institutes of Health, and by the generous support of Hilda Schwartz. The authors thank J.D. Eastman and D.M. Reid for editorial review and preparation of the manuscript.

References

1 Irving, C.S., Schoeller, D.A., Nakamura, K-I., Baker, A.L. & Klein P.D. The aminopyrine breath test as a measure of liver function. A quantitative description of its metabolic basis in normal subjects *J. Lab. Clin. Med.* 1982, **100**, 356–373.

2 Irving, C.S., Thomas, M.R., Malphus, E.W., Marks, L., Wong, W.W., Boutton, T.W. & Klein, P.D. Lysine and protein metabolism in young women: subdivision based on the novel use of multiple stable isotope labels. *J. Clin. Invest.* 1986, **77**, 1321–1331.

3 MacLean, W.C. Jr., Fink, B.B., Schoeller, D.A., Wong, W.W. & Klein, P.D. Lactose assimilation by full term infants. Relation of ^{13}C and H_2 breath tests with fecal ^{13}C excretion. *Pediatr. Res.* 1983, **17**, 629–633.

4 Shulman, R.J., Wong, W.W., Irving, C.S., Nichols, B.L. & Klein, P.D. Utilization of dietary cereal by young infants. *J. Pediatr.* 1983, **103**, 23–28.

5 Watkins, J.B., Schoeller, D.A., Klein, P.D., Ott, D.G., Newcomer, A.D. & Hofmann, A.F. ^{13}C-Trioctanoin: a nonradioactive breath test to detect fat malabsorption. *J. Lab. Clin. Med.* 1977, **90**, 422–430.

6 Sherr, H.P., Sasaki, Y., Newman, A., Banwell, J.G., Wagner, H.N. & Hendrix, T.R. Detection of bacterial deconjugation of bile salts by a convenient breath test. *N. Engl. J. Med.* 1971, **285**, 656–661.

7 Klein, P.D. & Klein, E.R. Applications of stable isotopes to pediatric nutrition and gastroenterology: measurement of nutrient absorption and digestion using ^{13}C. *J. Pediatr. Gastroenterol. Nutr.* 1985, **4**, 9–19.

8 Schoeller, D.A., Schneider, J.F., Solomons, N., Watkins, J.B. & Klein, P.D. Clinical diagnosis with the stable isotope ^{13}C in CO_2 breath tests: methodology and fundamental considerations. *J. Lab. Clin. Med.* 1977, **90**, 412–421.

9 Schoeller, D.A., Klein, P.D., Watkins, J.B., Heim, T. & MacLean, W.C. Jr. ^{13}C Abundances of nutrients and the effect of variations in ^{13}C isotopic abundance of test meal formulations for $^{13}CO_2$ breath tests. *Am. J. Clin. Nutr.* 1980, **33**, 2375–2385.

10 Kornberg, H.L. & Davies, R.E. Gastric urease. *Physiol. Rev.* 1955, **35**, 169–177.

11 Kornberg, H.L., Davies, R.E. & Wood, D.R. The activity and function of gastric urease in the cat. *Biochem. J.* 1954, **56**, 363–372.

12 Schoeller, D.A. & Klein, P.D. A microprocessor controlled mass spectrometer for the fully automated purification and isotopic analysis of breath CO_2. *Biomed. Mass Spectrom.* 1979, **6**, 350–355.

13 Haycock, G.B., Chir, B., Schwartz, G.J. & Wisotsky, D.H. Geometric method for measuring body surface area: a height–weight formula validated in infants, children and adults. *J. Pediatr.* 1978, **93**, 62–66.

14 Shreeve, W.W., Cerasi, E. & Luft, R. Metabolism of [2-^{14}C]pyruvate in normal, acromegalic and HGH-treated human subjects. *Acta Endocrinol.* 1970, **65**, 155–169.

15 Irving, C.S., Wong, W.W., Shulman, R.J., Smith, E.O. & Klein, P.D. ^{13}C bicarbonate kinetics in man: intra-vs interindividual variations. *Am. J. Physiol.* 1983, **245**, R190–R202.

16 Graham, D.Y., Klein, P.D., Evans, D.J., Evans, D.G., Alpert, L.C., Opekun, A.R. & Boutton, T.W. *Campylobacter pyloridis* detected by the ^{13}C-urea test. *Lancet* 1987, **i**, 1174–1177.

17 Evans, D.J. Jr., Evans, D.G., Graham, D.Y. & Klein, P.D. A specific and sensitive serologic assay for the detection of *Campylobacter pylori* infection. *Gastroenterology* 1988, **96**, 1004–1008.

18 Marshall, B.L. & Surveyor, I. Carbon-14 urea breath test for the diagnosis of *Campylobacter pylori* associated gastritis. *Clin. Sci.* 1988, **29**, 11–26.

19 Fromm, H. & Hofmann, A.F. Breath test for altered bile acid metabolism. *Lancet* 1971, **i**, 621–625.

20 Klein, P.D., Boutton, T.W., Hachey, D.L., Irving, C.S. & Wong, W.W. The use of stable isotopes in metabolism studies. *J. Anim. Sci.* 1986, **63**, 102–110.

21 Irving, C.S., Klein, P.D., Navratil, P.R. & Boutton, T.W. Measurement of $^{13}CO_2/^{12}CO_2$ abundance by nondispersive infrared heterodyne ratiometry as an alternative to gas isotope ratio mass spectrometry. *Anal. Chem.* 1986, **58**, 2172–2178.

22 Graham, D.Y., Klein, P.D., Opekun, A.R. & Boutton, T.W. Effect of age on the frequence of active *Campylobacter pylori* infection diagnosed by the [^{13}C] urea breath test in normal subjects and patients with peptic ulcer disease. *J. Infect. Dis.* 1988, **157**, 777–780.

23 Graham, D.Y., Klein, P.D., Opekun, A.R., Boutton, T.W., Evans, D.J., Evans, D.G., Alpert, L.C., Michaletz, P.A., Yoshimura, H.H. & Adam, E. Epidemiology of *Campylobacter pylori* infections: ethnic considerations. *Scand. J. Gastroenterol.* 1988, **23** (Suppl. 142), 9–13.

24 Graham, D.Y., Jordan, P.H., Opekun, A.R., David, I.A. & Klein, P.D. *Campylobacter pylori* is not *the* cause of duodenal ulcers. *Gastroenterology* 1988, **94**, A152.

25 Graham, D.Y., Klein, P.D., Opekun, A.R., Boutton, T.W., Evans, D.J., Evans, D.G., Alpert, L.C., Michaletz, P.A., Yoshimura, H.H. & Adam, E. Search for an effective and safe therapy for *Campylobacter pylori* infection. *Gastroenterology* 1988, **94**, A152.

26 Young, L.S. & Hewitt, W.L. Activity of five aminoglycoside antibiotics *in vitro* against Gram-negative bacilli and *Staphylococcus aureus*. *Antimicrob. Agents Chemother.* 1973, **4**, 617–625.

14 Campylobacter pylori and chronic gastritis

M.F. DIXON

Any attempt to evaluate the role of *Campylobacter pylori* in the pathogenesis of chronic gastritis has to take into account the heterogeneity that exists within chronic inflammatory disease in the stomach. Most histopathologists recognize chronic inflammation simply in terms of superficial or atrophic gastritis. The former term was originally defined as gastritis in which the inflammatory cell infiltrate is confined to the zone between the gastric pits or foveolae, while deeper extension accompanied by glandular atrophy constituted chronic atrophic gastritis [1]. However, there is confusion over the terminology when there is deep (transmucosal) inflammatory cell infiltration without glandular atrophy, and some authors have introduced the terms 'chronic interstitial' or 'pre-atrophic' gastritis for this category [2]. The latter epithets have not been widely accepted and because of these difficulties over nomenclature it is evident that there have been wide inconsistencies in the way that histopathologists have categorized chronic gastritis.

If we set aside semantic difficulties, it seems most appropriate in terms of functional consequences to grade chronic gastritis simply on the basis of the presence or absence of glandular atrophy, irrespective of the degree or depth of inflammatory cell infiltration. Thus, when there is no loss of glands the term *chronic superficial* is employed and when there is glandular loss then *chronic atrophic* is appropriate. This approach has recently been endorsed by Kekki *et al.* [3], but a more satisfactory terminology might be into 'non-atrophic' and 'atrophic' as advocated by Correa [4]. When glandular loss is severe the term 'gastric atrophy' has been employed[5] but this is not a distinct form and its use has little to commend it. While it is true that marked glandular atrophy in body mucosa is the classical finding in the gastritis of pernicious anaemia, this is not invariably the case [6] and gastric atrophy can also represent the end-stage of non-autoimmune gastritis [7].

As far as pathogenesis is concerned, the earliest attempt at classification was that proposed by Strickland & Mackay in 1973 [8]. They distinguished an autoimmune category characterized by involvement of body mucosa and the presence of anti-parietal cell antibodies in the serum (Type A), and the much more common, predominantly antral gastritis without associated antibodies, which they designated Type B. A further category, Type AB gastritis, was suggested by Glass & Pitchumoni [9] where there was patchy involvement of both antrum and body. AB gastritis is characterized by multifocal atrophy and intestinal metaplasia and carries an increased risk of gastric ulceration and the development of carcinoma.

While adding to our understanding of chronic gastritis, these attempts at pathogenetic classification have had little or no impact on histopathological practice. Currently most histopathologists will label biopsies containing increased numbers of chronic inflammatory cells as either *superficial* (CSG) or *atrophic* (CAG) gastritis (according to differing criteria). Biopsies showing no increase in inflammatory cells are almost invariably considered *normal*. It should be borne in mind when looking at previous data on the association between *C. pylori* and chronic gastritis that these are the diagnostic standards which will have been applied. This chapter will endeavour to demonstrate that there exists much wider heterogeneity within gastritis and that this has an important bearing on the relationship with *C. pylori*.

Prevalence of C. pylori in chronic gastritis

Following on from Warren & Marshall's first descriptions of *C. pylori* in chronic gastritis [10], there have been numerous studies from widely separated countries attesting to the very strong link between colonization and gastritis. This is particularly so when the gastritis has been designated 'active'. Active chronic gastritis is diagnosed when intraepithelial and interstitial neutrophil polymorphs are present in addition to lymphocytes and plasma cells (Fig. 14.1). A selection of studies from various countries which illustrate this link are summarized in Table 14.1 [11–17]. It can be seen that *C. pylori* positivity ranges from 70% to 92% in active chronic gastritis, but falls to as low as 13% in inactive cases.

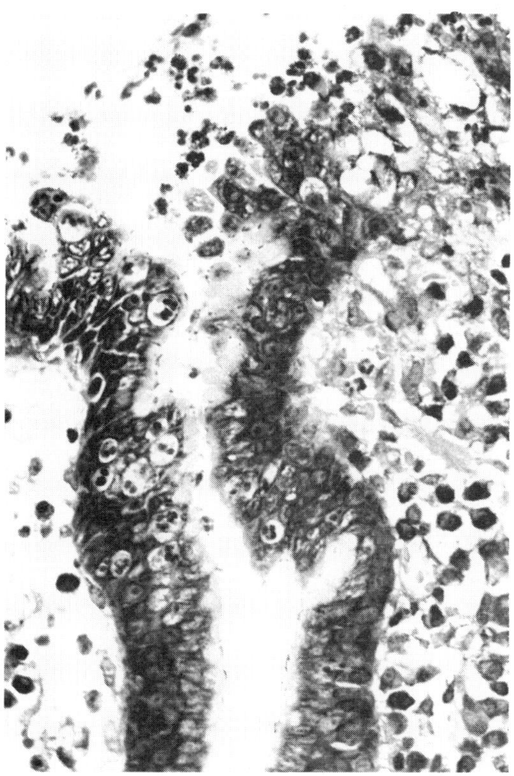

Fig. 14.1 Active chronic gastritis showing polymorph infiltration of the upper foveolar and surface epithelium in an area of *C. pylori* colonization. The organisms are seen at the surface and as dark granules within polymorphs in the exudate. (Modified Giemsa stain.)

While these figures give an overview of the association, they are not strictly comparable because the methods of detection of *C. pylori* differ and, as already argued, the criteria for histological diagnosis will vary. In a personal series, the present author has categorized antral biopsies from 322 consecutive patients undergoing endoscopy for dyspepsia and found that 178 exhibited chronic superficial or atrophic gastritis (Fig. 14.2). *Campylobacter pylori* were sought using the modified Giemsa stain in all cases. The breakdown of this group in terms of sub-type, activity, atrophy and *C. pylori* positivity is shown in Table 14.2.

It can be seen that active cases, be they superficial or atrophic, are almost invariably *C. pylori*-positive, and inactive cases exhibit higher prevalence rates than previous series. The higher prevalence is not a consequence of a higher overall detection rate of *C. pylori*. The overall prevalence in the present series, 173/322 (54%), is if anything lower than that found in previous studies on dyspeptic patients. The reason for the high positivity rate in chronic superficial and atrophic gastritis lies in the recognition of cases of reflux and lymphocytic gastritis, and a more consistent diagnosis of normality.

Overall the prevalence data point to a very close, if not causal, link between *C. pylori* and the predominantly antral form of chronic gastritis corresponding to Strickland & Mackay's Type B [8]. It therefore seems appropriate to retain the term Type B to signify a *bacteria-associated* form, as distinct from the *autoimmune* Type A and other recently-identified varieties of chronic gastritis [18].

Type A gastritis and C. pylori

Type A gastritis is characterized histologically

Table 14.1 *Campylobacter pylori* and activity in chronic gastritis

Study group	Active			Inactive		
	n	*C. pylori* positive	%	*n*	*C. pylori* positive	%
Goodwin *et al.* (1986) [11] Perth, Western Australia	51	47	(92%)	23	8	(35%)
Morris *et al.* (1986) [12] Auckland, New Zealand	28	23	(82%)	7	1	(14%)
Rawles *et al.* (1986) [13] Baltimore, USA	23	16	(70%)	47	7	(15%)
Wyatt *et al.* (1986) [14] Leeds, UK	109	95	(87%)	35	13	(37%)
Andersen *et al.* (1987) [15] Hillerod, Denmark	74	67	(91%)	16	2	(13%)
Fiocca *et al* (1987) [16] Pavia, Italy	239	213	(89%)	124	91	(73%)
Jiang *et al.* (1987) [17] Shanghai, China	49	44	(90%)	72	36	(50%)

by glandular atrophy and an absence of parietal cells from the fundic mucosa, a picture constituting severe atrophic gastritis. Such atrophy is usually, but not invariably, accompanied by intestinal metaplasia. Antral biopsies may be normal or show chronic gastritis but to a lesser degree than that exhibited by the body mucosa.

At an early stage in our studies on *C. pylori*, we found a much reduced prevalence of the organism in both body and antral mucosa when compared with age- and sex-matched controls with peptic ulcer [19]. This observation has been reinforced by the findings of Flejou *et al.* [20], who identified very few positive cases even when the antrum showed chronic gastritis. The most plausible conclusions to draw from these studies are that *C. pylori* does not have a role in the pathogenesis of Type A gastritis and that the finding of antral gastritis in these patients does not represent co-existence of Type B and Type A

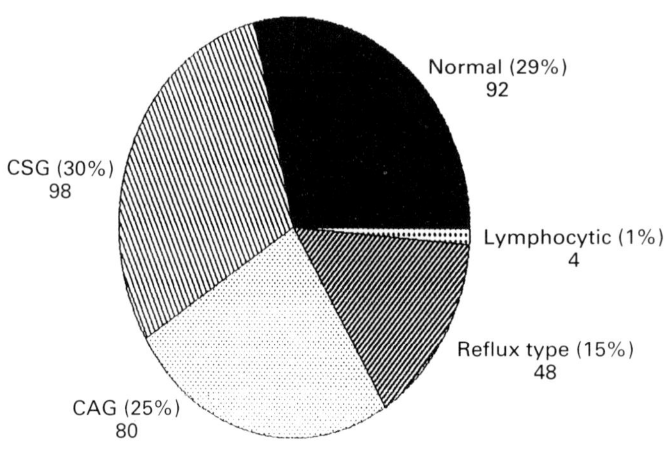

Fig. 14.2 Histological diagnoses in 322 consecutive patients undergoing oesophago-gastroscopy for dyspeptic symptoms.

Table 14.2 Characteristics of patients with chronic superficial and chronic atrophic gastritis

Condition	*n*	Active	*C. pylori* positive	Inactive	*C. pylori* positive
Chronic superficial gastritis	98	69	68 (99%)	29	25 (86%)
Chronic atrophic gastritis					
mild	58	50	50	8	6
moderate	19	18	18	1	1
severe	3	1	1	2	0
	80	69	69 (100%)	11	7 (64%)

gastritis. Nevertheless, it remains a possibility that the mixed bacterial flora of the hypochlorhydric stomach is hostile to *C. pylori* so that with the development of severe atrophy, organisms previously present in the stomach have been eliminated.

'New' forms of gastritis

Lymphocytic gastritis

The finding that the vast majority of cases of active chronic superficial gastritis are *C. pylori*-positive led us to review negative cases to determine whether or not there were any distinctive features. This review brought to light several cases where the surface and foveolar epithelium was infiltrated by unusually large numbers of lymphocytes — so-called 'lymphocytic gastritis' [21]. Lymphocytic gastritis has a characteristic endoscopic appearance frequently described as 'varioliform' and displaying nodular prominence of the gastric mucosa, the nodules being surmounted by shallow aphthoid erosions. This striking appearance is usually confined to the body part of the stomach but can be diffuse.

In contrast to Type B gastritis, lymphocytic gastritis shows a far lower prevalence of *C. pylori* positivity [22]. Recently Wallez *et al.* (who first recognized this entity) found only 8% positivity amongst 102 cases of lymphocytic gastritis [23]. However, this does not rule out a role for *C. pylori* in this condition; some of our negative cases exhibited serological evidence of previous *C. pylori* infection [22]. Furthermore, Flejou *et al.* [24] found on follow-up of *C. pylori*-positive patients that two of those who became negative revealed the histological picture of lymphocytic gastritis.

We have previously demonstrated that the intra-epithelial lymphocytes in lymphocytic gastritis are almost exclusively T cells [22]. More recently, using frozen material, we have established that these are CD8-positive suppressor cells. These findings suggest that the condition might arise from a perturbed immune response in the gastric mucosa analogous to that found in the small intestine in coeliac disease.

Reflux gastritis

Increasing familiarity with biopsies from the post-operative stomach led us to distinguish a further type of chronic gastritis related to entero-gastric reflux[25]. This 'reflux gastritis' results from persistent injury to the surface epithelium and is characterized by compensatory foveolar hyperplasia and a vascular and exudative response evidenced by capillary dilatation, congestion and lamina propria oedema. There is no notable increase in inflammatory cells. At a later stage increased numbers of smooth muscle fibres and collagen bundles appear in the lamina propria.

The pathogenesis of reflux gastritis involves

Table 14.3 Major forms of chronic gastritis

Diagnosis	Pathogenesis	Aetiology
Type A	Autoimmunity	?
Type B	Bacterial infection	*C. pylori*
Type C	Chemical injury	Enterogastric reflux NSAIDs Alcohol

damage to the surface epithelium by bile acids and lysolecithin resulting in disruption of the mucous barrier and cell exfoliation [26]. Access of the bile constituents to the mucosa leads to degranulation of mast cells and a histamine-mediated vascular response [27]. Further damage to the surface epithelium ensues — possibly mediated by depressed prostaglandin synthesis, and back-diffusion of hydrogen ions may contribute to the mucosal injury.

Although we initially described the histological appearances in reflux as 'distinctive', this is only true in the sense that they characterize *chemical* injury to the stomach. Identical appearances can be seen in repeated injury by salicylates and other NSAIDs [28], and it is possible that alcohol could produce a similar response. Thus there is a need for a less restrictive term than 'reflux' to describe these appearances and it could be suggested that the designation 'Type C (chemical) gastritis' would be an appropriate and useful omnibus term for this category. A new classification which includes this revised concept of reflux gastritis as part of a broader group of chemical injuries to the stomach is given in Table 14.3.

Reflux gastritis is essentially *C. pylori*-negative. However, in the post-operative stomach one is not dealing with entero-gastric reflux in isolation. Prior to peptic ulcer surgery, all (or almost all) patients will have a *C. pylori*-positive chronic gastritis. Following operations which lead to reflux, the Type B gastritis will gradually give way to a *C. pylori*-negative reflux gastritis which will only be seen in a 'pure' form some years after surgery [29]. In the intact stomach, the reflux (or Type C) picture can only be recognized in *C. pylori*-negative patients. In *C. pylori*-positive cases the Type B gastritis associated with colonization will dominate the histological appearances. It might be that *C. pylori*-positive patients who also have duodeno-gastric reflux or take NSAIDs exhibit more foveolar hyperplasia, oedema and congestion than pure Type B gastritis. Such 'mixed' cases will be difficult to distinguish, however, and might only be revealed by detailed morphometry. A pure Type C picture was found in 15% of the 322 patients in our dyspepsia series (Fig. 14.2). The relative importance of 'spontaneous' duodeno-gastric reflux and NSAID ingestion in the production of Type C gastritis in the intact stomach is the subject of current study.

Morphological aspects of Type B gastritis

Histology

The histological appearances in Type B gastritis are well known, but there are some aspects which are worth emphasizing. For instance, it is conventional to think that infiltration of the superficial lamina propria by lymphocytes and plasma cells represents the earliest stage of Type B gastritis. However, in cases where there is only mild inflammatory cell infiltration (and therefore presumed to be early), one frequently sees a mid-zonal infiltrate with sparing of the superficial mucosa (Fig. 14.3). In terms of polymorph infiltration this has been borne out by morphometric methods which have confirmed maximal activity centred on the pit–isthmus region [30]. In attempting to link polymorph response to *C. pylori* colonization, this represents something of a paradox in that the number of organisms declines steadily from surface to foveolar neck. Thus, although there is an overall correlation between the density of *C. pylori* colonization and polymorph infiltration, when the exact topographical relations are considered there are obvious discrepancies between the two. It might be that the immature foveolar epithelium of the isthmic proliferative

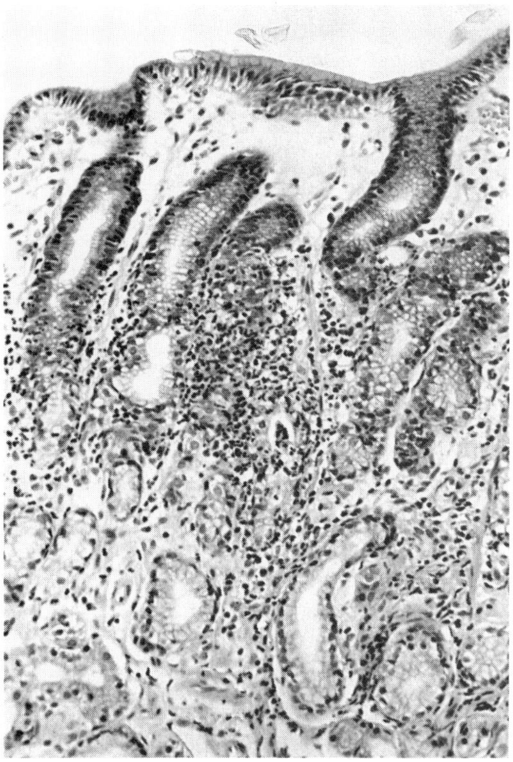

Fig. 14.3 'Early' chronic gastritis showing predominantly mid-zonal infiltration by lymphocytes and polymorphs centred on the pit–isthmus region. There is no increase of cells in the superficial lamina propria.

zone is more permeable to leucotactic factors, such as complement products, activated by *C. pylori* higher up the pits.

A further aspect of note is the finding of lymphoid follicles. These are more frequently seen when the inflammatory cell infiltrate extends through the full thickness of the mucosa and, by separating the glandular elements, may give a false impression of atrophy [4]. It is not widely appreciated that the finding of reactive lymphoid follicles is virtually confined to *C. pylori*-positive gastritis [31] (Fig. 14.4), and that they form a prominent feature of such gastritis in children [32].

Persistent chronic inflammation leads eventually to intestinal metaplasia. Initially, the metaplasia is of an incomplete small intestinal type, in which goblet cells secreting acidic sialomucins appear interspersed between the normal neutral mucin secreting gastric epithelium (Type II metaplasia). Later, the intervening cells become absorptive in character, there is villous transformation and the development of Paneth cells, so that the mucosa takes on an appearance closely resembling normal small intestine — complete metaplasia (Type I). Alternatively, the metaplasia may take on 'colonic' characteristics where both the goblet cells and the non-absorptive intervening cell secrete sulphomucins (Type III). *Campylobacter pylori* do not adhere to normal intestinal epithelium and in keeping with this, organisms are not found on areas of intestinal metaplasia in the stomach (Fig. 14.5). It has been suggested that this could explain the low prevalence of *C. pylori* in Type A gastritis [33], but metaplasia is rarely sufficiently widespread to preclude colonization.

With regard to the biological significance of intestinal metaplasia, the conventional view is that it represents 'faulty regeneration' of the gastric mucosa and that the change in differentiation, particularly with regard to Type III metaplasia, reflects a pre-malignant condition [34]. However, it is equally plausible that such metaplasia is an adaptive response to an altered microenvironment. One can speculate on one hand that the development of small intestinal epithelium could confer a biological advantage by preventing the adhesion of *C. pylori*. On the other, the acquisition of colonic characteristics is a reaction to the proliferation of faecal organisms in the hypochlorhydric stomach. Thus metaplasia can be viewed as a 'defence response' which could be an epiphenomenon with respect to the development of carcinoma and not constitute a pre-malignant condition *per se* [35]. Those conditions which provoke intestinal metaplasia also predispose to the formation of intra-gastric carcinogens. The development of carcinoma frequently follows in the train of metaplasia, but this does not prove causality.

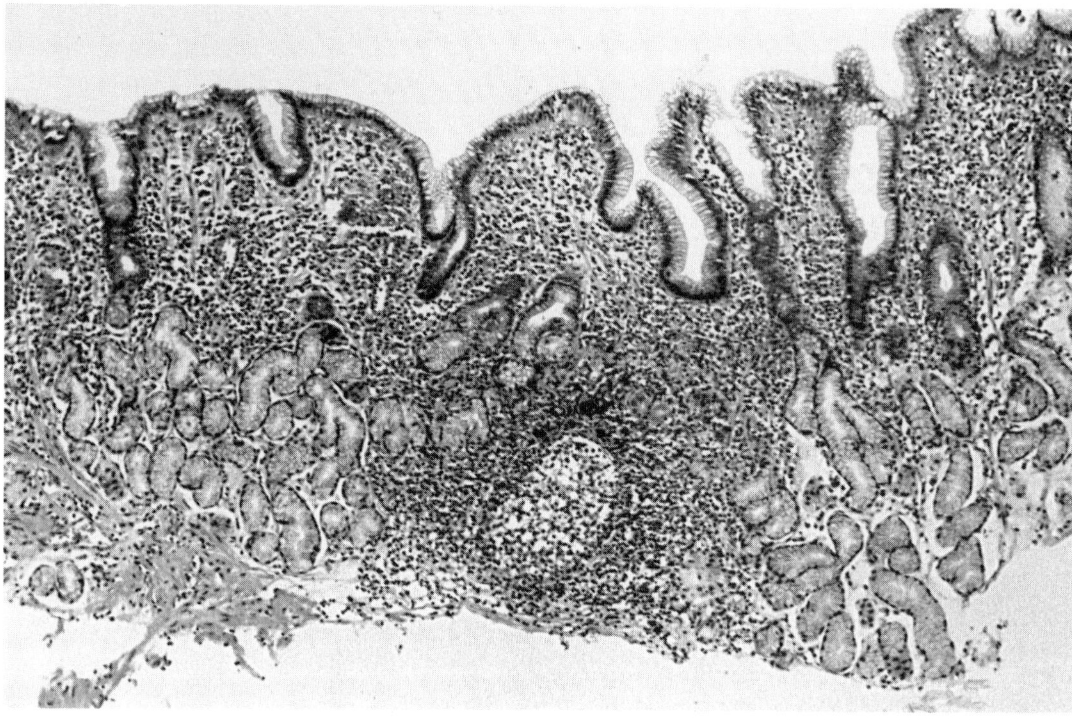

Fig. 14.4 *Campylobacter pylori*-positive chronic gastritis with lymphoid follicle formation. Whether this has resulted in separation of glands or is in an area of glandular atrophy is debatable and underlines the difficulties involved in categorizing biopsies as atrophic or not.

When malignancy develops in the *C. pylori*-positive stomach, the areas of high-grade dysplasia and intra-mucosal carcinoma are not colonized [17], presumably because of changes in the surface characteristics of the malignant cells.

Atrophy is recognized by glandular loss and separation of the remaining glands by fibrous tissue. In previous studies where the relationship between *C. pylori* and atrophy has been examined it has been found that positivity rates decline with increasing atrophy. This fall in positivity has been explained by some authors in terms of an altered environment 'hostile' to *C. pylori*. The two principal factors being suggested are intestinal-type mucus secreted by the metaplastic epithelium, and competition by the mixed bacterial flora colonizing the hypochlorhydric stomach [36]. This line of reasoning would be appropriate if all atrophic gastritis originated from *C. pylori*-positive chronic superficial gastritis. While no doubt this applies to the majority of cases, there is also a contribution from Type A and reflux gastritis which could have been *C. pylori*-negative from the outset. These interrelations are illustrated in Fig. 14.6. Thus, atrophy is the end result of more than one form of gastritis. It should be borne in mind that the histological picture of a 'quiescent' chronic atrophic gastritis might have been *C. pylori*-negative *ab initio* rather than assuming that eradication of organisms has been accompanied by a resolution of the inflammatory cell infiltrate.

Distribution

While the density of colonization by *C. pylori* varies between biopsies and even from area to

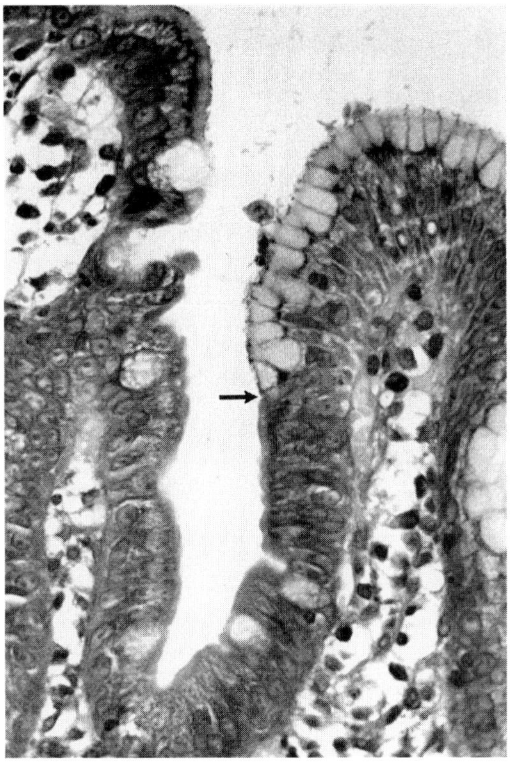

Fig. 14.5 Chronic gastritis with intestinal metaplasia. The heavy colonization of the normal gastric epithelium ends abruptly (at the arrow) with the change to intestinal type epithelium composed of absorptive cells and goblet cells (complete IM — Type I). (Modified Giemsa stain.)

area within the same biopsy, they are rarely completely absent in a biopsy from a 'positive' stomach. We have examined multiple biopsies (5–8) of antral and body mucosa from 50 patients, using the modified Giemsa stain to identify the organisms [37]. Of 38 *C. pylori*-positive patients, 24 had organisms in all their biopsies, and 10 in all but one. Four patients had small numbers of *C. pylori* in 1–3 biopsies. Thus we concluded that sampling errors rarely impair the histological detection of *C. pylori*, and that two antral biopsies are sufficient to establish *C. pylori* status.

Ultrastructure

Ultrastructural studies on *C. pylori* in chronic gastritis lend support to its pathogenic role, in that degenerative changes are frequently encountered in epithelial cells in close proximity to bacteria. Chen *et al.* [38] observed abundant phagolysosomes and depleted mucin content in cells showing intimate contact with *C. pylori*. Tricottet *et al.* [39] observed focal cell necrosis and partially digested organisms in heterophagosomes in foveolar cells. Both these studies revealed patchy loss of microvilli in regions of bacterial attachment, and bacteria were seen within widened intercellular gaps and in the canaliculi of parietal cells.

An early ultrastructural observation was the finding of so-called adhesion 'pedestals' at the site of *C. pylori* attachment to epithelial cells. These pedestals are similar to those marking the sites of attachment of enteropathogenic *E. coli* to intestinal epithelium [40]. Our observations indicate that proximity of *C. pylori* to the surface epithelium results in loss of microvilli from the plasma membrane.

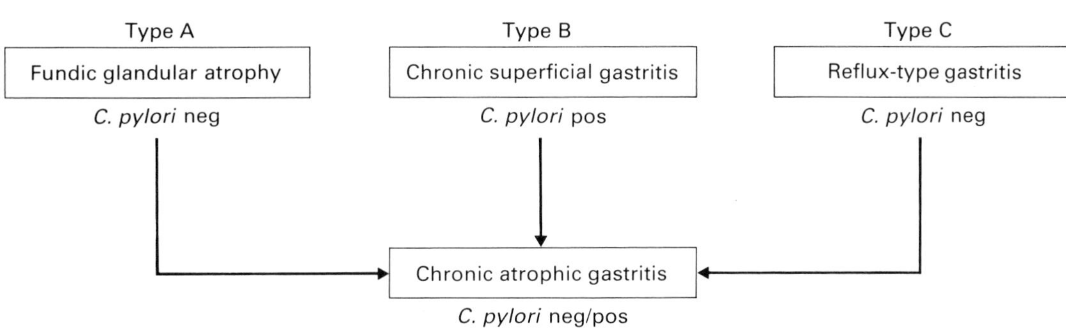

Fig. 14.6 Routes of development of chronic atrophic gastritis.

Initial adhesion follows the development of an electron dense 'fuzzy' or fibrillar layer between the organism and the membrane, and ruthenium red staining would support this being a glycoprotein ligand. The organism comes to abut the membrane which extrudes to form an elevated plateau and subsequently 'cups' the bacterium to form a fully-developed pedestal. The proportion of organisms exhibiting adhesion sites in *C. pylori*-positive antral gastritis varies from 0 to 41% (mean = 19%). Those showing full pedestal formation range from 0 to 21% (mean = 3.4%). The proportion of attached *C. pylori* is significantly correlated with the finding of degenerative changes in the surface epithelium seen on histological examination of parallel antral biopsies. It does not correlate with other indicators of disease activity such as polymorph infiltration of the lamina propria, mononuclear cell response and mucin depletion, but interestingly is inversely related to intraepithelial polymorph infiltration.

Natural history

As yet, there is very little data on the follow-up of *C. pylori*-associated gastritis. A study from Flejou *et al.* [24] reported in abstract form, examined the *C. pylori* status of patients one month to 10 years after initial diagnosis: thirteen of the 67 initially positive patients became negative, while 3 of the 41 initially negative became positive. Apart from treatment, they found that loss of organisms was associated with increasing intestinal metaplasia or occasionally with the development of lymphocytic gastritis.

However, if we assume that the majority of cases of chronic superficial gastritis will be *C. pylori*-positive, then there is an abundance of data on the natural history. Many of these data come from population studies in Scandinavia, and have consistently demonstrated that the prevalence of chronic gastritis rises steadily with age [41]. As seen in younger age groups the disease starts as a diffuse process affecting both antrum and body and the progress of the gastritis remains static until middle age [3]. Thereafter a small proportion (around 12% of the population) exhibits more rapid progression with the development of atrophy in divergent patterns so that clear-cut phenotypes emerge which correspond to the previously described [8] Type A (body > antrum), Type B (antrum > body) and AB (multifocal) [9]. Thus, it is argued, a few patients progress from a diffuse superficial gastritis through to the pernicious anaemia phenotype of severe body atrophic gastritis and anti-parietal cell antibodies. Genetic studies indicate that this occurs according to the inheritance of a simple dominant gene [3]. The role of *C. pylori* in this process has yet to be determined, but when it is recalled that very few patients with pernicious anaemia show *C. pylori* colonization, even when the antrum exhibits chronic superficial gastritis, it would seem unlikely that the organism is a pathogenetic factor in this disease.

Other cases of (presumed) *C. pylori*-positive antral superficial gastritis show progression to the atrophic stage. As indicated above, there is a decline in the prevalence of *C. pylori* in patients showing increasing atrophy, but this has not been adequately documented in longitudinal studies. Patients with the histological picture of quiescent antral atrophic gastritis will include some in whom the atrophy has resulted from repeated 'chemical' injury and who may never have been *C. pylori*-positive. Genetic studies on this subgroup with antral atrophy indicate a recessive Mendelian inheritance [3].

The conclusion to be drawn from these epidemiological studies is that there are various levels of genetic predisposition to gastritis mediated through mucosal defence and immunological mechanisms. This implies that the response to *C. pylori* infection is not uniform and will vary according to genetic susceptibility. Thus, many people exposed to infection might suffer a mild and transient acute gastritis in which the organisms are eliminated and the inflammatory response resolves. Other individuals are unable to eliminate the organism and a chronic gastritis ensues, a circumstance that is more likely to occur with increasing age. In the majority of patients the inflammation persists as a superficial gastritis showing no tendency to progress. However, a minority of patients — those who are genetically predisposed, undergo a progressive course with increasing glandular atrophy.

It seems unlikely that *C. pylori* has any role in producing the severe body atrophy of pernicious anaemia. The epidemiological data remain consistent with the view that this is a distinct autoimmune gastritis occurring in people who are predisposed by the inheritance of a dominant genetic abnormality. The relative importance of *C. pylori* and genetic factors, and the existence of other 'triggering' factors [42] in the pathogenesis of Type B gastritis remain to be determined.

References

1 Motteram, R. A biopsy study of chronic gastritis and gastric atrophy. *J. Pathol. Bact.* 1951, **63**,389–394.
2 Cheli, R., Santi, L., Ciancameria, G. & Canciani, G. A clinical and statistical follow-up study of atrophic gastritis. *Dig. Dis.* 1973, **18**, 1061–1066.
3 Kekki, M., Siurala, M., Varis, K., Sipponen, P., Sistonen, P. & Nevanlinna, H.R. Classification principles and genetics of chronic gastritis. *Scand. J. Gastroenterol.* 1987, **22**, (Suppl. 141), 1–28.
4 Correa, P. Chronic gastritis: a clinico-pathological classification. *Am. J. Gastroenterol.* 1988, **83**, 504–509.
5 Wood, I.J. & Taft, L.I. 1958. *Diffuse Lesions of the Stomach.* Arnold, London, p. 86.
6 Joske, R.A., Finckh, E.S. & Wood, I.G. Gastric biopsies. *Q. J. Med.* 1955, **24**, 269–294.
7 Cheli, R. & Giacosa, A. Chronic atrophic gastritis and gastric mucosal atrophy — one and the same. *Gast. Endoscopy* 1983, **29**, 23–25.
8 Strickland, R.G. & Mackay, I.R. A reappraisal of the nature and significance of chronic atrophic gastritis. *Dig. Dis.* 1973, **18**, 426–440.
9 Glass, G.B.J. & Pitchumoni, C.S. Atrophic gastritis. *Hum. Pathol.* 1975, **6**, 219–250.
10 Warren, J.R. & Marshall, B. Unidentified curved bacilli on gastric epithelium in active chronic gastritis. *Lancet* 1983, **i**, 1273–1275.
11 Goodwin, C.S., Armstrong, J.A. & Marshall, B.J. *Campylobacter pyloridis*, gastritis and peptic ulceration. *J. Clin. Pathol.* 1986, **39**, 353–365.
12 Morris, A., Arthur, J. & Nicholson, G. *Campylobacter pyloridis* infection in Auckland patients with gastritis. *N.Z. Med. J.* 1986, **99**, 353–355.
13 Rawles, J.W., Paull, G., Yardley, J.H., Hendrix, T.R., Rayich, W.J. Walters, L.L., Dick, J.D. Margolis, A. Gastric campylobacter-like organisms in a US hospital population. *Gastroenterology* 1986, **91**, 1599A.
14 Wyatt, J.I., Rathbone, B.J. & Heatley, R.V. Local immune response to gastric *Campylobacter* in non-ulcer dyspepsia. *J. Clin. Pathol.* 1986, **39**, 863–870.
15 Andersen, L.P., Holck, S., Povlsen, C.O., Elsborg, L. & Justesen, T. *Campylobacter pyloridis* in peptic ulcer disease. *Scand. J. Gastroenterol.* 1987, **22**, 219–224.
16 Fiocca, R., Villani, L., Turpini, R. & Solcia, E. High incidence of campylobacter like organisms in endoscopic biopsies from patients with gastritis, with or without peptic ulcer. *Digestion* 1987, **38**, 234–244.
17 Jiang, S.J., Liu, W.Z., Zhang, D.Z., Shi, Y., Xiao, S.D., Zhang, Z.H. & Lu, D.Y. Campylobacter like organisms in chronic gastritis, peptic ulcer and gastric carcinoma. *Scand. J. Gastroenterol* 1987, **22**, 553–558.
18 Wyatt, J.I. & Dixon, M.F. Chronic gastritis — a pathogenetic approach. *J. Pathol.* 1988, **154**, 113–124.
19 O'Connor, H.J., Axon, A.T.R. & Dixon M.F. Campylobacter-like organisms unusual in Type A (pernicious anaemia) gastritis. *Lancet* 1984, **ii**, 1091.
20 Flejou, P.F., Bahame, P., Smith, A., Stockbrugger, R.W., Rode, J. & Price, A.B. Pernicious anaemia and campylobacter like organisms: Is the gastric antrum resistant to colonization? *Gut* 1989, **30**, 60–64.
21 Haot, J., Delos, M., Wallez, L., Hardy, N., Lenzen, B. & Jouret-Mourin, A. Intraepithelial lymphocytes in inflammatory gastric pathology. *Acta Endoscop.* 1986, **16**, 61–67.
22 Dixon, M.F. Wyatt, J.I. Burke, D.A. & Rathbone, B.J. Lymphocytic gastritis — relationship to *Campylobacter pylori* infection. *J. Pathol.* 1988, **154**, 125–132.
23 Wallez, L., Weynaud, B. & Haot, J. Évaluation histologique de la présence d'organismes de type *Campylobacter* dans les gastrite lymphocytaires. *Acta Endoscop.* 1987, **17**, 41–50.
24 Flejou, J.F., Price, A.B. & Smith, A.C. A natural history of *Campylobacter pylori* in the stomach: a follow-up biopsy study. *Gastroenterology International* 1988, 1085A.
25 Dixon, M.F., O'Connor, H.J., Axon, A.T.R., King, R.F.J.G. & Johnston, D. Reflux gastritis: distinct histopathological entity? *J. Clin. Pathol.* 1986, **39**, 524–530.
26 Armstrong, D., Murphy, G.M. & Dowling, R.H. The relative gastrotoxicity of duodenal juice components; acute studies in the rat. *Gastroenterology* 1988, **94**, A13.

27 Rees, W. & Rhodes, J. Bile reflux in gastro-oesophageal disease. *Clin. Gastroenterol.* 1977, **6**, 179–200.

28 Laine, L., Marin-Sorensen, M. & Weinstein, W.M. The histology of gastric erosions in patients taking non-steroidal anti-inflammatory drugs (NSAIDs): a prospective study. *Gastroenterology* 1988, **94**, A247.

29 O'Connor, H.J., Dixon, M.F., Wyatt, J.I., Axon, A.T.R., Ward, D.C., Dewar, E.P. & Johnston, D. Effect of duodenal ulcer surgery and enterogastric reflux on *Campylobacter pyloridis*. *Lancet* 1986, **ii**, 1178–1181.

30 Hopwood, D. A histometric analysis of gastric biopsies from patients treated with Gastritex: a new drug active against acute or chronic gastritis. *J. Pathol.* 1988, **154**, 86A.

31 Wyatt, J.I. & Rathbone, B.J. Immune response of the gastric mucosa to *Campylobacter pylori*. *Scand, J. Gastroenterol.* 1988, **23** (Suppl. 142), 44–49.

32 Mahony, M.J., Wyatt, J.I. & Littlewood. J.M. *Campylobacter pylori* gastritis. *Arch. Dis. Child.* 1988, **63**, 654–655.

33 Thomas, J.M. Campylobacter-like organisms in gastritis. *Lancet* 1984, **ii**, 1217.

34 Filipe, M.I., Potet, F., Bogomoletz, W.V., Dawson, P.A., Fabiani, B., Chauveinc, P., Fenzy, A., Gazzard, B., Goldfain, D. & Zeegen, R. Incomplete sulphomucin-secreting intestinal metaplasia for gastric cancer. Preliminary data from a prospective study from three centres. *Gut* 1985, **26**, 1319–1326.

35 Ectors, N. & Dixon, M.F. The prognostic value of sulphomucin positive intestinal metaplasia in the development of gastric cancer. *Histopathology* 1986, **10**, 1271–1277.

36 Eastwood, G.L. Effect of pH on bile salt injury to mouse gastric mucosa: a light and electron microscope study. *Gastroenterology* 1975, **68**, 1456–1465.

37 Wyatt, J.I., Primrose, J. & Dixon, M.F. Distribution of *Campylobacter pylori* in gastric biopsies. *J. Pathol.* 1988, **155**, 350A.

38 Chen, X.G., Correa, P., Offerhaus, J., Rodriguez, E., Janney, F., Hoffmann, E., Fox, J., Hunter, F. & Diavolitsis, S. Ultrastructure of the gastric mucosa harbouring campylobacter-like organisms. *Am. J. Clin. Pathol.* 1986, **86**, 575–582.

39 Tricottet, V., Bruneval, P., Vire, O. & Camilleri, J.P. Campylobacter-like organisms and surface epithelium abnormalities in active, chronic gastritis in humans: an ultrastructural study. *Ultrastruct. Pathol.* 1986, **10**, 113–122.

40 Cantey, J.R., Lushbaugh, W.B. & Inman, L.R. Attachment of bacteria to intestinal epithelial cells in diarrhoea caused by *E. coli* strain RDEC-1 in the rabbit: stages and role of capsule. *J. Infect. Dis.* 1981, **143**, 219–230.

41 Villako, K., Kekki, M, Tamm, A., Tammur, R., Savisaar, E., Viirsalu, V. & Sipponen, P. Epidemiology and dynamics of gastritis in a representative sample of an Estonian urban population. *Scand. J. Gastroenterol.* 1982, **17**, 601–607.

42 Lee, A. & Hazell, S.L. *Campylobacter pylori* in health and disease: an ecological perspective. *Microbial Ecology in Health and Disease* 1988, **1**, 1–16.

15 Campylobacter pylori, duodenitis and duodenal ulceration

J.I. WYATT

'Peptic duodenitis', characterized by inflammation of the duodenal mucosa with polymorph infiltration and surface gastric metaplasia [1], is commonly seen in biopsies from the margin of duodenal ulcers, and also occurs in the absence of duodenal ulceration. Most workers believe that duodenitis and duodenal ulcer belong to the same pathophysiological spectrum [1, 2]. However, since duodenal biopsies are not routinely taken, duodenitis has received less attention than gastritis.

Patients with duodenal ulcer have chronic gastritis which diffusely involves the antral mucosa, often with mucosal lymphoid follicles; the body type mucosa is normal or only mildly inflamed, and does not show any atrophy of acid secreting glands [3]. This is the 'hypersecretory' pattern of gastritis described by Correa, which we now know to be very closely associated with *Campylobacter pylori* colonization [4].

C. pylori in patients with duodenal ulcer and duodenitis

The association between duodenal ulcer and *C. pylori* gastritis is very strong (Table 15.1) [5–14]. In our experience, 95% of duodenal ulcer patients have campylobacter-associated gastritis; others have found that in the few *C. pylori*-negative patients with duodenal ulcer there is often some different aetiological factor such as non-steroidal anti-inflammatory agents or pancreatitis to account for the duodenal ulcer [5, 15].

The association with duodenal ulcer may have two explanations — either that *C. pylori* is simply a marker of the type of gastritis that is present in duodenal ulcer patients, or that *C. pylori* plays a more direct role in causing the mucosal damage in the duodenum which may lead to ulceration.

Various pathogenic mechanisms have been proposed for *C. pylori* and it is not yet clear which is important *in vivo*; however, all require that the bacterium is closely associated with the mucosal surface for damage to occur. If *C. pylori* is to have a direct role in duodenal mucosal damage, colonization of the duodenal mucosa would therefore be necessary.

Several groups have shown that *C. pylori* is found in the duodenum in patients with duodenitis and duodenal ulcer. The reported frequency of this finding is between 29% and 100%, as shown in Table 15.2 [15–19]. In all these reports, *C. pylori* is only found in duodenal mucosa showing active inflammation, with areas of gastric type surface epithelium.

C. pylori and gastric metaplasia

Campylobacter pylori is unable to colonize small intestinal epithelium, as demonstrated by the sparing of intestinal metaplasia in chronic gastritis (Fig. 15.1) [20, 21]. In the same way, the distribution of *C. pylori* in the duodenum is restricted to foci of gastric epithelium in the duodenal mucosa (Fig. 15.2).

We studied histologically the relationship between campylobacter-associated gastritis, gastric metaplasia in the duodenum, and duodenal inflammation in 290 dyspeptic patients, including 23 with duodenal ulceration, and 13 who had had a highly selective vagotomy (HSV) for duodenal ulcer [17]. In this

Table 15.1 Prevalence of *C. pylori* in patients with duodenal ulcer

Patients studied	Number (%) with *C. pylori*	References	
70	63 (90%)	Marshall *et al.* (1985)	[5]
21	17 (80%)	Price *et al.* (1985)	[6]
32	25 (78%)	Booth *et al.* (1986)	[7]
36	35 (97%)	O'Connor *et al.* (1986)	[8]
37	28 (76%)	Marcheggiano *et al.* (1987)	[9]
14	12 (86%)	Jiang *et al.* (1987)	[10]
17	12 (70%)	Raskov *et al.* (1987)	[11]
66	61 (93%)	Coghlan *et al.* (1987)	[12]
36	36 (100%)	Rauws *et al.* (1988)	[13]
107	100 (93%)	Goodwin *et al.* (1988)	[14]

study, the proportion of the surface epithelium of the duodenum which showed gastric metaplasia was subjectively graded. The relationship between duodenal inflammation and gastric metaplasia is shown in Table 15.3. Active ('peptic') duodenitis was associated with gastric metaplasia and increased in frequency with increasing extent of metaplasia; however, a proportion of biopsies with metaplasia showed no inflammation. Thirty-four patients had active duodenitis. All of these patients had gastric metaplasia involving at least 5% of the duodenal surface and/or campylobacter gastritis, and in 30 (88%) both factors were present (Fig. 15.3). This was compatible with the hypothesis that duodenitis develops when the metaplastic epithelium is colonized by *C. pylori*.

In this series, *C. pylori* were observed in the duodenal biopsies in 55% of patients with duodenitis. This frequency is considerably lower than is seen in chronic gastritis, although reports do vary (Table 15.2). This may be for two resons: (i) unlike gastritis, duodenitis is a patchy lesion in the mucosa [22], and so sampling error will be important; (ii) *C. pylori* are sparser and smaller in the duodenum, often appearing coccoid [19, 23], and are thus less easily detected by histology.

Gastric epithelium in the duodenum

Since the presence of gastric epithelium in the duodenum is essential if *C. pylori* is to infect this site, the circumstances under which this epithelium occurs in the duodenum are important. Gastric epithelium is present in the duodenum in two situations — either over areas of heterotopic gastric body type mucosa, or as a metaplastic change in the duodenal epithelium. Gastric heterotopia, characterized by islands of fully developed fundic mucosa

Table 15.2 Frequency of *C. pylori* in duodenal biopsies from patients with duodenal ulcer and duodenitis

C. pylori +ve/total duodenal ulcer	*C. pylori* +ve/total duodenitis	*C. pylori* +ve/total normal	Reference
45/53	36/39	0/58	Johnson *et al.* (1986) [15]
8/11	–	–	Steer (1984) [16]
8/13	11/21	0/256	Wyatt *et al.* (1987) [17]
7/24	–	–	Bode *et al.* (1987) [18]
14/14	5/16	–	Caselli *et al.* (1988) [19]

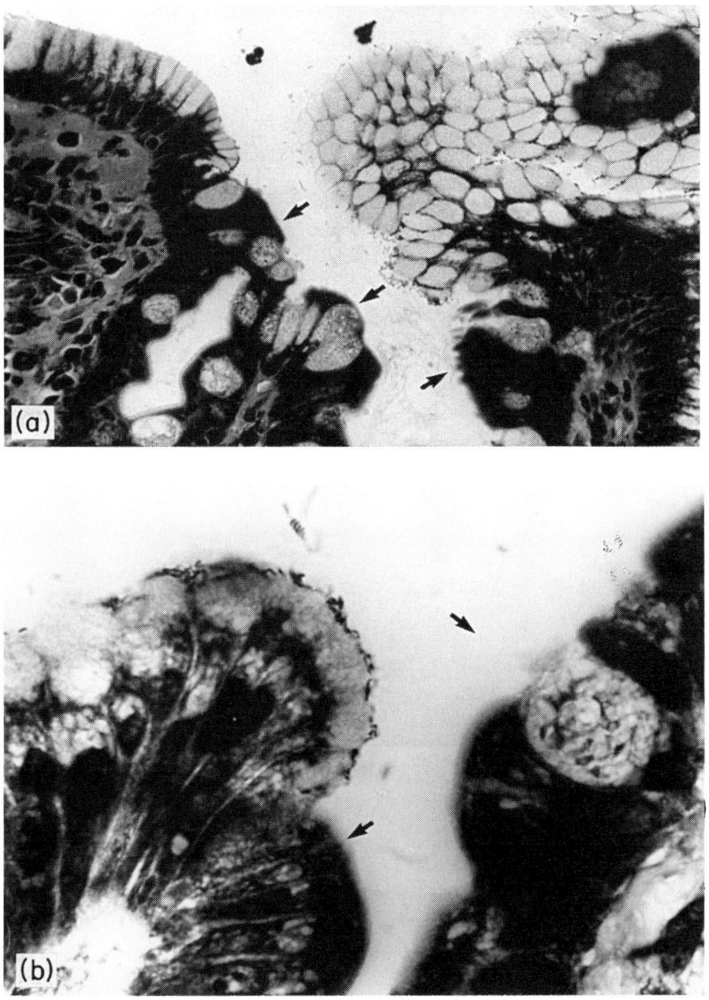

Fig. 15.1 Gastric mucosa with *C. pylori* colonization; the foci of intestinal metaplasia (arrows) are not colonized by bacteria. (Magnifications: (a) × 80; (b) × 198; modified Giemsa stain.)

which includes parietal and chief cells (Fig. 15.4) is rare, occurring in about 1–2% of duodenal biopsies. It is considered to be congenital and it is not generally associated with inflammation [17, 24]. Gastric metaplasia, however, is much more common, present in up to 64% of normal individuals [25], usually as small foci of gastric epithelial cells on the tips of villi (Fig. 15.5). It is more common and more extensive in patients with duodenal inflammation and ulceration, when it is present in up to 100% cases [19, 24, 26–28]. It is rare in children (5% in paediatric duodenal biopsies compared with 30% overall in biopsies from adults [29]) which argues against a congenital origin.

Gastric metaplasia probably represents a response to injury of the duodenal mucosa. This is supported by animal experiments, where gastric metaplasia developed during healing of surgically-produced defects in cats, and also in response to hyperacidity in the duodenum in cats, rats, pigs, and monkeys (Table 15.4).

In man, gastric metaplasia is seen in the duodenum only in patients who have a low fasting gastric juice pH [17], and has been correlated with maximal acid output by Pa-

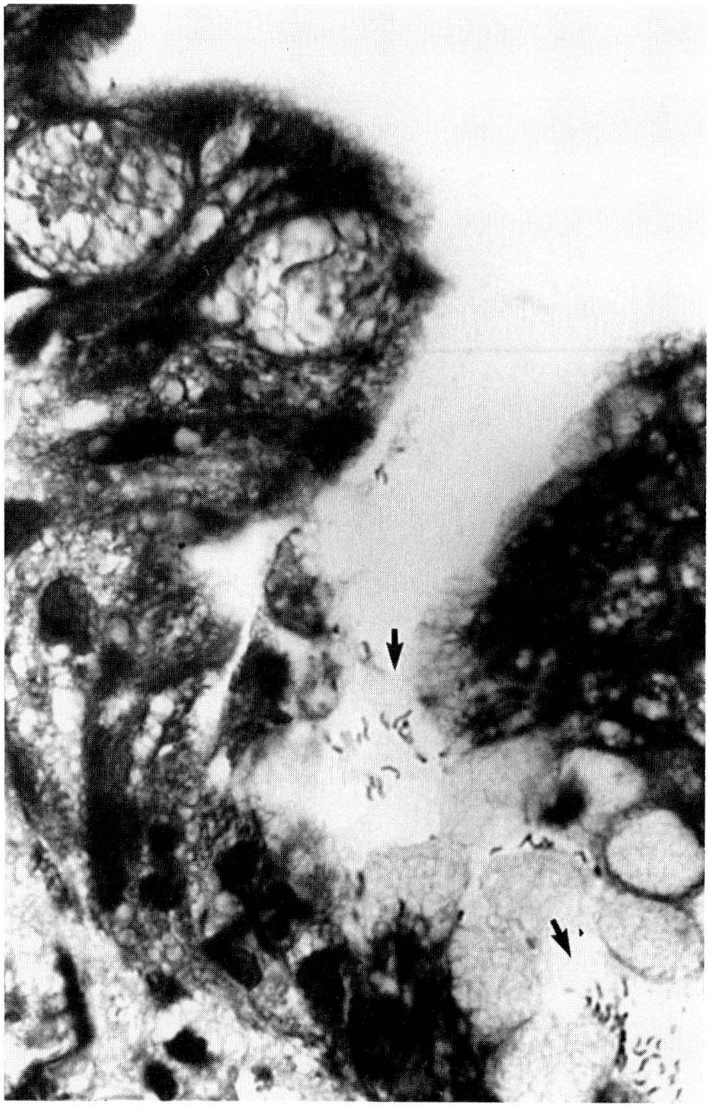

Fig. 15.2 *C. pylori* in the duodenum in a patient with duodenitis; bacteria are found only on areas of gastric metaplasia (arrows). (Modified Giemsa stain × 225.)

trick *et al.* [35]. It is extensive in patients with the Zollinger–Ellison syndrome [36]. It is not found in patients with atrophic gastritis and intestinal metaplasia affecting the acid-secreting gastric mucosa, and appears to be reduced after highly selective vagotomy [17]. It is not specifically a response to acid injury, since it is also occasionally seen in Crohn's disease, and ulcerative jejunitis [37, 38], but in the proximal duodenum, high acid load represents the most frequent potential cause of mucosal injury.

C. pylori in the pathogenesis of duodenal ulcer

Acid secretion by duodenal ulcer patients varies from normal to high levels. The physiology of acid secretion is disturbed in most ulcer patients who may have rapid gastric empty-

Table 15.3 Relationship between duodenal inflammation and gastric metaplasia in the duodenum

Gastric metaplasia	Duodenal histology		
	Normal	Chronic duodenitis	Active duodenitis
None	193	7	1
<5%	30	7	2
5–20%	11	6	17
20%	1	1	14
Totals	235	21	34

ing, nocturnal hypersecretion, or increased sensitivity to gastrin, all of which lead to an increased acid load entering the duodenum [39]. A minimal threshold of acid secretion is consistently present. Below this, duodenal ulceration does not occur [40].

In patients without campylobacter-associated gastritis, gastric metaplasia may represent a protective response of the duodenum to the acid load. However, if the patient becomes infected with *C. pylori*, the gastric metaplasia will also be colonized by *C. pylori* and become inflamed — the patient will have 'gastritis' in the duodenum. The inflamed mucosa may then be susceptible to ulceration, either because it is less able to withstand other ulcerogenic factors, or because severe inflammation itself leads to ulceration, perhaps as a result of an increased rate of cellular proliferation in duodenitis which cannot be sustained [41].

This hypothesis, concerning the sequence of events in the development of duodenitis and duodenal ulceration, accounts for the role of *C. pylori* and acid secretion in duodenal ulcer patients. However, other genetic and environmental factors are well recognized as predisposing to duodenal ulceration. These may potentially act at three levels: (i) in the development of gastric metaplasia, (ii) in determining host susceptibility to *C. pylori* and (iii) in causing ulceration of the inflamed duodenum. Future studies may clarify the interaction of these factors within a unifying concept of duodenal ulcer pathogenesis.

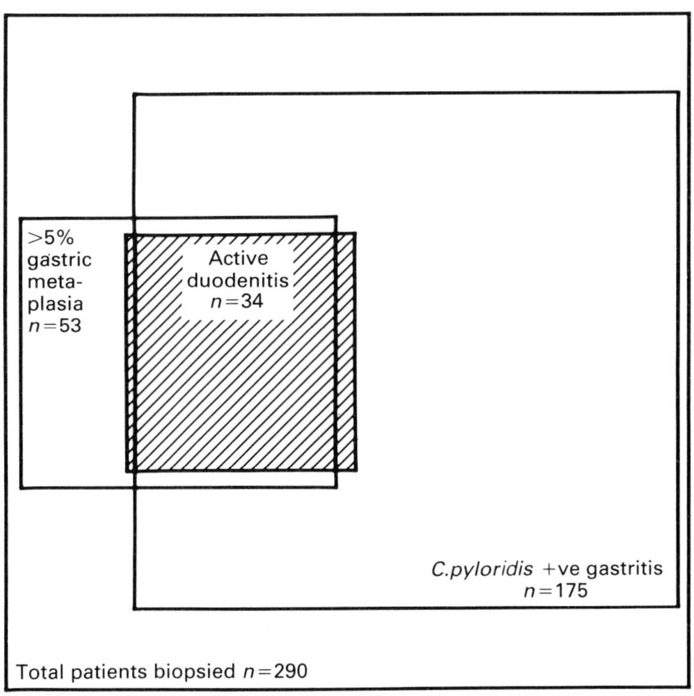

Fig. 15.3 Active duodenitis, gastric metaplasia in the duodenum and campylobacter-associated gastritis in 290 dyspeptic patients. (Reproduced with permission of the *Journal of Clinical Pathology* [17].)

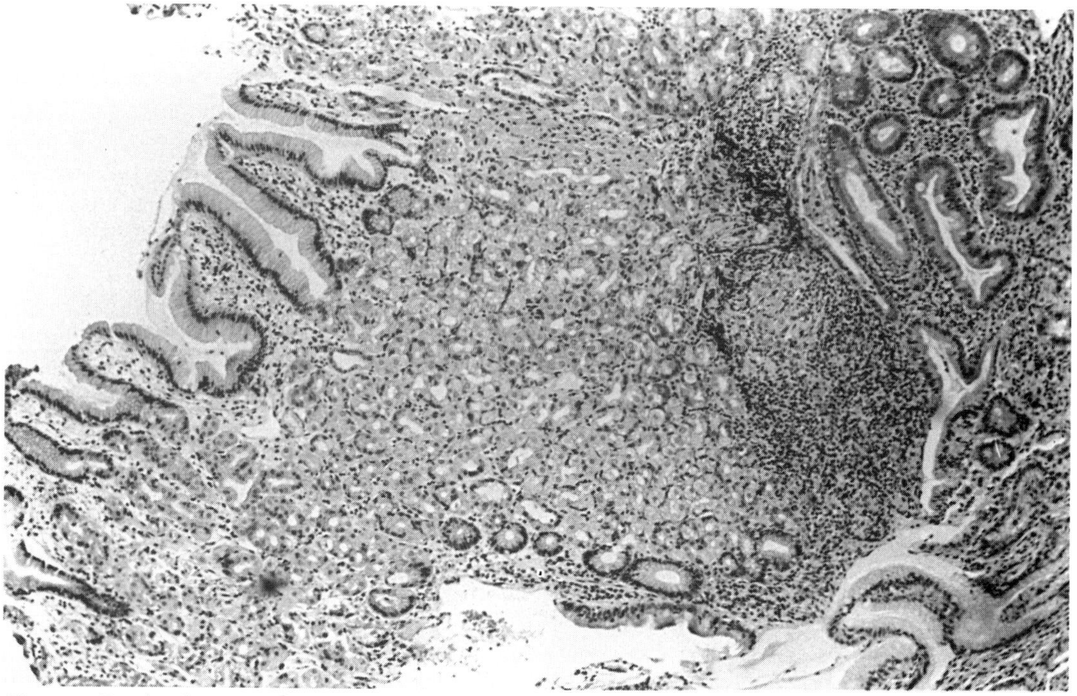

Fig. 15.4 Duodenal mocosa showing an area of gastric heterotopia, in which there are fundic-type gastric glands containing parietal and chief cells. (Haematoxylin & eosin, × 10.)

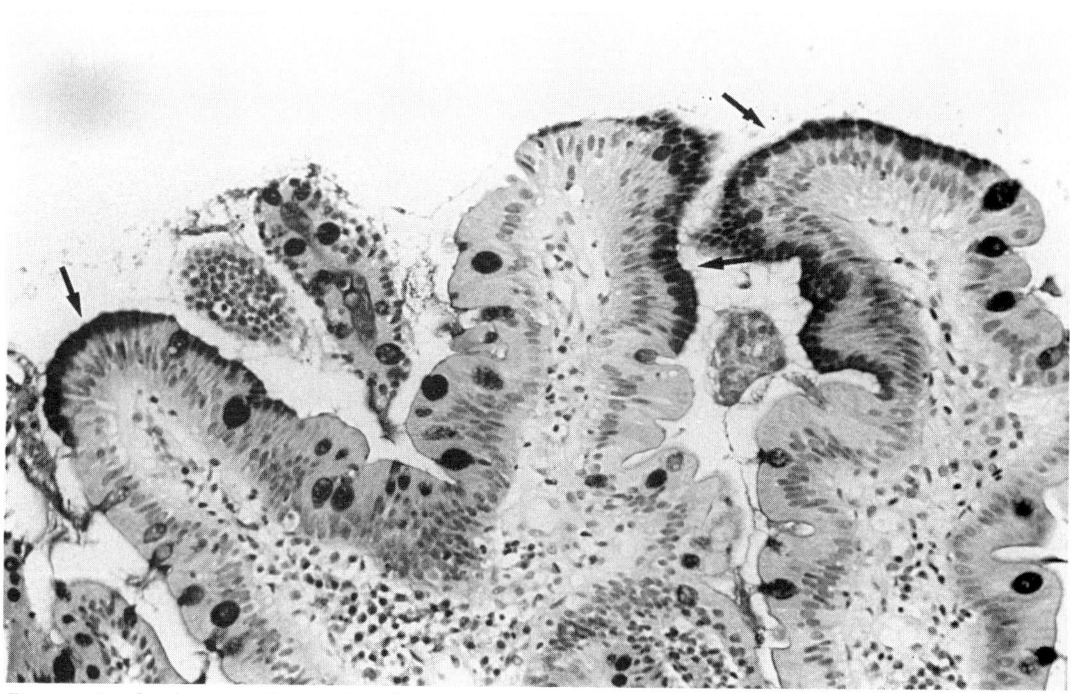

Fig. 15.5 Duodenal mucosa showing foci of gastric metaplasia involving the surface epithelium (arrows). The duodenal mucosa is otherwise normal. (Alcian blue/PAS, × 30.)

Table 15.4 Experimental production of gastric metaplasia in animal models

Animal	Mucosal injury	Reference
Cat	Surgically-produced mucosal defect	Florey & Harding (1935) [30]
Pig	Hyperacidity due to gastroduodenal fistula	Florey *et al.* (1939) [31]
Cat	Histamine-induced hyperchlorhydria	Rhodes (1964) [32]
Rat	Pharmacologically-induced hyperchlorhydria	Gaskin *et al.* (1975) [33]
Monkey	Stress-induced hyperchlorhydria	Natelson *et al.* (1977) [34]

References

1 Joffe, S.N., Lee, F.D. & Blumgart, L.H. Duodenitis. *Clin. Gastroenterol.* 1978, **7**, 635–650.

2 Thompson, W.O., Joffe, S.N., Robertson, A.G. *et al.* Is duodenitis a dyspeptic myth? *Lancet*, 1977, **ii**, 1197–1198.

3 Correa, P. Chronic gastritis: A clinico–pathological classification. *Am. J. Gastroenterol.* 1988, **83**, 504–509.

4 Chen, X.G., Correa, P., Offerhaus, J. *et al.* Ultrastructure of the gastric mucosa harbouring campylobacter-like organisms. *Am. J. Clin. Pathol.* 1986, **86**, 575–582.

5 Marshall, B.J., McGechie, D.B., Rogers, P.A. *et al.* Pyloric campylobacter infection and gastroduodenal disease. *Med. J. Australia* 1985, **142**, 439–444.

6 Price, A.B., Levi, J., Dolby, J.M. *et al. Campylobacter pyloridis* in peptic ulcer disease: microbiology, pathology and scanning electron microscopy. *Gut* 1985, **21**, 1183–1188.

7 Booth, L., Holdstock, G. MacBride, H. *et al.* Clinical importance of *Campylobacter pyloridis* and associated serum IgG and IgA antibody responses in patients undergoing upper gastrointestinal endoscopy. *J. Clin. Pathol.* 1986, **39**, 215–219.

8 O'Connor, H.J., Dixon, M.F., Wyatt, J.I. *et al.* Effect of duodenal ulcer surgery and enterogastric reflux on *Campylobacter pyloridis*. *Lancet* 1986, **ii**, 1178–1181.

9 Marcheggiano, A., Kannoni, C., Agnello, M. *et al.* Campylobacter-like organisms on the human gastric mucosa. Relation to type and extent of gastritis in different clinical groups. *Gastroenterol. Clin. Biol.* 1987, **11**, 376–381.

10 Jiang, S.J., Liu, W.Z., Zhang, D.Z. *et al.* Campylobacter-like organisms in chronic gastritis, peptic ulcer, and gastric carcinoma. *Scand. J. Gastroenterol.* 1987, **22**, 553–558.

11 Raskov, H., Lanng, K., Gaarslev, B. *et al.* Screening for *Campylobacter pyloridis* in patients with upper dyspepsia and the relation to inflammation of the human gastric antrum. *Scand. J. Gastroenterol.* 1987, **22**, 568–572.

12 Coghlan, J.G., Gilligan, D., Humphries, H. *et al. Campylobacter pylori* and recurrence of duodenal ulcers — a 12-month follow-up study. *Lancet* 1987, **ii**, 1109–1111.

13 Rauws, E.A.J., Langenberg, W., Houthoff, H.J. *et al. Campylobacter pyloridis*-associated chronic active antral gastritis; a prospective study of its prevalence and the effects of antibacterial and antiulcer treatment. *Gastroenterology* 1988, **94**, 33–40.

14 Goodwin, C.S., Marshall, B.J., Blincow, E.D. *et al.* Prevention of nitroimidazole resistance in *Campylobacter pylori* by co-administration of colloidal bismuth subcitrate: clinical and *in vitro* studies. *J. Clin. Pathol.* 1988, **41**, 207–210.

15 Johnson, B.J., Reed, P.I., Ali, M.H. Campylobacter-like organisms in duodenal and antral endoscopic biopsies: relationship to inflammation. *Gut* 1986, **27**, 1132–1137.

16 Steer, H.W. Surface morphology of the gastroduodenal mucosa in duodenal ulceration. *Gut* 1984, **25**, 1203–1210.

17 Wyatt, J.I., Rathbone, B.J., Dixon, M.F. *et al. Campylobacter pyloridis* and acid-induced gastric metaplasia in the pathogenesis of duodenitis. *J. Clin. Pathol.* 1987, **40**, 841–848.

18 Bode, G., Malfertheiner, P. & Ditschuneit, H. Pathogenetic implications of ultrastructural findings in *Campylobacter pylori* related gastroduodenal disease. *Scand. J. Gastroenterol.* 1988, **23**, (Suppl. 142), 25–39.

19 Caselli, M., Bovolenta, M.R., Aleotti, A. *et al.* Epithelial morphology of duodenal bulb and campylobacter-like organisms. *J. Submicrosc. Cytol. Pathol.* 1988, **20**, 237–242.

20 Thomas, J.M. Campylobacter-like organisms in gastritis. *Lancet* 1984, **ii**, 1217.

21 Wyatt, J.I. & Dixon, M.F. Chronic gastritis — a pathogenetic approach. *J. Pathol.* 1988, **154**, 113–124.

22 Hasan, M., Sircus, W. & Ferguson, A. Duodenal mucosal architecture in non-specific and ulcer-associated duodenitis. *Gut* 1981, **22**, 637–641.

23 Delmee, M., Debongnie, J.C., Warzee, P. *et al.* Une méthode originale di prélèvement bacteriologique pour l'étude de '*Campylobacter pylori*' dans l'ulcéré

duodénale. *Gastroenterol. Clin. Biol.* 1987, **11**, 550–553.

24 Shousha, S., Spiller, R.C. & Parkins, R.A. The endoscopically abnormal duodenum in patients with dyspepsia: biopsy findings in 60 cases. *Histopathology* 1983, **7**, 23–34.

25 Kreuning, J., Bosman, F.T., Kuiper, G. *et al.* Gastric and duodenal mucosa in 'healthy' individuals. An endoscopic and histopathological study of 50 volunteers. *J. Clin. Pathol.* 1978, **31**, 69–77.

26 James, A.H. Gastric epithelium in the duodenum. *Gut* 1964, **5**, 285.

27 Morrissey, S.M., Ward, P.M., Jayari, A.P. *et al.* Histochemical changes in mucosa in duodenal ulceration. *Gut* 1983, **24**, 909–913.

28 Gregory, M.A. & Spitaels, J.M. Variations in the morphology of villous epithelial cells within 8 mm of untreated duodenal ulcers. *J. Pathol.* 1987, **53**, 109–119.

29 Wyatt, J.I. & Rathbone, B.J. Gastric metaplasia in the duodenum and *Campylobacter pylori*. *Gastroenterol. Clin. Biol.* 1989, in press.

30 Florey, H.W. & Harding, H.E. The healing of artificial defects of the duodenal mucosa. *J. Pathol.* 1935, **40**, 211–218.

31 Florey, H.W., Jennings, M.A., Jennings, D.A. *et al.* The reactions of the intestine of the pig to gastric juice. *J. Pathol. Bact.* 1939, **49**, 105–123.

32 Rhodes, J. Experimental production of gastric epithelium in the duodenum. *Gut* 1964, **5**, 454–458.

33 Gaskin, R.J., Gad, A., Barros, A.A.J. *et al.* Natural history and morphology of secretagogue-induced ulcers in rats. *Gastroenterology* 1975, **69**, 903–910.

34 Natelson, B., Dubois, A. & Sodetz, F.J. Effect of multiple-stress procedures on monkey gastroduodenal mucosa, serum gastrin and hydrogen ion kinetics. *Dig. Dis. Sci.* 1977, **22**, 888–897.

35 Patrick, W.J.A., Denham, D. & Forrest, A.P.M. Mucous change in the human duodenum; a light and electron microscopic study and correlation. *Gut* 1974, **15**, 767–776.

36 Parrish, J.A. & Rawlins, D.C. Intestinal mucosa in the Zollinger–Ellison syndrome. *Gut* 1965, **6**, 286–289.

37 Lee, F.D. & Toner, P.G. 1980. *Biopsy Pathology of the Small Intestine*. Chapman & Hall, London.

38 Whitehead, R. 1984. *Mucosal Biopsy of the Gastrointestinal Tract*, 3rd edn. W.B. Saunders, Philadelphia.

39 Lam, S-K. Pathogenesis and pathophysiology of duodenal ulcer. *Clin. Gastroenterol.* 1984, **13**, 447–472.

40 Baron, J.H. Current views on pathogenesis of peptic ulcer. *scand. J. Gastroenterol.* 1982, **17** (Supp. 80), 1–10.

41 Bransom, C.J., Boxer, M.E., Palmer, K.R. *et al.* Mucosal cell proliferation in duodenal ulcer and duodenitis. *Gut* 1981, **22**, 277–282.

16 Campylobacter pylori, gastric ulceration and the post-operative stomach

H.J. O'CONNOR AND A.T.R. AXON

Over the past five years, the flurry of research activity into *Campylobacter pylori* and its disease associations has focused largely on the *C. pylori*-positive stomach and less attention has been paid to the clinical significance of finding a gastric mucosa devoid of the organism. At about the time *C. pylori* was rediscovered, we coincidentally became aware that alkaline reflux gastritis could produce a characteristic histological picture. The discovery that reflux gastritis was also a form of chronic gastritis largely negative for *C. pylori* prompted us to study the effects of peptic ulcer surgery and Roux-en-Y biliary diversion on the organism and latterly to assess possible differences between *C. pylori*-positive and *C. pylori*-negative gastric ulcers. The results of these studies and their clinical implications form the basis of this chapter.

Reflux gastritis

Although the concept of alkaline reflux gastritis has gained widespread acceptance as an endoscopic and clinical entity [1], there is little awareness that the syndrome may be accompanied by a distinctive histological picture. Recently, we became increasingly impressed by the incidence of certain histological features seen in biopsy specimens from patients, usually after gastric surgery, who have clinical and endoscopic evidence of bile reflux. The principal histological feature in such cases is the finding of elongation, tortuosity and hypercellularity of the gastric pits (foveolar hyperplasia), often giving the mucosa a villous appearance. Other changes usually accompany this hyperplasia including vasodilatation and capillary congestion in the superficial lamina propria, oedema in the lamina propria, and a paucity of both acute and chronic inflammatory cells. If reflux gastritis is truly a separate pathological entity with a distinctive histological picture, a radically different pattern of association with *C. pylori* to that found in 'conventional' chronic gastritis might also be expected.

To corroborate these observations and to test whether or not the histological features correlate with the degree of reflux of alkaline

Table 16.1 Details of patients studied

Group	*n*	Age (years)		Sex (M:F)	Length of follow-up after operation (years)	
		Mean	Range		Mean	Range
Normal endoscopy	19	41.7	22–75	10:9		
Duodenal ulcer	16	44.4	21–76	9:7		
Gastric ulcer	25	62.9	36–83	12:13		
Combined duodenal and gastric ulcer	15	54.5	37–77	6:9		
Post-operative patients*	23	56.9	26–78	19:4	14.8	1–40

* 8 Billroth I partial gastrectomy, 7 Billroth II partial gastrectomy, 6 truncal vagotomy and gastroenterostomy, 2 truncal vagotomy and antrectomy.

enteric contents into the stomach, a total of 98 patients (Table 16.1) underwent gastric biopsy, together with measurement of pH and total bile acid concentration [2] (measured in 72 patients) in their fasting gastric juice [3, 4].

Histological assessment

The histological sections were stained with haematoxylin & eosin and examined by a pathologist who was unaware of the patients' endoscopic findings, gastric pH or bile acid concentrations. A score from 0 (normal or absent) to 3 (severe) was allotted for each of the following histological features according to its severity: foveolar hyperplasia, oedema in the lamina propria, and vasodilatation and congestion of the lamina propria. Acute and chronic inflammatory cells were graded separately and given scores which reflected their paucity — that is, 0 (severe increase) to 3 (absence of polymorphs or normal or reduced numbers of chronic inflammatory cells). A composite reflux score was calculated for each patient by simply adding the scores allotted for each of the individual variables (minimum score, 0; maximum score, 15) (Fig. 16.1).

A further section was taken from each set of biopsies and stained by the Warthin–Starry method for *C. pylori*. These sections were examined by another pathologist, who was unaware of the findings in the routine sections and the other investigations.

Reflux scores, gastric pH and bile acid concentrations

Table 16.2 shows the mean reflux scores, fasting gastric pH and bile acid concentrations

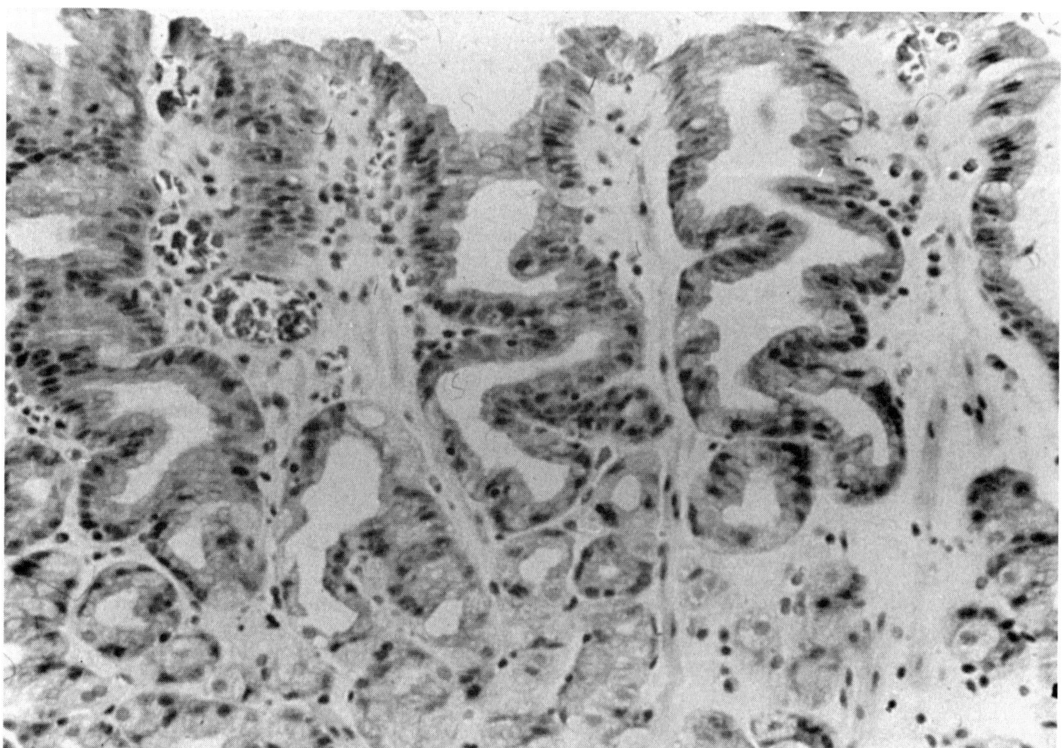

Fig. 16.1 Biopsy specimen showing marked foveolar hyperplasia (grade 3), moderate oedema in the lamina propria (grade 2), severe capillary congestion (grade 3), paucity of chronic inflammatory cells (grade 3), and absence of polymorphs (grade 3). (Haematoxylin & eosin, × 161.)

Table 16.2 Reflux scores, fasting intragastric pH and bile acid concentrations in the patient groups

Group	Reflux score (mean ± S.D.)	pH	Bile acid concentrations (mmol/l)
Normal endoscopy	9.4 (2.7)	2.1 (0.5)	0.10 (0.32)
Duodenal ulcer	7.9 (2.1)	1.9 (0.5)	0.11 (0.25)
Gastric ulcer	8.9 (3.1)	2.7 (1.3)	0.51 (2.04)
Combined duodenal and gastric ulcer	8.4 (3.6)	2.0 (0.4)	0.11 (0.22)
Post-operative patients	12.0 (3.2)	6.0 (1.8)	5.88 (6.67)

in the five groups of patients. Reflux scores were highest in the post-operative patient group and lowest in the duodenal ulcer group. The mean reflux score was significantly higher ($P < 0.01$) in the post-operative group compared with all of the other groups. Mean pH and bile acid concentrations were also significantly higher ($P < 0.01$) in the post-operative group. We found a highly significant correlation between hypochlorhydria (pH > 4) and increased bile acid concentrations (> 1 mmol/l) in the stomach, which remained significant when patients with intestinal metaplasia ($n = 27$) (another variable which may be associated with raised gastric pH) were excluded from analysis (χ^2, $P < 0.001$).

A significant correlation was found between severe change in each of the histological variables and both hypochlorhydria and increased concentrations of bile acids in the stomach. Furthermore, when the composite reflux scores were compared with gastric pH and degree of bile reflux, there was a significant association between a reflux score > 10 and both hypochlorhydria and high bile acid concentrations ($P < 0.005$).

Reflux scores and C. pylori

C. pylori was detected in 64 of the 98 patients (65%) and the distribution of the positive cases is shown in Table 16.3. Considerably fewer post-operative patients were positive compared with patients in the other clinical groups. Of the *C. pylori*–negative cases, three (one normal endoscopy, one duodenal ulcer, one post-operative) were considered to have normal histological findings, and were excluded from further analysis.

We found a highly significant difference in the distribution of *C. pylori*-positive and *C. pylori*-negative cases between patients with high (> 10) and low (⩽ 10) reflux scores. Of the 31 patients who were negative, 24 had a score of > 10 compared with only 12 of 64 in the positive group (χ^2, $P < 0.0001$). Reflux scores (mean ± S.D.) were also significantly

Table 16.3 *C. pylori* status and the patient groups

Group	*n*	*C. pylori*-positive	
		n	%
Normal endoscopy	19	12	63
Duodenal ulcer	16	14	88
Gastric ulcer	25	20	80
Combined duodenal and gastric ulcer	15	12	80
Post-operative patients	23	6	26

higher in the negative group compared with the positive group (11.7 ± 2.3 versus 8.2 ± 2.5; $P = 0.0004$).

To exclude the possibility that observed differences in reflux gastritis scores between the negative and positive groups might be accounted for solely on the basis of differences in the intensity of the inflammatory cell infiltrate, a modified reflux score was calculated for each patient by adding the scores allotted for foveolar hyperplasia, lamina propria oedema, and vasodilatation and congestion. This modified score, reflecting the epithelial and vascular components of the inflammatory reaction, was also significantly higher in the *C. pylori*-negative group.

Gastric pH, bile acid concentrations and C. pylori

When the relation between *C. pylori* status and fasting intragastric pH and bile acid concentrations was assessed, there was a highly significant association between the absence of *C. pylori* and both hypochlorhydria and high bile acid levels in the stomach (Table 16.4). Furthermore, intragastric pH (mean ± S.D.) was significantly higher in the negative group (4.3 ± 2.3) than in the positive group (2.5 ± 1.5) ($P = 0.001$), as were bile acid concentrations (3.76 ± 5.86 versus 0.73 ± 2.90; $P = 0.004$).

This study showed that a histological picture quite distinct from previously described categories of gastritis [5], comprising foveolar hyperplasia, oedema and capillary congestion of the lamina propria, and a relative paucity of acute and chronic inflammatory cells, is strongly associated with raised intragastric bile acid levels and increased pH. Using these histological criteria we found a highly significant difference in *C. pylori* status between patients with reflux scores above and below 10. Thus, the much lower incidence of *C. pylori* in patients, who by histological and other criteria are considered to have alkaline reflux into the stomach, supports the hypothesis that reflux is a distinct aetiological factor and that the resultant gastritis is unrelated to *C. pylori* infection.

It is important for histopathologists to recognize reflux gastritis because the changes may be confused with pre-malignant dysplasia. There is a tendency for histopathologists to interpret severe foveolar hyperplasia in the absence of inflammatory cell infiltration (as occurs in reflux gastritis) as dysplasia and the same changes in the presence of inflammatory cells as 'reactive' or 'regenerative'. Thus, a failure to appreciate the nature of the mucosal response to reflux may well underlie the high incidence of low-grade dysplasia reported in some studies on the post-operative stomach [6–9]. Recent studies documenting the strong association between *C. pylori* and 'conventional' chronic gastritis have claimed that most biopsy specimens negative for *C. pylori* are normal [10–13]. We suspect that the reluctance to diagnose gastritis in the absence of an inflammatory cell infiltrate means that features of reflux gastritis have been previously overlooked, or regarded as part of the 'nor-

Table 16.4 Gastric pH, bile acid concentrations and *C. pylori* status

Status	Gastric pH*		Bile acid concentrations (mmol.l)**	
	>4	<4	>1	<1
C. pylori-negative	16	15	11	11
C. pylori-positive	7	57	4	46

* $\chi^2 = 18.831$, $P < 0.001$.
** $\chi^2 = 16.340$, $P < 0.001$.

mal' spectrum, and that a proportion of these so-called normal biopsy specimens have reflux gastritis.

Finally, the treatment of reflux gastritis centres on the use of cytoprotective [14] or bile-binding agents [15] or surgical biliary diversion [16], whereas for type B chronic gastritis treatment with agents designed to eradicate *C. pylori* would now seem more logical. Indeed, eradication of *C. pylori* is increasingly used as an end-point in controlled studies in peptic ulceration and non-ulcer dyspepsia [17, 18]. It will therefore be important to distinguish *C. pylori*-positive chronic gastritis from reflux gastritis in assessing response to treatment in future trials.

C. pylori and the post-operative stomach

The early recognition that bismuth compounds were bactericidal to *C. pylori* [19], and that persistence of *C. pylori*-related gastritis might account for the frequent relapse of ulcers following H_2-antagonist therapy, probably explains why so little attention has been paid to the effects of peptic ulcer surgery on *C. pylori*. Prompted by our findings in relation to reflux gastritis we explored the effects of gastric surgery on *C. pylori* in a series of 89 duodenal ulcer patients, some of whom were untreated and others who had undergone either a conservative highly selective vagotomy (HSV) or a more radical operation which bypassed or removed the pylorus (Table 16.5) [20].

At endoscopy, at least two gastric mucosal biopsies were taken from within 5 cm of the pylorus from untreated patients and those who had undergone HSV, and from within 5 cm of the stoma from patients in the other groups. In 71 (80%) patients, fasting gastric contents were aspirated for measurement of total bile acid concentration [2]. The biopsy specimens were assessed 'blind' for the presence of *C. pylori* by one pathologist and independently scored for severity of reflux gastritis by another using the histological grading system described above.

Campylobacter pylori was detected in 65 (73%) of the 89 patients and the distribution of positive cases in the five diagnostic groups is shown in Fig. 16.2. The likelihood of not finding *C. pylori* rose with time which had elapsed since Billroth type I partial gastrectomy (BIPG), Billroth type II partial gastrectomy (BIIPG) and truncal vagotomy and gastro-enterostomy (TVGE), but not HSV. Of the 12 patients who had had their gastric resection or anastomosis < 10 years ago, 8 were *C. pylori*-positive compared with only 8 of 26 patients who had been operated on > 10 years ago (χ^2, $P < 0.005$).

Reflux scores, bile acid concentrations and diagnostic groups

Reflux scores and bile acid concentrations in

Table 16.5 Details of patients studied

Group	*n*	Age (years)		Sex (M:F)	Length of follow-up after operation (years)	
		Mean	Range		Mean	Range
Untreated duodenal ulcer	35	43.4	21–76	22:13		
Highly selective vagotomy	16	49.0	31–68	12:4	3.9	0.5–10
Billroth I partial gastrectomy (BIPG)	9	54.0	34–69	8:1	15.0	2.5–40
Billroth II partial gastrectomy (BIIPG)	17	58.5	26–81	12:5	20.3	7–33
Truncal vagotomy and gastro-enterostomy (TVGE)	12	53.6	34–67	11:1	10.8	2–22

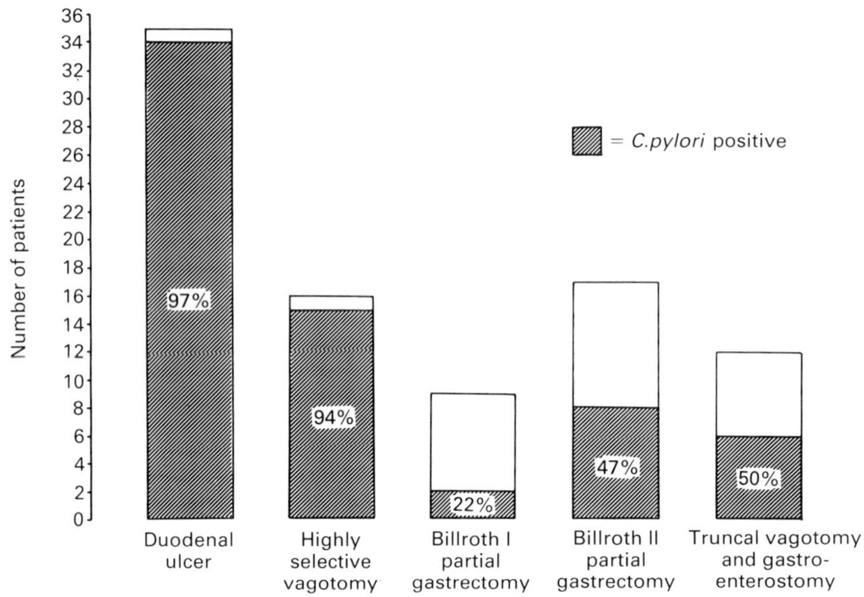

Fig. 16.2 Effect of peptic ulcer surgery on *C. pylori*: distribution of *C. pylori*-positive cases in the diagnostic groups.

the diagnostic groups are shown in Figs. 16.3 and 16.4, respectively. The untreated group and the HSV group had similar reflux scores and bile acid levels. Reflux scores were significantly higher in the BIPG, BIIPG, and TVGE groups than in either the untreated or HSV groups. Bile acid concentrations were significantly higher in the BIPG and BIIPG groups than in either the untreated or HSV groups. Bile acid levels after TVGE were also signifi-

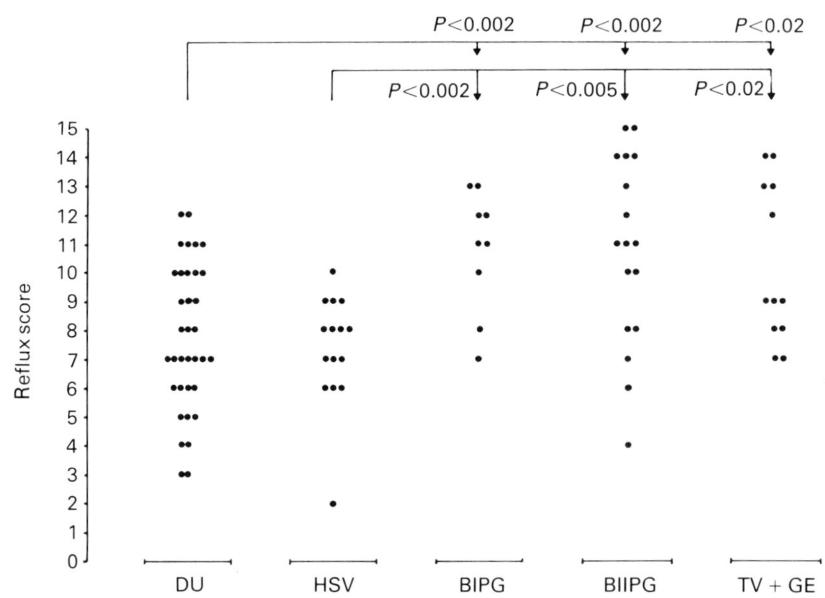

Fig. 16.3 Reflux scores and the diagnostic groups.

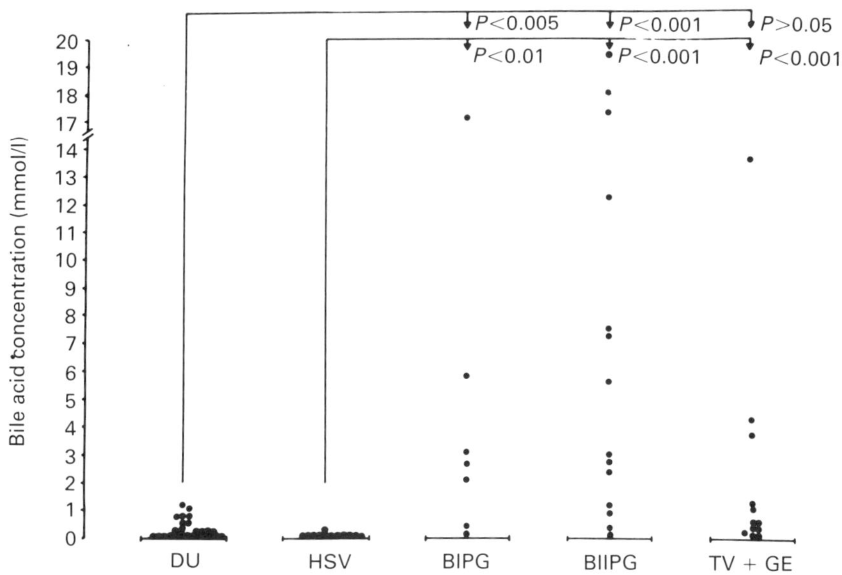

Fig. 16.4 Bile acid concentrations and the diagnostic groups.

cantly higher than after HSV and showed a non-significant trend towards higher values than in untreated duodenal ulcer patients.

In agreement with previous findings, reflux scores and bile acid levels were significantly higher in the *C. pylori*-negative patients than in the *C. pylori*-positive patients and there was a significant correlation between high bile acid concentrations (> 1 mmol/l) and severe gastritis (reflux score > 10).

These results show that the prevalence of *C. pylori* in duodenal ulcer patients is lower after partial gastrectomy or TVGE compared with HSV. Indeed the presence of *C. pylori* was an almost universal finding in the HSV and untreated duodenal ulcer patients. The fact that there is less enterogastric reflux after HSV than after operations that destroy or bypass the pylorus may be relevant [21].

Previous histological studies on gastritis in the post-operative stomach have found a discrepancy between the degree of 'conventional' chronic (type B) gastritis and objective tests of bile reflux [22–24]. Dewar *et al.* [25] observed notable differences in the bile acid levels in samples of fasting gastric juice from various post-operative groups but could not find significant differences in the degree of type B gastritis in their biopsy specimens. We postulate that these discrepancies stem from the fact that *C. pylori* and not entero-gastric reflux is causally related to type B gastritis. Our results also suggest that duodenal ulcer patients may undergo a transition from *C. pylori*-related gastritis to reflux gastritis after operations which resect or bypass the pylorus. This conclusion can be criticized on the grounds that the results are not based on longitudinal studies in individual patients and we did not assess whether the mucosal abnormalities were reversible after effective diversion of upper intestinal contents away from the gastric remnant.

To answer these criticisms and further develop the hypothesis we undertook a detailed study of gastric mucosal changes in individual patients who had undergone peptic ulcer surgery followed by Roux-en-Y (RY) biliary diversion. We studied 24 patients (21 male, 3 female) with a mean age of 49 years (range, 29–68 years) who underwent RY biliary diversion as treatment for symptomatic bile re-

Table 16.6 Original operative procedures

Operation	*n*
Billroth I partial gastrectomy	9
Billroth II partial gastrectomy	3
Truncal vagotomy and pyloroplasty	5
Truncal vagotomy and antrectomy	5
Gastroenterostomy	2

Table 16.7 Length of efferent limb used at Roux-en-Y diversion

	Length (cm)						
	25	35	45	50	55	100	Not stated
n	1	13	1	5	1	1	2

flux following an initial operation for peptic ulcer. All patients had scintigraphic [26] and/or biochemical[27] evidence of entero-gastric reflux. The indication for the original operation was duodenal ulcer in 19 (79%), gastric ulcer in 4 and combined gastric and duodenal ulceration in 1. The details of the original operative procedures are presented in Table 16.6. The interval from the initial operation to RY diversion ranged from 1 to 38 years (mean, 7.1 years). The length of the RY efferent limb was noted at operation in 22 patients (Table 16.7). All patients were followed up after RY diversion for 0.8–12.3 years (mean, 4.9 years).

Histological sections were obtained from endoscopic gastric biopsies and from specimens resected at operation. All patients underwent upper gastrointestinal endoscopy before and after RY diversion. The interval to endoscopy after the operation ranged from 0.8 to 9.3 years (mean, 4.2 years). At endoscopy, at least two gastric mucosal biopsies were taken from within 5 cm of the stoma. Histological sections were available from the original partial gastrectomy specimen in 12 patients (50%). Of the remaining 12 patients in the study, seven had not undergone gastric resection as part of their initial operation and we were unable to locate sections from the original specimen in the other five. Histological sections were available from prestomal tissue resected at the time of RY diversion in 14 patients (58%). The histological sections were examined 'blind' for the presence of *C. pylori* and specifically assessed for severity of reflux gastritis as described above.

Campylobacter pylori was detected in 10 of the 12 (83%) original partial gastrectomy specimens. Only 13 of the 24 patients studied (54%) were *C. pylori*-positive before RY diversion; this rose to 22 (92%) after biliary diversion ($P = 0.024$). In the nine patients who reverted from being *C. pylori*-negative to *C. pylori*-positive, recolonization of the gastric remnant was detected from 0.5 to 9 years (mean, 4.4 years) after RY diversion. All but one of the 13 patients who were *C. pylori*-positive before RY diversion remained positive after the operation. The median reflux score rose from 6 in the original partial gastrectomy specimens to 11 before RY diversion, falling again to 6 after biliary diversion (Fig. 16.5). Furthermore, high reflux scores (> 10) were found in 13 patients before RY diversion but were not detected in any of the original gastrectomy specimens ($P = 0.002$) or after biliary diversion ($P = 0.00003$).

There was a significant correlation between the fall in reflux scores after biliary diversion and the length of the efferent limb used at RY surgery (Spearman $r = 0.628$, $P = 0.004$), but not with time after operation.

The characteristic sequence of histological changes associated with partial gastrectomy for peptic ulcer followed by RY biliary diversion is illustrated in Fig. 16.6a–c.

A curious feature of several previous studies on the outcome of diversionary surgery has been the failure to demonstrate a decrease in the degree of histological gastritis, despite improvement in symptoms and virtual elimination of reflux as documented by biochemical analysis and/or scintigraphy [8, 24, 28–30]. This discrepancy may result from the fact that all these studies used standard Whitehead criteria [5] to evaluate gastritis.

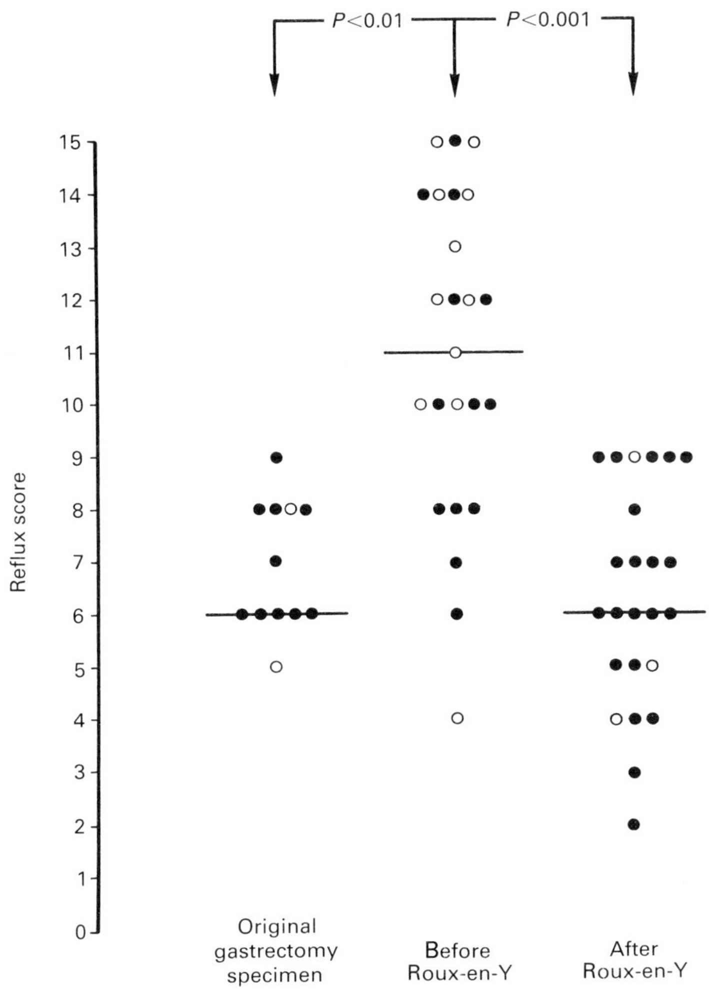

Fig. 16.5 Reflux scores in the original partial gastrectomy specimens and in gastric biopsies before and after Roux-en-Y diversion. ○ = *C. pylori*-negative; ● = *C. pylori*-positive; — = median.

The striking sequence of gastric mucosal changes we have found in individual patients related to RY diversion strongly supports the hypothesis that reflux gastritis is indeed a distinct form of chronic gastritis that defies evaluation by criteria previously applied to type B gastritis.

This study shows that *C. pylori*, apparently eradicated after peptic ulcer operations that increase enterogastric reflux, may reappear in the gastric remnant following effective biliary diversion. Furthermore, our results suggest that, associated with changes in *C. pylori* status and entero-gastric reflux, the predominant pattern of chronic gastritis in peptic ulcer patients may change first from type B gastritis to reflux gastritis after the initial ulcer operation and then back to type B gastritis after biliary diversion. It remains to be determined whether recolonization of the gastric remnant after RY diversion results from recrudescence of infection with the same strain of *C. pylori*, as occurs after unsuccessful antibiotic therapy [31], or from re-infection with a new strain.

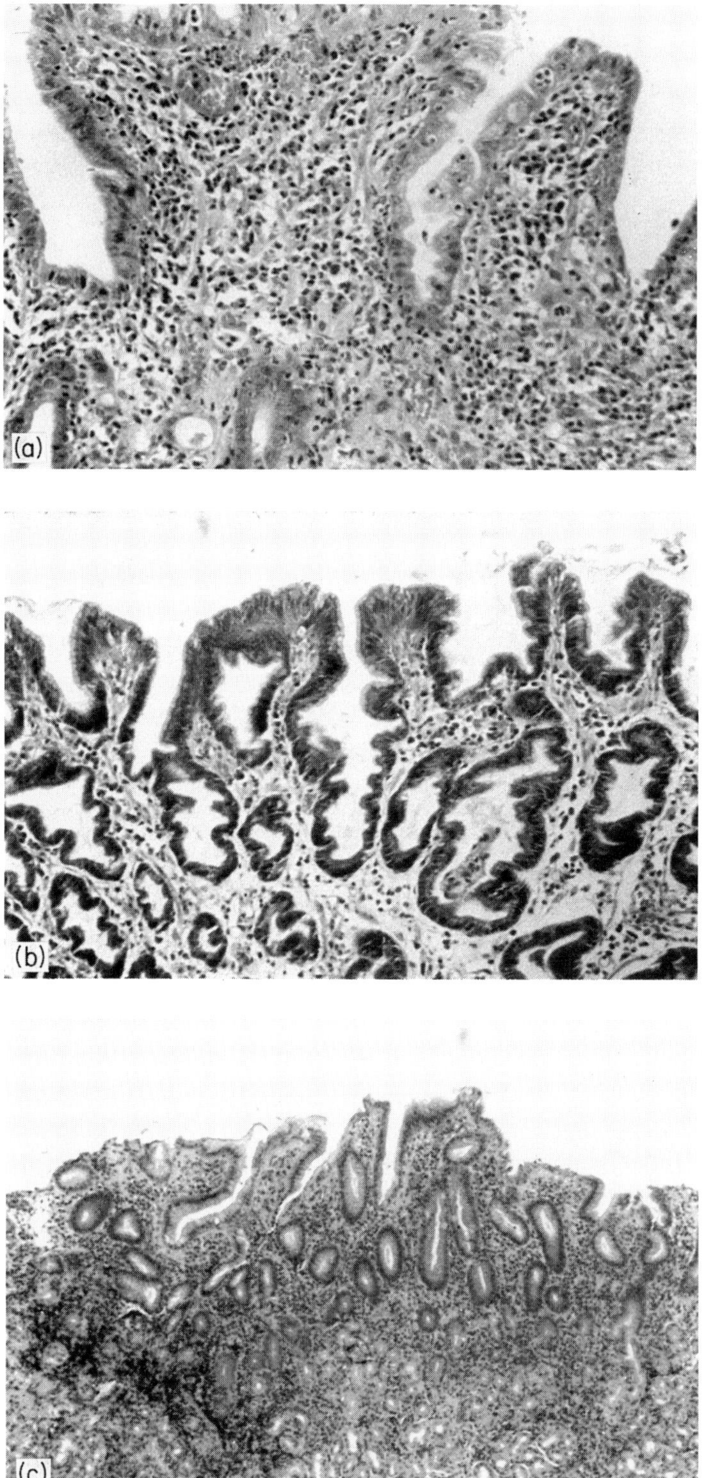

Fig. 16.6 (a) Operative specimen from a 56-year-old man who underwent Billroth II partial gastrectomy for duodenal ulceration. Reflux score = 6; *C. pylori*-positive. (Haematoxylin & eosin, × 69.) (b) Gastric mucosa resected from the same patient four years later at Roux-en-Y diversion. Reflux score = 11; *C. pylori*-negative. (Haematoxylin & eosin, × 69.) (c) Endoscopic biopsy from the gastric remnant seven years after Roux-en-Y diversion. Reflux score = 6; *C. pylori*-positive. (Haematoxylin & eosin, × 76.)

Prevalence of C. pylori in gastric ulceration

Of the 18 patients with benign gastric ulcer originally studied by Marshall & Warren [32], 14 (77%) were *C. pylori*-positive. Since then, several reports have documented the prevalence of the organism in gastric ulceration [12, 33–39] and the results are summarized in Table 16.8. The reported prevalence ranges from as low as 58% in the Finnish study by Niemala *et al.* [36] to 86% in the report from Shanghai by Jiang *et al.* [37]. On average, however, approximately 70% of gastric ulcer patients are colonized by *C. pylori*.

Relationship of ulceration to gastritis

Over the years, several histopathological studies have reported a strong association between gastritis and gastric ulceration [40–43]. The inflammatory changes are largely confined to the gastric antrum, irrespective of the site of the ulcer [42], and persist after ulcer healing [44]. Indeed, many authors hold the view that inflammatory changes precede ulceration and that gastritis is the fundamental change upon which other factors act to induce ulceration. The gastric mucosal barrier may be damaged in many ways and factors known to be associated with gastric ulceration include low socio-economic status [45], cigarette smoking [46], ingestion of aspirin and non-steroidal anti-inflammatory drugs [47], and duodeno-gastric reflux [48]. Both bile salts and lysolecithin are found in significantly greater quantity in the gastric aspirate of patients with gastric ulcer than in normal controls [48, 49].

The association between *C. pylori* and peptic ulcer disease is stronger for duodenal than for gastric ulceration, in that virtually all duodenal ulcer patients have *C. pylori*-positive antral gastritis whereas about 30% of gastric ulcer patients are consistently *C. pylori*-negative. To assess the significance of this finding in relation to the aetiology of gastric ulcer we undertook a study to evaluate possible differences between *C. pylori*-positive and *C. pylori*-negative gastric ulcers [35]. Fifty-four patients (mean age 59 years, range 36–78 years; 30 female) with untreated chronic benign gastric ulcer had at least two endoscopic biopsy specimens taken from uninvolved mucosa within 5 cm of the pylorus. The histological sections were examined 'blind' for the presence of *C. pylori* and assessed for severity of reflux gastritis, using the histological grading system described above. Thirty-nine patients (72%) (mean age 61 years, range 40–83; 21 female) were *C. pylori*-positive and 15 (mean age 56 years, range 36–78 years; 9 female) were negative. One patient in the negative group was taking oxytetracycline (to which *C. pylori* is sensitive [50]) at the time of the study and was therefore excluded from further analysis. Of the remaining 14 *C. pylori*-negative patients,

Table 16.8 Prevalence of *C. pylori* in gastric ulceration: previous studies

Reference	Patients studied (*n*)	*C. pylori*-positive *n*	%
Marshall & Warren (1984)[32]	18	14	77
McNulty & Watson (1984)[12]	8	5	63
Marshall *et al.* (1985)[33]	40	27	68
Buck *et al.* (1986)[34]	17	10	59
O'Connor *et al.* (1987)[35]	54	39	72
Niemala *et al.* (1987)[36]	33	19	58
Jiang *et al.* (1987)[37]	21	18	86
Feng & Wang (1988)[38]	28	24	86
Bedossa *et al.* (1988)[39]	71	50	70

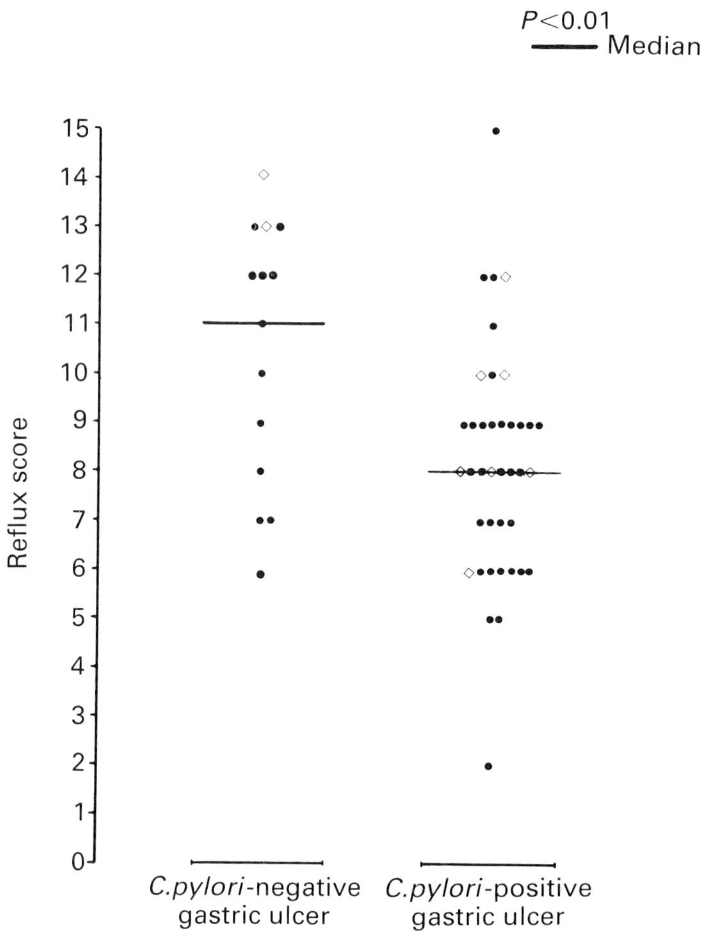

Fig. 16.7 Reflux scores in *C. pylori*-positive and *C. pylori*-negative gastric ulcer patients. ◇ = on non-steroid anti-inflammatory drugs.

8 had a reflux score > 10 compared with only 5 of the *C. pylori*-positive patients (χ^2, $P < 0.001$). Furthermore, *C. pylori*-negative patients had significantly higher reflux scores than positive patients ($P < 0.01$; Fig. 16.7). Seven patients in the *C. pylori*-positive group (18%) compared with two (13%) in the negative group were taking non-steroidal anti-inflammatory drugs at the time of the study ($P > 0.05$).

Thus, the features that we believe constitute the histological picture of reflux gastritis were more severe in *C. pylori*-negative than in *C. pylori*-positive patients, and colonization with the organism was uncommon in patients with a reflux score > 10. We speculate that the evolution of gastric ulcer in *C. pylori*-positive patients may be related in some way to *C. pylori*-induced gastric mucosal injury, whereas in *C. pylori*-negative subjects duodeno-gastric reflux may be an important contributory factor.

Acknowledgements

We are indebted to our colleagues Dr Mike Dixon, University Department of Pathology, General Infirmary, Leeds, and Dr Judith Wyatt, Department of Pathology, St James's University Hospital, Leeds, for their immense help throughout the conduct of these studies.

References

1 Ritchie, W.P. Alkaline reflux gastritis: a critical reappraisal. *Gut* 1984, **25**, 975–987.

2 Fausa, O. & Skalhegg, B.A. Quantitative determination of bile acids and their conjugates using thin layer chromatography and a purified α-hydroxysteroid dehydrogenase. *Scand. J. Gastroenterol.* 1974, **9**, 248–254.

3 Dixon, M.F., O'Connor, H.J. Axon, A.T.R., King, R.F.J.G. & Johnston, D. Reflux gastritis: distinct histopathological entity? *J. Clin. Pathol.* 1986, **39**, 524–530.

4 O'Connor, H.J., Wyatt, J.I., Dixon, M.F. & Axon, A.T.R. Campylobacter-like organisms and reflux gastritis. *J. Clin. Pathol.* 1986, **39**, 531–534.

5 Whitehead, R. 1985 *Mucosal Biopsy of the Gastrointestinal Tract*, 3rd edn. W.B. Saunders, Philadelphia.

6 Savage, A. & Jones, S. Histological appearances of the gastric mucosa 15–27 years after partial gastrectomy. *J. Clin. Pathol.* 1979, **32**, 179–186.

7 Geboes, K., Rutgeerts, P, Broeckaert, L., Vantrappen, G. & Desmet, V. Histologic appearances of endoscopic gastric mucosal biopsies 10–20 years after partial gastrectomy. *Ann. Surg.* 1980, **192**, 179–182.

8 Watt, P.C.H., Sloan, J.M. & Kennedy, T.L. Changes in gastric mucosa after vagotomy and gastrojejunostomy for duodenal ulcer. *Br. Med. J.* 1983, **287**, 1407–1410.

9 Thomas, W.E.G., Cooper, M.J., Mortenson, N.J.McC., Burton, P.A. & Davies, E.R. The clinical assessment of duodenogastric reflux by scintigraphy and its relation to histological changes in gastric mucosa. *Scand. J. Gastroenterol.* 1984, **19** (Suppl. 92), 195–199.

10 Langenberg, M.L., Tytgat, G.N.J., Schipper, M.E.I., Rietra, P.J.G.M. & Zanen, H.C. Campylobacter-like organisms in the stomach of patients and healthy individuals. *Lancet* 1984, **i**, 1348

11 Burnett, R.A., Forrest, J.A.H., Girdwood, R.W.A. & Fricker, C.R. Campylobacter-like organisms in the stomach of patients and healthy individuals. *Lancet* 1984, **i**, 1349.

12 McNulty, C.A.M. & Watson, D.M. Spiral bacteria of the gastric antrum. *Lancet* 1984, **i**, 1068–1069.

13 Jones, D.M., Lessells, A.M. & Eldridge, J. Campylobacter-like organisms on the gastric mucosa: culture, histological, and serological studies. *J. Clin. Pathol.* 1984, **37**, 1002–1006.

14 Buch, K.L., Weinstein, W.M., Hill, T.A., Eleshoff, J.D., Reedy, T.J., Tippoliti, A.F., Tedesco, F.J. & Singh, M. Sucralfate therapy in patients with symptoms of alkaline reflux gastritis. *Am. J. Med.* 1985, **79** (Suppl. 25), 49–54.

15 Nicolai, J.J., Speelman, P., Tytgat, G.N. & van der Stadt, J. Comparison of the combination of cholestyramine/alginate with placebo in the treatment of postgastrectomy biliary reflux gastritis. *Eur. J. Pharmacol.* 1981, **21**, 189–194.

16 Earlam, R. Bile reflux and the Roux-en-Y anastomosis. *Br. J. Surg.* 1983, **70**, 393–397.

17 Coghlan, J.G., Gilligan, D., Humphries, H., McKenna, D., Dooley, C., Sweeney, E., Keane, C. & O'Morain, C. *Campylobacter pylori* and recurrence of duodenal ulcers — a 12-month follow-up study. *Lancet* 1987, **ii**, 1109–1111.

18 Rokkas, T., Pursey, C., Uzoechina, E., Dorrington, L., Simmons, N.A., Filipe, M.I. & Sladen, G.E. *Campylobacter pylori* and non-ulcer dyspepsia. *Am. J. Gastroenterol.* 1987, **82**, 1149–1152.

19 McNulty, C.A.M., Gearty, J.C., Crump, B., Davis, M., Donovan, I.A., Melikian, V, Lister, D.M. & Wise, R. *Campylobacter pyloridis* and associated gastritis: investigator blind, placebo controlled trial of bismuth salicylate and erythromycin ethylsuccinate. *Br. Med. J.* 1986, **293**, 645–649.

20 O'Connor, H.J., Dixon, M.F., Wyatt, J.I., Axon, A.T.R., Ward, D.C., Dewar, E.P. & Johnston, D. Effect of duodenal ulcer surgery and enterogastric reflux on *Campylobacter pyloridis*. *Lancet* 1986, **ii**, 1178–1181.

21 Ward, D.C., Johnston, D., Belton, I.P. & Barker, M.C.J. Duodenogastric reflux decreases significantly after highly selective vagotomy. *Gut* 1986, **27**, A610–A611.

22 Keighley, M.R.B., Asquith, P. & Alexander-Williams, J. Duodenogastric reflux: a cause of gastric mucosal hyperaemia and symptoms after operations for peptic ulceration. *Gut* 1975, **16**, 28–32.

23 Hoare, A.M., McLeish, A., Thompson, H. & Alexander-Williams, J. Selection of patients for bile diversion surgery: use of bile acid measurement in fasting gastric aspirates. *Gut* 1978, **19**, 163–165.

24 Malagelada, J.R., Phillips, S.F., Higgins, J.A., Shorter, R.G., van Heerden, J.A. & Adson, M.A. A prospective evaluation of alkaline reflux gastritis: bile acid binding agents and Roux-en-Y diversion. *Gastroenterology* 1979, **76**, 1192.

25 Dewar, E.P., Dixon, M.F. & Johnston, D. Bile reflux and degree of gastritis after highly selective vagotomy, truncal vagotomy, and partial gastrectomy for duodenal ulcer. *World J. Surg.* 1983, **7**, 743–750.

26 Donovan, I.A., Fielding, J.W.L., Bradby, H., Sorgi, M. & Harding, L.K. Bile diversion after total gastrectomy. *Br. J. Surg.* 1982, **69**, 389–390.

27 Hoare, A.M., Keighley, M.R.B., Starkey, B. & Alexander-Williams, J. Measurement of bile acids in fasting gastric aspirates: an objective test for bile reflux after gastric surgery. *Gut* 1978, **19**, 166–169.

28 Malagelada, J.R., Phillips, S.F., Shorter, R.G., Higgins, J.A., Magrina, C., van Heerden, J.A. & Adson, MA. Postoperative reflux gastritis: pathophysiology and long-term outcome after Roux-en-Y diversion. *Ann. Intern. Med.* 1985, **103**, 178–183.

29 Schoon, I.M., Anderson, H., Faxen, A., & Olbe, L. Gastric bile acids before and after Roux-en-Y transposition for bile reflux gastritis and in asymptomatic controls. *Scand. J. Gastroenterol.* 1979, **14**, 969–974.

30 Mosimann, R., Loup, P., Fontolliet, C. & Mosimann, F. Postoperative reflux gastritis: results of surgical treatment. *Scand. J. Gastroenterol.* 1981, **16** (Suppl. 67), 237–239.

31 Langenberg, W., Rauws, E.A.J., Widjojokusomo, A, Tytgat, G.N.J. & Zanen, H.C. Identification of *Campylobacter pyloridis* isolates by restriction endonuclease DNA analysis. *J. Clin. Microbiol.* 1986, **24**, 414–417.

32 Marshall, B.J. & Warren, J.R. Unidentified curved bacilli in the stomach of patients with gastritis and peptic ulceration. *Lancet* 1984, **i**, 1311–1314.

33 Marshall, B.J., McGechie, D.B., Rogers, P.A. & Glancy, R.J. Pyloric *Campylobacter* infection and gastroduodenal disease. *Med. J. Australia* 1985, **142**, 439–444.

34 Buck, G.E., Gourley, W.K., Lee, W.K., Subramanyan, K., Latimer, J.M. & DiNuzzo, A.R. Relation of *Campylobacter pyloridis* to gastritis and peptic ulcer. *J. Infect. Dis.* 1986, **153**, 664–669.

35 O'Connor, H.J., Dixon, M.F., Wyatt, J.I., Axon, A.T.R., Dewar, E.P. & Johnston, D. *Campylobacter pylori* and peptic ulcer disease. *Lancet* 1987, **ii**, 633–634.

36 Niemala, S., Karttunen, T. & Lehtola, J. Campylobacter-like organisms in patients with gastric ulcer. *Scand. J. Gastroenterol.* 1987, **22**, 487–490.

37 Jiang, S.J., Lui, W.Z., Zhang, D.Z., Shi, Y., Xiao, S.D., Zhang, Z.H. & Lu, D.Y. Campylobacter-like organisms in chronic gastritis, peptic ulcer, and gastric carcinoma. *Scand. J. Gastroenterol.* 1987, **22**, 553–558.

38 Feng, Y-Y. & Wang, Y. *Campylobacter pylori* in patients with gastritis, peptic ulcer, and carcinoma of the stomach in Lanzhou, China. *Lancet* 1988, **i**, 1055–1056.

39 Bedossa, P., Poynard, T., Chaput, J-C. & Martin, E. A decade of *Campylobacter pylori*. *Lancet* 1988, **i**, 417–418.

40 Meyers, W.C. A study of gastric mucosa in various diseases affecting the upper part of the gastrointestinal tract. *Gastroenterology* 1948, **10**, 923–938.

41 Magnus, H.A. The pathology of peptic ulceration. *Postgrad. Med. J.* 1954, **30**, 131–136.

42 Schrager, J., Spink, R. & Mitra, S. The antrum in patients with duodenal and gastric ulcers. *Gut* 1967, **8**, 497–508.

43 Gear, M.W.L., Truelove, S.C. & Whitehead, R. Gastric ulcer and gastritis. *Gut* 1971, **12**, 639–645.

44 McIntyre, R.L.E. Piris, J. & Truelove, S.C. Effect of cimetidine on chronic gastritis in gastric ulcer patients. *Aust. N.Z. J. Med.* 1982, **12**, 106.

45 Langman, M.J.S. 1979. *The Epidemiology of Chronic Digestive Disease*. Edward Arnold, London.

46 Paffenberger, R.S., Wing, A.L. & Hyde, RT. Chronic disease in former college students. *Am. J. Epidemiol.* 1974, **100**, 307–315.

47 Duggan, J.M., Dobson, A.J., Johnson, H. & Fahey, P. Peptic ulcer and non-steroidal anti-inflammatory agents. *Gut* 1986, **27**, 929–933.

48 Rhodes, J., Bernardo, D.E., Phillips, S.F., Rovelstad, R.A. & Hofmann, A.F. Increased reflux of bile into the stomach in patients with gastric ulcer. *Gastroenterology* 1969, **57**, 241–252.

49 Johnson, A.G. & McDermott, S.J. Lysolecithin: a factor in the pathogenesis of gastric ulceration? *Gut* 1974, **15**, 710–713.

50 Lambert, J.R., Hansky, J., Davidson, A., Pinkard, K. & Stockman, K. Campylobacter-like organisms (CLO) — *in vivo* and *in vitro* susceptibility to antimicrobial and antiulcer therapy. *Gastroenterology* 1985, **88**, 1462.

17 Campylobacter pylori and gastric cancer

P. CORREA AND B. RUIZ

Multidisciplinary research carried out over several decades by teams of investigators in many countries has led to the hypothesis that most gastric cancers in humans are the end result of events related to multiple consecutive regressive phenotypic changes in the gastric epithelium [1]. In most human situations, many years elapse between the initial stages of the process and the final invasion and spread of the tumour.

Most of the information available on this pre-cancerous process was obtained before *Campylobacter pylori* was recognized as a factor in gastric disease. Only recently have attempts to understand the relationship between *C. pylori* and the pre-cancerous process been reported [2]. This relationship is explored in this chapter.

Types of gastric cancer and their precursors

Biologically, gastric carcinoma is not a homogeneous entity and considerable differences in the aetiological forces involved have been identified for the several types so far recognized. For one of the types, the so-called 'diffuse' or 'infiltrative' type of gastric carcinoma, our knowledge of the precursor stages is so meager as to make it impossible to describe them in any meaningful way [3–4]. For three other types of gastric cancer specific pre-cancerous lesions have been characterized.

Gastric carcinoma associated with the pernicious anaemia syndrome is preceded by diffuse corporal (Type A) atrophic gastritis [5]. Such gastritis appears unrelated to *C. pylori.*

In patients with previous gastrectomy the so-called 'stump carcinoma' is observed. Characteristically, these patients manifest 'reflux' gastritis, again apparently unrelated to *C. pylori* [6]. The precursor of stump carcinoma is the so-called 'gastritis cystica polyposa', present only at the anastomosis site itself and not observed away from it, where reflux gastritis is ubiquitous [7]. There are no grounds to implicate *C. pylori* in the pathogenesis of this disease complex.

In the region of the gastro-oesophageal junction, adenocarcinomas arise from either the cardiac gastric glands or the glands of Barrett's oesophagus. In some cases the exact site of origin is obscure. The precursors of carcinoma of the gastric cardiac glands have not been characterized. The precursors of carcinoma arising in Barrett's oesophagus have been fairly well characterized as metaplastic and dysplastic changes essentially identical to those described below for the 'epidemic' or 'intestinal' type of carcinoma. It should be mentioned at this point that both the cardiac region of the stomach and the epithelium in Barrett's oesophagus are frequently colonized by *C. pylori.*

In most human populations displaying high gastric cancer risk, the types of gastric cancer mentioned above are the least frequent. Most gastric carcinomas are morphologically of the 'intestinal' or 'expansive' type and have epidemiological characteristics which have made them deserve the label 'epidemic' type. Such carcinomas make their appearance in the midst of a precursor lesion called 'multifocal chronic atrophic gastritis' (MAG), sometimes also called 'environmental' or 'Type B' [5].

Multifocal chronic atrophic gastritis (MAG) as a cancer precursor

MAG as a nosologic entity covers a broad morphological spectrum as it involves progressively more extensive areas of the gastric mucosa. It has a characteristic topographic distribution and pattern of progression: in younger individuals, it appears around the area of the incisura angularis, the lesser curvature and the corpus–antrum junction as independent foci which coalesce and extend to the antrum and corpus. Atrophy of glands appears preceded by chronic inflammatory infiltrate in the superficial portion of the lamina propria and around groups of gastric glands, especially on the antrum. Focal acute inflammation (polymorphonuclear infiltrate) is occasionally seen in and around the epithelium of the foveolar and the gland necks.

Intestinal-type epithelium replaces the glands lost in the atrophic process, first resembling small intestine (Type I intestinal metaplasia) and later resembling colonic epithelium (Types II and III). In more advanced lesions mature intestinal phenotypes are replaced by immature types, which display the histological pattern of dysplasia, a lesion considered pre-cancerous [8–10].

Epidemiological and experimental studies have identified aetiological factors which play a role in the pre-cancerous process, either initiating and promoting its progress or retarding and blocking it. Factors which inhibit the pre-cancerous process are mainly antioxidants such as vitamin C, vitamin E and beta-carotene. Factors which aggravate the process are mostly irritants, such as excessive salt in foods [5]. These irritants not only induce gastric inflammation but also potentiate the effect of gastric carcinogens such as MNNG [11] by some poorly-understood mechanism. The pre-cancerous role of gastric irritants is substantiated by abundant epidemiological experimental literature [5].

The role of Campylobacter in MAG

The association between *C. pylori* and chronic gastritis has been reported in many countries but frequently the heterogeneity of chronic gastritis is not recognized and no indication is given as to the type of gastritis being reported. Table 17.1 shows the prevalence of *C. pylori* reported in case series specifically mentioning atrophic gastritis.

Table 17.1 Reported prevalence of *C. pylori* in atrophic gastritis

Reference	Number of cases (*n*)	Percentage
Fiocca *et al.* (1987)[12]	38	89.5
Le Bodic *et al.* (1987)[13]	12	75.0
Barbosa *et al.* (1988)[14]	7	71.4
Booth *et al.* (1986)[15]	34	70.6
Menge *et al.* (1987)[16]	31	67.7
Andersen *et al.* (1987)[17]	17	64.7
Price *et al.* (1985)[18]	13	61.5
Jones *et al.* (1986)[19]	17	52.9
Chen *et al.* (1986)[20]	30	46.7
Sethi *et al.* (1987)[21]	23	43.5

The reports show clearly that *C. pylori* is frequently associated with atrophic gastritis. It is well known that *C. pylori* does not colonize metaplastic epithelium, as shown in Fig. 17.1 in which the bacteria are very abundant around foveolar epithelial cells but are notoriously absent in the immediate vicinity of cells of intestinal phenotype.

Epithelial cell damage is associated with *C. pylori* colonization, as seen in Fig. 17.2 in which a marked decrease of cytoplasmic mucin is seen where colonization is severe. A certain degree of irregular polarity is also observed in the nuclei (Fig. 17.2). Cell damage is followed by cell repair, as seen in Fig. 17.3 in which the presence of the bacteria is associated with increase in size and number of nuclei as well as their displacement to upper portions of the cytoplasm. From epidemiological and histopathological evidence, it is therefore clear that *C. pylori* is very much part of the chronic atrophic gastritis spectrum.

MAG correlates very closely with the risk of gastric cancer: its prevalence is higher in populations at high cancer risk. We recently studied two series of gastric biopsies from subjects of high risk communities [22]. In Nariño,

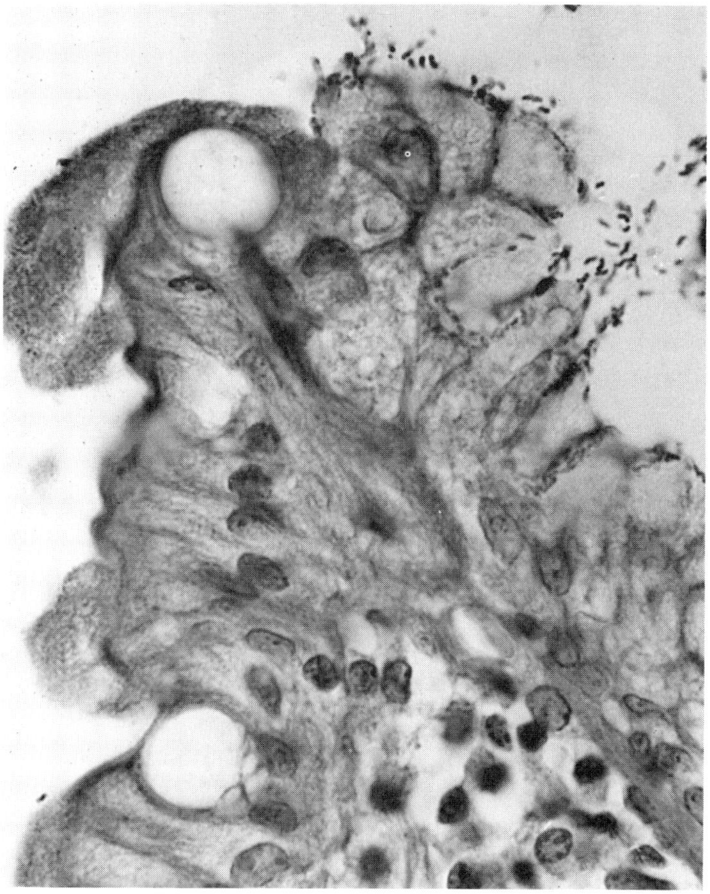

Fig. 17.1 Microphotograph of the area around the limit between a focus of intestinal metaplasia (left) and gastric mucosa lined by foveolar cells (right). Abundant *Campylobacter* organisms are seen on and between foveolar cells but not near the goblet cells or enterocytes of the metaplastic epithelium. (Warthin–Starry stain, × 800.)

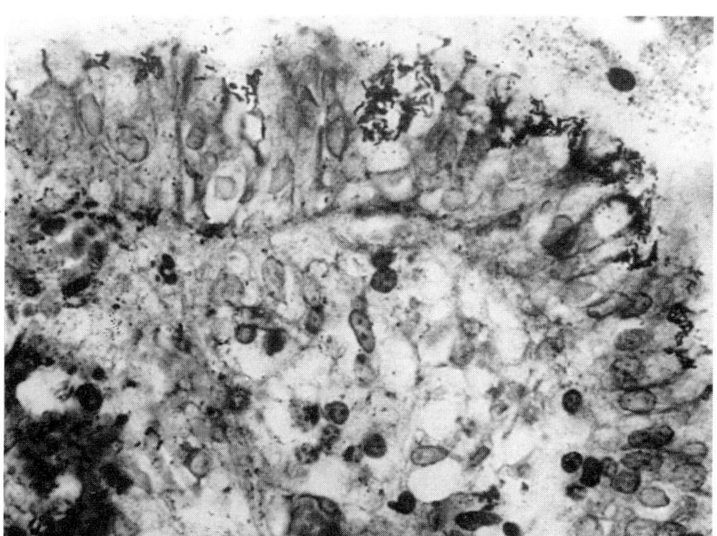

Fig. 17.2 Microphotograph of the gastric mucosa with acute and chronic inflammation associated with marked colonization by *C. pylori*. The bacteria are between foveolar cells which have lost much of their normal cytoplasmic mucin. The nuclei are somewhat enlarged and irregular in location and orientation. (Warthin–Starry stain, × 454.)

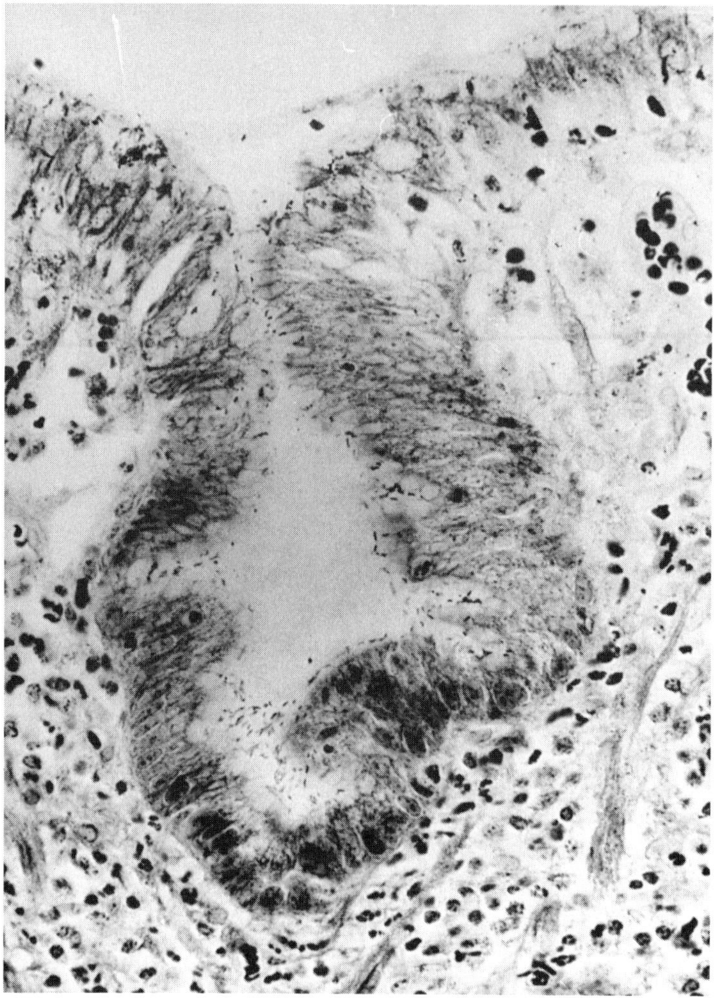

Fig. 17.3 Microphotograph of gastric foveolar with acute and chronic inflammation associated with *C. pylori* infection. Abundant organisms are seen in the foveolar lumen in the immediate vicinity of the epithelial cells. Cellular hyperplasia is manifested by increase in size and number of the nuclei of the foveolar cells. (Warthin–Starry stain, × 400.)

Colombia, the risk is extremely high (over 100/100 000 per year) and MAG is the predominant type of gastritis. In the New Orleans' black community, the risk is lower, but still relatively high (around 20/100 000) compared to the risk in whites (under 10/100 000), MAG is also frequent, followed by diffuse antral gastritis.

The prevalence of *C. pylori* in these populations as evaluated by the Warthin–Starry stain is seen in Table 17.2. In the MAG complex (chronic atrophic gastritis and intestinal metaplasia), the prevalence is around 94% in Colombia, compared to 64% in New Orleans. The higher rate in Colombia is significant, when compared with New Orleans. It should also be noted that, although small in number, all cases of dysplasia were colonized by *C. pylori* in the areas of remaining gastric foveolar epithelium.

In populations of low gastric cancer risk, MAG is very rare or non-existent. In such populations diffuse antral gastritis is usually predominant. This type of gastritis, however, is not known to be associated with increased gastric cancer risk. *Campylobacter pylori* is known to play a dominant role in diffuse antral gastritis and infection is not therefore a

Table 17.2 *C. pylori* in gastric biopsy material in two locations. Values represent the percentage of *C. pylori*-positive biopsies to the total biopsies in each case

Biopsy	Colombia (*N* = 147)		New Orleans (*N* = 173)	
Normal	0/1	0.0%	0/15	0.0%
Superficial gastritis	33/38	87.0%	12/14	85.7%
Diffuse antral gastritis	8/8	100.0%	29/34	85.3%
Chronic atrophic gastritis	64/66	97.0%	39/57	68.4%
Intestinal metaplasia	26/28	93.0%	30/51	58.8%
Dysplasia	6/6	100.0%	2/2	100.0%

cause of gastric cancer in isolation. Its role, if any, seems dependent on the previous existence of MAG.

Campylobacter and dysplasia

As shown in Table 17.2, *C. pylori* is very frequently associated with gastric dysplasia. Preliminary results of treatment of these patients with bismuth salts have reported dramatic histological and clinical improvement and drastic reduction in cellular regeneration, atypia and inflammatory infiltrate [23]. Although these results need confirmation, it appears that anti-*Campylobacter* therapy may play a role in reducing the intensity of, and perhaps also reverting, dysplastic changes.

Campylobacter and gastric carcinoma

There have been some reports of association of *C. pylori* with gastric carcinoma, as shown in Table 17.3.

It is well known that *C. pylori* does not colonize cancer cells and it follows that the prevalence in cancer cases represents colonization in the remaining foveolar epithelium of the stomach. In a way these findings reinforce the observation that non-cancerous mucosa is frequently colonized by *C. pylori*.

We recently made a special search for campylobacter-like organisms utilizing Warthin–Starry and Gimenez stains in 20 gastrectomy specimens with gastric carcinoma. In 8, campylobacter-like organisms were identified but the negative results may reflect bacterial

Table 17.3 *Campylobacter* prevalence in gastric carcinoma

Reference	Method*	Number of cases (*n*)	Percentage positive
Cheng *et al.* (1987)[24]	ELISA	19	73.7
Marshall *et al.* (1985)[25]	Gram, culture	5	80.0
Cortes *et al.* (1987)[26]	W.S., Giemsa	14	71.4
Gilman *et al.* (1987)[27]	W.S.	20	65.0
Lambert *et al.* (1985)[28]	W.S.	24	62.5
Rathbone & Wyatt (1988)[2]	Giemsa	52	46.1
Von Wolffen *et al.* (1988) [29]	Gram, culture	13	40.0
Jiang *et al.* (1987) [30]	Gram, culture	28	25.0
Feng & Wang (1988) [31]	W.S., culture	19	21.0
Rollanson *et al.* (1984) [32]	H.E., W.S.	?	19.0

*W.S. = Warthin–Starry; H.E. = haematoxylin & eosin.

autolysis because such specimens are not fixed immediately after resection. This may explain some of the discordant results observed in Table 17.3.

Campylobacter and the pre-cancerous process

Agents involved in the aetiology of cancer are customarily classified mainly as either initiators or promoters. Initiators are genotoxic agents which may induce damage to DNA on short-term exposure and which can be complete carcinogens when administered in multiple doses through a prolonged period. *Campylobacter pylori* is present for many years in patients with chronic diffuse antral gastritis and probably plays a causal role in the disease. Since the condition has not been shown to increase the risk of gastric cancer, it follows that *C. pylori* cannot be assigned an initiator or genotoxic role.

Promoters, on the other hand, effectively increase the risk of cancer only when applied repeatedly to cells which have been previously exposed to an initiator. Promoters may act by increasing cell replication, thereby making cells more vulnerable to the genotoxic effects of initiators, which require cell replication for their interaction with DNA. That type of role has been documented for other gastric irritants such as excessive salt [33]. As shown in Figs. 17.2 and 17.3, glands in which *C. pylori* are present display increased nuclear activity. As an irritant of the gastric mucosa and as an inducer of excessive cell replication, it is plausible to assume that *C. pylori* may play the role of a cancer promoter. In the setting of diffuse antral gastritis such promoting effect is not evident, presumably because of lack of initiators. In the setting of multifocal atrophic gastritis such irritation and excessive replication may result in promotion of the carcinogenesis process, presumably because genotoxic agents are also present.

Acknowledgement

This work was supported by Grant #R01-AI-25590 of the National Institutes of Health.

References

1 Correa, P. A human model of gastric carcinogenesis. *Cancer Res.* 1988, **48**, 3554–3560.

2 Rathbone, B. & Wyatt, J. 1988. *Campylobacter pylori* and precancerous lesions. In Reed, P.I. & Hill, M.J. (eds.) *Gastric Carcinogenesis*, Excerpta Medica, Amsterdam, pp. 132–144.

3 Lauren, P. The two main types of gastric carcinoma: diffuse and so-called intestinal type carcinoma. An attempt at a histoclinical classification. *Acta Path. Microbiol. Scand.* 1965, **64**, 31–49.

4 Ming, S.C. Gastric carcinoma. A pathologic classification. *Cancer* 1977, **39**, 2475–2485.

5 Correa, P. Chronic gastritis: A clinico-pathologic classification. *Am. J. Gastroenterol.* 1988, **83**, 504–509.

6 O'Connor, H.J., Wyatt, J.I., Dixon, M.F. *et al.* Campylobacter like organisms and reflux gastritis. *J. Clin. Pathol.* 1986, **39**, 531–534.

7 Littler, E.R. & Gleibermann, E. Gastritis cystica polyposa: gastric mucosal polyposa at gastroenteroctomy site, with cysts and infiltrative hyperplasia. *Cancer* 1972, **29**, 205–209.

8 Heilman, K.K. & Hopker, W.W. Loss of differentiation in intestinal metaplasia in cancerous stomachs. A comparative study. *Pathol. Res. Pract.* 1979, **164**, 249–258.

9 Silva, S. & Filipe, M.I. Intestinal metaplasia and its variants in gastric carcinoma of Portuguese subjects: a comparative analysis of biopsy and gastrectomy material. *Hum. Pathol.* 1986, **17**, 988–995.

10 Cuello, C., Correa, P., Zarama, G. *et al.* Histopathology of gastric dysplasias. *Am. J. Surg. Pathol.* 1979, **3**, 491–500.

11 Takahashi, M., Kokuho, T., Furukawa, F. *et al.* Effect of high salt diet on rat gastric carcinogenesis induced by MNNG. *Gann* 1983, **74**, 28–34.

12 Fiocca, R., Villani, L., Turpini, R. & Salcia, E. High incidence of campylobacter-like organisms in endoscopic biopsies from patients with gastritis with or without peptic ulcer. *Digestion* 1987, **38**, 234–244.

13 Le Bodic, M.F., Barre, P. & Freland, C. *Campylobacter pylori* et muqueuse gastrique: Étude histologique, bacteriologique et résultats préliminaires d'une enquête epidemiologique dans la région nantaise. *Gastroenterol. Clin. Biol.* 1987, **11**, 543–549.

14 Barbosa, A.J., Queiros, D.M., Mendes, E.N. *et al.* Im-

munocytochemical identification of *Campylobacter pylori* in gastritis and correlation with culture. *Arch. Pathol. Lab. Med.* 1988, **112**, 523–525.

15 Booth, L., Holdstock, G., MacBide, H. *et al.* Clinical importance of *Campylobacter pyloridis* and associated serum IgG and IgA antibody responses in patients undergoing upper gastrointestinal endoscopy. *J. Clin. Pathol.* 1986, **39**, 215–219.

16 Menge, H., Warrehman, M., Joy, V. *et al. Campylobacter pylori* in Magen, Duodenum und Kolon gastroenterologischer Patienten. *Dtsch med. Wochenschr.* 1987, **112**, 1403–1407.

17 Andersen, L.P., Holck, S., Povlsen, C.O. *et al. Campylobacter pyloridis* in peptic ulcer disease. I. Gastric and duodenal infection caused by *C. pyloridis*: Histopathologic and microbiologic findings. *Scand. J. Gastroenterol.* 1987, **22**, 219–224.

18 Price, A.B., Levi, J., Dolby, J.M. *et al. Campylobacter pyloridis* in peptic ulcer disease: microbiology, pathology and scanning electron microscopy. *Gut* 1985, **26**, 1183–1188.

19 Jones, D.M., Eldridge, J., Fox, A.J., Sethi, P. & Whorwell, P.J. Antibody to the gastric campylobacter-like organisms (*Campylobacter pyloridis*). Clinical correlations and distribution in the normal population. *J. Med. Microbiol.* 1986, **22**, 57–62.

20 Chen, X.C., Correa, P., Offerhaus, J. *et al.* Ultrastructure of the gastric mucosa harboring campylobacter-like organisms. *Am. J. Clin. Pathol.* 1986, **86**, 575–582.

21 Sethi, P., Banerjee, A.K., Jones, D.M. *et al.* Gastritis and gastric campylobacter-like organisms in patients without peptic ulcers. *Postgrad. Med. J.* 1987, **63**, 543–545.

22 Correa, P., Muñoz, N., Cuello, C. *et al.* The role of *Campylobacter pylori* in gastroduodenal disease. *Prog. Surg. Pathol.*, in press.

23 Leon-Baura, R., Recavarren, S., Ramirez, A. *et al.* Reversal of gastric mucosal dysplasia associated with *Campylobacter pylori* using oral bismuth therapy. *Gastroenterology*, 1988, **94**, A256.

24 Cheng, S.C.W., Sanderson, C.R., Waters, T.E. *et al. Campylobacter pyloridis* in patients with gastric carcinoma. *Med. J. Australia* 1987, **147**, 202–203. (Letter)

25 Marshall, B.J., McGechie, D.B., Rogers, P.A. & Glancy, R.J. *Campylobacter* infection and gastroduodenal disease. *Med. J. Australia* 1985, **142**, 439–444.

26 Cortes, A., Carmona, F., Carrascal, E. *et al. Campylobacter pyloris* en biopsia de mucosa gastrica. *Colombia Medica* 1987, **18**, 101–109.

27 Gilman, R.H., Leon-Baura, R., Ramirez-Ramos, A. *et al. Campylobacter pyloridis* fails to colonize sites of adenocarcinoma but not adjacent non-cancerous tissue in patients with gastric cancer. *Gastroenterology*, 1987, **92**, 1406. (Abstract)

28 Lambert, J.R., Hamsky, J., Eaves, E.K. *et al.* Campylobacter-like organisms in human stomach. *Gastroenterology* 1985, **88**, 1463. (Abstract)

29 Von Wulffen, H., Grote, H.J., Gatermann, S. *et al.* Immunoblot analysis of immune response to *Campylobacter pylori* and its clinical associations. *J. Clin. Pathol.* 1988, **41**, 653–659.

30 Jiang, S.J., Liu, W.Z., Zhang, D.Z. *et al.* Campylobacter-like organisms in chronic gastritis, peptic ulcer and gastric carcinoma. *Scand. J. Gastroenterol.* 1987, **22**, 553–558.

31 Feng, Y. & Wang, Y. *Campylobacter pylori* in patients with gastritis, peptic ulcer and carcinoma of the stomach in Lanzhou, China. *Lancet*, 1988, **i**, 1055. (Letter)

32 Rollanson, T.P., Stone, J. & Rhodes, J.M. Spiral organisms in endoscopic biopsies of the human stomach. *J. Clin. Pathol.* 1984, **37**, 23–26.

33 Charnley, G. & Tannenbaum, S.R. A flow cytometric analysis of the effect of sodium chloride on gastric cancer risk in the rat. *Cancer Res.* 1985, **45**, 5608–5616.

18 Ultrastructure of Campylobacter pylori in vivo

H.W. STEER

The microenvironment of the bacterium *Campylobacter pylori* is unique, displaying a number of unusual features necessitating the conclusion that only a highly specialized bacterium could survive at such a site. Electron microscopy has helped to define this microenvironment aiding in determining the characteristics of the bacterium and its role in any pathological process. To examine the ultrastructural features of this microenvironment, the *in vivo* appearance of *C. pylori* will be considered prior to its relationship with its surroundings.

In vivo appearance of C. pylori

The shape of *C. pylori in vivo* can vary from spiral to kidney-shaped to a coccal form. Spiral and kidney-shaped forms have been observed with both the transmission and scanning electron microscopes [1, 2], confirming previous observations made with the light microscope. It has not been possible to define the coccal shape with transmission electron microscopy because of possible cross-cutting of the bacteria but this form of bacterium has been observed with the scanning electron microscope [3]. Evidence is lacking about any possible pattern of distribution of the various forms of *C. pylori* or any relationship between the different forms. The coccal form is seen in older bacteriological cultures and has been considered degenerate, although this concept may have to be reconsidered in the light of recent evidence that the coccal form is capable of regrowth.

The size of *C. pylori* has been reported to vary in patients with chronic duodenal ulceration depending upon the site at which the bacterium is located. Thus, *C. pylori* are larger in antral biopsies compared with those biopsies taken from the edge of the duodenal ulcer [3].

Campylobacter pylori have 4–6 unipolar sheathed flagella which have terminal bulbs [4]. The flagella are attached to the bacterial body by a spherical body or basal granule. The terminal bulbs are infrequently observed *in vivo* because of their diameter but occasional terminal bulbs, flagella and basal granules are seen (Figs. 18.1 and 18.2). There are two types of terminal bulbs seen. The larger terminal bulbs (Fig. 18.1) have a diameter of 0.28 μm with a sheathed flagellum, which has an overall diameter of only 67 nm and a basal granule diameter of only 50 nm. The substructure of the larger terminal bulb includes an organized array of microtubules (seen in cross-section in Fig. 18.1) with a diameter of 17 nm. The precise arrangement of these microtubular structures cannot be determined from the *in vivo* studies but in cross-section they are separated by a space of 20–25 nm. It appears that some of these microtubules open onto the surface of the terminal bulb. This profile of the terminal bulb is larger than that described by Goodwin *et al.* [4]. The substructure of this smaller terminal bulb (Fig. 18.2) differs from that already described and this type of terminal bulb has a diameter of only 58 nm. The flagella attached to these smaller terminal bulbs have a diameter of 25 nm. These differences in the flagella and terminal bulbs may represent a dimorphism as far as the bacteria are concerned and may reflect different dynamic states of the bacteria.

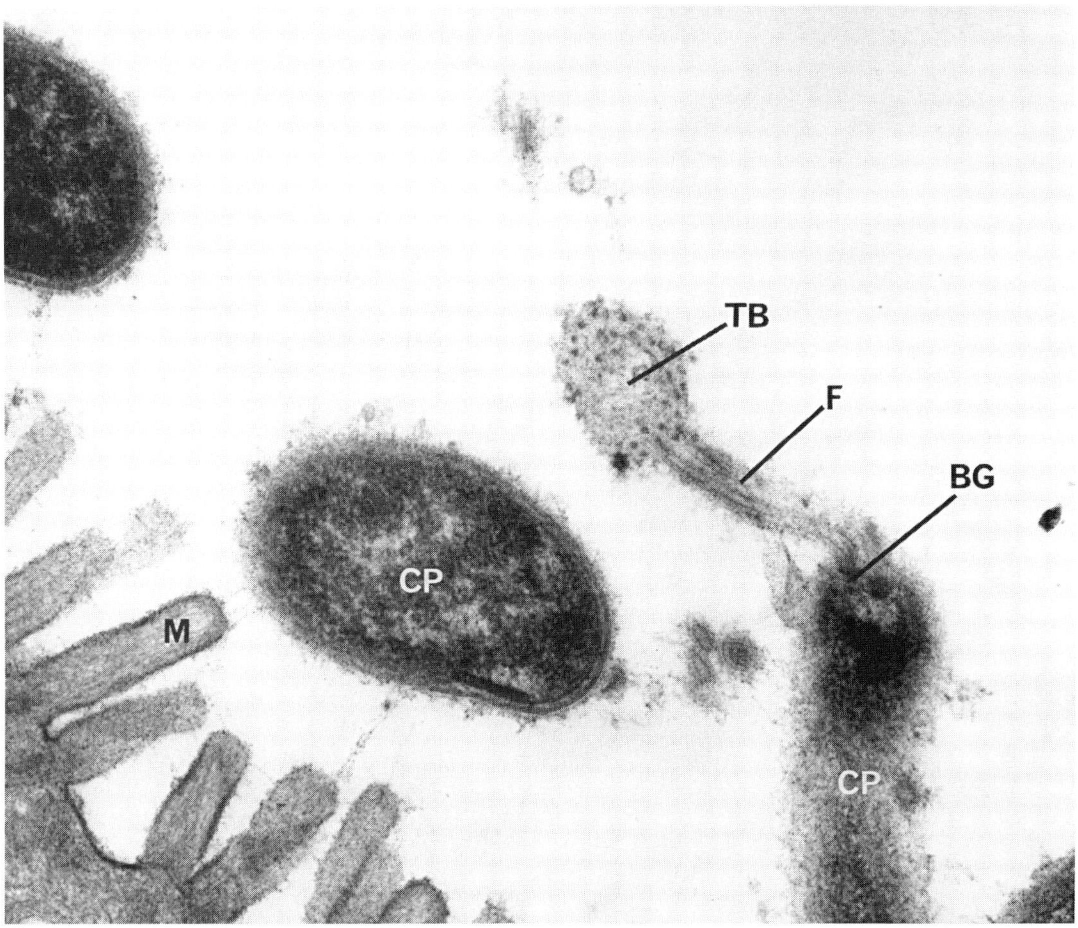

Fig. 18.1 *C. pylori* (CP) are related to the microvillous surface (M) of the gastric epithelium. A terminal bulb (TB) at the end of a flagellum (F) is seen together with two basal granules (BG). (Transmission electron micrograph, ×45 600.)

Campylobacter pylori and its environment

The localization of *C. pylori* is limited to cells derived from the gastric-type mucosa [1–5]. They are never related to intestinal-type mucosal cells. The bacteria are most frequently found deep to the layer of mucus overlying the epithelial cells and are therefore not apparent unless this layer of mucus is removed [1]. Some *C. pylori* are present in the overlying gastric mucus and at this site tend to be aggregated in 'cave-like' structures [3]. Some of these bacteria have been identified in the gastric juice of patients when the appropriate bacteriological analyses have been performed.

The distribution of the epithelial-related bacteria is 'patchy' with frequent clusters of bacteria apparent [6]. *Campylobacter pylori* are most frequently found near the neck of gastric mucosa although bacteria have been seen in the depths of gastric glands and even in the canaliculi of parietal cells [7]. The *C. pylori* in the duodenum are always related to areas of gastric metaplasia. The reasons for this specificity of mucosal relationship includes the differences in the physico-chemical properties of the mucus at the sites of intestinal metaplasia where the goblet cells produce an acidic mucus [8]. Differences in the antigenic recog-

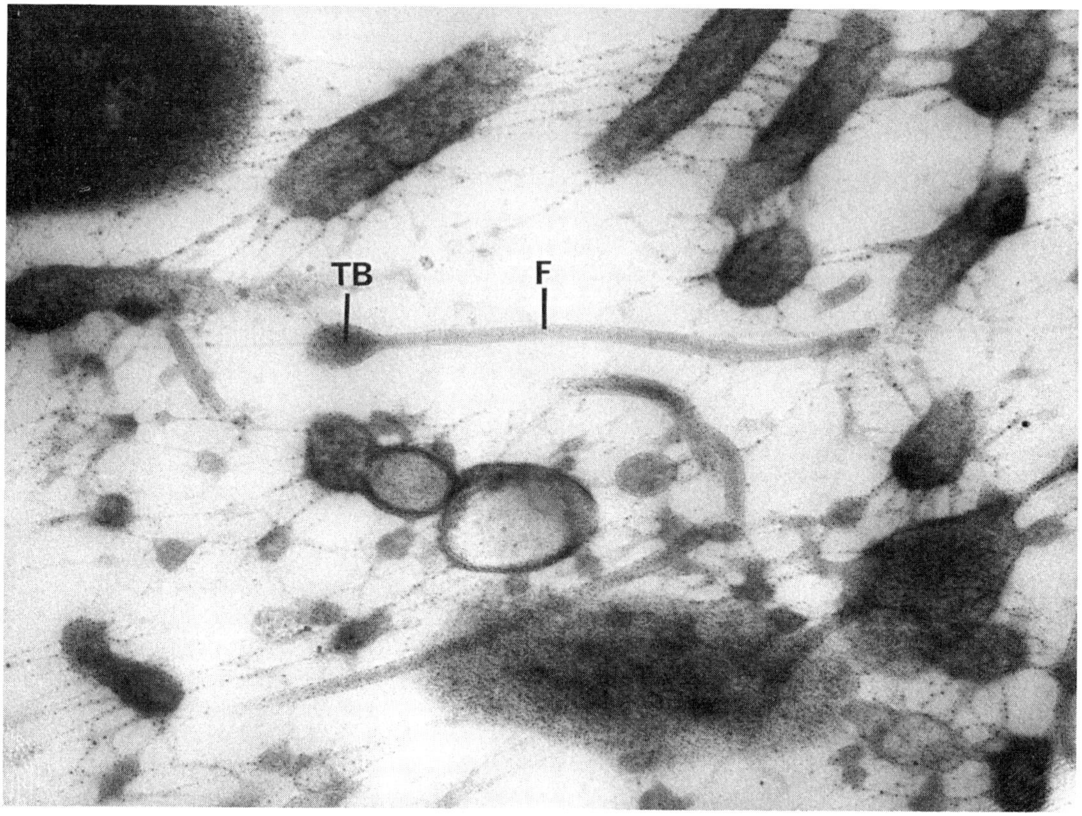

Fig. 18.2 The flagellum (F) and an attacked terminal bulb (TB) from a *C. pylori*. (Transmission electron micrograph, × 73 000.)

nition and expression of the two types of cell surface could influence bacterial adhesion but if this is a criterion for their absence at areas of intestinal metaplasia, it would be reasonable to expect to find non-adherent *C. pylori* at these sites. However, even non-adherent *C. pylori* are not found at the areas of intestinal metaplasia, making this possibility untenable.

Campylobacter pylori can be related to any part of the luminal surface of gastric-mucus secreting cells but are most frequently sited in the 'grooves' at the junction of individual epithelial cells (Figs. 18.3 and 18.4). The epithelial cells related to bacteria have a decreased number of surface microvilli [1, 3, 4]. The siting of the bacteria in the 'grooves' between neighbouring epithelial cells results in their close proximity to the intercellular junctions [1, 3, 4, 9]. The localization of the bacteria at these sites, seen most clearly with the scanning electron microscope, is unlikely to be artefactual resulting from removal of the overlying mucus, because such a localization has been confirmed with the transmission electron microscope. This siting must confer benefits for individual bacteria and may enable the bacteria to avoid the principal flow of mucus from the cell surface. Mucus is secreted from the gastric mucus-secreting cells by a process of merocrine secretion with the mucus passing from the apex of individual cells into the overlying mucus layer. Bacteria located at the apex of actively secreting mucus cells would be continually forced into the mucus layer in transit to other hostile parts of the gastro-intestinal tract. This may

Fig. 18.3 The gastric epithelial surface in a normal stomach showing individual epithelial cells (EP) covered with microvilli. The 'grooves' (G) between the surface of individual cells is clearly seen. (Scanning electron micrograph, × 5936.)

account for the aggregates of bacteria present in the overlying mucus [3] and those *C. pylori* isolated from the gastric juice. It is possible that large numbers of bacteria are lost in this way, so that those bacteria observed in the 'grooves' at the epithelial surface would be more appropriately termed the 'resident' bacterial population.

Most *C. pylori* are found between the epithelial cell surface and the overlying mucus barrier. The reason for this localization may be related to the pH gradient across the gastric mucus barrier [10]. At the epithelial cell surface many *C. pylori* are adherent to the epithelial surface. This adherence is a characteristic of bacterial pathogenicity and can be so intimate that they are only separated by a distance of 10 nm. Adhesion may take one of three forms. Most frequently the bacterial surface and the epithelial cell surface are separated by part of the glycocalyx [6]. Such initial adhesion may involve non-specific interactions before a more specific process occurs. The accompanying adhesive mechanisms result in a thickening of the cell membrane [3, 4] forming an adhesive pedestal. Adhesion is usually between the bacterial body rather than the flagellae or terminal bulbs. Whether the different forms of adhesion have any pathogenic significance has not been investigated.

The zona occludens together with the zona adherens and the macula adherens are specialized intercellular junctions vital to the integrity of the epithelial cell surface. The continuity of this layer remains in spite of the

Fig. 18.4 The epithelial surface of gastric mucus-secreting cells from the pre-pyloric area of a patient with chronic duodenal ulceration. The number of surface microvilli is decreased. The *C. pylori* (CP) are numerous and present in the 'grooves' (G) between individual epithelial cells (EP). Coccal (C), kidney-shaped (K) and spiral (S) forms of *C. pylori* are seen. (Scanning electron micrograph, × 3907.)

different and often highly specialized cells of the gastro-intestinal tract. Disruption of this epithelial cell continuity either by premature cell death or disintegration of the intercellular junction will result in a sequence of events leading to epithelial ulceration. The location of *C. pylori* near the zona occludens (Fig. 18.5) enables these bacteria to be ideally sited to disrupt these junctions (Fig. 18.6) and participate in the processes of ulceration. There is ultrastructural evidence for the disruption of those intercellular junctions near *C. pylori* and the siting of *C. pylori* in the intercellular spaces [3, 7, 9]. Whether this disruption of the intercellular junctions results from the presence of the bacteria and what are the mechanisms for this disruption have yet to be elucidated.

Ultrastructural evidence has also been produced for the intracellular invasion of *C. pylori*. These bacteria have been described in mucus granules, in lysosomes and closely related to such structures [3, 11, 12]. The frequency of such *C. pylori* invasion has been reported to be 10% in patients with active duodenal ulceration [3]. Whether such invasion is a characteristic of *C. pylori*, an alteration to the *C. pylori* when subjected to appropriate conditions, to a characteristic of other species of *Campylobacter*, or the cross-cutting of ultra-thin sections, is debatable. *In vitro*, *C. pylori* are characteristically non-invasive [13] and contrast with other *Campylobacter*.

Campylobacter pylori are found in areas of acute and chronic gastritis. They are most

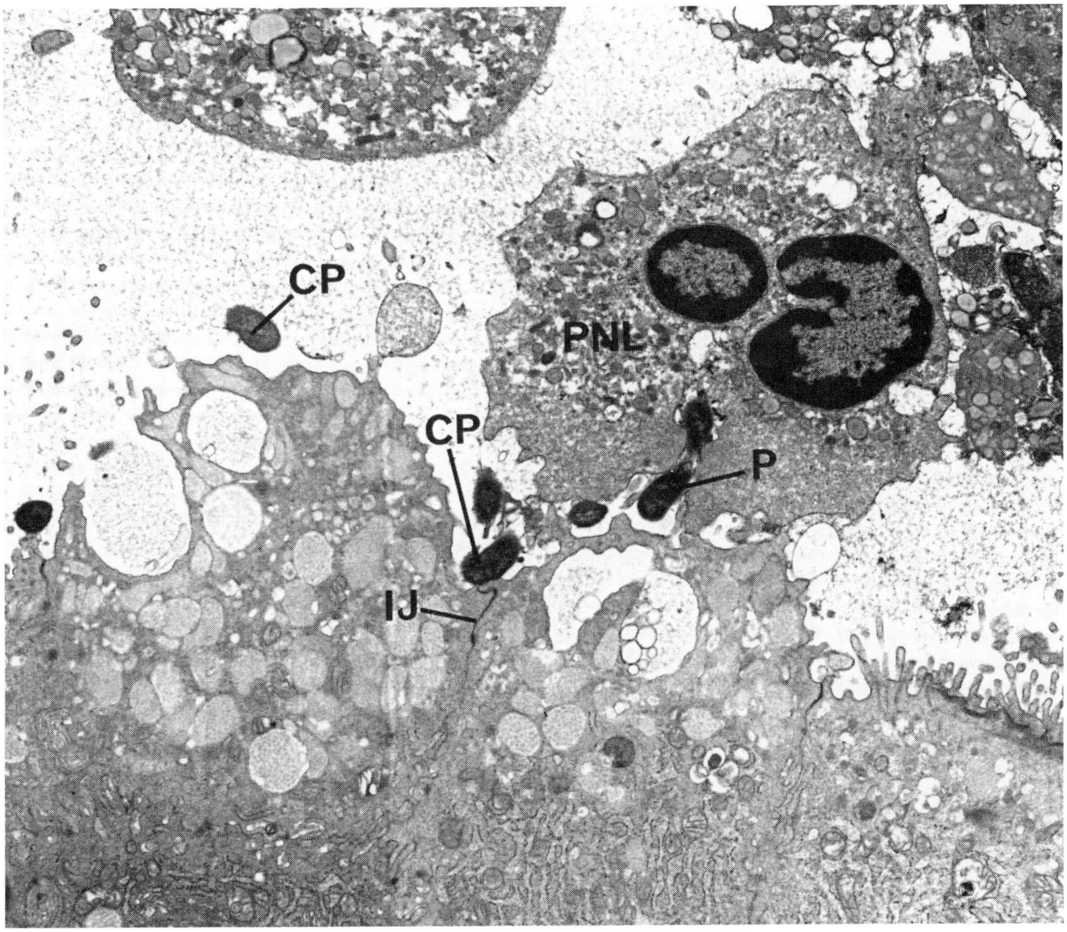

Fig. 18.5 *C. pylori* (CP) and a polymorphonuclear leucocyte (PNL) are closely related to gastric mucus-secreting cells. A bacterium is being phagocytosed (P) by the polymorphonuclear leucocyte. The close relationship between the *C. pylori* and the intercellular junctions (IJ) is apparent. (Transmission electron micrograph, × 6350.)

numerous in those areas of active chronic gastritis [5, 6] when bacteria are frequently associated with migrating polymorphonuclear leucocytes (Fig. 18.7). At such sites there is a positive correlation between the number of intraepithelial polymorphonuclear leucocytes and the number of bacteria [8]. This correlation is not absolute [6] and depends upon the site from which the biopsy has been taken. Thus, at the edge of a benign ulcer, there is a disproportionate number of intraepithelial polymorphonuclear leucocytes. At the edge of the ulcer, the protective mucus barrier is incomplete. This enables the gastric/duodenal fluid to reach the epithelial surface, resulting in a hostile environment for *C. pylori* survival but acts as a potent stimulator of polymorphonuclear leucocyte migration. The nature of the chemotactic factors responsible for the migration of polymorphonuclear leucocytes at the sites of active chronic gastritis have not been fully elucidated. After migrating from the subepithelial layer through the overlying epithelium, phagocytosis of *C. pylori* by polymorphonuclear leucocytes does, however, occur [3–6, 14]. A bacterium in the process of being phagocytosed by an intraluminal polymor-

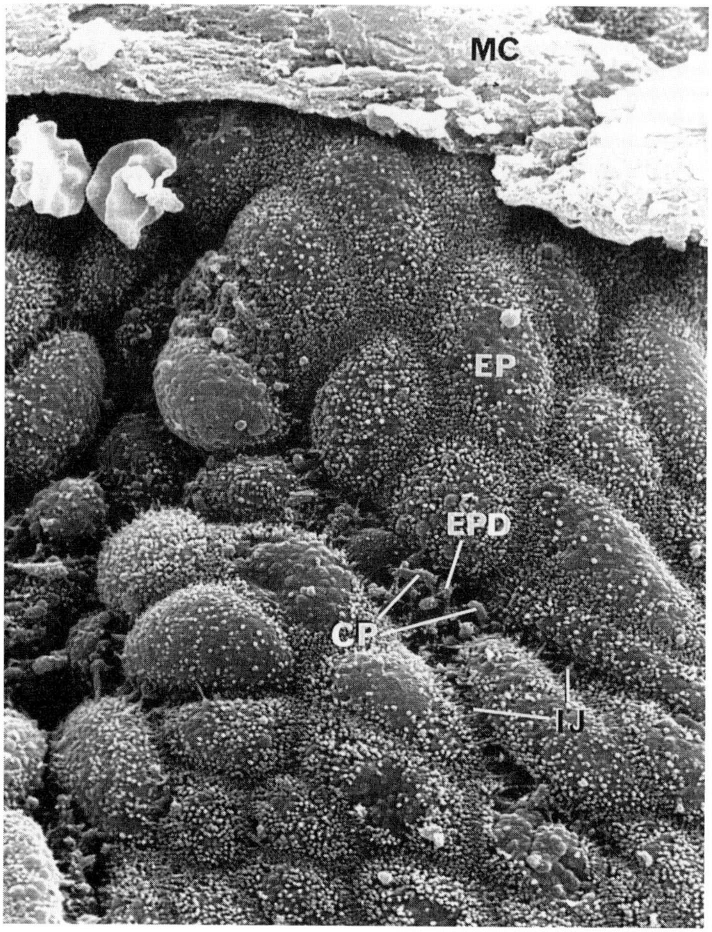

Fig. 18.6 The epithelial surface of a pre-pyloric biopsy from a patient with chronic duodenal ulceration. The microvilli are decreased over the surface of individual epithelial cells (EP). The surface of one epithelial cell is grossly disorganized (EPD) by either secreting mucus granules or the disease process. Mucus (MC) overlies part of the surface. *Campylobacter pylori* (CP) are seen in the 'groove' between epithelial cells. The intercellular junction (IJ) between some epithelial cells appears to be disrupting and at two sites a gap is present between individual epithelial cells. (Scanning electron micrograph, × 1477.)

phonuclear leucocyte is seen in Fig. 18.5. This phagocytosis is less frequent that might be anticipated from the number of migrating polymorphonuclear leucocytes and bacteria. Moreover, with the large numbers of migrating polymorphonuclear leucocytes it is perhaps surprising that the natural defence mechanisms are not more successful in eradicating the bacteria.

The immune response to *C. pylori* infections differs qualitatively when examining patients with and without ulceration [15] but in both circumstances there is a significant increase in systemic IgG, IgG^1, IgG^4 and IgA. This response is typical of a chronic infection. Immunohistological studies have demonstrated that the bacteria are coated with IgA, IgG and IgM [16]. The proportion of bacteria coated with IgA exceeds that of the other immunoglobulins. If large numbers of *C. pylori* are solely coated with IgA this would account for the limited phagocytosis observed. Significant numbers (89%) of bacteria are however coated with IgM and/or IgG, so that it would be reasonable to expect a greater proportion of the bacteria to have undergone phagocytosis.

The pathogenic role of *C. pylori* would involve an interference with the cytoprotective mechanisms of gastric integrity. Whether the *C. pylori* directly interferes with the function of the epithelial cells has not been observed in those epithelial cells in close proximity to the bacteria. Changes which have been described are the dissolving of mucous granules result-

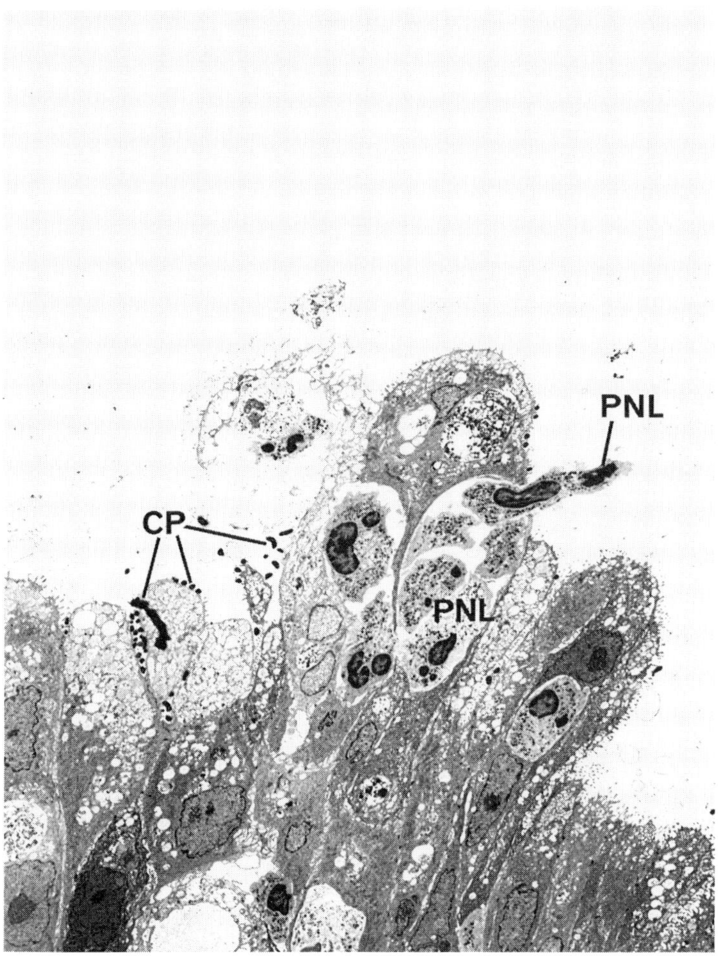

Fig. 18.7 The surface epithelium of a pre-pyloric biopsy from a patient with chronic duodenal ulceration. Numerous *C. pylori* (CP) are related to the surface epithelium and many polymorphonuclear leucocytes (PNL) are migrating through the epithelium with one polymorphonuclear leucocyte passing into the gastric lumen. (Transmission electron micrograph, × 894.)

ing in translucent empty structures [3], a decrease in the mucus content of infected cells [3–5], a qualitative alteration in the mucus resulting in an increase in its sialic acid-rich glycoprotein content [3] and disruption of the intercellular junctions [7]. Changes observed in parietal cells have also been attributed to *C. pylori* and these changes include the presence of lipid droplets, intracellular oedema and the occurrence of laminated bodies [7].

The association of *C. pylori* with acute and chronic gastritis is well established. The presence of *C. pylori* and the accompanying cellular changes are integral components of the observations associated with peptic ulceration. Many of the features associated with *C. pylori* are accepted characteristics of pathogenic bacteria. Therefore, it would be reasonable to conclude that *C. pylori* must be one of the many factors implicated in the aetiology of benign peptic ulceration.

References

1 Steer, H.W. Surface morphology of the gastroduodenal mucosa in duodenal ulceration. *Gut* 1984, 25, 1203–1210.
2 Phillips, A.D., Hine, K.R., Holmes, G.K.T. & Woodings, D.F. Gastric spiral bacteria. *Lancet* 1984, ii, 100–101.
3 Bode, G., Malfertheiner, P. & Ditschuneit, H. Patho-

genetic implications of ultrastructural findings in *Campylobacter pylori* related gastroduodenal disease. *Scad. J. Gastroenterol.* 1988, **23** (Suppl. 142), 25–37.

4 Goodwin, C.S., Armstrong, J.A. & Marshall, B.J. *Campylobacter pyloridis*, gastritis and peptic ulceration. *J. Clin. Pathol.* 1986, **39**, 353–365.

5 Steer, H.W. & Colin-Jones, D.G. Mucosal changes in gastric ulceration and their response to carbenoxolone sodium. *Gut* 1975, **16**, 590–597.

6 Steer, H.W. Ultrastructure of cell migration through the gastric epithelium and its relationship to bacteria. *J. Clin. Pathol.* 1975, **28**, 639–646.

7 Chen, X.G., Correa, P., Offerhaus, E., Fox, J., Hunter, F. & Diavolitsis, S. Ultrastructure of the gastric mucosa harboring campylobacter-like organisms. *Am. J. Clin. Pathol.* 1986, **86**, 575–582.

8 Steer, H.W. The gastroduodenal epithelium in peptic ulceration. *J. Pathol.* 1985, **146**, 355–362.

9 Droy-Lefais, M.T., Megraud, F., Lamouliatte, H., Geraud, G. & Jutteau, J. 1987. *Campylobacter pylori* and gastric mucosal lesions: a scanning electron microscopy study. In Kaijser, B. & Falsen, E. (eds.) *Campylobacter IV: Proceedings of the 4th International Workshop on Campylobacter Infections*, University of Göteborg, Göteborg, pp. 392–393.

10 Williams, S.E. & Turnberg, L.A. Demonstration of a pH gradient across mucus adherent to rabbit gastric mucosa: evidence for a 'mucus–bicarbonate' barrier. *Gut* 1981, **22**, 94–96.

11 Meyrick-Thomas, J., Poynter, D., Gooding, C., Woodings, D.F., Selway, S., Cook, A.R., Hill, M.J. & Misiewicz, J.J. Gastric spiral bacteria. *Lancet* 1984, **ii**, 100.

12 Lee, W.K., Gourley, W.K., Buck, G.E. & Subramanyam, K. A light and electron microscopic study of a campylobacter-like bacteria inhabiting the human stomach. *Gastroenterology* 1985, **88**, 1470.

13 Steer, H.W. & Newell, D.G. 1985. Mucosa-related bacteria in benign peptic ulceration. In Pearson, A.D., Skirrow, M.B., Lior, H. & Rowe, B. (eds.) *Campylobacter III: Proceedings of the 3rd International Workshop on Campylobacter Infections, Ottawa, Canada.*

14 Shousha, S., Bull, T.B. & Parkins, R.A. Gastric spiral bacteria. *Lancet* 1984, **ii**, 101.

15 Steer, H.W., Hawtin, P.R. & Newell, D.G. An ELISA technique for the serodiagnosis of *Campylobacter pyloridis* infection in patients with gastritis and benign duodenal ulceration. *Serodiagnosis and Immunology* 1987, **1**, 253–259.

16 Wyatt, J.I., Rathbone, B.J. & Heatley, R.V. Local immune response to gastric *Campylobacter* in non-ulcer dyspepsia. *J. Clin. Pathol.* 1986, **39**, 863–870.

19 Campylobacter pylori and non-ulcer dyspepsia

T.M. SHALLCROSS, B.J. RATHBONE AND R.V. HEATLEY

In the UK, the incidence of dyspepsia resulting in a general practice consultation is about 1% per year [1], probably accounting for one-fifth to one-quarter of all general practice consultations [2]. In Sweden 2% of all outpatient consultations and 31% of gastroenterology referrals between 1978 and 1981 were for 'gastritis' — which is probably synonymous with dyspepsia [3], while in Denmark the prevalence of dyspepsia in a community survey was 27% [4].

Definitions are hard to standardize in a field where seemingly simple terms have very different meanings even to experienced doctors [5]. A recent international working party [6] defined non-ulcer dyspepsia (NUD) as 'upper abdominal or retrosternal pain, discomfort, heartburn, nausea, vomiting, or other symptoms considered to be referable to the proximal alimentary tract, and lasting for more than 4 weeks, unrelated to exercise, and for which no focal lesion or systemic disease can be found responsible'. Inevitably, broad definitions may cover a wealth of pathology, and the number of patients ultimately labelled as having NUD rather than a more specific diagnosis will depend in part on the fervour of the investigator and of course on the tenacity of the patient.

Talley & Piper in a study of 327 patients with 'NUD' [7] suggested that in over three-quarters of patients a definite diagnosis could be reached, although nearly one-quarter remained undiagnosed. The main diagnoses found were irritable bowel syndrome, gastro-oesophageal reflux, aerophagy and gallstones. The term 'essential dyspepsia' was suggested to cover the undiagnosed group, and one can presume that with increasing knowledge and an expanding number of tests, the size of this group will steadily decline.

Unfortunately, this confusion of terminology makes comparison of the findings of different non-ulcer dyspepsia studies difficult, since the proportion of patients with 'essential dyspepsia' may vary considerably between different centres and clinicians. Also, because 'essential dyspepsia' itself must include more than one pathology, the difficulties in identifying a small subgroup (with, for instance, *Campylobacter pylori* as a cause) are considerable.

The 'rediscovery' of *Campylobacter pylori* led to hopes not only of a new approach to peptic ulcers, but also to the possibility of an explanation for 'essential dyspepsia' [8]. Three main questions can be asked about the role of this organism in NUD: firstly, is colonization by *C. pylori* more common in patients with NUD than in an asymptomatic population?; secondly, are there specific symptoms, or a symptom profile that typifies NUD associated with *C. pylori*?; and thirdly, does eradication of *C. pylori* lead to resolution or reduction of dyspeptic symptoms?. This review examines the current status of *C. pylori* in the aetiology of non-ulcer dyspepsia, and evaluates whether these questions can yet be answered.

Chronic gastroduodenitis and upper gastro-intestinal symptoms

Historically, acid and pepsin have been thought of as the main aggressors to the gastric mucosa, often aided by other self-inflicted irritants such as alcohol, tobacco and spices. These supposedly cause inflammation of the gastric mucosa with resultant pain — hence the confusion of terminology between the 'touch of gastritis' blamed for acute upper gastro-intestinal symptoms, the endoscopist's gastritis (varying from slight hyperaemia, to the florid picture of a red angry mucosa with haemorrhages and erosions), and the pathologist's classification based on cellular infiltrate and architectural changes. Many of the early writers attributed quite specific though diverse symptoms to gastritis [9], but unfortunately the simple concept that a red mucosa must be painful does not appear to hold true.

Because *C. pylori* has only been identified since 1983, earlier work on NUD cannot be applied directly to questions about the bacterium. However, the strong association between *C. pylori* and antral gastritis and duodenitis should allow some information to be gleaned from studies linking symptoms with gastric histology. Unfortunately, before the widespread availability of fibre-optic upper gastrointestinal endoscopy, most gastric biopsies were taken using the Wood's tube or a similar device, producing untargeted biopsies mainly from the gastric body or fundus [10, 11]. Since gastritis of the body or fundus is much less sensitive and specific than antral gastritis as a predictor of *C. pylori* colonization, only studies using antral biopsies are of use in this context. Furthermore, 'gastritis' is often used to refer to the endoscopic appearance of the mucosa rather than to the histological changes, which is unsatisfactory given the wide observer variation and lack of an accepted classification for the macroscopic features seen. Not surprisingly, just as *C. pylori* is present in an appreciable proportion of asymptomatic individuals, so is gastritis. Kreuning *et al.* [12] found antral gastritis in 36% of 50 normal volunteers (mean age 33 years), and chronic duodenitis in 12%. Cheli *et al.* [13] have shown antral or fundal gastritis in 46% and 73% of asymptomatic Italians and Hungarians respectively (two groups of 100 subjects, age range 20–79 years). Comparable results have been reported in 10 medical students [14].

Comparative studies are hard to find, but neither Cheli *et al.* [15] nor Villako *et al.* [16] were able to show a significant association between gastritis and dyspepsia. Toukan *et al.* [17] compared the density of inflammatory infiltrate in gastroduodenal biopsies from 31 NUD patients with biopsies taken from a group of asymptomatic age-matched controls. The patients showed significantly higher neutrophil counts in the gastric body, antrum and duodenum than controls. Some subsequent workers have linked *C. pylori* with a neutrophil infiltrate [8, 18] but this finding is disputed by others [19].

Several groups have suggested a link between heartburn or reflux symptoms and gastritis. Earlam *et al.* [20], in a study of patients undergoing surgery for duodenal ulcer, found that all patients with heartburn also had gastritis, and 62% of those with gastritis had heartburn. However, as would be expected, 94% of the subjects had gastritis, and although the 2 without gastritis did not experience heartburn, this cannot be considered significant. Fink *et al.* [21], in a study of 23 patients with gastro-oesophageal reflux (diagnosed by typical symptoms, positive Bernstein test and histology), found antral gastritis in 78%, compared to only 10% in a group of 20 normal volunteers. Unfortunately, the mean age of the patients was 53 years and that of the volunteers 29 years, the obvious conclusion being that both antral gastritis and reflux are more common with increasing age. However, the study did also show a significant correlation between the severity of gastritis judged histologically and the magnitude of delayed gastric emptying of a semi-solid isotope-labelled meal, suggesting a link with symptoms.

One American group in a study of 28 patients with NUD [22] showed a highly significant correlation between heartburn and the histological findings of chronic fundal gastritis, acute antral gastritis and chronic antral gastritis. The age difference between the two groups (heartburn and no heartburn) was less than 10 years.

Duodenitis is also strongly linked with both antral gastritis and *C. pylori* [23, 24]. There is evidence that it may form part of a disease spectrum in which duodenal ulcer also occurs [25]. Opinion is divided as to whether duodenitis on its own can cause symptoms [26, 27], although the number of people who fit this diagnostic group is small.

Undoubtedly asymptomatic chronic gastritis is very common. There is, however, at present little hard evidence of a clear subgroup in which it causes symptoms, although there is some support for a link with heartburn. These studies therefore do not provide a definite answer to questions about the symptomatology of *C. pylori* infection.

Chronic C. pylori infection and non-ulcer dyspepsia

Most of the initial interest in *C. pylori* was directed at the association with gastritis and peptic ulcers. Consequently, little information concerning the symptoms found in *C. pylori*-positive and -negative individuals was recorded. More recently, attention has turned towards symptomatology. This is particularly with the availability of non-invasive tests to detect *C. pylori* infection, and with fairly effective treatment which at least temporarily clears the organism from the gastric mucosa. In reviewing these studies it is perhaps easiest to categorize them according to the methods used to detect *C. pylori* and relate this to symptoms.

Prevalence studies

The majority of studies reporting the prevalence of *C. pylori* infection in different groups of patients have not included detailed symptomatic assessments. However, if *C. pylori* is responsible for many of the symptoms of NUD, then it should be possible to demonstrate a higher frequency of infection in these patients, than in normal controls.

Unfortunately, there are considerable difficulties in making valid comparisons between NUD patients and normals. The prevalence of *C. pylori* infection has been shown to rise steadily with increasing age, being present in less than 10% of people under 20 years of age, but rising to over 40% by the age of 50 years, and possibly much higher in old age [28–31]. The age profile of the control group is therefore vital if comparable data are to be obtained. Not only is age important, but there may also be marked ethnic variations in the frequency of infection, with rates of 60% in Chinese aged 20–39 years [29], and 46% in Indians aged 20–29 years [32]. Beg *et al.* [33] report rates of over 90% in Indians and Fijians presenting for endoscopy, and Gutlerrez found *C. pylori* in 87% of 15 normal volunteers in Colombia [34]. Matching the ethnic origins of study and control groups could cause considerable problems in multi-racial societies, and so far few reports have included mention of the races studied.

We do not yet know what other environmental factors may significantly affect *C. pylori* carriage rates (for instance social class, size of families, religious dietary taboos). Until more information is available, reliable comparisons between groups may not be possible.

With these obstacles in view, what interpretation can be made of the studies published so far? Two different methods can be used to look at prevalence in relation to symptoms: either large non-invasive population studies, or endoscopic studies. These will be considered separately.

NON-INVASIVE POPULATION STUDIES

Non-invasive population studies can now be carried out, using either serology or the ^{13}C- or ^{14}C-breath tests to screen large numbers of subjects for *C. pylori* infection, and a search made for a correlation with symptoms assessed by interview or self-administered questionnaire. Obviously, a firm diagnosis of NUD cannot be made by this method, since between 3% and 23% of the population can be expected to have peptic ulcers [35–37] and only 25 % of people with 'dyspepsia' will have no other diagnosis after proper investigation [7]. Nevertheless, this type of study may yield useful information.

Wyatt *et al.* [31], using a self-administered questionnaire to study 247 blood donors, found that only the occurrence of previous investigations for dyspepsia in those aged over 40 years was associated with *C. pylori* infection as assessed by ELISA of serum; 64 % of those with positive serology gave no history of clinically-significant dyspepsia.

Using a passive haemagglutination test, Marshall *et al.* [38] studied 292 blood donors who had never been diagnosed as having had peptic ulcer disease. The frequency of *C. pylori* in subjects who had either been investigated for dyspepsia with a barium meal or endoscopy, or who had taken antacid in the preceding seven days, was twice that of the remainder of the group (28 % versus 14 % — a highly significant difference).

Skoglund & Whalen [39] carried out a serological survey of 121 subjects who also completed a self-assessment questionnaire detailing antacid use. A strong positive correlation was shown between the quantity of antacid used and *C. pylori* colonization assessed by ELISA.

These reports do suggest some association between *C. pylori* infection and 'dyspepsia' using a broad definition, but do not answer more specific questions about NUD, nor define a symptom profile. Detailed information will be hard to obtain using this type of study without devoting considerable resources.

Endoscopic prevalence studies

Tables 19.1–19.3 list the frequency of *C. pylori* found by different workers in patients with NUD, and normal volunteers. Of the combined studies, only that by Rauws *et al.* [41] shows a significant difference in frequency of *C. pylori* between NUD patients and controls. The average age of the controls, however, was 18 years lower than that of patients. Interestingly, the Colombian group [34] found a higher frequency of infection in their normal volunteers!

We have examined the age-related frequency of *C. pylori* determined histologically, in patients attending a dyspepsia clinic with NUD (defined as abdominal symptoms referable to the upper gastrointestinal tract, in the absence of ulceration seen at endoscopy, or other likely cause of the symptoms). These results have been compared with the frequency of infection as determined by ELISA, in a group of randomly-selected blood donors from the same city (Fig. 19.1). Infection was more common in patients compared to blood donors at all ages, although this failed to reach significance in two of the groups.

In most reports, the frequency of infection in NUD is of the order of 50 %. Many of these patients are 'middle aged', when the prevalence of *C. pylori* in the European or American population also approaches 50 %. The conclusion must be that at present there is no overwhelming evidence that *C. pylori* is more common in patients undergoing endoscopy for NUD, although a number of studies suggest that this is the case. However, it must

Table 19.1 Frequency of *C. pylori* infection in studies including both non-ulcer dyspepsia and normal volunteer groups

Reference	Country	Non-ulcer dyspepsia group			Normal volunteers		
		n	Mean age (years)	(%) *C. pylori* positive	*n*	Mean age (years)	(%) *C. pylori* positive
Pettross *et al.* (1988) [40]	USA	97	47	43	15	29	13
Rauws *et al.* (1988) [41]	Netherlands	240	56	70	34	38	20
Gutlerrez *et al.* (1988) [34]	Colombia	34	?	79	15	?	87

Table 19.2 Frequency of *C. pylori* colonization in non-ulcer dyspepsia

Reference	Country	*n*	Mean age (years)	(%) *C. pylori* positive
Rouvroy *et al.* (1987) [42]	Rwanda (Africa)	173	34	71
Blomberg *et al.* (1988) [82]	Sweden	103	56	45
Jiang *et al.* (1987) [43]	China	136	49	74
Marshall & Warren (1984) [8]	Australia	69	55	51
Schnell & Schubert (1988) [44]	USA	30	?	33

be remembered that a good proportion of asymptomatic normal volunteers have *C. pylori*-positive antral gastritis, making a strong link with NUD less likely.

Comparative endoscopic studies (symptomatology)

Marshall & Warren [8] in their original study of 100 patients found that the only symptom which correlated with the presence of *C. pylori* was 'burping', and this association remained even after exclusion of patients with peptic ulcer. Marshall subsequently suggested [48] that the most severe cases of 'nervous dyspepsia' (with symptoms including substernal and epigastric burning pain, reflux, burping, epigastric distension and periodic nausea) have *C. pylori* infection and gastritis, which responds to antibacterial therapy. However, this is anecdotal evidence only.

In a recent study by Loffeld *et al.* [49], a detailed history was taken from 109 dyspeptic patients using a questionnaire. Patients with peptic ulcers, reflux oesophagitis, active duodenitis, previous gastrectomy, carcinoma, and who had recently used steroids, non-steroidal anti-inflammatory drugs (NSAIDs) or antibiotics were not included. During endoscopy, gastric antral biopsies were taken for histology and culture of *C. pylori*. On the basis of histology, the patients were divided into two groups: normals and those with gastritis. The normal group was further divided into those with and those without *C. pylori*. No significant differences in symptoms were found between the groups. In contrast to most studies however, a high proportion of the patients with normal histology showed colonization by *C. pylori* (66%), although in low numbers as assessed by histology. The reason for this is not obvious.

Rokkas *et al.* [50] studied 55 patients with dyspepsia of at least 6 months duration, and who were not taking NSAIDs; 15 normal controls were also studied. The only symptom which showed a significant association with the presence of *C. pylori* was post-prandial

Table 19.3 Frequency of *C. pylori* infection in studies of normal volunteers only (figures to nearest interger)

Reference	Country	*n*	Mean age (years)	(%) *C. pylori* positive
Barthel *et al.* (1988) [45]	USA	20	29	20
Langenberg *et al.* (1984) [46]	Netherlands	25	27	24
Peterson *et al.* (1986) [47]	USA	18	33	50*

*Prevalence of 61% if fundus rather than antrum is considered.

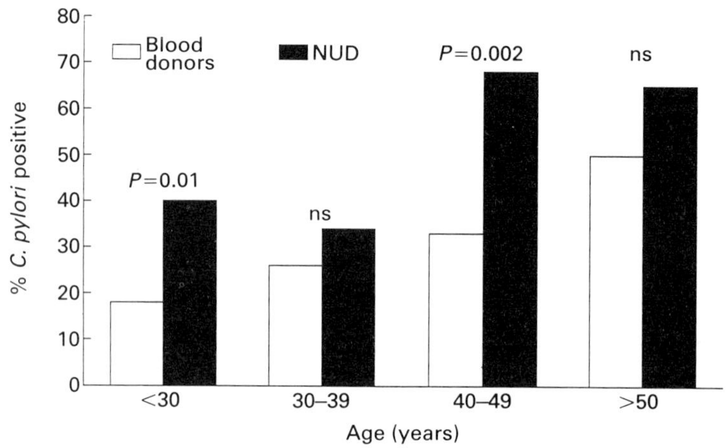

Fig. 19.1 Frequency of *C. pylori* in NUD patients and blood donors.

bloating. The frequency of *C. pylori* in the NUD group was 45.5%, significantly greater than the 13.3% frequency in the control group. The mean age of the *C. pylori*-positive, -negative, and control groups was comparable (39.3–42.6 years), but the frequency of infection in the control group was considerably lower than that found in most of the volunteer studies (see Tables 19.1–19.3). The applicability of this result to the general population must therefore be questioned.

In a study of 149 consecutive patients undergoing gastroscopy, Borsch *et al.* [51] found no significant difference in symptoms between those with or without *C. pylori*. There were, however, differences in the use of ulcer healing drugs, smoking habits, and antibiotic use between the two groups which may have affected symptoms. It is not stated whether patients with peptic ulcer were included in the series.

Andersen *et al.* [52] showed a significant difference in only two symptoms between *C. pylori*-positive and -negative groups in 33 patients with NUD. In the *C. pylori*-positive group, symptoms which had been present for more than 5 years and symptoms which varied throughout the year were more common.

No clear symptom profile emerges from these studies, with each group finding different associations. This may be partly due to the problem of different workers using varying symptom definitions and also because a number of chance associations will be found when multiple symptoms are analysed. It strongly suggests, however, that *C. pylori* does not cause specific symptoms.

Treatment studies

If *C. pylori* is a cause of NUD, then eradication of the organism from the gastric mucosa would be expected to result in an improvement in dyspeptic symptoms. A number of treatment studies evaluating changes in symptoms have now been published.

An early study of treatment aimed at eradicating *C. pylori* in patients with NUD was reported by McNulty *et al.* in 1986 [53]: 50 subjects were evaluated after being randomized to receive bismuth salicylate, matched placebo, or erythromycin suspension; 11/12 of the symptomatic patients in the group cleared of *C. pylori* reported an improvement in symptoms compared with 21/32 in the group with persistent infection — this difference did not reach statistical significance. There was no association between the resolution of gastritis following treatment, and improvement of symptoms. As the authors point out, the method used for assessing changes in symptoms (better or worse than before treatment) may be less reliable than absolute scoring systems [54].

Several other studies using bismuth have appeared, mainly in abstract form. Lambert *et al.* [55] evaluated 54 patients in a randomized, controlled study, and showed a significant improvement in day and night pain, and of nausea/vomiting in subjects cleared of bacteria. The study was double blind (in so far as this is possible using bismuth). Borody *et al.* [56] in an uncontrolled trial of bismuth plus an antibiotic, reported an improvement in epigastric burning/pain, post-prandial abdominal distension, and nausea, following treatment. A further double-blind study from the same centre [57] evaluated a two-week course of bismuth in 43 patients. Significant improvements were shown in pain/burning, abdominal distension, and nausea in the bismuth group. However, no correlation was noted between either clearance of bacteria and symptoms, nor improvement of gastritis and symptoms.

Rokkas *et al.* [58, 59] report a significantly better symptomatic response to bismuth compared with placebo in a double-blind study of 40 patients with NUD. Again, no details of symptomatic improvement related to bacterial clearance are given. Also, at variance with most studies of NUD, no improvement in symptoms was shown by the placebo group.

None of these studies is really adequate to demonstrate whether eradication of *C. pylori* leads to an improvement in the symptoms of NUD. They do, however, give some positive encouragement. Unfortunately, the use of bismuth as a therapeutic agent to eradicate *C. pylori* has some inherent problems. Although bismuth has been shown to have a bactericidal effect on *C. pylori in vitro*, and is effective in clearing the bacteria from the gastric mucosa with a resultant improvement in gastritis, it also has a number of other local effects in the stomach which may be important either in improving symptoms, or in healing gastritis. Bismuth binds to mucus glycoproteins and may reduce acid attack on the gastric mucosa [60]. It inhibits pepsin both *in vitro* and *in vivo* [59], stimulates prostaglandin synthesis by the gastric mucosa [61] and inhibits peptic degradation of epidermal growth factor [62]. It is therefore possible that the bactericidal properties are of minor importance in its mode of action. There also remains the problem that a truly double-blind study is difficult to perform with an agent that produces such an obvious change in the stool.

In order to demonstrate conclusively that eradication of *C. pylori* alone can improve both gastritis and symptoms, studies using agents which clear the bacteria but have no direct effect on the mucosa are needed. Glupczynski *et al.* [63] carried out a double-blind placebo-controlled trial of amoxycillin suspension (1 g twice daily for 8 days) in 45 patients with NUD and *C. pylori*-positive antral gastritis: 91% of the treatment group were cleared of bacteria, with resolution of active antral gastritis in 68%, although chronic gastritis was not affected. There was a greater improvement in the overall symptom score of those cleared of bacteria compared with those not cleared. This was small however, and not statistically significant. Neither was a significant improvement shown in individual symptoms between these two groups. It is possible that a longer period of treatment or a different dosage would have been more effective in eradicating bacteria or improving gastritis and symptoms. A Peruvian double-blind placebo-controlled study [64] of furazolidone or nitrofurantoin similarly showed good clearance of bacteria, and significant improvement of gastritis, but failed to show a consistent improvement in symptoms between the control and active therapy groups.

These studies have been disappointing in not providing a clear answer to some of the questions about NUD. Certainly they have not shown a clear role for *C. pylori*, but nor have they been so dramatically unsuccessful as to make this organism an unlikely candidate for a place in NUD. Obviously, further studies of treatment in relation to symptoms are needed.

C. pylori and NSAID-associated dyspepsia

Dyspepsia is common in patients taking NSAIDs, but the reasons for this are unclear [65, 66]. Upadhyay *et al.* [67], in a study of 14 patients referred for endoscopy and 38 volunteers, all of whom were taking NSAIDs, found a significant positive correlation between the presence of *C. pylori* and the severity of dyspeptic symptoms: 5/26 bacteria-positive patients and 3/26 -negative patients had peptic ulcers. Symptomatic data are not however provided on these patients specifically, and so no definite conclusions about NUD can be drawn.

A similar study from New Zealand [68] of 22 patients and 12 asymptomatic volunteers showed no correlation between symptoms and the presence of bacteria. However, only a small number of patients was studied, and only 41% experienced dyspepsia, making a significant result less likely. In an ongoing study, we have investigated 40 patients with dyspepsia referred by general practitioners and rheumatologists, and who had been taking an NSAID regularly until 4 weeks before endoscopy. The age-related frequency of *C. pylori* was no different from that found in a group of 405 blood donors from the same geographical area. Because of the high rate of peptic ulceration, only 13 *C. pylori*-negative patients and 9 of those who were positive had non-ulcer dyspepsia (this was not statistically significant).

We have also carried out a survey of 97 randomly-selected patients attending a rheumatology clinic. A structured questionnaire was completed at interview, and *C. pylori* colonization assessed by ELISA of anti-*C. pylori* antibodies in serum. Both current dyspepsia (defined as pain, discomfort or burning in the upper abdomen or lower chest) and a past history of dyspepsia were more common in the *C. pylori*-positive patients. This was observed taking the group as a whole, and also looking at the subgroups taking NSAIDs with or without rheumatoid arthritis, but the difference was not statistically significant. Greater numbers might reveal a significant difference, although of course a diagnosis of NUD cannot be made by questionnaire, and peptic ulceration is common in these patients.

C. pylori and dyspepsia in children

Serological studies suggest that *C. pylori* is uncommon in children [28] with a prevalence $< 10\%$ in most series. A retrospective review of consecutive biopsies taken at gastroscopy in one children's hospital [69] showed *C. pylori* in 11% of 53 patients. Abdominal pain was the main indication for endoscopy in 24 of these children, but no indication is given of the presenting complaint in those who were *C. pylori*-positive; 4/6 *C. pylori*-positive children had peptic ulcers. A prospective study by the same group [70] showed a frequency of 10.4% in 67 children, abdominal pain being the main reason for endoscopy in 39; 5/7 of those with *C. pylori* had duodenal ulcers. Again it is not possible from the data presented to correlate symptoms with *C. pylori* status. However, these results do not suggest that *C. pylori* is a major cause of non-ulcer dyspepsia in children, since the frequency of *C. pylori* in those without ulcers was only 6.3%. It was not stated how many of these had abdominal pain however.

Thomas *et al.* [71] studied 51 consecutive symptomatic patients aged 5–16 years undergoing gastroscopy. These were compared serologically with 51 age-matched hospital controls with no gastro-intestinal symptoms, and 150 healthy schoolchildren; 4% of the hospital controls and 5% of the schoolchildren had significant antibody titres to *C. pylori*, compared with 20% of the group undergoing gastroscopy. An unusually high sensitivity and specificity (both 100%) was reported for the ELISA. A link between *C. pylori* infection and abdominal pain is certainly suggested, but unfortunately details of the endoscopic diagnoses are not given, and so no firm conclusions can be drawn about NUD.

Cadranel *et al.* [72] have reported the results of endoscopies on 114 children with ages up to 16 years. Approximately 73% of those with chronic epigastric pain were *C. pylori*-positive, compared to only 25% of those with other diagnoses (excluding peptic ulcer). It is not possible to extract exact figures from the data provided. However, it must be noted that 34% of this sample were from Africa (mainly North Africa), and the prevalence of *C. pylori* colonization is high in the adult population from these areas. This fact may have biased the results, and it is not clear into which diagnostic groups these African children fitted.

Once again, the data available do not allow a positive answer to be given about the role of *C. pylori* in childhood NUD, but is seems likely that there is a subgroup of children with chronic abdominal pain whose symptoms are related to this infection.

Acute infection and non-ulcer dyspepsia

There is now substantial evidence that acute infection with *C. pylori* can produce the clinical picture of a 'viral gastro-enteritis' with marked upper abdominal symptoms. Both studies [73, 74] in which volunteers have ingested *C. pylori* have reported short-lived upper abdominal symptoms such as epigastric pain, nausea, vomiting, and epigastric fullness or bloating. However, the symptoms appear to have been self-limiting even though infection persisted in one case, with no suggestion that chronic dyspepsia has resulted.

Epidemic hypochlorhydria associated with acute active gastritis is now well documented [75–80], and in some cases *C. pylori* has been identified as the cause. In a number of these episodes a mild systemic illness accompanied by dyspeptic symptoms has been reported, but once again these have been self-limiting, of short duration, and continuing dyspepsia has not been noted.

Interestingly, none of three children who presented with an acute *C. pylori*-positive gastritis and protein-losing enteropathy was noted to have had abdominal pain during the illness [81].

Conclusion

There seems little doubt that acute *C. pylori* infection can be symptomatic in some people, although it does not always cause symptoms. It is unlikely that many cases of acute infection will be encountered in hospital practice, and in general practice these cases will cause few problems. Certainly, there is no reason to attempt treatment of such cases at present.

Currently, there is little convincing evidence that *C. pylori* is a major factor in the pathogenesis of NUD, although the large number of reports showing some correlation do suggest that it has a role, at least in a subgroup of patients.

The high frequency of colonization by *C. pylori* in asymptomatic people, coupled with the considerable changes in the rate with age, ethnic group, and perhaps other as yet unidentified factors, have made the results of prevalence studies difficult to interpret and often of questionable value.

The diverse actions of bismuth compounds makes most of the treatment studies carried out so far of limited value in defining the role of *C. pylori* in any symptomatic improvement.

There is a need for large studies of the prevalence of *C. pylori* in NUD patients, with careful definition of the selection criteria, and strict attention to the comparability of a control group. Carefully controlled treatment studies seem more likely to provide a definitive answer, but at the moment these are hampered by the lack of an agent with specific anti-*C. pylori* efficacy, which will produce long-term eradication.

It must be concluded that at present there is no good indication to treat *C. pylori*-associated dyspepsia with anti-bacterial drugs, except within the confines of clinical trials. Nevertheless, many clinicians will continue to be convinced, from anecdotal and personal experience, that there is a subgroup of dyspeptics whose problems are caused by *C. pylori*, and in whom specific therapy is both justified and effective.

References

1 Gear, M.W.L. & Barnes, R.J. Endoscopic studies of dyspepsia in a general practice. *Br. Med. J.* 1980, **280**, 1136–1137.

2 Heatley, R.V. & Rathbone, B.J. Dyspepsia: A dilemma for doctors. *Lancet* 1987, **ii**, 779–782.

3 Adami, H.O., Agenas, I., Gustavsson, S., Loof, L., Nyberg, A., Nyren, O. & Tyllstrom, J. The clinical diagnosis of 'gastritis'. *Scand. J. Gastroenterol.* 1984, **19**, 216–219.

4 Tibblin, G. Introduction to the epidemiology of dyspepsia. *Scand. J. Gastroenterol.* 1985, Supplement 109, 29–33.

5 Knill-Jones, R.P. A formal approach to symptoms in dyspepsia. *Clin. Gastroenterol.* 1985, **14**, 517–529.

6 Anon. Management of dyspepsia: Report of a working party. *Lancet* 1988, **i**, 576–579.

7 Talley, N.J. & Piper, D.W. The association between non-ulcer dyspepsia and other gastrointestinal disorders. *Scand. J. Gastroenterol.* 1985, **20**, 896–900.

8 Marshall, B.J. & Warren, J.R. Unidentified curved bacilli in the stomach of patients with gastritis and peptic ulceration. *Lancet* 1984, **i**, 1311–1315.

9 Cheli, R., Perasso, A. & Giacosa, A. 1987. Symptomatology. In *Gastritis: A Critical Review*. Springer-Verlag, Berlin, Heidelberg. pp. 159–167.

10 Shiner, M. & Doniach, I. A study of X-ray negative dyspepsia with reference to histologic changes in the gastric mucosa. *Gastroenterology* 1957, **32**, 313–324.

11 Roberts, D.M. Chronic gastritis, alcohol, and non-ulcer dyspepsia. *Gut* 1972, **13**, 768–774.

12 Kreuning, J., Bosman, F.T., Kuiper, F., Van der Wal, A.M. & Lindeman, J. Gastric and duodenal mucosa in 'healthy' individuals. *J. Clin. Pathol.* 1978, **31**, 69–77.

13 Cheli, R., Simon, L., Aste, H., Figus, I.A., Nicold, G., Bajtai, A. & Puntoni, R. Atrophic gastritis and intestinal metaplasia in asymptomatic Hungarian and Italian populations. *Endoscopy* 1980, **12**, 105–108.

14 Myren, J. & Serck-Hanssen, A. Gastroscopic observations related to bioptical histology in healthy medical students. *Scand. J. Gastroenterol.* 1975, **10**, 353–355.

15 Cheli, R., Perasso, A. & Giacosa, A. Dyspepsia and chronic gastritis. *Hepato-gastroenterology* 1983, **30**, 21–23.

16 Villako, K., Ihamaki, T., Tamm. A & Tammur, R. Upper abdominal complaints and gastritis. *Ann. Clin. Res.* 1984, **16**, 192–194.

17 Toukan, A.U., Kamal, M.F., Amr, S.S., Arnaout, M.A. & Abu-Romiyeh, A.S. Gastroduodenal inflammation in patients with non-ulcer dyspepsia. A controlled endoscopic and morphometric study. *Dig. Dis. Sci.* 1985, **30**, 313–320.

18 Wyatt, J.I., Rathbone, B.J. & Heatley, R.V. Local immune response to gastric *Campylobacter* in non-ulcer dyspepsia. *J. Clin. Pathol.* 1986, **39**, 863–870.

19 Price, A.B. Histological aspects of *Campylobacter pylori* colonization and infection of gastric and duodenal mucosa. *Scand. J. Gastroenterol.* 1988, **23**, 21–24.

20 Earlam, R.J., Amerigo, J., Kakavoulis, T. & Pollock, D.J. Histological appearances of oesophagus, antrum and duodenum and their correlation with symptoms in patients with a duodenal ulcer. *Gut* 1985, **26**, 95–100.

21 Fink, S.M., Barwick, K.W., DeLuca, V., Sanders, F.J., Kandathil, M. & McCallum, R.W. The association of histologic gastritis with gastroesophageal reflux and delayed gastric emptying. *J. Clin. Gastroenterol.* 1984, **6**, 301–309.

22 Volpicelli, N.A., Yardley, J.H. & Hendrix, T.R. The association of heartburn with gastritis. *Dig. Dis.* 1977, **22**, 333–339.

23 Wyatt, J.I., Rathbone, B.J., Dixon, M.F. & Heatley, R.V. *Campylobacter pyloridis* and acid induced gastric metaplasia in the pathogenesis of duodenitis. *J. Clin. Pathol.* 1987, **40**, 841–848.

24 Johnston, B.J., Reed, P.I. & Ali, M.H. Campylobacter like organisms in duodenal and antral endoscopic biopsies: relationship to inflammation. *Gut* 1986, **27**, 1132–1137.

25 Sircus, W. Duodenitis: A clinical, endoscopic and histopathologic study. *Q. J. Med.* 1985, **56**, 593–600.

26 Greenlaw, R., Sheahan, D.G., DeLuca, V., Miller, D., Myerson, D. & Myerson, P. Gastroduodenitis, a broader concept of peptic ulcer disease. *Dig. Dis. Sci.* 1980, **25**, 660–672.

27 Thomson, W.O., Joffe, S.N., Robertson, A.G., Lee, F.D., Imrie, C.W. & Blumgart, L.H. Is duodenitis a dyspeptic myth? *Lancet* 1977, **i**, 1197–1198.

28 Jones, D.M., Eldridge, J., Fox, A.J., Sethi, P. & Whorwell, P.J. Antibody to the gastric campylobacter-like organism ('*Campylobacter pyloridis*') — clinical correlations and distribution in the normal population. *J. Med. Microbiol.* 1986, **22**, 57–62.

29 Graham, D.Y., Klein, P., Opekun, A.R. & Boutton, T.W. Effect of age on the frequency of active *Campylobacter pylori* infection diagnosed by the [^{13}C]-urea breath test in normal subjects and patients with

peptic ulcer disease. *J. Med. Microbiol.* 1988, **157**, 777–780.

30 Berkowicz, J. & Lee, A. Person-to-person transmission of *Campylobacter pylori*. *Lancet* 1987, **ii**, 680–681. (Letter).

31 Wyatt, J.I., Rathbone, B.J., Heatley, R.V. & Losowsky, M.S. *Campylobacter pylori* and history of dyspepsia in healthy blood donors. *Gut* 1988, **29**, A706–A707.

32 Graham, D.Y., Klein, P.D., Opekun, A.R, Boutton, T.W., Evans, D.J. & Evans, D. Epidemiology of *Campylobacter pylori* infection: Ethnic considerations. *Scand. J. Gastroenterol.* 1988, Supplement 142, 9–13.

33 Beg, F., Oldmeadow, M., Morris, A., Miller, M. & Nicholson, G. *Campylobacter pylori* infection in patients undergoing endoscopy in Fiji. *N.Z. Med. J.* 1988, **101**, 140–141.

34 Gutlerrez, D., Sierra, F., Gomez, M.C. & Camargo, H. *Campylobacter pylori* in chronic environmental gastritis and duodenal ulcer patients. *Gastroenterology* 1988, **94**, A163.

35 Ihamaki, T., Varis, K. & Siurala, M. Morphological, functional and immunological state of the gastric mucosa in gastric carcinoma families. *Scand. J. Gastroenterol.* 1979, **14**, 801–812.

36 Araki, S. & Goto, Y. Peptic ulcer in male factory workers: a survey of prevalence, incidence, and aetiological factors. *J. Epidemiol. Comm. Health* 1985, **39**, 82–85.

37 Jorde, R. & Burhol, P.G. Asymptomatic peptic ulcer disease. *Scand. J. Gastroenterol.* 1987, **22**, 129–134.

38 Marshall, B.J. 1988. *Campylobacter pylori*: Addressing the controversies . In Menge, H., Gregor, M. & Tytgat, G.N.J. (Eds.) *Campylobacter pylori*. Springer-Verlag, Berlin, Heidelberg. 235–245.

39 Skoglund, M.L. & Whalen, J.W. Correlation of *C. pylori* and upper GI symptoms. *Gastroenterology* 1988, **94**, A430.

40 Pettross, C.W., Appleman, M.D., Cohen, H., Valenzuela, J.E. & Chandrasoma, P. Prevalence of *Campylobacter pylori* and association with antral mucosal histology in subjects with and without upper gastrointestinal symptoms. *Dig. Dis. Sci.* 1988, **33**, 649–653.

41 Rauws, E.A.J., Langenberg, W., Houthoff, H.J., Zanen, H.C. & Tytgat, G.N.J. *Campylobacter pyloridis*-associated chronic active antral gastritis. *Gastroenterology* 1988, **94**, 33–40.

42 Rouvroy, D., Bogaerts, J., Nsengiumwa, O., Omar, M., Versailles, L. & Haot, J. *Campylobacter pylori*, gastritis, and peptic ulcer disease in central Africa. *Br. Med. J.* 1987, **295**, 1174.

43 Jiang, S.J., Liu, W.Z., Zhang, D.Z., Shi, Y., Xiao, S.D., Zhang, Z.H. & Lu, D.Y. Campylobacter-like organisms in chronic gastritis, peptic ulcer, and gastric carcinoma. *Scand. J. Gastroenterol.* 1987, **22**, 553–558.

44 Schnell, G.A. & Schubert, T.T. Relation of CLO test to clinical diagnosis. *Gastroenterology* 1988, **94**, A410.

45 Barthel, J.S., Westblom, T.U., Havey, A.D., Gonzalez, F. & Everett, E.D. Gastritis and *Campylobacter pylori* in healthy asymptomatic volunteers. *Arch. Intern. Med.* 1988, **148**, 1149–1151.

46 Langenberg, M.L., Tytgat, G.N.J., Schipper, M.E.I., Reitra, P.J.G.H. & Zanen, H.C. Campylobacter-like organisms in the stomach of patients and healthy individuals. *Lancet* 1984, **i**, 1348–1349. (Letter).

47 Peterson, W.L., Lee, E.L. & Feldman, M. Gastric campylobacter-like organisms in healthy humans: correlation with endoscopic appearance and mucosal histology. *Gastroenterology* 1986, **90**, 1585.

48 Marshall, B.J. Nervous Dyspepsia. *Med. J. Australia* 1985, **142**, 704. (Letter).

49 Loffeld, R.J.L.F., Potters, H.V.P.J., Arends, J.W., Stobberingh, E. & Flendrig, J.A. *Campylobacter* associated gastritis in patients with non-ulcer dyspepsia. *J. Clin. Pathol.* 1988, **41**, 85–88.

50 Rokkas, T., Pursey, C., Uzoechina, E., Dorrington, L., Simmons, N.A. & Fillipe, M.I. *Campylobacter pylori* and non-ulcer dyspepsia. *Am. J. Gastroenterol.* 1987, **11**, 1149–1152.

51 Borsch, G., Wegener, M., Schmidt, G., Sandmann, M., Adamek, R. & Reitemeyer, E. Prospective analysis of clinical and histologic factors associated with *Campylobacter pylori* colonization. *Gastroenterology* 1988, **94**, A44.

52 Andersen, L.P., Elsborg, L. & Justesen, T. *Campylobacter pylori* in peptic ulcer disease. III. Symptoms and paraclinical and epidemiologic findings. *Scand. J. Gastroenterol.* 1988, **23**, 347–350.

53 McNulty, C.A.M., Gearty, J.C., Crump, B., Davis, M., Donovan, I.A. & Melikian, V. *Campylobacter pyloridis* and associated gastritis: investigator blind, placebo controlled trial of bismuth salicylate and erythromycin ethylsuccinate. *Br. Med. J.* 1986, **293**, 645–649.

54 Nyren, O., Adami, H.O., Bates, S. Bergstrom, R., Gustavsson, S., Loof, L. & Sjoden, P. Self-rating of pain in non-ulcer dyspepsia. *J. Clin. Gastroenterol.* 1987, **9**, 408–414.

55 Lambert, J.R., Borromeo, M., Korman, M.G. & Hansky, J. Role of *Campylobacter pyloridis* in non-ulcer dyspepsia — A randomised controlled trial. *Gastroenterology* 1987, **92**, 1488.

56 Borody, T., Daskalopoulos, G., Brandl, S., Carrick, J. & Hazell, S. Dyspeptic symptoms improve following eradication of gastric *Campylobacter pyloridis*. *Gastroenterology* 1987, **92**, 1324.

57 Borody, T., Hennessy, W., Daskalopoulos, G., Carrick, J. & Hazell, S. Double blind trial of De-Nol in non-ulcer dyspepsia associated with *Campylobacter pyloridis* gastritis. *Gastroenterology* 1987, **92**, 1324.

58 Rokkas, T., Pursey, C., Simmons, N.A., Filipe, M.I. & Sladen, G.E. Non-ulcer dyspepsia and colloidal bismuth subcitrate therapy: the role of *Campylobacter pyloridis*. *Gastroenterology* 1987, **92**, 1599.

59 Rokkas, T. & Sladen, G.E. Bismuth: Effects of gastritis and peptic ulcer. *Scand. J. Gastroenterol.* 1988, Supplement 142, 82–86.

60 Tasman-Jones, C., Maher, C., Thomsen, L., Lee, S.P. & Vanderwee, M. Mucosal defences and gastroduodenal disease. *Digestion* 1987, Supplement 2, 1–7.

61 Konturek, S.J., Brzozowski, T., Drozdowicz, D. & Bielanski, W. 1988. Gastroprotective and ulcer healing properties of bismuth salts. In Menge, H., Gregor, M., Tytgat, G.N.J. & Marshall, B.J. (eds.) *Campylobacter pylori*. Springer-Verlag, Berlin, Heidelberg. pp. 184–194.

62 Slomiany, B.L., Bilski, J., Sarosiek, J., Liau, Y.H. & Slomiany, A. Colloidal bismuth subcitrate (De-Nol) inhibits peptic degradation of epidermal growth factor. *Gastroenterology* 1988, **94**, A431.

63 Glupczynski, Y., Burette, A., Labbe, M., Deprez, C., De Reuck, M. & Deltenre, M. *Campylobacter pylori*-associated gastritis: A double-blind placebo-controlled trial with amoxycillin. *Am. J. Gastroenterol.* 1988, **83**, 365–372.

64 Gilman, R., Leon-Barua, R., Ramirez-Ramos, A., Morgan, D., Recavarron, S. & Spira, W. Efficacy of nitrofurans in the treatment of antral gastritis associated with *Campylobacter pyloridis*. *Gastroenterology* 1987, **92**, 1405.

65 Larkai, E.N., Lacey-Smith, J., Lidsky, M.D. & Graham, D.Y. Gastroduodenal mucosa and dyspeptic symptoms in arthritic patients during chronic nonsteroidal anti-inflammatory drug use. *Am. J. Gastroenterol.* 1987, **82**, 1153–1158.

66 Katzka, D.A, Sunshine, A.G. & Cohen, S. The effect of nonsteroidal anti-inflammatory drugs on upper gastrointestinal tract symptoms and mucosal integrity. *J. Clin. Gastroenterol.* 1987, **9**, 142–148.

67 Upadhyay, R., Howatson, A., McKinlay, A., Danesh, B.J.Z., Sturrock, R.D. & Russell, R. *Campylobacter pylori* associated gastritis in patients with rheumatoid arthritis taking nonsteroidal anti-inflammatory drugs. *Br. J. Rheumatol.* 1988, **27**, 113–116.

68 Doube, A. & Morris, A. Nonsteroidal anti-inflammatory drug-induced dyspepsia — is *Campylobacter pylori* implicated? *Br. J. Rheumatol.* 1988, **27**, 110–112.

69 Drumm, B., O'Brien, A., Cutz, E. & Sherman P. *Campylobacter pyloridis*-associated primary gastritis in children. *Pediatrics* 1987, **80**, 192–195.

70 Drumm, B., Sherman, P., Cutz, E. & Karmali, M. Association of *Campylobacter pylori* on the gastric mucosa with antral gastritis in children. *N. Eng. J. Med.* 1987, **316**, 1557–1561.

71 Thomas, J., Eastham, E.J., Elliot, T.S.J., Dobson, C.M. & Jones, D.M. *Campylobacter pylori* gastritis in children — a common cause of symptoms? *Gut* 1988, **29**, A707.

72 Cadranel, S., Glupczinsky, Y., Labbe, M. & De Prez, C. 1988. *Campylobacter pylori* in children. In Menge, H., Gregor, M., Tytgat, G.N.J. & Marshall, B.J. (eds.) *Campylobacter pylori*. Springer-Verlag, Berlin, Heidelberg. pp. 110–117.

73 Marshall, B.J., Armstrong, J.A., McGechie, J.A. & Glancy, R.J. Attempt to fulfil Koch's postulates for pyloric *Campylobacter*. *Med. J. Australia* 1985, **142**, 436–439.

74 Morris, A. & Nicholson, G. Ingestion of *Campylobacter pyloridis* causes gastritis and raised fasting gastric pH. *Am. J. Gastroenterol.* 1987, **82**, 192–199.

75 Graham, D.Y., Smith, J.L., Alpert, L.C. & Yoshimura, H.H. Epidemic achlorhydria is not viral but is caused by *Campylobacter pyloridis*. *Gastroenterology* 1987, **92**, 1412.

76 Hazell, S.L., Carrick, J., Edwards, P. & Lee, A. Acute infection with *Campylobacter pylori* can mimic gastric cancer. *Gastroenterology* 1988, **94**, A178.

77 Ramsay, E.J., Carey, K.V., Peterson, W.L., Jackson, J.J., Murphy, F.K. & Read, N.W.T. Epidemic gastritis with hypochlorhydria. *Gastroenterology* 1979, **76**, 1449–1457.

78 Goodwin, C.S., Armstrong, J.A. & Marshall, B.J. *Campylobacter pyloridis*, gastritis, and peptic ulceration. *J. Clin Pathol.*, 1986, **39**, 353–365.

79 Gledhill, T., Leicester, R.J., Addis, B., Lightfoot, N., Barnard, J. & Viney, N. Epidemic hypochlorhydria. *Br. Med. J.* 1985, **290**, 1383–1386.

80 Wiersinga, W.M. & Tytgat, G.N. Clinical recovery owing to target parietal cell failure in a patient with Zollinger–Ellison syndrome. *Gastroenterology*, 1977, **73**, 1413–1417.

81 Hill, I.D., Sinclair-Smith, C., Lastovica, A.J., Bowie, M.D. & Emms, M. Transient protein losing enteropathy associated with acute gastritis and *Campylobacter pylori*. *Arch. Dis. Child.* 1987, **62**, 1215–1219.

82 Blomberg, B., Jarnerot, G., Kjellander, J., Danielsson, D. & Kraaz, W. Prevalence of *Campylobacter pylori* in an unselected Swedish population of patients referred for gastroscopy. *Scand. J. Gastroenterol.* 1988, **23**, 358–362.

20 Campylobacter pylori in paediatric populations

M.J. MAHONY AND J.M. LITTLEWOOD

Paediatricians frequently see children who have recurrent abdominal pain and dyspeptic symptoms of nausea and vomiting. Their approach to such patients has been influenced by the pioneering work of the late John Apley in the 1950s and 1960s [1]. Apley surveyed 1000 well schoolchildren and 100 children referred to hospital because of recurrent abdominal pain. The children were all aged 3 years or over and recurrent abdominal pain was defined as three or more episodes of abdominal pain over a period of at least 3 months. In well schoolchildren the overall prevalence of recurrent abdominal pain was 10.8%. In the children with recurrent abdominal pain there was often a history of abdominal pain in siblings and parents and a family history of peptic ulcer disease was present in 10%, compared to 3% in unaffected children. Most children with abdominal pain were aged between 5 and 8 years and there were later peaks at 9 years in girls and at 14 years in boys. The pain was typically peri-umbilical and the further the pain was localized from the umbilicus the more often an organic cause was found. An organic cause was found in only 8 of the 100 children investigated. Apley concluded that recurrent abdominal pain was largely 'non-organic' in nature and he emphasized the importance of psychosomatic factors. Follow-up of 60 children revealed that the abdominal pain continued into early adult life in 23, and one had a perforated duodenal ulcer. Hence Apley's well-known aphorism 'Little bellyachers grow up to be big bellyachers'.

Apart from this psychosomatic approach, recognized organic causes of recurrent abdominal pain in children include constipation [2], lactose intolerance [3], food intolerance [4] and more rarely Crohn's disease [5] and pancreatitis [6].

The availability of small-diameter fibreoptic endoscopes in recent years has meant that more children have been investigated by upper gastro-intestinal endoscopy [7]. The ability to visualize and biopsy the upper gastro-intestinal tract has led to the increased recognition of conditions such as oesophagitis, peptic ulcer disease, gastro-duodenal Crohn's disease and eosinophilic gastroenteritis in children presenting with abdominal pain and upper gastro-intestinal symptoms. There has been little published work on gastritis in children, most reports describing acute gastritis following drug ingestion and chronic gastritis associated with pernicious anaemia.

Paediatric gastroenterologists, especially those who perform upper gastrointestinal endoscopy, followed the description of *Campylobacter pylori*-associated gastritis in 1983 [8] and subsequent literature with great interest. The first reports of *C. pylori*-associated gastritis in children appeared in 1986 [9–11] and there followed a rapidly expanding paediatric literature on the subject with reports from many different parts of the world.

We summarize here the present knowledge of *C. pylori* gastro-duodenal disease in children based on published reports and our own experience.

Prevalence of Campylobacter pylori in children

As *C. pylori* has been diagnosed mainly by examination of histological specimens obtained at upper gastrointestinal endoscopy, estimates of the prevalence of *C. pylori* in symptomatic children are based on reported upper GI endoscopy series [12–15]. These studies obviously contain highly selected patients but do give some indication of the occurrence of the organism in this selected paediatric population. More objective information on general prevalence is available from three serological surveys [16–18] performed on non-symptomatic healthy children.

Retrospective surveys

We reviewed 111 upper GI endoscopies performed on children in our unit between 1981 and 1987; on re-examination of 38 antral biopsies, where endoscopy had been performed for abdominal pain, *C. pylori* colonization was found in 8 (21%) of 10 (26.3%) biopsies showing antral gastritis [12]. One hundred and seventy nine upper gastro-intestinal endoscopies performed at Toronto Hospital for Sick Children in the years 1984 and 1985 were reviewed in another study [13]; of 53 antral biopsies which were available for study, 10 (18.8%) had primary antral gastritis and 6 (11%) of these latter biopsies had *C. pylori* colonization. Oderda *et al.* in Turin carried out a retrospective study of 669 gastric biopsies obtained at upper gastro-intestinal endoscopy performed between 1983 and 1987; 58 (8.6%) biopsies showed antral gastritis and 41 (6.1%) had *C. pylori* colonization (G. Oderda, personal communication).

Prospective studies

Cadrenal *et al.* [14] performed 137 endoscopies in 114 children aged 2 months to 16 years (mean 6.7 years). Indications for endoscopy were recurrent vomiting, chronic abdominal pain, haematemesis, follow-up of Nissen's fundoplication, gastro-duodenal ulcers and oesophageal caustic injuries. *Campylobacter pylori* was found on histology and culture in 36/114 (31.6%) children. Duodenal ulcer was present in two children and both were *C. pylori*-positive. When 67 children who had upper gastro-intestinal endoscopy and antral biopsy were studied by Drumm *et al.* [15], primary antral gastritis associated with *C. pylori* colonization was found in seven children (10.4%), and duodenal ulcer was found, in addition, in five patients with *C. pylori* on the antral mucosa. In a study of 51 children endoscoped for upper gastrointestinal symptoms reported by Thomas *et al.* [16], *C. pylori*-associated antral gastritis was found in 20%. We have studied 40 children, aged 2–16 years, by upper gastro-intestinal endoscopy and antral biopsy. *Campylobacter pylori* was identified on histology of antral biopsies in nine (22.5%), five of whom showed active chronic gastritis, three chronic gastritis and one showed lymphoid hyperplasia only without gastritis. No duodenal or gastric ulcers were found. In another study *C. pylori* was cultured from six out of 25 (24%) children endoscoped, and four of these patients had chronic gastritis with duodenal ulcer in one case [9].

Serological studies

Jones *et al.* [17] performed a large population survey for *C. pylori* antibodies on sera from 607 healthy individuals without gastro-intestinal symptoms, using complement fixation and ELISA techniques: 109 of those studied were children aged 6–14 years, and *C. pylori* antibodies were found in one case only. Of 151 healthy individuals aged 15–20 years *C. pylori* antibodies were present in nine (6%). *Campylobacter pylori*-specific antibodies were detected using a sensitive ELISA method in 5% of 150 well schoolchildren, and 4% of 51 children in hospital with non-gastro-enterological symptoms [16]. In another study IgG antibodies specific to *C. pylori* were measured by ELISA in 37 adult peptic ulcer patients, 50 healthy adult laboratory workers and 31 children under 10 years of age. The peptic ulcer patients had significantly raised serum IgG antibodies to *C. pylori* compared with the laboratory workers and the children. The children had lower antibody levels than the healthy adults, suggesting that *C. pylori* is acquired later in life [18].

From these studies, it appears that *C. pylori* infection is rare in apparently healthy children. However, in children with upper gastro-intestinal symptomatology sufficient to warrant investigation by endoscopy, *C. pylori*-associated gastritis is found in a significant proportion ranging from 6% to 31%.

Clinical features

Age

Campylobacter pylori is principally found in older children. In most published reports, affected children have been in the 8–16 year age group. One study [14] includes two children aged 3 and 4 years, respectively, and we have found *C. pylori* in two children aged 2 and 5 years, respectively. *Campylobacter pylori* gastritis associated with protein-losing enteropathy involving three children in a younger age group (2–4 years) [19] has been reported.

Sex ratio

Although not all published reports give details of the children's sex, there does not appear to be a significant difference between male and female children affected.

Ethnic origin

Campylobacter pylori has been reported in children from Australia [9], United States [11], United Kingdom [12, 16], Belgium [14], Canada [15], Italy [20] and Poland [21]. Cadrenal reported a high incidence among North African children living in Brussels [14].

Abdominal pain

Recurrent abdominal pain has been the main presentation in children in whom *C. pylori* has been found on upper gastro-intestinal endoscopic biopsies [14, 15]. It is usually epigastric and suggestive of peptic ulcer disease, but can also be peri-umbilical [11]. The site of abdominal pain in nine children with *C. pylori* gastritis diagnosed by the present authors was epigastric in four, right-sided para-umbilical in three, peri-umbilical in one and retrosternal in one case.

Vomiting

Recurrent vomiting alone, or in association with abdominal pain, is a common presentation. In our children, it was a prominent symptom in 4/9 cases with *C. pylori* gastritis, often accompanied by nausea, headache and lethargy.

Haematemesis

Haematemesis was a major complaint in three of five children with *C. pylori* gastritis described in one report; gastric erosions were found in one case and duodenal ulcer in another [11].

Gastro-duodenal ulcer

Association with endoscopically proven gastric and duodenal ulcers has been reported frequently. Most series [9, 12, 14, 15, 22] of affected children have had a proportion of children with duodenal ulcer ranging from 1.75% to 29%. A family history of duodenal ulcer is also frequently found, both in children with duodenal ulcer and those with gastritis. Gastric ulcers are less frequent but were present in 18% in one series [22]. In one report of *C. pylori*-associated gastritis in five children [11], duodenal ulcer was present in three and gastric ulcer in one case.

As in adult patients [23] *C. pylori* has been associated with recurrence of peptic ulcer in children following treatment with H_2-receptor blockers. Oderda *et al.* [24] carried out a five-year follow-up study on 31 children (20 males, age range 3–14 years) with endoscopically-diagnosed duodenal ulcer (23 cases) and antral ulcer (eight cases). All were treated with ranitidine (5–10 mg/kg/day) for 8 weeks and ulcer healing was achieved in 93 %. After 1–3 years, 14 children (45 %), including 12 males, showed a recurrent ulcer crater on endoscopy. *Campylobacter pylori* was present on the antral mucosa in 12 (85 %) of the ulcer relapse patients, and in five (29 %) of the non-relapse patients.

Anaemia

Iron-deficiency anaemia (haemoglobin = 4.0 g %) was the presentation in one of our patients, and one patient in another series [16] also presented with iron-deficiency anaemia.

Protein-losing enteropathy

Hill *et al.* [19] reported 3 young children aged 2, 3 and 4 years who presented with a transient protein-losing enteropathy (PLE) (confirmed by ^{51}Cr-labelled albumin studies) associated with acute gastritis and the presence of *C. pylori* on the gastric mucosa. Stool culture and microscopy was negative for bacterial pathogens and parasites. Gastroscopy showed severe gastritis mainly affecting the body of the stomach with swollen gastric rugae in two cases. Histology showed active gastritis and presence of *C. pylori*, which were also cultured in two cases, and normal duodenal histology. In two children, treatment was with erythromycin. All three children recovered from PLE and this occurred more rapidly in the two treated children. Follow-up endoscopy showed a normal gastric mucosa. Histology of gastric body biopsies was normal with clearance of *C. pylori* and *C. pylori* culture was negative. This is the only report of an association between acute gastritis, *C. pylori* and PLE. Transient PLE has been described in children after viral gastroenteritis, and Hill's study [19] lacked detailed virological investigations. In this report, two of the patients had an antecedent diarrhoeal illness and could have had viral gastroenteritis. Detailed virological studies to exclude a viral cause for PLE will be needed on similar patients to firmly establish a causal relationship between *C. pylori*, gastritis and PLE.

Diagnosis

Upper gastro-intestinal endoscopy

In children with antral gastritis a distinctive macroscopic antral nodularity or pseudopolypoid appearance (Fig. 20.1) has been described [11, 14, 22]. This is not usually found in adult patients. Antral erythema may be present, but the endoscopic appearance can be normal. Antral biopsy is therefore essential for diagnosis. The duodenum is also inspected for ulceration and biopsied. Antral biopsies are taken for histology and culture, and may also be examined by scanning electron microscopy studies and the urease test (CLO test).

Detailed endoscopic findings have been reported in two studies. In Oderda's study [22] of 44 children aged 5–17 years who were *C. pylori*-positive on antral biopsy, 17 (44 %) had antral nodularity, 27 (61 %) antral hyperaemia, 8 (18 %) gastric ulcer and duodenal ulcer in 13 (29 %). Additional endoscopic findings were duodenitis in 11 (25 %) and oesophagitis in 26 (59 %) which was associated with Barrett's metaplasia in two (4.5 %). In Cadrenal's study [14] 36 of 114 children were positive for *C. pylori*. In the 36 *C. pylori*-positive children, the endoscopic appearance was normal in five (14 %), while 19 (52 %) had antral nodularity. Hiatus hernia was present in nine children and seven were *C. pylori*-positive, three of whom had Barrett's oesophagus and *C. pylori* was isolated on biopsy of the cardia in these three children. In addition *C. pylori* was identified in gastric biopsies of five children undergoing follow-up endoscopy for caus-

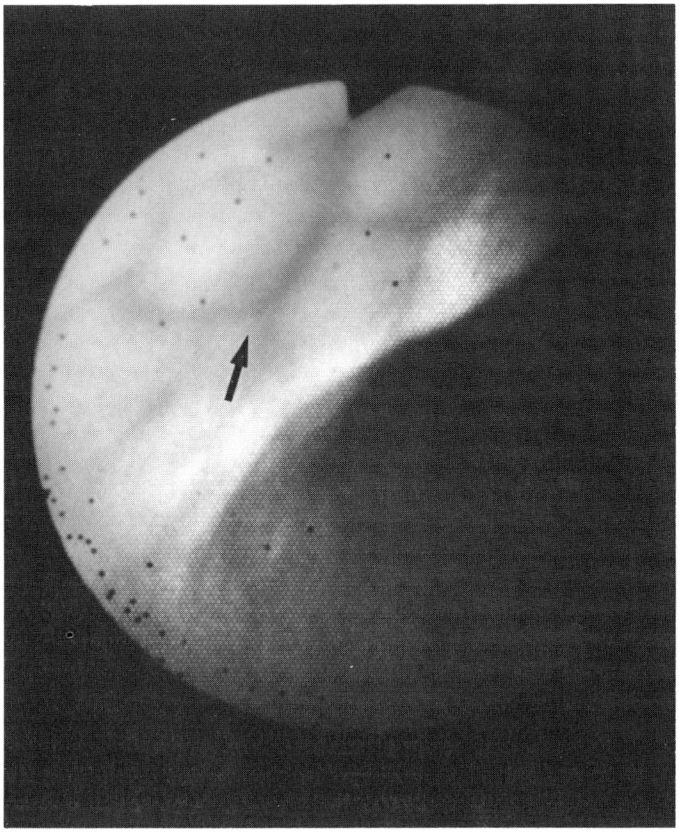

Fig. 20.1 Endoscopic appearance of antral nodularity (arrow) in a child with *C. pylori* gastritis.

tic oesophagitis which occurred one year previously, but not in two children endoscoped within one month of suffering caustic oesophagitis. In nine children positive for *C. pylori* studied by the present authors, the endoscopic features were antral nodularity in five, antral erythema in two, deformed pylorus in one and normal mucosa in one case.

In contrast to these series, Drumm *et al.* [15] found a poor correlation between endoscopic and histological findings. Gross inflammation was noted on endoscopy in only two of ten children with primary antral gastritis.

Histology

Campylobacter pylori colonization is detected histologically by a number of stains. The Warthin–Starry silver stain was the first described, but in our laboratory a modified Giemsa stain is successful in identifying *C. pylori* [25]. Some investigators have also used scanning electron microscopy to detect *C. pylori* [16]. Chronic gastritis is a histological diagnosis based on a diffuse chronic inflammatory cell infiltrate of the lamina propria of the gastric mucosa. The presence of neutrophils within the glandular epithelium associated with degenerative changes on epithelial cells indicates 'active' chronic gastritis [26].

Drumm *et al.* [15] have described detailed histological findings. Antral biopsies were obtained from upper gastro-intestinal endoscopies in 67 children: 49 biopsies were histologically normal, and negative for *C. pylori* on silver staining and culture; 18 showed evidence of gastritis. In eight of these patients, an identifiable cause for the gastritis was present from the clinical history or histology: Crohn's disease in three, eosinophilic gastroenteritis in two, acetylsalicylic acid in two and corticosteroids in one child. These were termed 'secondary antral gastritis'. These biopsies were all negative for *C. pylori* on Warthin–Starry staining and on culture. In ten (15%) biopsies there was gastritis for which no cause could be determined from the clinical history or histology. These were termed 'primary antral gastritis'; *C. pylori* was identified by Warthin–Starry stain and by culture in seven of the ten biopsies. Duodenal ulcer was diagnosed endoscopically in five patients and *C. pylori* was cultured from the gastric antrum in all five; histological examination of the gastric antrum in four of these showed gastritis with *C. pylori* colonization although *C. pylori* was not found in the duodenal biopsies. Drumm *et al.* [15] found no correlation between the endoscopic appearance and the histological findings.

In our experience lymphoid follicles are a prominent histological feature of paediatric biopsies with *C. pylori*-associated antral gastritis and correlate well with the presence of antral nodularity on endoscopy.

Gastric histology from 79 children studied by the present authors showed chronic gastritis in 19, of whom 16 were *C. pylori*-positive by histology. The three *C. pylori*-negative cases had unusual histological patterns of gastritis, characteristic of 'lymphocytic gastritis' (two cases) and 'reflux gastritis' (one case). Histologically, the inflammatory infiltrate of campylobacter-associated gastritis seen in children appears to differ in some ways from that in adults. The chronic inflammatory cell component tends to be less dense in children, while lymphoid follicles (seen in 8/16 cases) are often a prominent feature. In adults, campylobacter-associated gastritis usually shows an 'active' polymorph component, but in children this is less common. Gastric biopsies from 49 children were entirely normal. However, a further 11 cases showed one or two mucosal lymphoid aggregates without other features of chronic gastritis — a situation rarely encountered in adults. In 2 cases, they were histologically *C. pylori*-positive. The relationship of this finding to the natural history of the infection requires further investigation.

Microbiology

Drumm *et al.* [15] cultured antral biopsies for *C. pylori* using blood agar and chocolate agar in Skirrow's medium under microaerophilic conditions. All seven patients who had histological evidence of *C. pylori*-associated primary gastritis on Warthin–Starry stain grew *C. pylori* on culture. Cadrenal *et al.* [27] isolated *C. pylori* by culture in 30/175 patients but *C. pylori* was identified by histological staining (haematoxylin & eosin) in 22 biopsies only. Eastham *et al.* [28] reported positive antral biopsy culture for *C. pylori* in 4/7 patients with histological active chronic gastritis associated with *C. pylori*.

Urease test

Gastric urease assay as a marker of *C. pylori* infection was evaluated in a study of 25 paediatric patients aged 2–18 years and undergoing upper gastro-intestinal endoscopy for investigation of epigastric pain, nausea and vomiting and haematemesis [29]. Gastric biopsies were placed in Stuart urease test broth and incubated at room temperature (a colour change from yellow to pink in this test broth indicates the presence of urease): 8 (32%) biopsies were positive for *C. pylori* on histology and culture, and all demonstrated urease activity. None of the urease-negative biopsies had *C. pylori*, giving 100% sensitivity and specificity for the urease test. In another report [15], however, the urease test was positive in only 3 out of 6 patients with *C. pylori* antral gastritis.

Serology

We have measured serum IgG antibodies spe-

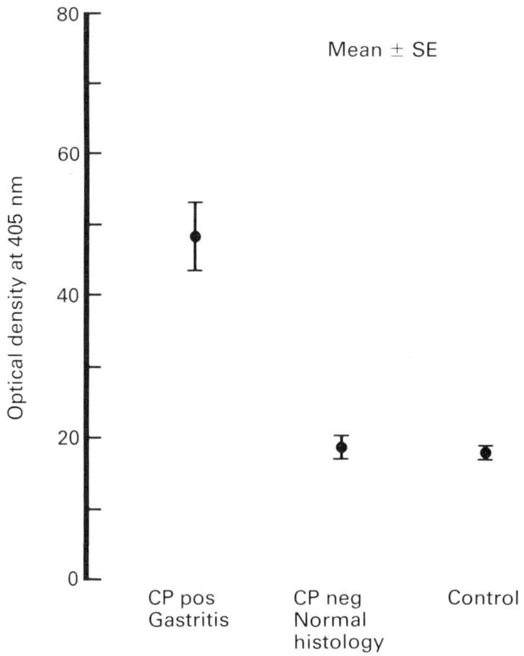

Fig. 20.2 Serum IgG antibody levels to *C. pylori*.

cific to *C. pylori* in 36 children (age range 2–15 years, median 12.5 years, and including 14 males) who were investigated for *C. pylori*-associated gastro-duodenal disease by upper gastro-intestinal endoscopy and biopsy, and in 16 control children (age range 1.5–15 years, median 11.5 years, and including 6 males). The control children were orthopaedic inpatients undergoing treatment for fractures and scoliosis and none had gastro-intestinal symptoms. *Campylobacter pylori*-specific IgG antibodies were assayed by an indirect ELISA technique using a soluble bacterial antigen preparation and quantitated by optical density measurement at 405 nm. *Campylobacter pylori* IgG antibody titres (mean optical density ± S.E.) were 48 ± 9.6 in 9 children who had *C. pylori*-positive gastritis, 18 ± 3.5 in 27 *C. pylori*-negative normal-histology children, and 17.3 ± 2.4 in the 16 control children. There was a significant difference between *C. pylori*-positive and *C. pylori*-negative children ($P < 0.05$), and between *C. pylori*-positive and control children ($p < 0.01$) by the Wilcoxon rank sum non-parametric test, (Fig. 20.2).

Thomas *et al.* [16] assayed *C. pylori*-specific IgG antibodies by ELISA in 51 children investigated for upper gastro-intestinal symptoms by endoscopy and gastric biopsy. Serum IgG antibodies specific to *C. pylori* were present in 'significant titres' (> 1 : 1600) in 20%, all of whom had *C. pylori* colonization, and chronic gastritis was present in all but two cases. Only 4% of 51 age-matched hospital controls had significant *C. pylori* antibody titres and in all cases these were < 1 : 1600.

Treatment

Children with *C. pylori* gastritis and associated duodenal ulceration have been treated with colloidal bismuth subcitrate and ampicillin. Follow-up data are limited at present. Potential toxicity of bismuth in paediatric patients has been a major concern because of the description of an encephalopathy in patients taking long-term bismuth [30].

We have treated five patients (aged 10–16 years) with *C. pylori*-associated primary antral gastritis with tripotassium dicitratobismuthate (De-Nol) at 240 mg b.d. for 2 months and ampicillin 500 mg q.d.s. for 2 weeks. Symptoms cleared in four children and follow-up endoscopy, performed 3–6 weeks after completion of treatment, revealed eradication of *C. pylori* from antral biopsies in all four cases; three antral biopsies showed normal mucosa, and the other residual gastritis without neutrophils; one patient remained symptomatic and follow-up histology showed persistence of *C. pylori* and chronic gastritis. Serum-specific *C. pylori* IgG antibody titres, measured 6–12 months after completion of treatment, were lower in the four children who responded to treatment (Fig. 20.3). Serum bismuth levels were measured in all patients and all were in the non-toxic range (3–19 μg/l). Neurotoxicity becomes more likely at serum levels > 50–100 μg/l [30]. No patient developed signs of bismuth toxicity,

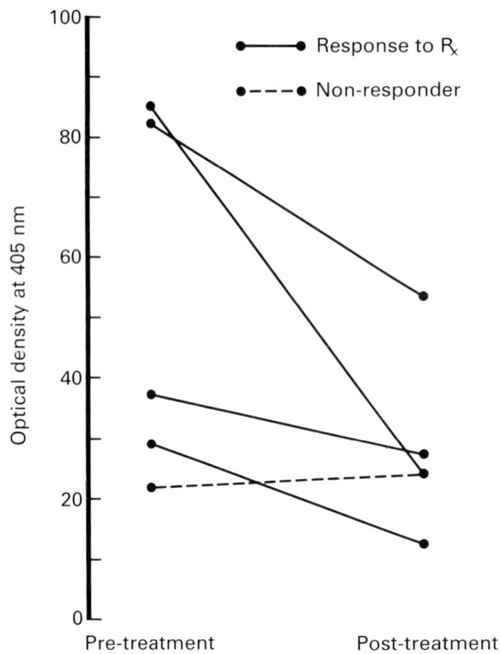

Fig. 20.3 Serum IgG antibody levels to *C. pylori* before and after treatment.

and biochemical monitoring of renal and hepatic function was normal during and after treatment.

Eastham & Elliott treated six patients with *C. pylori* gastritis using De-Nol for 2 months and ampicillin (250/500 mg/q.d.s.) for 2 weeks [28]. De-Nol at 480 mg/day in divided dose was given to children > 12 years of age and 240 mg/day to those < 12 years. All experienced resolution of symptoms, eradication of *C. pylori* and histological improvement. Socha *et al.* [21] treated ten children, five with *C. pylori* gastritis and five with *C. pylori*-associated duodenal ulcer with De-Nol for 4 weeks. *Campylobacter pylori* colonization was reduced from 100% to 30%. Bonamico *et al.* [20] treated six children with *C. pylori*-positive antral gastritis using ampicillin (five cases for 4 weeks and one case for 2 weeks). Resolution of symptoms and eradication of *C. pylori* from biopsy specimens taken 15 days after discontinuation of therapy occurred only in those treated for 4 weeks.

Conclusions

The emergence of *C. pylori* gastritis as an identifiable cause of recurrent abdominal pain and upper gastro-intestinal symptoms in childhood represents an important advance in paediatric gastro-enterology. Chronic active gastritis is rare in children but is strongly associated with *C. pylori* colonization. Available estimates based on limited numbers of serological surveys suggest that *C. pylori* infection is rare in asymptomatic healthy children. Among symptomatic children undergoing upper gastro-intestinal endoscopy, *C. pylori* infection has been demonstrated in various series in between 6% and 31% of children. The infection occurs mainly in older children in the 8–16 year age group, but has been found in children as young as 2 years. Information on treatment responses and long-term follow-up data is scant, although there are encouraging reports of small numbers of treated children.

Like other investigators, paediatric researchers have not identified a reservoir of infection or the mode of spread. A family history of peptic ulcer disease is frequently found in affected children and this may suggest a common exposure to a source of infection in families.

Paediatric *C. pylori* disease differs from adult disease in that recognizable endoscopic abnormalities are commonly found in the gastric antrum which has a nodular appearance, gastric and duodenal ulcers are less common, and lymphoid follicles are more frequent on histology. Serological evidence of *C. pylori* infection in asymptomatic children is rare (< 5%) as compared to the adult population where 50% of people > 50 years have raised levels of *C. pylori* antibodies [17], and affected children appear to be more often symptomatic than adults.

As childhood infection is more likely to represent the early stages of infection with *C. pylori*, the study of *C. pylori* in children and follow-up into adult life is likely to contribute to our knowledge of the natural history and pathogenic role of this organism in gastroduodenal pathology.

References

1 Apley, J. 1975. *The Child with Abdominal Pains*, 2nd edn. Blackwell Scientific Publications, Oxford.
2 Dimson, S.B. Transit time related to clinical findings in children with recurrent abdominal pain. *Pediatrics* 1971, **47**, 666–674.
3 Blumenthal, I., Kelleher, J. & Littlewood, J.M. Recurrent abdominal pain and lactose intolerance. *Br. Med. J.*, 1981, **282**, 2014–2015.
4 Minford, A.M.B., MacDonald, A. & Littlewood, J.M. Food intolerance and food allergy in children: a review of 68 cases. *Arch. Dis. Child.* 1982, **57**, 742–747.
5 O'Donoghue, D.P. & Dawson, A.M. Crohn's disease in childhood. *Arch. Dis. Child.* 1977, **52**, 627–632.
6 Forbes, A., Leung, J.W.C. & Cotton, P.B. Relapsing acute and chronic pancreatitis. *Arch. Dis. Child.* 1984, **59**, 927–934.
7 Miller, V. & Doig, C.M. Upper gastrointestinal tract endoscopy. *Arch. Dis. Child.* 1984, **59**, 1100–1102.
8 Warren, J.R. & Marshall, B.J. Unidentified curved bacilli on gastric epithelium in active chronic gastritis. *Lancet* 1983, **i**, 1273–1275.
9 Hill, R., Pearman, J., Worthy, P., Carusso, V., Goodman, S. & Blincow, E. *Campylobacter pyloridis* in children. *Lancet* 1986, **i**, 387.
10 Cadrenal, S., Goossens, H., de Boeck, M., Malengreau, A., Rodesch, P. & Butzler, J.P. *Campylobacter pyloridis* in children. *Lancet* 1986, **i**, 735–736.
11 Czinn, S.J., Dahms, B.B., Jacobs, G.H., Kallen, B. & Rothstein, F.C. Campylobacter-like organisms in association with symptomatic gastritis in children. *J. Pediatr.* 1986, **109**, 80–83.
12 Mahony, M.J., Wyatt, J.I. & Littlewood, J.M. *Campylobacter pylori* gastritis. *Arch. Dis. Child.* 1988, **63**, 654–655.
13 Drumm, B., O'Brien, A., Cutz, E. & Sherman, P. *Campylobacter pyloridis*-associated primary gastritis in children. *Pediatrics* 1987, **80**, 192–195.
14 Cadrenal, S., Glupczynsky, Y., Labbe, M. & de Prez, C. 1988. *Campylobacter pylori* in children. In Menge, H., Gregor, M., Tytgat, G.N.J. & Marshall, B.J. (eds.) *Campylobacter pylori*. Springer-Verlag, Berlin, Heidelberg. pp. 110–115.
15 Drumm, B., Sherman, P., Cutz, E. & Karmali, M. Association of *Campylobacter pylori* on the gastric mucosa with antral gastritis in children. *N. Engl. J. Med.* 1987, **316**, 1557–1561.
16 Thomas, J.E., Eastham, E.J., Elliott, T.S.J. & Nelson, R. *Campylobacter pylori* gastritis in children — a common cause of symptoms. *Gut* 1988, **29**, A707.
17 Jones, D.M., Eldridge, J., Fox, A.J., Sethi, P. & Whorwell, P.J. Antibody to the gastric campylobacter-like organism (*Campylobacter pyloridis*) — clinical correlation and distribution in the normal population. *J. Med. Microbiol.* 1986, **22**, 57–62.
18 Kaldor, J., Tee, W., McCarthy, P., Watson, J. & Dwyer, B. Immune response to *Campylobacter pyloridis* in patients with peptic ulceration. *Lancet* 1985, **i**, 921.
19 Hill, I.D., Sinclair-Smith, C., Lastovica, A.J., Bowie, M.D. & Emms, M. Transient protein losing enteropathy associated with acute gastritis and *Campylobacter pylori*. *Arch. Dis. Child.* 1987, **62**, 1215–1219.
20 Bonamico, M., Chiesa, C., Carpino, F., Pacifico, L., Magliocca, F.M., Pitzalis, G., Bosco, D., Castronuovvo, R., Midulla, M. & Signoretti, A. *Campylobacter pylori* in children — preliminary study. *Pediatr. Res.* 1988, **24**, 416. (Abstract)
21 Socha, J., Celinska-Cedro, D., Teisseyre, M. & Rasinski, A. *Campylobacter pyloridis* in children. *Pediatr. Res.* 1988, **24**, 416. (Abstract)
22 Oderda, G., Forni, M., Dell'olio, D. & Ansaldi, N. *Campylobacter pylori* (CP) and upper G.I. endoscopic and pathological changes in children. *Pediatr. Res.* 1988, **24**, 417. (Abstract)
23 Coghlan, J.G., Gilligan, D., Humphries, H., McKenna, D. & Dooley, C. *Campylobacter pylori* and recurrence of duodenal ulcer — a 12 month follow-up study. *Lancet* 1987, **ii**, 1109–1111.
24 Oderda, G., Farina, L. & Ansaldi, N. Peptic ulcer in children: 5 years follow-up after ranitidine therapy. *Pediatr. Res.* 1988, **24**, 417. (Abstract)
25 Gray, S.F., Wyatt, J.I. & Rathbone, B.J. Simplified techniques for identifying *Campylobacter pyloridis*. *J. Clin. Pathol.* 1986, **39**, 1279–1280.
26 Whitehead, R., Truelove, S.C. & Gear, M.W.L. The histological diagnosis of chronic gastritis in fibreoptic biopsy specimens. *J. Clin. Path.* 1972, **25**, 1–11.
27 Cadrenal, S., Rodesch, P., Di Lorenzo, C., De Prez, C., Goossens, H. & Glupczynsky, Y. Chronic gastritis in childhood: role of *Campylobacter pylori*. *Pediatr. Res.* 1987, **22**, 109. (Abstract)
28 Eastham, E.J. & Elliott, T.S.M. *Campylobacter pyloridis* in children. *Arch. Dis. Child.* 1987, **62**, 652. (Abstract)
29 Czinn, S.J. & Carr, H. Rapid diagnosis of *Campylobacter pyloridis*-associated gastritis. *J. Pediatr.* 1987, **110**, 569–570.
30 Bader, J.P. The safety profile of De-Nol. *Digestion* 1987, **37** (Suppl. 2), 53–59.

21 Campylobacter pylori and spontaneous hypochlorhydria

R.H. HUNT

Acid is secreted by the stomach of a wide variety of animals and man. Although acid secretion in man is not essential to life, it is important for the activation of pepsinogens and initiation of protein digestion, and to augment dietary iron and calcium absorption. The secretion of gastric acid is an important primary bactericidal barrier to protect the highly specialized enterocyte of the small bowel from ingested bacteria [1, 2]. The presence of gastric acid also modulates the secretion of gastrin through a feedback mechanism and this may be important in controlling the trophic effects of this hormone [3].

Bactericidal effects of gastric acid

Low pH is principally responsible for the bactericidal properties of gastric juice [4]. Exogenous bacteria introduced into the stomach at $\leqslant$ pH 3.0 are destroyed within 15 min, and some bactericidal activity is observed up to pH 4.0. The intragastric bacterial count is normally $< 10^5$ organisms/ml [5], rising to $> 10^5$ organisms/ml when acidity is reduced [6]. The bacterial count also increases temporarily during treatment with antisecretory drugs such as the H_2-receptor antagonists, or omeprazole [7], while nitrites and *N*-nitroso compounds also increase transiently [6, 8].

Reduction of acid predisposes to infection with a variety of bacteria, including *Salmonella, Shigella, Vibrio cholerae, Brucella* and parasites such as *Giardia* and *Strongyloides* [1]. Previous gastric surgery predisposes a patient to the risk of *Salmonella* infection [9], which is directly related to the reduction in gastric acid secretion. Gastric acid secretion decreases with age [10] and this is probably also responsible for the increased risk of severe *Salmonella* infection seen in patients >60 years [11].

Consequences of reduced acid secretion

The prevalence of achlorhydria varies widely but is high in less-developed countries, where it is commonly encountered with malnutrition [12] and is associated with a higher incidence of cholera.

It has been suggested that typhoid may induce achlorhydria [13], but it seems more likely that hypochlorhydria predisposes to typhoid infection with increasing severity [14].

Patients with pernicious anaemia and achlorhydria have an increased risk of gastric carcinoma [15, 16], which may also be true for prolonged acid reduction following gastric surgery [17]. The hypothesis that reduced acidity results in gastric colonization by nitrate-reducing bacteria proposes an increase in nitrites and *N*-nitroso compounds, which are, in some models, carcinogenic [18]. However, although hypoacidity predisposes to bacterial overgrowth and nitrate generation it does not appear to enhance nitrosation which is maximal at low pH, suggesting a chemical rather than bacterial nitrosation. These results are contrary to the nitrosamine hypothesis of gastric carcinogenesis which has never been shown in man.

Suppression of gastric acidity is accompanied by an increase in circulating gastrin [19, 20] which is trophic to the gastric mucosa. A spectrum of changes in gastric mucosal morphology has been described in animals treated with several potent H_2-receptor antagonists or omeprazole. These changes range from adenocarcinoma in the rat antrum and hyperplasia and dysplasia or carcinoma in the rat forestomach to hyperplasia of ECL cells and the development of gastric carcinoids [21]. These changes may be principally due to structural characteristics of the drug molecule rather than to reduced acidity *per se*. However, the associated hypergastrinaemia will be associated with a trophic stimulus to the gastro-intestinal mucosa [22]. The possible effects of hypergastrinaemia as a trophic stimulus on gastric mucosal cell dysplasia and malignant change is not known. Gastrin potentiates the growth of rat gastric cancer cell lines *in vitro* [23], however, and xenotransplanted human gastric cancer in male mice [24]. High, unremitting levels of gastrin seen in patients with pernicious anaemia and atrophic gastritis [25] have little or no trophic response on the mucosa. However, in the presence of nitrate-reducing bacteria, an organ-specific carcinogen and gastric mucosal disease such as gastritis, the gastric epithelium may be more sensitive to neoplastic and trophic stimuli.

Spontaneous hypochlorhydria

Spontaneous hypochlorhydria is a well-recognized but relatively rare event, which has usually been reported as a result of the development of achlorhydria in previously healthy individuals during the course of gastric secretion studies. Clinical investigators performing repeated tests of gastric acid secretion have observed prolonged periods of basal anacidity [26, 27] following what had previously been a measurable basal acid output. There are also a number of reports in the literature describing the development of achlorhydria in previously healthy individuals [28–31] and in patients with the Zollinger–Ellison syndrome [32–34]. Furthermore, two reports have described this phenomenon in a cluster of patients: Ramsay *et al.* [35] reported hypochlorhydria in 17/37 healthy volunteers participating in studies of food-stimulated gastric acid secretion, while Gledhill *et al.* [36] observed 4/5 previously healthy subjects who developed hypochlorhydria during a study of intragastric acidity following antacid treatment.

Hirschowitz *et al.* [28] described the development of anacidity occurring in a 23-year-old male medical student who was undergoing a study of the effects of corticosteroids on gastric acid secretion. Just prior to the development of achlorhydria the subject complained of non-specific epigastric pain and nausea. Spiro & Schwartz [29] described seven subjects in whom basal gastric acid was $> \text{pH}\ 6$, and all subjects had superficial gastritis and did not respond to histamine. Waterfall [30] described the development of histamine and pentagastrin-fast achlorhydria in himself during a series of gastric secretory studies. Biopsies of the gastric mucosa showed atrophic gastritis with the presence of chronic inflammatory cells and a reduced number of parietal cells, while another patient with a duodenal ulcer developed a similar reduction in gastric acid secretion, but did not have a biopsy performed. In the study reported by Desai *et al.* [31], seven patients with duodenal ulcer and three healthy volunteers underwent repeated gastric secretion studies. At the commencement of the study all the subjects had a gastric acid output within the normal range and a spontaneous decrease in the maximal acid output to histamine, ranging from 27 to 72%, was subsequently noted. All patients remained asymptomatic and in five cases the reduction in acid output lasted for > 4 months. Only two subjects had gastric biopsy performed and these showed changes consistent with atrophic gastritis.

The occurrence of spontaneous achlorhydria in patients with Zollinger–Ellison syndrome has also been clearly described [32, 34]. Desai & Anita [33] described a 53-year-old male patient with probable Zollinger–Ellison syndrome in whom they observed a spontaneous decrease in basal acid secretion from 40–50 to 0.6–1 mmol/hour and of histamine-stimulated peak acid output from 60–80 to 0.4–0.8 mmol/hour. This occurred in association with the development of histological confirmed chronic gastritis, and an apparent reduction in the total number of parietal cells. In the patient described by Lawrie *et al.* [32], a 48-year-old male with the Zollinger–Ellison syndrome, treatment with anticholinergic drugs was accompanied by the development of total achlorhydria which was attributed to the treatment.

Wiersinga & Tytgat [34] described a 29-year-old male patient with Zollinger–Ellison syndrome who experienced an intercurrent illness characterized by malaise, nausea, vomiting and mild pyrexia, at which time the patient spontaneously reported relief of his symptoms of pain and diarrhoea. A marked reduction in both the basal and stimulated gastric acid output was observed with almost anacidity. Histology of the gastric mucosa was considered relatively normal before the development of the intercurrent illness, but subsequently showed marked chronic inflammation for up to 8 months.

These reports all show that healthy individuals, during the course of gastric secretory studies, or patients with Zollinger–Ellison syndrome who remained under close medical supervision, developed spontaneous hypo- or achlorhydria. Thus, the true incidence of this phenomenon might be much greater in the community setting, since these reports could represent chance observations made in highly selected individuals. The occasional clustering of cases suggested in the reports by Spiro & Schwartz [29], Waterfall [30] and Desai *et al.* [31] raised the possibility that a transmissible agent might be involved in the pathogenesis of this condition. The histological changes of gastritis have often been reported in these studies, and in the Zollinger–Ellison cases reported by Desai & Anita [33], and Wiersinga & Tytgat [34]. While the histological changes are non-specific they could, nevertheless, have been caused by an infective agent. The possibility that an infective agent might be responsible is supported by evidence from the observations of an outbreak of spontaneous hypochlorhydria in the subjects reported by Ramsey *et al.* [35] and Gledhill *et al.* [36]. Both these studies described isolated outbreaks of achlorhydria in subjects participating in gastric secretory studies. Ramsey *et al.* [35] found hypochlorhydria in 17/37 healthy volunteers undergoing measurements of gastric acid secretion after stimulation by food. Gastric juice was frequently sampled from the stomach, and pH estimated using a common combined glass electrode, following which the gastric juice was returned to the stomach. As the pH electrode was not sterilized between measurements, these authors suggested that an infective agent might have been transmitted via a contaminated electrode, although no specific agent was identified. In the study reported by Gledhill *et al.* [36], which also employed a similar protocol and a common electrode to measure intra-gastric acidity following antacid treatment, 4/6 previously healthy subjects developed hypochlorhydria, following a transient illness with nausea, vomiting, and abdominal pain. All subjects initially had a normal secretory profile but mean basal and peak acid outputs were 0 and 2.3 mmol/hour 1 month after the onset of illness, and 1.5 and 27 mmol/hour at 8 months follow-up; two of the subjects were followed at 18 months when mean basal and peak acid output was 3.9 and 33.5 mmol/hour. The basal and peak acid outputs at 1, 8 and 18 months after onset of illness are shown in Table 21.1 for the subjects studied. At 1 month following the illness, the four subjects also underwent a 24-hour intra-gastric acidity study. Figure 21.1 shows the mean hourly pH values in the four subjects who developed spontaneous hypochlorhydria and pH values in 8 control subjects studied earlier. All mean hourly pH values in the subjects with spontaneous hypochlorhydria were higher than those in the controls.

Table 21.1 Basal and peak acid outputs (mmol/hour) at 1, 8, and 18 months after onset of illness [36]. See text

Case No.	Level of output	Period after onset of illness (months)		
		1	8	18
1	Basal	0	2.6	2.6
	Peak	0.3	31.4	31.8
2	Basal	0	1.2	5.2
	Peak	8.8	29.4	35.2
3	Basal	0	0.8	*
	Peak	0	20.2	*
4	Basal	0	†	†
	Peak	1.2	†	†

* Refused further studies.
† Working overseas.

Schilling tests performed in three subjects at 8 months showed diminished retention of vitamin B_{12}.

All subjects had blood taken for viral serology 2 days and 2 weeks after the condition was recognized. Complement fixation tests were performed against the following antigens: influenza A, influenza B, adenovirus, cytomegalovirus, and *Mycoplasma pneumoniae*. One month after the onset of hypochlorhydria, the patients were admitted for 24 hours to study intra-gastric acidity (see Fig. 21.1), bacterial counts and concentrations of nitrite and *N*-nitroso compounds. Of the 48 samples of gastric juice examined, 47 had a bacterial growth $> 10^5$ organisms/ml, and 46 had a growth of nitrate-reducing bacteria $> 10^5$ organisms/ml. Mean intra-gastric nitrite concentrations were ten times higher than in a group of eight healthy controls but both the mean total and mean stable *N*-nitroso compound concentrations were not appreciably different from those of controls. Gastric juice was aspirated with a sterile syringe and transferred to a universal container

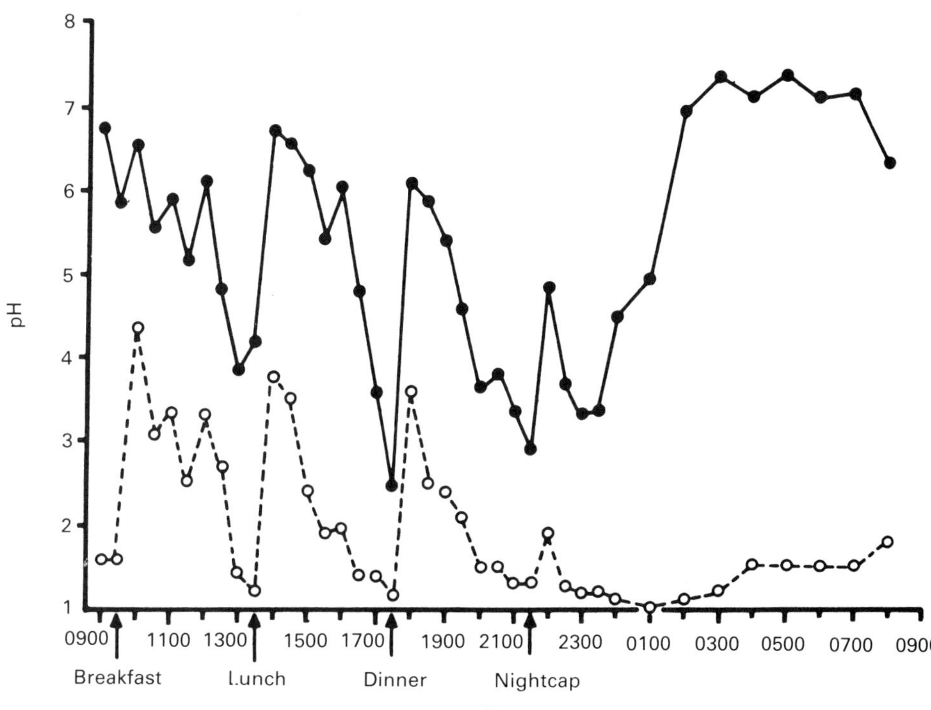

Fig. 21.1 Mean hourly pH values in 4 subjects who developed spontaneous hypochlorhydria (——) [36] and in 8 subjects studied earlier (– – – –).

sterilized with gas; mucus and food were broken up and bacteria dispersed by vortex mixing. Samples were placed inside an anaerobic chamber where all subsequent manipulations were performed. The gastric juice was diluted (down to 10^{12}) in pre-reduced brain/heart infusion broth, and added to indonitrite broth in micro-titre plates. The total numbers of bacteria and nitrate-reducing bacteria were estimated by the most probable number method after anaerobic incubation at 36.5°C for 5 days. In addition, one sample from each patient was examined to identify the individual bacterial species present. Haematological and biochemical screenings performed before, and 2 weeks after, the onset of illness showed unchanged results, and viral serology detected no rise in titres to influenza A and influenza B, adenovirus, cytomegalovirus, or *Mycoplasma pneumoniae*. Bacteria isolated by culture included *Neisseria, Corynebacterium, Streptococcus, Staphylococcus, Acinetobacter, Lactobacillus, Fusobacterium, Bacteroides, Veillonella* and *Bifidobacterium*, all of which are considered normal oral flora, and had presumably been swallowed. Coliforms were not isolated.

Initial endoscopy in all subjects following illness showed a macroscopically normal stomach; however, histology showed active superficial gastritis in all subjects affected. The inflammatory response was most noticeable in the gastric antrum, and in some subjects clusters of polymorphs were present within the gastric glands. Cultures of gastric biopsy specimens in baboon kidney, HEP_2, and MRC_5 cell lines did not show the presence of any virus. Endoscopy and biopsy was repeated after a further 8 months. Histology had returned to normal in one patient, although the three other subjects showed features of chronic gastritis with plasma cell and lymphocytic infiltration replacing the polymorphs seen in the earlier specimens. A further patient still had active chronic gastritis after 18-month follow-up. Thus, all patients in this study had superficial gastritis which became chronic in three cases, and in one patient moderate titres of parietal-cell antibodies developed suggesting that this condition might progress to atrophic gastritis.

Although community transmission was considered a possibility by these authors, serological screening and electron microscopy of gastric biopsy specimens failed to show an infective cause. Transmission of an unidentified enteric pathogen via a contaminated pH electrode was therefore suspected.

Spontaneous hypochlorhydria and Campylobacter pylori

The study reported by Gledhill *et al.* [36] was conducted between 1981 and 1982, at which time Warren [37] and Marshall [38] had not yet described the unidentified curved bacilli on the gastric epithelium of patients with active chronic gastritis, which has subsequently proved to be *Campylobacter pylori*.

The observation that hypo- or achlorhydria was associated with the voluntary ingestion of *C. pylori* in two otherwise healthy subjects [39, 40] suggests that *C. pylori* might have been the causative organism in the cluster studies reported by Ramsey *et al.* [35] and Gledhill *et al.* [36].

Subsequent to the discovery of *C. pylori*, Peterson *et al.* [41] tested serum samples taken during and after the illness in subjects reported by Ramsey *et al.* [35] for antibodies to *C. pylori*. They felt that the demonstration of organisms on mucosal biopsy would not be proof of infection because there is a high prevalence of *C. pylori* in normal fundic tissue of healthy subjects [42]. Acute and convalescent serum samples obtained at diagnosis and 6 months later from 12 patients were coded and assessed for the presence of IgG antibody against *C. pylori* using an enzyme-linked immunosorbent assay (ELISA). There was no significant difference between those subjects with gastritis and controls although the convalescent gastritis sera titres were significantly increased. Antibody titres to *C. pylori* rose during convalescence by at least one dilution in 9/12 patients with gastritis, suggesting that the organism played a role in the epidemic of gastritis.

The documentation of alterations in gastric acid secretion in the two volunteer studies reported [39, 40] are somewhat limited. When Marshall [39] ingested *C. pylori*, no formal studies of basal or stimulated acid secretion were undertaken but the vomitus did not taste acidic and was not acid when tested with litmus paper. In the study reported by Morris & Nicholson [40], basal acid pH was tested extensively before, during and subsequent to infection. Following ingestion of *C. pylori*, pH remained acidic until day 5 and by day 8 had risen to a value of 7.6, which was maintained until day 27, returning to 1.6 on day 29.

A further carefully-documented report of *C. pylori* infection as a cause of epidemic hypochlorhydria was reported by Graham *et al.* [43] in a 37-year-old previously healthy man having weekly gastric analyses with endoscopy and biopsies in a study of gastric adaptation to aspirin. In the second week of aspirin administration he developed nausea and epigastric discomfort and aspirin was discontinued but gastric analyses and biopsies were continued following the protocol. When compared to studies preceding the acute illness there was a transient 7.4-fold increase in basal acid, a 3.6-fold increase in mucus secretion and a 12-fold increase in gastric bleeding. During the second week of infection, basal acid secretion fell to zero and pepsin secretion was one-third of control. Endoscopy at the time of symptoms showed erosions in the gastric body and antrum as well as numerous haemorrhages and an acute ulcer was present in the antrum. Endoscopy 7 days later revealed that the gastric mucosa had almost completely recovered. Gastric biopsies were normal before and during the first 2 weeks of aspirin ingestion but at the time of the acute illness, when basal acid secretion was markedly increased, there was an acute inflammation of the antrum with many *C. pylori* present. At that time neither acute inflammation nor *C. pylori* were found in biopsies from the body of the stomach but 1 week later biopsies showed acute inflammation in the gastric body and antrum. Within 1 week, the features of chronic gastritis were present and 2 years later persistent chronic gastritis with *C. pylori* was still present. This case showed that an increased basal acid and pepsin secretion occurred in response to the acute infection with *C. pylori* and this was followed by a phase of hypochlorhydria and subsequent recovery. Stimulated acid secretion in response to gastrin or histamine was not undertaken.

Acid secretion and serum gastrin levels have been evaluated in 36 individuals with *C. pylori* confirmed by Warthin–Starry stain [44]: 19 patients, 10 with gastric ulcer and nine with duodenal ulcer had normal gastric secretion and two patients with gastric ulcer were hypochlorhydric. Median basal acid output was higher for those with duodenal ulcer at 38 mmol/hour than gastric ulcer at 28 mmol/hour. *Campylobacter pylori* was found in six normogastrinaemic patients with elevated acid output and in one gastrinoma patient with marked acid hypersecretion. Histological chronic gastritis was present in all subjects and 29 had active chronic gastritis. While the authors conclude that *C. pylori* can survive in a wide range of acid conditions, this study gives no data to suggest prolonged hypochlorhydria in patients infected with *C. pylori*. Thus if infection with *C. pylori* is indeed associated with diminished or absent acid secretion the majority of patients appear to recover their acid secretory status.

Possible inhibitory effects of infection on gastric secretion in man

The mechanism by which hypochlorhydria occurs remains unclear. In the study reported by Ramsey *et al.* [35] careful histology failed to demonstrate a reduction in parietal cell mass, and in the case reported by Graham *et al.* [43] parietal cell morphology and numbers appeared normal. Ramsey *et al.* [35] also confirmed that the reduction in gastric acid secretion was real and not the result of an increased rate of hydrogen-ion back-diffusion across the damaged gastric mucosa.

Bacterial infections which cause fever have frequently been associated with a marked reduction in acid secretion. Histamine fast achlorhydria was found in patients suffering from a variety of bacterial infections including typhoid, paratyphoid, pulmonary tuberculosis, bronchopneumonia and lung abscess [45]. This was reported from a region of China associated with a low natural incidence of achlorhydria.

In 106 febrile patients with infection, body temperature was positively correlated with the degree of suppression of acid secretion [46]. The basal secretion of acid was reduced to approximately one-third of the expected normal, and the prevalence of total achlorhydria was increased 8-fold. Gastric secretion returned to normal levels in 90% of patients after eradication of the infection.

In a group of 325 patients with pulmonary tuberculosis of varying degrees of severity, there was a steady increase in the prevalence of achlorhydria with advancement of the disease [47].

It is interesting to note that in 1883 William Beaumont had observed a reduction in the gastric secretion of his gastric fistula patient Alexis St. Martin during a febrile illness [48]. In the presence of fever, he noted that '... the secretions become greatly vitiated, greatly diminished or entirely suppressed ...'.

It is possible that fever rather than infection *per se* is responsible for the suppression of acid output. A transient inhibition of gastric secretion has been produced in man by artificially raising body temperature to 38–39° C in a heating cabinet [49].

There is some evidence that parasitic infections in man cause a reduction in acid secretion. Infection with the fish tapeworm *Diphyllobothrium latum* is more common in association with hypochlorhydria [50], but there is some evidence to suggest that the infection itself may suppress acid output [51] through an unknown mechanism. Infection with *Trypanosoma cruzei*, the causative agent in Chagas' disease, is associated with reduction in basal and stimulated acid output [52]. This may be because of the gastric parasympathetic denervation caused by the infecting organism which reduces the responsiveness of the parietal cell to physiological stimuli. Reduced gastric acid secretion has also been documented in patients infected with the hookworm *Ancylostoma duodenale* [53]. The parasite is confined to the small intestine and the mechanism whereby it influences gastric secretion is unknown.

The effects of viral infection on gastric secretion in man have not received much attention, although a virus-derived peptide after vaccinia infection appears to be able to suppress acidity in laboratory animals [54]. As nausea and vomiting are frequent and non-specific symptoms of viral infection in man, it might be expected that certain viruses would produce quantifiable changes in gastric secretion or motility. Ingestion of two strains of parvovirus by human volunteers produced no alteration in either basal or stimulated acid output [55], but markedly slowed gastric emptying. The two viruses used do not cause any inflammation of the gastric mucosa [56], and this situation is quite different from that of epidemic hypochlorhydria in which there is diffuse gastritis [35, 36].

In the case of *C. pylori*, hypochlorhydria may result from the production of a bacterial toxin acting directly or indirectly on acid secretion. We have recently reported the effect of *C. pylori* on acid production, assayed using the ^{14}C-aminopyrine uptake technique in isolated guinea-pig parietal cells [57]. *Campylobacter pylori* caused a reduction in basal acid secretion of about 80% while histamine-stimulated acid secretion was reduced to 50% of maximal stimulation within 15 min of the addition of *C. pylori*. Only a partial return of responsiveness to histamine was seen when *C. pylori* was washed from the preparation. Electron microscopy showed some *C. pylori* remaining attached to parietal cells. It was clear from these studies that while *C. pylori* does reduce basal and histamine-stimulated acid production in this preparation, the effect is only partially reversible. The reduction in acid secretion is therefore unlikely to be solely the result of cytotoxicity or mechanical damage caused by *C. pylori*.

References

1 Howden, C.W. & Hunt, R.H. The relationship between gastric secretion and infection. *Gut* 1987, **28**, 96–107.

2 Cook, G.C. Infective gastroenteritis and its relationship to reduced gastric acidity. *Scand. J. Gastroenterol.* 1980, **20** (Suppl. 111), 17–22.

3 Penston, J. & Wormsley, K.G. Achlorhydria: hypergastrinaemia: carcinoids — a flawed hypothesis? *Gut* 1987, **28**, 488–505.

4 Giannella, R.A., Broitman, S.A. & Zamchek, N. Gastric acid barrier to ingested microorganisms in man: studies *in vivo* and *in vitro*. *Gut* 1972, **13**, 251–256.

5 Anonymous. Bacteria in the stomach. *Lancet* 1981, **ii**, 906–907.

6 Milton-Thompson, G.J., Lightfoot, N.F., Ahmet, Z., Hunt, R.H. *et al.* Intragastric acidity, bacteria, nitrite, and N-nitroso compounds before, during and after cimetidine treatment. *Lancet* 1982, **i**, 1091–1095.

7 Sharma, B.K., Santana, I.A., Wood, E.C. *et al.* Intragastric bacterial activity and nitrosation before, during and after treatment with omeprazole. *Br. Med. J.* 1984, **289**, 717–719.

8 Meyrick-Thomas J., Misiewicz, J.J., Cook, A.R., Hill, M.J. *et al.* Effect of one year's treatment with ranitidine and of truncal vagotomy on gastric contents. *Gut* 1987, **28**, 726–738.

9 Waddell, W.R. & Kunz, L.J. Association of *Salmonella* enteritis with operations on the stomach. *N. Engl. J. Med.* 1956, **255**, 555–559.

10 Grossman, M.I., Kirsner, J.B. & Gillespie, I.E. Basal and histalog-stimulated gastric secretion in control subjects and in patients with peptic ulcer. *Gastroenterology* 1963, **45**, 14–26.

11 Gorbach, S.L. 1983. In Sleisenger, M.H. & Fordran, J.S. (eds.) *Gastrointestinal Disease: Pathophysiology, Diagnosis, Management.* W.B. Saunders, Philadelphia. pp. 925–965.

12 Abdou, S. Susceptibility to cholera. *Lancet* 1948, **i**, 903–904.

13 Bhalla, F. Vij, J.C., Anand, B.S. *et al.* Gastric acid secretion in patients with typhoid fever. *Gut* 1985, **26**, 491–494.

14 Howden, C.W. & Hunt, R.H. Gastric secretion in patients with typhoid. *Gut* 1985, **26**, 1387.

15 Mosbech, J. & Vidback, A. Mortality from and risk of gastric carcinoma among patients with pernicious anaemia. *Br. Med. J.* 1950, **ii**, 390.

16 Blackburn, E.K. Callender, S.T., Dacie, J.V. *et al.* Possible association between pernicious anaemia and leukaemia: a prospective study of 1625 patients with a note on the very high incidence of stomach cancer. *Int. J. Cancer* 1963, **3**, 163–170.

17 Caygill, C.P., Hill, M.J., Kirkham, J.S. *et al.* Mortality from gastric cancer following gastric surgery for peptic ulcer. *Lancet* 1986, **i**, 929–931.

18 Ruddell, W.S.J., Bone, E.S., Hill, M.J. & Walters, C.L. Pathogenesis of gastric cancer in pernicious anaemia. *Lancet* 1978, **i**, 521–523.

19 Walsh, J.H. & Grossman, M.I. Gastrin. *N. Engl. J. Med.* 1975, **292**, 1324–1332.

20 Lanzon-Miller, S., Pounder, R.E., Hamilton, M.R. *et al.* Twenty-four-hour intragastric acidity and plasma gastrin concentration before and during treatment with either ranitidine or omeprazole. *Aliment. Pharmacol. Ther.* 1987, **1**, 239–251.

21 Yamada, T. Gastric acid suppression: potential longterm consequences. *Postgrad. Med.* 1985 (November supplement), 67–75.

22 Dembinski, A.B. & Johnson, L.R. Growth of pancreas and gastrointestinal mucosa in antrectomized and gastrin treated rats. *Endocrinology* 1979, **105**, 769–773.

23 Kabori, O., Vuillot, M.T. & Martin, F. Growth responses of rat stomach cancer cells to gastro-entero-pancreatic hormones. *Int. J. Cancer*, 1982, **30**, 65–67.

24 Ohkura, H., Hanafusa, K., Maruyama, K. *et al.* Gastrin-enhanced tumor growth of a xenotransplantable human gastric carcinoma in nude mice. *Jpn. J. Clin. Oncol.* 1980, **10**, 255–264.

25 Lanzon-Miller, S., Pounder, R.E., Hamilton, M.R. *et al.* Twenty-four-hour intragastric acidity and plasma gastrin concentration in healthy subjects and patients with duodenal or gastric ulcer, or pernicious anaemia. *Aliment. Pharmacol. Ther.* 1987, **1**, 225–237.

26 Makhlouf, S.M., McManus, J.P.A. & Card, W.I. A comparative study of the effects of gastrin, histamine, histalog and mechothane on the secretory capacity of the human stomach in normal subjects over 20 months. *Gut* 1965, **6**, 525–534.

27 Baron, J.H. Gastric secretion in a healthy man 1949–1969. *Lancet* 1970, **ii**, 547–549.

28 Hirschowitz, B.I., Streeten, D.H.P., London, J.A. *et al.*

A steroid-induced gastric ulcer. *Lancet,* 1956, **ii**, 1081–1083.

29 Spiro, H.M. & Schwartz, R.D.L. Superficial gastritis: a cause of temporary achlorhydria and hyperpepsinemia. *N. Engl. J. Med.* 1958, **259**, 682–684.

30 Waterfall, W.E. Spontaneous decrease in gastric secretory response to humoral stimuli. *Br. Med. J.* 1969, **iv**, 459–461.

31 Desai, H.G., Zaveri, M.P. & Anita, F.P. Spontaneous and persisting decrease in maximal acid output. *Br. Med. J.* 1971, **ii**, 313–315.

32 Lawrie, R.S., Williamson, A.W.R. & Hunt, J.N. Zollinger–Ellison syndrome treated with poldine methyl methosulphate. *Lancet* 1962, **i**, 1002–1004.

33 Desai, H.G. & Anita, F.P. Spontaneous achlorhydria with atrophic gastritis in the Zollinger–Ellison syndrome. *Gut* 1969, **10**, 935–939.

34 Wiersinga, W.M. & Tytgat, G.N.J. Clinical recovery due to target parietal cell failure in a patient with Zollinger–Ellison syndrome. *Gastroenterology* 1977, **73**, 1413–1417.

35 Ramsey, E.J., Carey, K.V., Peterson, W.L. *et al.* Epidemic gastritis with hypochlorhydria. *Gastroenterology* 1979, **76**, 1449–1457.

36 Gledhill, T., Leicester, R.J., Addis, B., Lightfoot, N., Barnard, J., Viney, N., Darkin, D. & Hunt, R.H. Epidemic hypochlorhydria. *Br. Med. J.* 1985, **289**, 1383–1386.

37 Warren, J. Unidentified curved bacilli on gastric epithelium in active chronic gastritis. *Lancet* 1983, **i**, 1273.

38 Marshall, B. Unidentified curved bacilli on gastric epithelium in active chronic gastritis. *Lancet* 1983, **i**, 1273–1275.

39 Marshall, B.J., Armstrong, J.A., McGechie, D.B. *et al.* Attempt to fulfill Koch's postulates for pyloric *Campylobacter. Med. J. Australia* 1985, **142**, 436–439.

40 Morris, A. & Nicholson, G. Ingestion of *Campylobacter pylori* causes gastritis and raised fasting gastric pH. *Am. J. Gastroenterol.* 1987, **82**, 192–199.

41 Peterson, W.L., Lee, E. & Skogland, M. The role of *Campylobacter pyloridis* in epidemic gastritis with hypochlorhydria. *Gastroenterology* 1987, **92**, 1575.

42 Peterson, W.L., Lee, E. & Feldman, M. Gastric campylobacter like organisms in healthy humans; correlation with endoscopic appearance and mucosal histology. *Gastroenterology* 1986, **90**, 1585.

43 Graham, D.Y., Alpert, L.C., Smith, J.L. & Yoshimura, H.H. Iatrogenic *Campylobacter pylori* infection is a cause of epidemic achlorhydria. *Am. J. Gastroenterol.* 1988, **83**, 974–980.

44 Brady, C.E., Hadfield, T.L. & Hyatt, J.R. Acid secretion and serum gastrin levels in individuals with *Campylobacter pylori. Gastroenterology* 1988, **94**, 923–927.

45 Berglund, H. & Chang, H.C. Transitory character of the achlorhydria during fever demonstrated by the histamine test. *Proc. R. Soc. Exp. Biol. Med.* 1929, **26**, 422–423.

46 Chang, H.C. Gastric secretion in fever and infectious diseases. *J. Clin. Invest.* 1933, **12**, 155–169.

47 Kruger, A.L. Gastric acidity in pulmonary tuberculosis. *Am. J. Dig. Dis.* 1943, **10**, 111–114.

48 Beaumont, W. 1883. *Experiments and Observations on the Gastric Juice and the Physiology of Digestion.* Harvard University Press, Boston. p. 107.

49 Bandes, J., Hollander, F. & Bierman, W. The effect of physically induced pyrexia on gastric acidity. *Gastroenterology* 1948, **10**, 697–707.

50 Giannella, R.A., Broitman, S.A. & Zamchek, N. Influence of gastric acidity on bacterial and parasitic enteric infections. A perspective. *Ann. Intern. Med.* 1973, **78**, 271–276.

51 Salokannell, J. Intrinsic factor in tapeworm anaemia. *Acta Med. Scand.* 1970, **188** (Suppl. 517), 1–51.

52 Padovan, W., Godoy, R.A., Meneghelli, U.G., *et al.* Acid and pepsin secretion in chronic Chagas' disease patients in response to graded doses of pentagastrin and bethanechol. *Digestion* 1982, **23**, 48–56.

53 Pimparkar, B.D., Sharma, P., Satoskar, R.S. *et al.* Anaemia and gastrointestinal function in ancylostomiasis. *Postgrad. Med. J.* 1982, **28**, 51–63.

54 Aonuma, S., Koihama, Y., Enmi, K. *et al.* Studies on anti-ulcerogenic protein in inflamed rabbit skin tissues. III. Anti-ulcerogenic peptide obtained from tissues infected with vaccinia virus. *Yakugaku Zasshi* 1984, **104**, 362–373.

55 Meeroff, J.C., Schreiber, D.S., Trier, J.S. *et al.* Abnormal gastric motor function in viral gastroenteritis. *Ann. Intern. Med.* 1980, **92**, 370–373.

56 Widerlite, L., Trier, J.S., Blacklow, N.R. *et al.* Structure of the gastric mucosa in acute infectious nonbacterial gastroenteritis. *Gastroenterology* 1975, **68**, 425–430.

57 Defize, J., Goldie, J. & Hunt, R.H. Effect of *Campylobacter pylori* on acid production by isolated guinea pig parietal cells. *Gut* 1988, **29**, A1435.

22 Campylobacter pylori: human ingestion studies

A. MORRIS AND G. NICHOLSON

The observation that curved bacilli were almost always present in gastric biopsies showing active chronic gastritis [1] led Australian workers to grow and characterize the organism currently known as *Campylobacter pylori* [2, 3]. Despite their conviction that the organism caused histological gastritis, they were still concerned that the organism could be an opportunistic colonizer of previously abnormal mucosa [4]. In an effort to test the hypothesis of cause and effect, Dr Barry Marshall conducted an ingestion study in July 1984 which was published in 1985 [4].

Marshall ingestion study

At 1 month before challenge, the subject underwent endoscopy and antral, body and duodenal biopsies were taken for histology and culture. Endoscopic findings were normal. No histological abnormalities were present and *C. pylori* was neither seen nor cultured. The same day that the initial biopsies were taken *C. pylori* was isolated from a 66-year-old male with non-ulcer dyspepsia and this isolate had antibiotic susceptibility tests performed before being freeze-dried.

At 3 hours before challenge, gastric acidity was reduced with 600 mg of cimetidine (Smith, Kline & French). The challenge dose was a suspension of a 3-day-culture of *C. pylori*, approximately 10^9 colony-forming units, in 10 ml of alkaline peptone water (pH 8.0).

Apart from borborygmi in the 24 hours following ingestion, the subject was asymptomatic for 1 week. On day 7 epigastric fullness was felt after the evening meal. The following day early-morning hunger occurred and a small amount of mucus was vomited. There was no fever or diarrhoea but headaches, irritability and 'putrid' breath were experienced.

Endoscopy was repeated on day 10, when antral biopsies were obtained for culture and histology. Histological gastritis was present and was characterized by polymorphonuclear leukocytes within the lamina propria and on the mucosal surface. A reduction in the amount of intracellular mucus was noted. Organisms were seen adhering to the epithelial surface in silver-stained and electron microscopic sections. *Campylobacter pylori* was recovered.

Early-morning hunger continued and on day 14 endoscopy was repeated. Biopsies were taken and evidence of persisting infection was expected. However, the histological gastritis had diminished, no polymorphonuclear leukocytes were present and only a mild accumulation of mononuclear cells remained. The mucus content of the epithelial cells had increased. No organisms were seen in any biopsy. Tinidazole (500 mg b.d. for 7 days) was begun on day 14 and symptoms completely resolved within 24 hours of therapy. No seroconversion was detected when sera were tested in a passive haemagglutination assay [4]. Long-term follow-up has been performed by this volunteer and no evidence of infection has been found on endoscopy and biopsy and he remains seronegative (Marshall, personal communication).

This experiment was the first direct evidence that *C. pylori* could colonize normal gastric mucosa. More importantly, colonization was associated with the histological changes typically seen in infected patients. It supported the hypothesis that *C. pylori* is a cause of superficial active chronic gastritis.

Morris ingestion study [5]

The aims of this challenge experiment were to confirm Marshall's observations, to study fasting gastric pH in a prospective fashion in a challenged subject and to relate any pH changes to gastric histopathology. It was thought that if infection persisted this would provide the opportunity to document seroconversion and investigate the effect of therapy.

pH recordings

Intragastric pH was measured by aspirating gastric contents through a nasogastric tube. Gastric aspirates were taken between 8.00 a.m. and 8.30 a.m. after overnight fasts. A nasogastric tube was positioned several centimetres proximal to the pylorus under X-ray screening. The tube was passed to the same distance for each aspiration. Gastric fluid was aspirated and a period of 10–15 min was allowed before a second volume was collected. When no further fluid could be aspirated a period of 10–20 min was allowed before the final gastric specimen was aspirated. The pH of the final aspirate was determined within 10 min of collection using a glass electrode.

A summary of the experiment is shown in Table 22.1. Days are described as either before or after the day of ingestion leading to infection. The results of subsequent gas–liquid chromatography of gastric aspirates are shown in Table 22.2. Results of long-term follow-up are shown in Table 22.3.

Clinical illness

Review of the literature prior to ingestion suggested that the likelihood of experiencing a clinical illness was approximately 60% [4, 6–8]. In the study of Ramsey *et al.* [6], 9/17 patients with hypochlorhydria, probably caused by *C. pylori* infection, experienced symptoms of mild-to-moderate epigastric pain, four experienced nausea and two experienced vomiting. The illness lasted 1–4 days. No fever was noted subjectively by any subject. Myalgia was experienced in one case [6]. The patient in the report of Wiersinga & Tytgat had an illness with severe malaise, nausea, vomiting, perspiration and fever up to 38.5° C, but no abdominal pain was experienced and the illness lasted a week [7]. Marshall experienced epigastric fullness, hunger, irritability, headaches, halitosis and one episode of vomiting.

The illness experienced by Morris was more severe than expected from this literature review. The epigastric pain came in waves and at its height the subject could not sleep and walked the floor for hours attempting to get relief. The vomitus on both occasions contained fluid and food ingested up to 8 hours before. The symptoms were such that high intestinal obstruction was suspected but erect and supine abdominal X-rays were normal.

Increased fasting pH

Fasting pH was normal before and for up to 6 days after ingestion. The rise in fasting pH coincided with the development of histological gastritis in the body of the stomach. This sequence has also been observed by Graham *et al.* [9]. Raised fasting pH persisted until doxycycline treatment was commenced. It appeared that this antibiotic treatment had suppressed infection allowing a return of gastric acidity. However, normal fasting pH was observed after stopping the antibiotic when infection and histological gastritis of the fundus were still present (biopsy 5, day 67). In the study of Ramsey *et al.* hypochlorhydria lasted some time, 53–235 days (mean 125 days, median 112 days) before resolving [6]. Hypochlorhydria was still present at 11 months in Wiersinga & Tytgat's patient [7]. In the Ramsey *et al.* study, maximum depression in peak acid output occurred early in infection

Table 22.1 Summary of events in experimental *C. pylori* infection[†]

Day(s)*	Comments
−87 to −2	pH always < 2; pH range 1.0–1.7; 23 recordings
−71	*Biopsy 1*. Normal antral and fundal histology. *C. pylori* not isolated
−41	pH 1.7; challenge with the Marshall strain of *C. pylori* fails
−30	*Biopsy 2*. Normal antral and fundal histology. *C. pylori* not isolated
0	pH 7.5; ingestion of local strain of *C. pylori*
1	pH 1.1; no symptoms
2	Evening: occasional mild epigastric cramping
3	Afternoon: attacks of moderate epigastric pain. Evening: no sleep
4	Morning: two episodes of vomiting, pain more severe Evening: episodes of pain decreasing in frequency and severity
5	pH 1.2. *Biopsy 3*. Antrum: active gastritis, *C. pylori* isolated, positive Gram stain. Fundus: normal histology, no *C. pylori* isolated, negative Gram stain
6–7	Epigastric sensation of hunger, bloated sensation after small amounts of food or drink
8	pH 7.6; *C. pylori* recovered from gastric aspirate
11	pH 7.6; *Biopsy 4*. Both antrum and fundus have active chronic gastritis; *C. pylori* isolated from each site
12–19	pH 7.8–7.9; *C. pylori* isolated from gastric aspirates
25	pH 5.9; *C. pylori* isolated from gastric aspirate. Doxycycline (100 mg b.d.) begun
27	pH 7.4; gastric aspirate culture negative
29	pH 1.6
32–53	pH 1.0–1.8; 4 readings; gastric aspirates culture negative
60	Doxycycline stopped
61	pH 1.3; *C. pylori* isolated from gastric aspirate
67	pH 1.1. *Biopsy 5*. Continuing antral and fundal gastritis, *C. pylori* isolated from each site. Bismuth subsalicylate (600 mg q.i.d.) begun
97	Bismuth subsalicylate stopped
103	*Biopsy 6*. Only a mild residual chronic inflammatory infiltrate present at both sites. Negative Gram stains. *C. pylori* not isolated

*Days described as either before or after the day of ingestion leading to infection.
[†]Table reproduced from Morris, A & Nicholson, G. Ingestion of *Campylobacter pyloridis* causes gastritis and raised fasting gastric pH. *American Journal of Gastroenterology* 1987, **82**, 192–199, by The American College of Gastroenterology with kind permission of the copyright holder.

and the level of its reduction paralleled the severity of the histological gastritis in the body [6]. In the ingestion experiment, return to normal fasting pH was not associated with any change in the histological gastritis where active chronic gastritis persisted. How altered pH or hypochlorhydria occur and, more interestingly, how they apparently resolve in the presence of continuing infection and gastritis are not understood at this time.

Table 22.2 Gas–liquid chromatography of gastric aspirates

Day	Comments
−34	No volatile fatty acids detected
−2	No volatile fatty acids detected
8	Acetic, proprionic and isobutyric acids detected
18	Proprionic acid detected
25	Acetic and isobutyric acids detected

Gas–liquid chromatography

In Osler's description of acute gastritis, quoted in detail by Marshall *et al.* [4], the presence of acetic and butyric acids was noted in patients suffering from clinical illnesses similar to that associated with *C. pylori*. Five gastric aspirates were analysed by gas–liquid chromatography, two taken before infection and three taken after infection (Table 22.2). Gastric asp-

Table 22.3 Long-term follow-up of experimental *C. pylori* infection

Day(s)	Biopsy No.	Comments*
103	6	Only mild chronic gastritis remains after bismuth subsalicylate treatment
357	7	Antral and fundal biopsies show ACG; *C. pylori* isolated
374–386		Course of norfloxacin 400 mg b.d.
392	8	Antral and fundal biopsies show ACG; cultures overgrown but Gram stain of biopsies shows *C. pylori*
435	9	Antral and fundal biopsies show ACG; *C. pylori* isolated
463	10	Continuing antral and fundal ACG; *C. pylori* isolated
634–652		Amoxycillin–clavulanic acid (500 mg b.d.) for otitis media and soft-tissue infections
817	11	Continuing antral and fundal ACG; *C. pylori* isolated
964	12	Continuing antral and fundal ACG; *C. pylori* isolated

*ACG = active chronic gastritis.

irates obtained after *C. pylori* infection contain volatile fatty acids not present before infection. It is not known if these fatty acids are the result of metabolism of *C. pylori* or of other organisms.

Serology

Serology was originally performed by a local enzyme-linked immunosorbent assay (ELISA) using whole organisms as antigen [10], and an IgG seroconversion was demonstrated [5]. Subsequently, early serum samples were tested in a different ELISA assay [11] by Perez-Perez *et al.* A clear IgM seroconversion, followed much later by both IgA and IgG responses, was observed (Morris & Blaser, unpublished observations).

Long-term follow-up

The results of long-term follow-up are shown in Table 22.3. Relapse of infection and histological gastritis were detected at biopsy 7 taken 254 days after apparent eradication of infection with bismuth subsalicylate. Infection has persisted for more than 2 years 8 months with histological gastritis continuing in antral and fundal biopsies of both greater and lesser curves.

Bismuth subsalicylate was chosen because at that time it was the agent with the best eradication rate [12]. However, ten of the patients apparently cleared of *C. pylori* by bismuth subsalicylate have been followed and eight have relapsed (McNulty, personal communication). It therefore appears that bismuth subsalicylate has a similar relapse rate to colloidal bismuth subcitrate [13].

Concluding comments

Both ingestion studies have shown that *C. pylori* is capable of colonizing human gastric mucosa and causing histological gastritis. Seroconversion has been demonstrated in one volunteer in whom infection has persisted. The observations made in the volunteer studies [4, 5] are comparable with other instances of documented acute *C. pylori* infection [6–9].

Although more needs to be learned about the early stages of *C. pylori* infection in man and their sequelae, there can be little doubt that *C. pylori* is a cause of active chronic gastritis. Even if the organism were to have little other clinical importance, its ability to cause histological gastritis solves one previously ill-understood aspect of gastric histopathology.

References

1 Warren, J. Unidentified curved bacilli on gastric epithelium in active chronic gastritis. *Lancet* 1983, **i**, 1273.

2 Marshall, B., Royce, H., Annear, D. *et al.* The original isolation of *Campylobacter pyloridis* from human gastric mucosa. *Microbios Lett.* 1984, **25**, 83–88.

3 Marshall, B. & Goodwin, C. Revised nomenclature of *Campylobacter pyloridis*. *Int. J. Systematic Bacteriol.* 1987, **37**, 68.

4 Marshall, B., Armstrong, J., McGechie, D. & Glancy, R. Attempt to fulfil Koch's postulates for pyloric *Campylobacter*. *Med. J. Australia* 1985, **142**, 436–439.

5 Morris, A. & Nicholson, G. Ingestion of *Campylobacter pyloridis* causes gastritis and raised fasting gastric pH. *Am. J. Gastroenterol.* 1987, **82**, 192–199.

6 Ramsey, E., Carey, K., Peterson, W., *et al.* Epidemic gastritis with hypochlorhydria. *Gastroenterology* 1979, **76**, 1449–1457.

7 Wiersinga, W. & Tytgat, G. Clinical recovery owing to target parietal cell failure in a patient with Zollinger–Ellison syndrome. *Gastroenterology* 1977, **73**, 1413–1417.

8 Peterson, W., Lee, W. & Skoglund, M. The role of *Campylobacter pyloridis* in epidemic gastritis with hypochlorhydria. *Gastroenterology* 1987, **92**, 1575. (Abstract)

9 Graham, D., Alpert, L., Smith J. & Yoshimura, H. Iatrogenic *Campylobacter pylori* infection is a cause of epidemic achlorhydria. *Am. J. Gastroenterol.* 1988, **83**, 974–980.

10 Morris, A., Nicholson, G., Lloyd, G., Haines, D., Rogers, A. & Taylor, D. Seroepidemiology of *Campylobacter pyloridis*. *N.Z. Med. J.* 1986, **99**, 657–659.

11 Perez-Perez, G., Dworkin, B., Chodos, J. & Blaser, M. *Campylobacter pylori* antibodies in humans. *Ann. Intern. Med.* 1988, **109**, 11–17.

12 McNulty, C., Gearty, J., Crump, B. *et al.* *Campylobacter pyloridis* and associated gastritis: investigator blind, placebo controlled trial of bismuth salicylate and erythromycin ethylsuccinate. *Br. Med. J.* 1986, **293**, 645–649.

13 Rauws, E., Langenberg, W., Mouthoff, M., Zanen, H. & Tytgat, G. *Campylobacter pyloridis*-associated chronic active antral gastritis. A prospective study of its prevalence and the effects of antibacterial and antibiotic treatment. *Gastroenterology* 1988, **94**, 33–40.

23 The prevalence of Campylobacter pylori in human populations

B. DWYER, J. KALDOR, W. TEE AND K. RAIOS

Although *Campylobacter pylori* has been recovered from the gastric mucosa of dyspeptic people on all inhabited continents, there are relatively few studies of the prevalence of infection in the general populations from which they have been drawn. The public health importance of the outcomes of chronic infection with *C. pylori* will only be properly understood when the prevalence of infection is known.

Prevalence in disease

For the moment, the most extensively endoscopically-studied population groups are those from Western Europe, New Zealand, Australia, and the United States. These studies have concentrated on the prevalence of infection with *C. pylori* in symptomatic patients undergoing endoscopy, thereby introducing a biased estimate of prevalence. In all groups there is clearly a very high rate of infection in patients undergoing endoscopy for dyspepsia (40–60%) and in patients with duodenal ulcer the rate is 85–95%, while in benign gastric ulcer the organism is found in about 70%. Lower isolation rates in some series may reflect the difficulties many experience when first attempting to culture this organism.

Prevalence in general populations

Prevalence at autopsy

There would appear to be only one contemporary study of the prevalence of *C. pylori* in gastric material examined post-mortem [1]. Morris *et al.* sought the organism in the gastric mucosa using Gram stains of mucous scrapings, tests for urease activity and histology. Infection was present in 12/20 (60%) Polynesians and in 6/31 (20%) Caucasians. There was a significant difference in frequency between these two groups, which were of approximately equal age distribution.

Prevalence studies using endoscopy

The histological review of endoscopic specimens obtained from asymptomatic, mostly young, adult, medical students in the Netherlands showed that infection in this predominantly male group was about 25% [2]. A prospective study from Cleveland in the United States using culture as well as histology showed a comparable prevalence [3]. However, histology was more successful than culture in this study. A study on morbidly obese white women (apparently with a high proportion of immigrants) undergoing intra-gastric balloon treatment in Sydney (Australia) for appetite control also showed a similar rate of 18% [4].

The defects of these studies are that they have been performed on selected (asymptomatic) volunteers, they appear to have a preponderance of males (with the exception of the Sydney study) and because of the small numbers involved they do not show an adequate distribution across age, let alone through geographical location, social class, cultural or occupational groups. Although clearly useful, this method of study is unlikely to continue to provide additional useful information due to the logistic difficulties involved in attaining sample sizes representative of more general population groups in the countries involved.

Prevalence studies using isotopically labelled urea

The elegant simplicity of detecting the presence of *C. pylori* infecting the gastric mucosa by the rapid hydrolysis of ingested urea to ammonia and carbon dioxide is very appealing. The exhaled fraction of labelled carbon dioxide can then be detected by a ß-counter if ^{14}C-labelled urea is used, or by mass spectroscopy if ^{13}C-urea is used. Neither method has been widely adopted and although the non-radioactive method would be preferred, the cost of the material and special instrumentation is likely to limit this method for some time to come [5–7]. The reliability of the results from both methods is likely to be influenced by recent antibiotic therapy. There are no reports describing the relationship between antibody status, urea hydrolysis breath test detection and culture. However, the two methods have been each validated by comparison with culture and/or histology.

To date there have only been two studies on prevalence using this method in normal (asymptomatic) subjects, both from Houston, Texas [7, 8]. The first study recruited from two population groups, American citizens in Houston and visiting Chinese scientists. In Americans there was no difference in infection rates for the sexes. The composition, by sex, of the Chinese was not stated. The Americans and Chinese were recruited from groups with backgrounds in the professions. The frequency of infection rose with age in the Americans, from 24% in the age group 20–39 years, to 36% in the age group 40–50 years, to 82% in the age group 60–84 years. A limited yet striking contrast was found in the Chinese: all subjects were in the age group 20–39 years and 60% were infected.

The second study may well include some subjects reported in the first study. For comparative purposes only the results on Indian and Mexican nationals are discussed here. The subjects were resident in the United States for periods of up to five years, all were in the age range 20–29 years, but the precise distribution by gender was not given. All were from the middle or upper socio-economic classes. Mexican nationals showed a similar rate of infection to US nationals (33%), while Indian nationals showed a higher rate again (46%).

Prevalence studies using serological methods

Sero-epidemiological methods have gained wide acceptance as valid tools for measuring the prevalence of many infections in a variety of populations and subgroups. For the study of the prevalence of infections with *C. pylori* most workers have used ELISA assays because of their convenience. Serology only identifies antibodies which ordinarily are not necessarily indicative of current infection. In the case of this organism, most workers would accept that the possession of antibodies to *C. pylori* is probably indicative of continuing infection because of the known chronicity of infection. One of the earliest surveys was carried out on 'patients without gastrointestinal conditions, healthy individuals from health screening clinics, healthy normal children and adults' from Manchester, England [9]. In this survey a close correlation was found between the results obtained by ELISA and those obtained by complement fixation. However, the ELISA assay was considered slightly more sensitive.

The earliest survey of prevalence in volunteer (unpaid) blood donors was that reported from Western Australia [10]. Marshall *et al.* reported that there was a rising age-specific gradient in positive serological responses to *C. pylori* as measured by reverse passive haemagglutination. They found that positive titres regressed with age and sex. Children (age undefined), had the lowest rates of infection, 11% for males and 5% for females. Adults showed the following patterns: 23% of males and 5% of females < 30 years had antibody, while 29% of males and 25% of females > 30 years had antibody.

While this survey was useful, the work from Morris and his colleagues clearly showed that there may be distinct differences in antibody rates between peoples of different cultures or races living in the same area [11]. Using sera from volunteer blood donors, Morris *et al.* showed that there was a gradient in seroprevalence which was age related. The overall rate was 36%. They also demonstrated that in donors of unknown race, 15% of sera were positive, while for Maoris the rate was 21% (this difference was not significant); there were, however, significant differences for Cook Islanders (39%), Samoans (44%) and Tongans (70%). Although precise details were not provided, it was stated that the higher seropositivity amongst these Polynesian blood donors was not due to a preponderance of older subjects.

Morris *et al.* also examined sera from a number of agricultural and meat-processing occupations and with the exception of Auckland meatworkers, freezing works veterinary surgeons and pest control officers, there were no significant deviations from the rate for donors of unknown race. The 80 meatworkers showed a seropositivity rate of 41%. This is in contrast to a recent survey of northern Victorian meatworkers where the seropositivity rate was 11% in 64 meatworkers (unpublished data, Dwyer *et al.*).

Further confirmation of an age related rise in seropositivity has been recently shown by Hirschl, using sera from healthy children and blood donors from Austria [12]. Blaser's group from Colorado has also recently reported the results of their serological studies on healthy people [13]. Antibody was absent in children < 9 years, and then rose with age, until in people > 40 years antibody was present in over half. The Sydney group of Lee & Hazell [14] have also shown a rise in seropositivity with age, with an interesting and unexplained biphasic form. Antibody was present in about 6% of blood donors aged 18–22 years, rising to over 20% in donors aged 23–27 years, before falling in the next two successive 4-year age cohorts, and then successively rising again in the next three 4-year age cohorts.

Lee & Hazell [14] also quote data of Carrick which show distinct differences in detection rates of *C. pylori* in endoscoped dyspeptic patients. Anglo-Saxons had the lowest positivity rate, 81/213 (38%), Italians the next with 54/83 (67%), then Greeks with 50/57 (87%) and Arabs had the highest rate of infection with 19/21 (90%).

At Fairfield Hospital in Melbourne, Australia, there has arisen an almost unique opportunity to study sera from a variety of population groups immigrant to Australia. These people present to a central health screening clinic to receive a limited clinical examination, immunization advice, and screening and treatment for syphilis, tuberculosis and intestinal parasites [15]. The immigrants are seen within several days of arrival in Australia. We have studied a number of different population groups who have availed themselves of this service [15, 16]. The serological data are summarized in Table 23.1. For many of these groups there are still either no data or only limited data on the prevalence of duodenal ulcer disease.

The salient serological features are that Vietnamese show a pattern intermediate to that for both Melbourne and rural Victorian blood donors. Both El Salvadoreans and Ethiopians show rates of infection which are higher than these other groups at each age group, but which do not rise with age. There is limited information which indicates that the peak age for complicated ulcer in some groups of Ethiopians is in the years 20–30 [17]. This correlates to some degree with our serological observations.

We have also studied the prevalence of *C. pylori*-specific IgG in a group of Aborigines from northern Australia [18]. It has been widely recognized that Aborigines from northern and north-western Australia rarely, if ever, develop duodenal or gastric ulcer [19]. We considered it important to determine

Table 23.1 Distribution of sera with IgG antibodies specific to *Campylobacter pylori* in various age and ethnic groups. (Numerators indicate the numbers positive by ELISA while denominators indicate the numbers tested in each group)

Group	Total (%)	Age groups (years)					
		10–19	20–29	30–39	40–49	50–59	>60
Melbourne (Australia) blood donors	23/154 (15%)	2/19	0/32	3/24	6/36	10/33	2/10
Rural Victoria (Australia) blood donors	49/171 (29%)	5/30	7/30	5/30	5/29	17/32	10/20
Manchester, England*	126/607 (21%)	9/238	12/64	10/53	6/37	11/28	77/165
New Zealand blood donors**	175/480 (36%)	56/175	54/165	42/93	19/35	4/12	0/3
Northern Australian Aborigines	2/274 (1%)	0/106	0/80	0/43	0/27	2/18	NT
Southern Australian Aborigines	142/285 (50%)	29/68	36/78	33/61	27/46	10/23	7/9
Vietnamese	35/190 (18%)	NT	6/49	11/51	8/47	4/18	6/25
El Salvadoreans	30/75 (40%)	NT	6/14	4/16	8/17	8/16	4/12
Ethiopians	32/74 (43%)	NT	8/17	19/46	5/11	NT	NT

*Modified from Jones *et al.* [9].
**Modified from Morris *et al.* [11].
NT = not tested.

whether this group had serological evidence of infection with *C. pylori*, because its presence would cast doubt on the hypothesis that this organism was important in the aetiology of these two conditions, especially duodenal ulcer. The data indicate that infection is rare in this population group (< 1%).

By way of contrast, we also studied a broad cross-section of people of Aboriginal descent from a major river valley of south-eastern Australia. These people live in urban and semi-urban conditions, often side-by-side with non-Aboriginal families. Duodenal ulcer is known to occur in this group, but details of its frequency are not available. The contrast with seropositivity rates for Melbourne blood donors and for rural Victorian blood donors was indeed striking. Aborigines from this group showed an overall rate of seropositivity of 50%, which appeared in teenagers (the youngest age group studied) and maintained this rate across all age groups. Rural Victorian blood donors (almost entirely non-Aboriginal) showed a higher rate of seropositivity than Melbourne donors although the differences were not substantial until the donors were over 60 years.

More recently, we have studied over 1200 sera from two indigenous populations in Papua New Guinea. One population is from a coastal location with a low prevalence of duodenal ulcer, the other from a highland population with a high prevalence of duodenal ulcer. Preliminary analysis of the data indicates that approximately 2% of the coastal population has antibody, while over 20% of the highland population has antibody.

Conclusions

The linkage between *C. pylori*, antral inflammatory (type B) gastritis and duodenal ulcer is now so strong that there is a need to re-examine the well-described variations in prevalence of duodenal ulcer in various population groups in relation to the prevalence of infection with *C. pylori*. There are two major problems for such an exercise. Firstly, obtaining direct cultural or histological evidence of infection by mass screening using upper gastro-intestinal endoscopy is impractical for all but a few population groups. Indirect evidence of infection using isotopically-labelled urea as a marker for the presence in the stomach of urease-producing bacteria, presumably *C. pylori*, is likely to remain too expensive for application in large scale surveys for the foreseeable future. Thus assays for serum antibody, usually IgG, specific to *C. pylori* would appear to have the practical edge over these other techniques. Secondly, almost all reported studies on prevalence of infection have not indicated the frequency of duodenal ulcer in the communities studied. In fact there are major methodological problems to be overcome in order to determine the actual rates of duodenal ulcer in various populations. For the moment, the best estimates will be made from clinical records.

At first sight the prevalence of infection due to *C. pylori* would appear to be linked in some way to the degree of industrialization and economic development of the populations studied. In the general populations of Western Europe, United States of America, Australia and New Zealand there would appear to be an overall prevalence of 15–25%. In Vietnamese, the prevalence is at a similar level, while in Ethiopians and El Salvadoreans, Samoans and Cook Islanders, the prevalence is of the order of 40% and in Chinese and Tongans may be even higher (60% and 70%, respectively). However, industrialization, or rather a lack of it, is not necessarily predictive of infection rates when one considers the prevalence data in northern Australian Aborigines or populations in Papua New Guinea.

Sero-epidemiological study of populations under a variety of conditions may well provide important insights into understanding the mode and conditions for transmission of the infection. If one accepts that there really is sufficient evidence now to implicate *C. pylori* as a necessary component in the causation of duodenal ulcer then we will also have to find satisfactory explanations for the following phenomena:

1 Why do some intermingling populations from different ethnic groups have differing rates of presentation with duodenal ulcer? For example the Indians of Fiji appear to have rates of duodenal ulcer at least twice that seen in ethnic Fijians [20]. Beg and co-workers have shown that *C. pylori* is present with equal frequency in both Fijian and Indian patients with duodenal ulcer from this country [21].

2 Why has there been a decline in mortality and morbidity for duodenal ulcer in all West European nations, the United States and Japan through this entire century and why has this been accompanied by a birth cohort effect, the brunt of which was carried through life by people born at the end of the 19th century [22–24]?

3 Is there a continuing risk through life for exposure and then chronic infection with *C. pylori*? The serological data from Europe, Australia, New Zealand and the United States would be consistent with this, although it may be that this risk is now declining. Certainly the risks of encounter would appear to be vastly different from those seen with *Campylobacter jejuni* which has a trimodal distribution of infections through life in those countries.

4 How do peaks in age-related prevalence of duodenal ulcer disease, both complicated and uncomplicated, relate to patterns of *C. pylori* seropositivity? In other words is presentation with duodenal ulcer just a function of chronicity of infection or are there other determinants to be considered?

The answers to these questions may well provide us with good public health approaches to an infectious disease which appears to have declined in frequency (but by no means disappeared!) in some countries even before we recognized its presence. We now need to define the determinants for acquisition of infection, for persistence of infection, for transmission and for outcome. These are just as important to study as the mode(s) of transmission, the effects of host and microbial genetics, variations in virulence and the nature of the immune response. With a better understanding of these factors, then the greater the likelihood that strategies can be devised to deal with this organism and its consequences. Clearly an organism with such a high prevalence of infection and such close links with gastritis and duodenal ulcer needs to be recognized more widely.

Acknowledgements

We are indebted to Dr Michael Oldmeadow for his kind permission to use information gathered by himself and his colleagues.

References

1 Morris, A., Maher, K., Thomsen, L., Miller, M., Nicholson, G. & Tasman-Jones, C. Distribution of *Campylobacter pylori* in the human stomach obtained at postmortem. *Scand. J. Gastroenterol.* 1988, **23**, 9–13.

2 Langenberg, M.L., Tytgat, G.J.N., Schipper, M.E.I., Rietra, P.J.G.M. & Zanen, H.C. Campylobacter-like organisms in the stomach of patients and healthy individuals. *Lancet* 1984, **i**, 1348.

3 Barthel, J.S., Westblom, T.U., Harvey, A.D., Gonzalez, F. & Everett, E.D. Gastritis and *Campylobacter pylori* in healthy, asymptomatic volunteers. *Arch. Intern. Med.* 1988, **148**, 1149–1151.

4 Carrick, J., Hazell, S., Ralston, M., & Borody, T. The incidence of *Campylobacter pyloridis* gastritis in healthy volunteers. *Aust. N.Z. J. Med.* 1987, **17**, 137.

5 Marshall, B.J. & Surveyor, I. Carbon-14 urea breath test for the diagnosis of *Campylobacter pylori* associated gastritis. *J. Nucl. Med.* 1988, **29**, 11–16.

6 Graham, D.Y., Klein, P.D., Evans, D.J., Evans, D.G., Alpert, L.C., Opekun, A.R. & Boutton, T.W. *Campylobacter pylori* detected non-invasively by the ^{13}C-urea breath test. *Lancet* 1987, **i**, 1174–1177.

7 Graham, D.Y., Klein, P.D., Opekun, A.R. & Boutton, T.W. Effect of age on the frequency of active *Campylobacter pylori* infection diagnosed by the [^{13}C] urea breath test in normal subjects and patients with peptic ulcer disease. *J. Infect. Dis.* 1988, **157**, 777–789.

8 Graham, D.Y., Klein, P.D., Opekun, A.R., Boutton, T.W., Evans, D.J., Evans, D.G., Alpert, L.C., Michaletz, P.A., Yoshimura, H.H. & Adam, E. Epidemiology of *Campylobacter pylori* infection: ethnic considerations. *Scand. J. Gastroenterol.* 1988, **23** (Suppl. 142), 9–13.

9 Jones, D.M., Eldridge, J., Fox, A.J., Sethi, P. & Whorwell, P.J. Antibody to the gastric campylobacter-like organism (*'Campylobacter pyloridis'*) — clinical correlations and distribution in the normal population. *J. Med. Microbiol.* 1986, **22**, 57–62.

10 Marshall, B.J., McGechie, D.B., Francis, G.J. & Utley, P.J. Pyloric *Campylobacter* serology. *Lancet* 1984, **ii**, 281.

11 Morris, A., Nicholson, G., Lloyd, G., Haines, D., Rogers, A. & Taylor, D. Seroepidemiology of *Campylobacter pyloridis*. *N.Z. Med. J.* 1986, **99**, 657–659.

12 Hirschl, A.M. Frequency of occurrence of *Campylobacter pylori* and analysis of the systemic and local immune response. *Zbl. Bakt. Hyg. (A)*, 1987, **266**, 526–542.

13 Perez-Perez, G.I., Dworkin, B.M., Chodos, J.E. & Blaser, M.J. *Campylobacter pylori* antibodies in humans. *Ann. Intern. Med.* 1988, **109**, 11–17.

14 Lee, A. & Hazell, S.L. *Campylobacter pylori* in health and disease: An ecological perspective. *Microb. Ecol. Health Dis.* 1988, **1**, 1–16.

15 Ryan, N., Plackett, M. & Dwyer, B. Parasitic infections of refugees. *Med. J. Australia* 1988, **148**, 491–494.

16 Dwyer, B., Kaldor, J., Tee, W., Marakowski, E. & Raios, K. Antibody response to *Campylobacter pylori* in diverse ethnic groups. *Scand. J. Infect. Dis.* 1988, **20**, 349–350.

17 Tsega, E. & Mengistu, M. Peptic ulcer in Ethiopia: A prospective analysis of 200 endoscopically proven cases. *Trop. Geogr. Med.* 1985, **37**, 150–155.

18 Dwyer, B., Sun, N., Kaldor, J., Tee, W., Lambert, J., Luppino, M. & Flannery, G. Antibody response to *Campylobacter pylori* in an ethnic group lacking peptic ulceration. *Scand. J. Infect. Dis.* 1988, **20**, 63–68.

19 Bateson, D.M. Duodenal ulcer — does it exist in Australian Aborigines? *Aust. N.Z. J. Med.* 1976, **17**, 137–139.

20 Parshu, R. Peptic ulcer in Fiji — a retrospective study. *Fiji Med. J.* 1975, **3**, 148–153.

21 Beg, F., Oldmeadow, M., Morris, A., Miller, M. & Nicholson, G. *Campylobacter pylori* infection in patients undergoing endoscopy in Fiji. *N.Z. Med. J.* 1988, **101**, 140–141.

22 Sonnenberg, A. Geographic and temporal variation

in the occurrence of peptic ulcer disease. *Scand. J. Gastroenterol.* 1985, **20** (Suppl. 110), 11–24.

23 Susser, M. Period effects, generation effects and age effects in peptic ulcer mortality. *J. Chron. Dis.* 1982, **35**, 29–40.

24 Sonnenberg, A. & Müller, H. Cohort and period effects in peptic ulcer mortality from Japan. *J. Chron. Dis.* 1984, **37**, 699–704.

24 Evidence for person-to-person spread of Campylobacter pylori

H.M. MITCHELL, A. LEE AND T.D. BOHANE

As yet, no natural animal host has been found for *Campylobacter pylori*. Indeed, laboratory experiments have shown that this organism will not colonize the stomachs of animals other than the germ-free piglet or other primates, e.g. monkeys and chimpanzees. The hypothesis has been put forward that the human is the natural host for *C. pylori* and that this organism has adapted itself to the peculiar ecological niche of the human stomach [1]. If this is indeed the case then one would have to postulate person-to-person spread as the most likely means of transmission of *C. pylori*. The volunteer studies of Barry Marshall [2] and Arthur Morris [3] certainly show that infection can be acquired by ingestion. Evidence of person-to-person transmission is found in the study of Ramsay *et al.* [4], who reported an outbreak of epidemic gastritis amongst a group of healthy volunteers participating in ongoing experiments of gastric acidity. This outbreak of gastritis is now accepted to have been caused by *C. pylori* presumably transmitted from person to person via the pH electrode that was being used to measure stomach acidity [5]. Thus it is clear that *C. pylori* can be acquired by mouth and can be transmitted via an inanimate object, however the evidence for person-to-person spread in the natural setting is limited. Studies are described below that have been carried out in this area and these include institutional spread, family studies, and spread amongst endoscopy staff. The major aim of this work has been to study groups of the population in which person-to-person spread would be highly likely to occur and to determine if there is an increased number of persons infected with *C. pylori* in this group.

As it is impractical to use endoscopy for epidemiological studies, an alternative accurate predictor of *C. pylori* infection is required. Serology has been the method of choice for the majority of epidemiological studies particularly using the enzyme-linked immunosorbent assay (ELISA).

Validation of the ELISA

Any serological test that is used to predict active *C. pylori* infection has to be extensively validated. The results of our studies described below have been obtained using an ELISA with a sonicate of whole cells as the antigen. The test was validated on a group of 189 endoscopy patients of whom 121 were known to be *C. pylori*-positive by histological examination of biopsy specimens and 68 subjects who were known to be *C. pylori*-negative on histology. The ELISA was found to be highly predictive of active *C. pylori* infection associated with histological active chronic gastritis with a sensitivity of 100% and a specificity of 94% [6]. The cut-off point for active infection by *C. pylori* was an optical density (O.D.) reading at 405 nm of > 1.0. All sera with an O.D. of between 1.0 and 1.4 were absorbed with *C. jejuni* prior to a second ELISA to increase the sensitivity of the test. There has been much discussion in the literature on which particular antigen is appropriate for *C. pylori* serology. Indeed it has been suggested that whole cell sonicates may not be useful [7]. However, as long as methods used have been validated as above, the choice of antigen will not influence the result.

Institutional spread of C. pylori

Infectious diseases which are spread from person to person by close contact are commonly found to have a higher incidence in institutions for the intellectually handicapped. This is due to the fact that personal hygiene is often lacking and close personal contact often occurs. This phenomenon was demonstrated for hepatitis B virus in an Australian institution where an increased incidence of infection was found in the inmates compared to that of the normal population [8]. The sera (which had been carefully stored at −70°C) used in this hepatitis study were kindly made available to us by Professors C. Boughton and R. Hawkes. Antibody to *C. pylori* was measured using the ELISA and an accurate assessment of *C. pylori* infection in this institution was made. These results have been published previously [9] (see Table 24.1). The control groups selected were either normal blood donors or children whose sera were being routinely screened for complaints not related to the gastro-intestinal tract of Guillain–Barré syndrome.

Table 24.1 Incidence of *C. pylori* infection and *C. jejuni* antibody in the normal population and in institution residents in New South Wales, Australia

Age (years)	Normal (blood donors)	Institution residents	*P*
*Incidence of C. pylori infection**			
3–17	1/31 (3.2%)	11/31 (35%)	<0.005
18–32	41/262 (15.6%)	19/20 (66%)	<0.0005
33–47	66/360 (18.3%)	18/24 (75%)	<0.0005
>48	49/145 (33.8%)	12/14 (86%)	<0.0005
All	157/798 (19.7%)	60/98 (61%)	<0.0005
*Incidence of C. jejuni antibody***			
3–17	3/31 (10%)	4/31 (13%)	<0.7
18–32	45/262 (17.2%)	10/29 (35%)	<0.025
33–47	63/360 (17.5%)	10/24 (42%)	<0.05
>48	26/145 (17.9%)	6/14 (43%)	<0.005
All	137/798 (17.2%)	30/98 (31%)	<0.005

*Incidence of persons with an O.D. >1.0 for *C. pylori* using an ELISA validated as being predictive of active infection.
**The *C. jejuni* results showed the incidence of persons with O.D. >0.7, i.e. presumably evidence of past infection.

The results show that the residents of the institution became infected with *C. pylori* a great deal earlier in life than the normal control population and at all age groups a much higher proportion were infected. Antibody to *C. jejuni*, the intestinal pathogen, was also found to be increased in the residents consistent with an increased spread of gut-borne pathogens. Absorption studies showed that the *C. pylori* antibody was specific and therefore all those shown to be serologically positive for *C. pylori* would be found to have an active infection with *C. pylori* on endoscopy. These results are consistent with the hypothesis of person-to-person spread of *C. pylori*.

Family studies

Family studies have shown a high incidence of active *C. pylori* infection in consanguineous contacts of children infected with *C. pylori*. In general, children tend to have closer personal contact with family members than with those outside the family unit. Therefore it appeared possible that if direct transmission of *C. pylori* was involved, the families of index cases of *C. pylori*-infected children would have a higher incidence of active *C. pylori* infection than the general population. The data on four families published in a letter to *The Lancet* were consistent with this hypothesis [10]. This study has now been extended to 10 families and we present these new data below. Histological information on the index cases is shown in Table 24.2. All patients showed evidence of active chronic gastritis in either the stomach antrum, or body, or both, and all the patients except one were very heavily infected with the organism. In the ELISA [6] O.D. readings of >1 indicate active infection with *C. pylori*. The data are shown in Table 24.3 and as can be seen from these results, a remarkably high incidence of *C. pylori* infection existed in this group. Consistent with the high predictive

Table 24.2 Histological findings in index patients, in family study*

Site	Family number and age of patient (years)									
	1 ($7\frac{1}{2}$)	2 ($8\frac{1}{2}$)	3 (14)	4 (16)	5 (14)	6 (7)	7 (16)	8 (11)	9 (14)	10 (11)
Oesophagus										
Histology	±	NA	++	NS	NS	+	NA	NA	NS	±
C. pylori	–	NA	+	–		–	NA	NA	–	–
Stomach (body)										
Histology	++	+	+	++	++	+	+	++	+	++
C. pylori	+	–	+	++	+++	+	+++	+++	+++	++
Stomach (antrum)										
Histology	++	NA	++	++	++	+	+	+	++	++
C. pylori	+++	+++	+++	++	+	+	+++	+++	+++	++
Duodenum										
Histology	NA	+	NA	++	NS	+	++	++	±	NS
C. pylori	NA	–	NA	–		–	–	–	–	–

*NA = not available; NS = no significant abnormality; – = negative, ± = equivocal, + = few, ++ = moderate and +++ = large numbers of *C. pylori* or, on histology, cells (plasma cells, mononuclear cells, polymorphs, or lymphocytes).

value of the ELISA, 100% of the index cases had positive serology. Of the 33 consanguineous contacts of these children, 27 (81%) were actively infected with *C. pylori*, which compares with 21% in an age-matched control population of Australian blood donors and children. The finding that such a high percentage of family members was actively infected with *C. pylori* is further evidence of person-to-person spread. However, it should be pointed out that there is one potential flaw in this interpretation. It has been shown that there are marked differences in the *C. pylori* infection rates in different ethnic groups. In the same community in which the above study was done, Dr Jane Carrick (personal communication) has shown that in a series of 424 adult dyspeptic patients 81/213 (38%) of Anglo-Saxons have *C. pylori* infection compared to 54/83 (67%) of Italians, 50/57

Table 24.3 IgG absorbance levels* in patients and direct family members

Family member	Family number									
	1	2	3	4	5	6	7	8	9	10
Patient	1.65	1.42	1.55	1.9	1.16	1.07	1.5	1.6	1.45	1.6
Father	> 2.00	0.940	1.23	1.76	1.56	1.62	1.73	1.6	1.71	1.85
Mother	0.77	1.220	1.67	1.92	0.63	1.85	1.08	1.6	1.72	1.60
Sister (1)	0.84	0.510		1.8	1.30	–	–	–	–	1.70
Sister (2)		–	–	–	–	–	–	–	–	–
Brother (1)	0.68	–	> 2.00	–	–	–	1.77	–	1.18	1.25
Brother (2)	–	–	–	–	–	–	1.41	–	–	1.55
Paternal grandmother	–	1.6	–	–	–	–	–	–	–	–
Maternal grandmother	–	–	–	–	–	–	–	–	1.55	–

*Optical density reading (O.D.) at 405 nm of ELISA using *C. pylori* as the antigen. An O.D. > 1.00 has been shown to be predictive of active *C. pylori* infection.

(87%) Greeks and 20/21 (90%) Arabs. The families in the study described above were from different ethnic groups, therefore the control population used was not ideal. More information is required on families in which there has been no history of stomach disease and which are culturally matched but these data are difficult to obtain. Despite the cautions noted above, the family data are still very suggestive of person-to-person spread.

There are conflicting data in the literature with regard to transmission of *C. pylori* in family members. Work presented by Jones *et al.* [11] studied the household contacts of 40 patients in whom *C. pylori* had been cultured from biopsy specimens. They found only 3/26 (12%) consanguineous household contacts to be *C. pylori*-positive while 8/38 (21%) non-consanguineous were positive. The difference between these results and our own could be that in Jones's work the index cases were adults with an average age of 48 years whereas our cases were children from 7–16 years old. One could expect close physical contact to be greater when a child is the index case compared with an adult. Which family member would have been the first infected is not possible to determine.

Langenberg *et al.* [12] showed the same restriction endonuclease pattern in *C. pylori* isolated from a husband and wife which is consistent with person-to-person transmission of this organism. Majewski & Goodwin [13] on the other hand found isolates from a husband and wife to be different as were isolates from a brother and sister. However, the fact that these workers found in pairs of isolates from eleven individuals that the second isolate was markedly different in six subjects raises some doubt about the methodology used. There is an urgent need for more restriction endonuclease fingerprinting in family groups.

Infection amongst endoscopy clinic staff

The evidence presented above gives no information concerning the actual process of person-to-person spread. While faecal-to-oral spread could be suggested by the institutional data, this is considered to be unlikely as *C. pylori* has never been isolated from the stools of gastritis patients and therefore faecal material would not be expected to be infectious. The most likely route is via stomach secretions possibly by regurgitation or vomitus although there is no evidence for this. Marshall [14] assumed from the descriptions of acute gastritis in the 1920 edition of Sir William Osler's *Textbook of Medicine*, in which vomiting was a common symptom and the vomitus was shown to be of non-acid pH, that the cases were active *C. pylori* infection. If this was so then vomitus presumably contained the organism. Craige [15] in the *Textbook of Veterinary Medicine* has also shown that non-*C. pylori* natural spiral colonizers of the canine stomach mucus can be discharged in the vomitus of dogs with gastritis.

To examine the likelihood of stomach secretions playing a part in person-to-person transmission, the incidence of active *C. pylori* infection in a group of gastroenterologists who perform endoscopies was studied because this group is more likely to come in contact with infected gastric secretions. At present, there is limited information on the incidence of *C. pylori* amongst gastroenterologists and the data available are conflicting. We have carried out a study of the incidence of active *C. pylori* infection in 33 gastroenterologists compared with that in control groups of blood donors, endoscopy sisters and general practitioners. We have found a statistically significant difference in the incidence of active *C. pylori* infection, i.e. 51% in the gastroenterologists compared to 28% in the general practitioners, 19% in the endoscopy sisters and 21% in the blood donor controls. This does not agree with the results of Morris *et al.* [16] who found no significant difference between 25 gastroenterologists and the normal population. Rawles *et al.* [17] also examined the incidence of active *C. pylori* infection in endoscopy staff and found a higher frequency of antibody to *C. pylori* compared with blood-bank controls; however this difference was shown to be not statistically significant. Clearly there is a need for much more work on endoscopy staff before this point can be clarified.

Animals as a source of C. pylori

The only report to date that would be consistent with an animal source for *C. pylori* is the serological study of Morris *et al.* [18]. This study compared the incidence of *C. pylori* infection in various occupational groups, including freezer workers, veterinarians and Auckland meatworkers, to normal blood donor controls. ELISA results showed a significantly raised incidence in these two occupational groups. When using an ELISA, it is essential to use a carefully validated method or misleading conclusions can be reached. This was demonstrated in our recent study of the sero-epidemiology of *C. pylori* in Papua New Guinea. Knowing that there is a high incidence of *C. jejuni* infection in children in Papua New Guinea, we were concerned that our figures could be a result of the known cross reactivity between the flagellar antigen of *C. pylori* and *C. jejuni* antibody [19]. Absorption of the sera with the middle range response (i.e. O.D. 1.0–1.4) in our ELISA showed that a high percentage of the Papua New Guinean sera absorbed out proving this concern to be true. However, this did not occur in the Australian blood donor sera where the level of *C. jejuni* antibody is very low. This explained why our test was so predictive of *C. pylori* infection in the Australian population without the absorption step, whereas in the Papua New Guinean studies absorption with *C. jejuni* is essential. If a high *C. pylori* O.D. is present, i.e. > 1.4, then absorption with *C. jejuni* makes little or no difference even if *C. jejuni* antibody is present. Thus in any population where a high level of contact with *C. jejuni* is likely to occur and in which whole cells or whole cell sonicates are being used in the *C. pylori* serological assay, all marginally-positive sera should be absorbed with *C. jejuni*. This applies to the group of veterinarians and meatworkers examined by Morris as it is known that a high carriage rate of *C. jejuni* exists in farm animals. If Morris is correct and there is this occupational association with *C. pylori* infection, then the hypothesis of the human as the main reservoir of *C. pylori* and person-to-person transmission will need to be reappraised.

Conclusion

On balance the evidence to date supports person-to-person transmission of *C. pylori*. However, more work is required to increase the limited amount of data available at present. It is important to gather statistically significant numbers of subjects, to choose suitable groups of individuals for study, to validate the serological method used and to have age and culture matched controls.

Acknowledgement

We would like to thank Nerissa Lee for considerable help in the preparation of this manuscript.

References

1 Lee, A. & Hazell, S.L. *Campylobacter pylori* in health and disease: An ecological perspective. *Micro. Ecol. Health. Dis.* 1988, **1**, 1–16.

2 Marshall, B.J., Armstrong, J.A., McGechie, D.B. & Clancy, R.J. Attempt to fulfill Koch's postulates for pyloric *Campylobacter*. *Med. J. Australia* 1985, **142**, 436–439.

3 Morris, A. & Nicholson, G. Ingestion of *Campylobacter pyloridis* causes gastritis and raised fasting pH. *Am. J. Gastroenterol.* 1987, **82**, 192–199.

4 Ramsay, E.J., Carey, K.V., Peterson, W.L. *et al.* Epidemic gastritis with hypochlorhydria. *Gastroenterology*, 1979, **76**, 1449–1457.

5 Blaser, M.J. Gastric *Campylobacter*-like organisms, gastritis, and peptic ulcer disease. *Gastroenterology* 1987, **93**, 371–383.

6 Mitchell, H.M., Lee, A., Berkowicz, J. & Borody, T. The

use of serology to diagnose active *Campylobacter pylori* infection. *Med. J. Australia*, **149**, 604–609.

7 Newell, D.G. Identification of the outer membrane proteins of *C. pyloridis* and antigen cross reactivity between *C. pyloridis* and *C. jejuni*. *J. Gen. Microbiol.* 1987, **133**, 163–170.

8 Boughton, C.R., Hawkes, R.A., Schroeter, D.R. & Harlor, J.A. The epidemiology of hepatitis B in a residential institution for the mentally retarded. *Aust. N.Z. J. Med.* 1976, **6**, 521–529.

9 Berkowicz, J. & Lee, A. Person to person transmission of *Campylobacter pylori*. *Lancet* 1987, **ii**, 680–681.

10 Mitchell, H.M., Bohane, T.D., Berkowicz, J., Hazell, S.L. & Lee, A. Antibody to *Campylobacter pylori* in families of index children with gastrointestinal illness due to *Campylobacter pylori*. *Lancet* 1987, **ii**, 681–682. (Letter)

11 Jones, D.M., Eldridge, J. & Whorwell, P.J. Antibodies to *Campylobacter pyloridis* in household contacts of infected patients. *Br. Med. J.* 1987, **294**, 615.

12 Langenberg, W., Rauws, E.A.J., Widjojokusumo, A., Tytgat, G.N.J. & Zanen, H.C. Identification of *Campylobacter pyloridis* isolates by restriction endonuclease DNA analysis. *J. Clin. Microbiol.* 1986, **24**, 414–417.

13 Majeweski, S.I.H. & Goodwin, S.C. Restriction endonuclease analysis of the genome of *Campylobacter pylori* with a rapid extraction method: Evidence for considerable genomic variation. *J. Infect. Dis.* 1988, **157**, 465–471.

14 Marshall, B.J. *Campylobacter pyloridis* and gastritis. *J. Infect. Dis.* 1986, **153**, 650–657.

15 Craige, J. 1959. In Hoskins, H.P., Lacroix, J.V. & Mayer, K. (eds.) *Canine Medicine*. 2nd edn. American Veterinary Publications, Santa Barbara, California.

16 Morris, A., Lloyd, G. & Nicholson, G. *Campylobacter pyloridis* serology among gastroendoscopy clinic staff. *N.Z. Med. J.* 1986, **99**, 820–821.

17 Rawles, J.W., Harris, M.L., Paul, G. *et al.* Antibody to *Campylobacter pyloridis* in endoscopy personnel, patients and controls. *Gastroenterology* 1987, **92**, 1589. (Abstract)

18 Morris, A., Nicholson, G., Lloyd, G., Haines, D., Rogers, A. & Taylor, D. Seroepidemiology of *Campylobacter pyloridis*. *N.Z. Med. J.* 1986, **89**, 657–659.

19 Lee, A., Logan, S.M. & Trust, T.J. Demonstration of a flagellar antigen shared by a diverse group of spiral shaped bacteria that colonize intestinal mucus. *Infect. Immun.* 1987, **55**, 828–831.

25 Possible pathogenic mechanisms in Campylobacter pylori infection

B.J. RATHBONE AND R.V. HEATLEY

The very strong association between *Campylobacter pylori* and Type B chronic gastritis and duodenal ulcer disease has been confirmed worldwide [1]. Human ingestion studies (either by accident or design), treatment and animal studies all suggest a pathological role for *C. pylori* in chronic gastritis. Furthermore, increasing evidence supports this organism as playing an important role in duodenal ulcer disease.

Campylobacter pylori colonizes the human gastric mucosa usually without any competition from other organisms. Although no similar organisms are commonly found in humans, spiral, motile, strongly urease-positive micro-aerophilic organisms are well documented colonizing mucus-secreting epithelia in other mammals such as the cat. The urease activity, spiral shape and gas requirements appear to be important factors enabling the organisms to survive in a viscous, poorly-oxygenated environment. In many animals the colonization appears commensal, but in humans the colonization is virtually always associated with gastric inflammation.

Possible mechanisms in gastric inflammation

Bacterial adherence

One of the major characteristics differentiating enteropathic from non-enteropathic *Escherichia coli* is the ability of the former to adhere to receptors on the epithelial surface [2]. Such adherence protects the bacteria from peristalsis and results in high concentrations of toxins locally at the epithelial cell surface.

The adherence of *C. pylori* to gastric epithelium by pedestals very similar to those found with *E. coli* has been reported [3]. Such pedestals are seen relatively uncommonly; however, other ultrastructural changes are common. Intracellular oedema occurs together with intracellular mucin depletion. Where bacteria are numerous, often in the intercellular regions, microvilli are depleted or absent, together with the organized cores of microfilaments which normally contribute to the cytoskeleton of the apical cytoplasm [3]. Disruption of the intercellular junctions has been demonstrated [4], as has the presence of the organisms in parietal cell canaliculi [4]. *Campylobacter pylori* have been seen in neutrophil phagocytic vacuoles [5] and recently they have been reported to be covered by epithelial microvilli and/or engulfed within phagocytotic vacuoles or in close contact with foci of cell necrosis containing cytoplasmic debris [6]. Within these areas partially autolysed organisms have been seen. Following elimination of *C. pylori* by bismuth salts the ultrastructural changes regress [3]. Although these ultrastructural abnormalities strongly implicate *C. pylori* as a pathological organism, they tell us little of the underlying cytopathic mechanism.

Biochemical factors

One well-recognized method for gut bacteria to exert a cytopathic effect is via toxin production. Mattsby-Baltzer & Goodwin have presented data showing that *C. pylori* lipopolysaccharide contained lipid A, a class of substance responsible for the endotoxic properties of a wide range of Gram-negative bacteria [7].

Leunk *et al.* [8] have studied the effect of broth culture filtrates on mammalian cell lines. They found that 45% of the bacterial strains tested produced cytopathic effects. These effects were abolished by rabbit antisera (against whole or sonicated *C. pylori*), but not by antiserum against a lipopolysaccharide fraction. Clearly the significance of these findings requires further work.

Enzymatically, *C. pylori* has a wide range of enzymes [9], a number of which may play a role in pathogenesis. The most notable is urease, the activity of which is so marked as to enable colonization to be diagnosed by a urea breath test or a biopsy urease test. It has been postulated that the ammonia produced envelops the bacteria, protecting it from acidic damage. Ammonia is certainly toxic to cells in tissue culture and recent animal studies have shown ammonia to cause a dose-dependent decrease in gastric mucosal potential difference, together with increasing gastric epithelial damage [10, 11].

Hazell & Lee [12] have postulated that the urease activity at the epithelial surface is a major factor in pathogenesis, with the rapid hydrolysis of urea creating a mucosal urea gradient and a transmucous ammonia gradient resulting in an increase of epithelial surface pH. This in turn, they suggest, might affect mucosal charge gradient, paracellular permeability and epithelial cell Na^+/K^+-ATPase leading to back-diffusion of H^+ ions and concomitant hypochlorhydria. This may then in extreme circumstances result in gastric ulcer formation. There is little direct evidence to support this hypothesis and the finding of gastric urease-positive spirals in other animals without inflammation, is against the urease being of prime pathogenic importance [13]. As this enzyme is seen in all the spirals colonizing mammalian upper gastro-intestinal mucus-secreting epithelia, it would appear to play an important role in bacterial colonization or survival. Studies using urease blockers may help define the role of this enzyme.

As yet little work has been directed to the effects of ammonia production on the mucus–bicarbonate barrier and epithelial and parietal-cell metabolism. Attempts to measure epithelial pH have however been carried out and demonstrate a reduced epithelial pH in gastritis which suggests that the mucus–bicarbonate barrier is being compromised.

One important mechanism by which *C. pylori* might be pathogenic is by direct alteration of the gastric mucus gel. Increased cell turnover and epithelial cell mucus depletion are well recognized features of chronic gastritis. *In vitro* studies of gastric mucus glycoprotein synthesis and secretion suggest an increased turnover of mucus glycoprotein in *C. pylori*-associated chronic gastritis [14]. Whether this is purely secondary to the increased epithelial cell turnover found in chronic gastritis is unclear. *Campylobacter pylori* has been demonstrated to have proteolytic enzymes capable of degrading porcine gastric mucus [15]. Interestingly, the optimum pH for this reaction is pH 7. The same group has also reported that the mucus in peptic ulcer patients is less viscous than in normals, has a higher proportion of low-molecular-weight mucin, less mucin polymer and a diminished ability to retard H^+ diffusion [16, 17].

If, as seems likely, *C. pylori* compromises the mucus–bicarbonate barrier, the gastric epithelium would then become exposed to the conventional aggressive luminal factors, acid, pepsin, bile and perhaps also drugs. These agents singly, or in combination, may then be the factors resulting in mucosal inflammation.

Immunological factors

Non-autoimmune chronic gastritis (Type B gastritis) is essentially a chronic inflammatory response, with or without polymorph infiltration and architectural damage. The very strong association between *C. pylori* and chronic gastritis in the absence of any other bacteria raises the possibility that the inflammatory response seen in chronic gastritis is a response to *C. pylori* antigens.

In the initial report documenting the association between *C. pylori* and chronic gastritis, it was noted that the organism was particularly associated with a polymorph response (active gastritis) [18]. Most workers have confirmed this association and in active gastritis the neutrophils appear predominantly between epithelial cells, being most numerous in the deep portion of the gastric pit [19]. Relatively few neutrophils are seen in the gastric gland lumen, and the finding of phagocytosed *C. pylori* has been documented [5]. It is currently unclear whether the polymorph response is elicited by the organisms, or whether the environment of an actively inflamed mucosa is more favourable for the growth of *C. pylori*.

Soon after the original *C. pylori* isolation it was recognized that there was a systemic IgG and IgA antibody response. This systemic response can be used to diagnose *C. pylori* colonization and is particularly useful in population studies [20]. Following clearance of the organisms it appears that the antibody titres decrease [21].

Studies on local gastric humoral responses prior to the isolation of *C. pylori* demonstrated that, with the development of histological gastritis, plasma cells increase in number and the epithelium of the isthmic zone shows secretion of IgA, together with lysozyme [22], and lactoferrin [23].

Initial studies on local *C. pylori* humoral responses concentrated on identifying organism-specific antibodies in gastric juice. Measurable titres of *C. pylori*-specific IgA and IgM were demonstrated in a proportion of subjects with *C. pylori*-associated chronic gastritis [24]. The absence of a demonstratable gastric juice IgG response was expected as this antibody would be rapidly broken down in the acidic gastric juice. No specific IgA, IgM or IgG antibodies were found in the gastric juice of *C. pylori*-negative, histologically-normal patients.

Proof that gastric-juice immunoglobulins were gastric in origin has been provided by short-term organ culture studies, where only *C. pylori*-positive biopsies produced high titres of *C. pylori*-specific antibodies [25]. Plasma cell counts were assessed in the same patients and the density of the plasma cell infiltrate was found to correlate with the specific immunoglobulin titre [26]. The data thus suggest that a proportion of the plasma cell infiltrate, which characterizes chronic gastritis, is responsible for the humoral response to *C. pylori* antigens.

Similar organ culture studies have been carried out using duodenal tissue and demonstrate a duodenal humoral response in *C. pylori*-positive patients [27]. The duodenal response was most marked in those patients with active duodenitis, and such patients have previously been shown to have duodenal *C. pylori* colonization on areas of gastric metaplasia [28, 29].

Antibody coating of *C. pylori* in tissue sections from patients with chronic gastritis has been investigated using an immunoperoxidase technique [19]. IgG, IgA and IgM coating has been demonstrated. However, it is unclear whether this coating is by specific or non-specific immunoglobulins. An interesting finding in this study was the failure of bacteria deep in the gastric pits to be coated, despite being in a location where a high concentration of IgA might be expected.

In vitro studies have demonstrated *C. pylori* to be sensitive to antibody-dependent, complement-mediated bactericidal activity of serum [30]. In the presence of serum opsonins the bacteria were phagocytosed and efficiently killed by polymorphs. The relevance of these studies to the situation at the mucosal surface, where the concentration of complement is unknown and IgA is the major antibody class, is unclear. There is the possibility that the bacterial coating is protective to the organism or *C. pylori* has the ability to produce an IgA protease.

Previous work on gastric cell-mediated immunity has concentrated on Type A autoimmune chronic gastritis, where T-cell subset and functional studies have given conflicting results [31, 32]. Recent studies on Type B gastritis have demonstrated that the inflamed mucosa, unlike normal gastric mucosa, is HLA-DR positive [33, 34] and thus theoretically may be capable of antigen presentation.

In Type B gastritis, T cells are increased in both the epithelial compartment and lamina propria. Studies on T-cell subsets give conflicting results with some workers reporting the majority of epithelial T cells to be of the suppressor/cytotoxic phenotype [34]. Others have, however, reported an increase in the T-helper phenotype [35]. In this latter study, the T-helper cells from patients with gastritis showed increased expression of CD7, a T-cell stimulation marker.

One particularly interesting group of gastritic patients from an immunological point of view are those with lymphocytic gastritis. This is a histologically identifiable condition where the epithelial compartment has a great excess of T cells. In a recent study, *C. pylori* was detected histologically in less than half of the cases. A serological response to *C. pylori* was, however, demonstratable in four out of six patients who were histologically negative [36]. This suggests that either the organism was present in small numbers and not detected, or that there had been a recent infection by *C. pylori*. Lymphocytic gastritis may thus represent an effective immune response to *C. pylori* which is capable of clearing the organism.

The studies of cell-mediated immunity are thus limited at present, and there is no information regarding the role of macrophages which are greatly increased in Type B gastritis.

Gastric autoantibodies have been recognized for many years, particularly in association with pernicious anaemia. In Type B chronic gastritis, IgG autoantibodies to gastrin cells have been demonstrated in a small proportion of patients [37]. In relation to *C. pylori*, the only data suggesting possible autoimmune mechanisms come from the raising of monoclonal antibodies. At least one monoclonal antibody to *C. pylori* cross-reacts with healthy gastric epithelium, suggesting autoimmunity may be possible [38].

Campylobacter pylori and duodenal inflammation

Campylobacter pylori is never found in association with intestinal-type epithelium. The only time that *C. pylori* colonizes the duodenum is when there is gastric metaplasia [39, 40]. The presence of duodenal gastric metaplasia appears to be related to low duodenal pH [41–43]. In a recent study Wyatt *et al.* demonstrated that duodenal gastric metaplasia only occurred in patients with fasting gastric juice pH <2.5, and that *C. pylori*, when present on gastric metaplasia, was always associated with polymorphs [28]. The same group demonstrated that the combination of significant duodenal gastric metaplasia and *C. pylori*-associated gastritis identified those patients with active duodenitis reliably with a sensitivity of 88% and a specificity of 98%. The association was equally strong whether an ulcer crater was present or not. Such data suggest that both acid/pepsin and *C. pylori* are important in duodenitis and duodenal ulcer disease and that these two conditions have a common pathogenesis. The mechanisms leading to *C. pylori*-associated duodenitis are probably similar to those responsible for gastric mucosal damage.

Conclusions

Campylobacter pylori certainly appears to be causally related to chronic gastritis, but as yet we have little information regarding its mechanisms of pathogenicity. There is evidence that the mucus–bicarbonate barrier is compromised. Furthermore, the inflammatory response seen in Type B chronic gastritis certainly seems to be directed at least in part at *C. pylori* antigens. This chronic immune response may well be damaging but other bacterial factors could also adversely affect the normal mucosal defences. Differing pathogenic mechanisms may have changing significance at various phases of colonization and in individual patients. Different bacterial strains may also have variable pathogenic potential [44]. In contrast to chronic gastritis, *C. pylori* colonization (and associated inflammation) in duodenal ulceration appears to be a necessary prerequisite upon which focal ulceration may intermittently occur, presumably due to the interplay of other important factors.

There is clearly much to be learnt regarding *C. pylori* and gastro-duodenal disease. For example, why in the face of a continuing active immune response directed at least in part at *C. pylori* does colonization persist? Why do some individuals such as those reported by Marshall *et al.* [45] eliminate their acute infection, whereas others reported by Morris *et al.* [46] go on to develop chronic colonization?

Campylobacter pylori gastro-intestinal colonization is unique to humans, in that a specialized epithelial niche is infected without competition from other bacteria and then results in a long-standing infection in association with chronic inflammation presumably lasting for years. One must assume that the bacteria gain benefit from the associated chronic inflammation. As the system is uncomplicated by other bacteria it provides a good model for studying bacteria–epithelial interactions. Such knowledge may be particularly relevant in the wider understanding of epithelial microbial ecology, bacterial pathogenicity and gut defence mechanisms.

References

1 Rathbone, B.J., Wyatt, J.I. & Heatley, R.V. *Campylobacter pyloridis*—A new factor in peptic ulcer disease? *Gut* 1986, **27**, 635–641.

2 Old, D.C. Bacterial adherence. *Med. Lab. Sci.* 1985, **42**, 78–85.

3 Goodwin, C.S., Armstrong, J.A. & Marshall, B.J. *Campylobacter pyloridis*, gastritis, and peptic ulceration. *J. Clin. Pathol.* 1986, **39**, 353–365.

4 Chen, X.G., Correa, P., Offerhaus, J., Rodriguez, E., Janney, F., Hoffmann, E., Fox, J. & Hunter, F. Diavolitsis. Ultrastructure of the gastric mucosa harboring campylobacter-like organisms. *Am. J. Clin. Pathol.* 1986, **86**, 575–582.

5 Shousha, S., Bull, T.B. & Parkins, R.A. Gastric spiral bacteria. *Lancet* 1984, , 101.

6 Tricottet, V., Bruneval, P., Vire, O. & Camilleri, J.P. Campylobacter-like organisms and surface epithelium abnormalities in active, chronic gastritis in humans: an ultrastructural study. *Ultrastruct. Pathol.* 1986, **10**, 113–122.

7 Mattsby-Baltzer, I. & Goodwin, C.S. 1988. Lipid A in *C. pylori*. In Kaijser, B. & Falsen, E. *Campylobacter IV*. University of Göteborg, Göteborg. Abstract No. 232.

8 Leunk, R.D., Johnson, P.T., David, B.D., Kraft, W.G. & Morgan, D.R. 1988. Identification of cytotoxic activity produced by *Campylobacter pyloridis*. In Kaijser, B. & Falsen, E. *Campylobacter IV*. University of Göteborg, Göteborg. Abstract No. 97.

9 Megraud, F., Bonnet, M., Garnier, M. *et al.* Characterization of *Campylobacter pyloridis* by culture, enzymatic profile, and protein content. *J. Clin. Microbiol.* 1985, **22**, 1007–1010.

10 Murakami, M., Yoo, J.K., Mizuno, M., Saita, H., Inada, M. & Miyake, T. Effects of ammonia, urea, and urease on the rat gastric mucosa. *Gastroenterology* 1987, **92**, 1544.

11 Murakami, M., Yoo, J.K., Mizuno, M., Saita, H., Inada, M. & Miyake, T. Role of gastric ammonia, urea and urease in gastric mucosal lesions in azotemia. *Gastroenterology* 1987, **92**, 1545.

12 Hazell, S.L. & Lee, A. *Campylobacter pyloridis*, urease, hydrogen ion back diffusion, and gastric ulcers. *Lancet* 1986, **ii**, 15–17.

13 Tompkins, D.S., Wyatt, J.I., Rathbone, B.J. & West, A.P. The characterization and pathological significance of gastric campylobacter-like organisms in the ferret: a model for chronic gastritis? *Epidem. Inf.* 1988, **101**, 269–278.

14 Crabtree, J.E., Rathbone, B.J., Wyatt, J.I., Heatley, R.V. & Losowsky, M.S. *In vitro* mucus glycoprotein synthesis and secretion by gastric mucosa colonized with *Campylobacter pyloridis*. *Gut* 1987, **28**, A1409.

15 Slomiany, B.L., Bilski, J., Sarosiek, J., Murty, V.L.N., Dworkin, B., Van Horn, K., Zielenski, J. & Slomiany, A. *Campylobacter pyloridis* degrades mucin and undermines gastric mucosal integrity. *Biochem. Biophys. Res. Commun.* 1987, **144**, 307–314.

16 Slomiany, B.L., Bilski, J., Murty, V.L.N., Sarosiek, J., Dworkin, B., Van Horn, K. & Slomiany, A. *Campy-*

lobacter pyloridis degrades mucin and undermines gastric mucosal integrity. *Gastroenterology* 1987, **92**, 1645.

17 Sarosiek, J., Gabryelewicz, A. & Slomiany, B.L. Changes in the macromolecular organization and physical properties of gastric mucus with peptic ulcer. *Gastroenterology* 1987, **92**, 1615.

18 Warren, J.R. Unidentified curved bacilli on gastric epithelium in active chronic gastritis. *Lancet* 1983, **i**, 1273.

19 Wyatt, J.I., Rathbone, B.J. & Heatley, R.V. Local immune response to gastric *Campylobacter* in non-ulcer dyspepsia. *J. Clin. Pathol.* 1987, **39**, 863–870.

20 Newell, D.G. & Rathbone, B.J. The serodiagnosis of *Campylobacter pylori* infection. *Serodiag. Immunother.* 1989, **3**, 1–6.

21 Vaira, D., Holton, J., Cairns, S.R., Falzon, M., Polydorou, A., Dowsett, J.F., Salmon, P.R. Antibody titre to *Campylobacter pylori* after treatment for gastritis. *Br. Med. J.* 1988, **297**, 397.

22 Isaacson, P. Immunoperoxidase study of the secretory immunoglobulin system and lysozyme in normal and diseased gastric mucosa. *Gut.* 1982, **23**, 578–588.

23 Valnes, K., Brandtzaeg, P., Elgjo, K. & Stave, R. Specific and non-specific humoral factors in the epithelium of normal and inflamed gastric mucosa. *Gastroenterology* 1984, **86**, 402–412.

24 Rathbone, B.J., Wyatt, J.I., Worsley, B.W., Shires, S.E., Trejdosiewicz, L.K., Heatley, R.V. & Losowsky, M.S. Systemic and local antibody responses to gastric *Campylobacter pyloridis* in non-ulcer dyspepsia. *Gut* 1986, **27**, 642–647.

25 Rathbone, B.J., Wyatt, J.I., Tompkins, D. & Heatley, R.V. *In vitro* production of *Campylobacter pyloridis* specific antibodies by gastric mucosal biopsies. *Gut* 1986, **27**, A607.

26 Wyatt, J.I. & Rathbone, B.J. Immune response of the gastric mucosa to *Campylobacter pylori. Scand. J. Gastroenterol.* 1988, **23** (Suppl. 142), 44–49.

27 Crabtree, J.E., Rathbone, B.J., Shallcross, T.M., Wyatt, J.I., Heatley, R.V. & Losowsky, M.S. Duodenal secretion of *Campylobacter pylori* specific antibodies in patients with gastritis and duodenitis. *Gut* 1988, **29**, A1438.

28 Wyatt, J.I., Rathbone, B.J., Dixon, M.F. & Heatley, R.V. *Campylobacter pyloridis* and acid induced gastric metaplasia in the pathogenesis of duodenitis. *J. Clin. Pathol.* 1987, **40**, 841–848.

29 Johnson, B.J., Reed, P.I. & Ali, M.H. Campylobacter like organisms in duodenal and antral endoscopic biopsies: relationship to inflammation. *Gut* 1987, **27**, 1132–1137.

30 Pruul, H., Lee, P.C., Goodwin, C.S. & McDonald, P.J. Interaction of *Campylobacter pyloridis* with human immune defence mechanisms. *J. Med. Microbiol.* 1987, **23**, 233–238.

31 Rathbone, B.J. & Heatley, R.V. 1986. Gastritis. In Losowsky, M.S. & Heatley, R.V. (eds.) *Gut Defences in Clinical Practice.* Churchill-Livingstone, Edinburgh.

32 Wright, R. 1988. Role of autoimmunity in disease of the gastrointestinal tract and liver. In Heyworth, M.F. & Jones, A.L. (eds.) *Immunology of the Gastrointestinal Tract and Liver*, Raven Press, New York. pp. 193–217.

33 Engstrand, L., Scheynius, A., Grimelius, L., Schwan, A. & Gustavsson, L. Induced expression of class II transplantation antigens on gastric epithelial cells in patients with *Campylobacter pylori* positive gastric biopsies. *Gastroenterology* 1988, **94**, A115.

34 Papadimitriou, C.S., Ioachim-Velogianni, E.E., Tsianos, E.B. & Moutsopoulos, H.M. Epithelial HLA-DR expression and lymphocyte subsets in gastric mucosa in type B chronic gastritis. *Virchows Archiv (A) Pathol. Anat.* 1988, **413**, 197–204.

35 Rathbone, B.J., Wyatt, J.I., Trejdosiewicz, L.K., Heatley, R.V. & Losowsky, M.S. Mucosal T cell subsets in normal gastric antrum and *C. pylori* associated chronic gastritis. *Gut* 1988, **29**, A1438.

36 Dixon, M.F., Wyatt, J.I., Burke, D.A. & Rathbone, B.J. Lymphocytic gastritis — relationship to *Campylobacter pylori* infection. *J. Pathol.* 1988, **154**, 125–132.

37 Vandelli, C., Bottazzo, G.F., Doniach, D. *et al.* Autoantibodies to gastrin-producing cells in antral (type B) chronic gastritis. *New Engl. J. Med.* 1979, **300**, 1406–1410.

38 Rathbone, B.J. & Trejdosiewicz, L.K. 1988. A strain-restricted monoclonal antibody to *Campylobacter pyloridis*. In Kaijser, B. & Falsen, E. *Campylobacter IV.* University of Göteborg, Göteborg. Abstract No. 137.

39 Marshall, B.J., McGechie, D.B., Rodgers, P.A. & Glancy, R.J. Pyloric *Campylobacter* finds a volunteer. *Med. J. Australia.* 1985, **142**, 439–444.

40 Steer, H.W. Surface morphology of the gastro-duodenal mucosa in duodenal ulceration. *Gut* 1984, **25**, 1203–1210.

41 James, A.H. Gastric epithelium in the duodenum. *Gut* 1964, **5**, 285–294.

42 Patrick, W.J.A., Denham, D. & Forrest, A.P.M. Mucosa change in the human duodenum, a light and electron microscope study and correlation with disease and gastric acid secretion. *Gut* 1974, **15**, 767–776.

43 Rhodes, J. Experimental production of gastric epithelium in the duodenum *Gut* 1964, **5**, 454–458.

44 Burnie, J., Lee, W., McNulty, C. & Dent, J. Virulence of *Campylobacter* strains and degree of gastritis. *Lancet* 1988, **i**, 302.

45 Marshall, B.J., Armstrong, J.A., McGechie, D.B. & Glancy, R.J. Attempt to fulfil Koch's postulates for pyloric *Campylobacter. Med. J. Australia* (1985), **142**, 436–439.

46 Morris, A. & Nicholson, G. Ingestion of *Campylobacter pyloridis* causes gastritis and raised fasting gastric pH. *Am. J. Gastroenterol.* 1987, **82**, 192–199.

26 Bacteriological and pharmacological basis for the treatment of Campylobacter pylori infection

C.A.M. McNULTY

There is now strong evidence that *Campylobacter pylori* causes chronic active (Type B) gastritis and that this may predispose to peptic ulceration. Challenge of gnotobiotic piglets and two human volunteers has produced the characteristic gastritis [1–4]. Clearance of the organism with amoxycillin [5] or bismuth salts [6] leads to resolution of gastritis. Clinicians' interest in treating this infection has increased in parallel with the evidence for a causative role of *C. pylori* in gastro-intestinal disease. Initial results indicate that long-term eradication of this organism will be difficult and a search is now on for the most effective antimicrobial.

The first step in the search for a suitable antimicrobial to treat any infection is to determine the *in vitro* susceptibility of the causative organism to suitable therapeutic agents. The activity of a range of antimicrobial agents against *C. pylori* was first determined at neutral pH (Table 26.1) [7, 8]. *Campylobacter pylori* is highly sensitive to most of the B lactams except cefsulodin. Penicillin, ampicillin, cefuroxime, cefoxitin and cephalexin have minimum inhibitory concentrations for 90% of organisms (MIC_{90}) < 0.5 mg/l; flucloxacillin and aztreonam are moderately active with MIC_{90} of 2 mg/l. The macrolides and quinolones except amifloxacin ($MIC_{90} = 4$ mg/l) also have good activity with $MIC_{90} <$ 0.5 mg/l. Other agents that have good *in vitro* activity include nitrofurantoin ($MIC_{90} =$ 0.5 mg/l), gentamicin ($MIC_{90} = 0.25$ mg/l),

Table 26.1 MIC values of 18 antimicrobial agents for isolates of *C. pylori*

Antimicrobial agent	Isolates tested (*n*)	MIC (mg/l)		
		MIC_{50}	MIC_{90}	Range
Penicillin	70	0.015	0.03	0.002–0.06
Ampicillin	97	0.015	0.03	⩽0.008–0.12
Flucloxacillin	50	1	2	0.12–8
Cefuroxime	97	⩽0.12	⩽ 0.12	⩽0.12–0.25
Cefoxitin	70	0.12	0.12	0.015–0.5
Cephalexin	97	⩽0.25	0.5	⩽0.25–2
Cefsulodin	97	16	>16	8–>16
Aztreonam	97	1	2	0.06–8
Erythromycin	97	0.12	0.25	⩽0.008–0.25
Azithromycin	97	0.12	0.25	0.03–0.5
Ciprofloxacin	97	0.25	0.5	⩽0.06–0.5
Ofloxacin	21	0.12	0.25	0.12–0.25
Amifloxacin	97	4	4	1–8
Nitrofurantoin	96	0.5	0.5	⩽0.03–0.5
Furazolidone	96	0.12	0.25	⩽0.03–0.5
Gentamicin	70	0.12	0.25	0.06–0.5
Tetracycline	70	0.25	0.25	0.12–1
Rifampicin	21	0.5	1	0.12–2

tetracycline (MIC_{90} = 0.25 mg/l) and rifampicin (MIC_{90} = 1 mg/l). Metronidazole has moderate activity but about 25% of strains are less sensitive [9]. *Campylobacter pylori* is resistant to vancomycin, cefsulodin, trimethoprim and the sulphonamides.

The known therapeutic value and safety profile of the H_2-receptor antagonists, sucralfate, carbenoxolone and bismuth salts in peptic ulcer disease suggested that they may be active against *C. pylori*. However, only the bismuth salts have activity at concentrations attainable in the gastric mucosa (Table 26.2). The MIC_{90} of the H_2-receptor antagonists, carbenoxolone and sucralfate are > 1000 mg/l. Recent data suggest that the anti-ulcer drug sofalcone, used in Japan, also has activity with an MIC_{90} of 50 mg/l. If the antibacterial activity of sofalcone is confirmed in clinical trials, this agent may become a suitable alternative to the bismuth salts and other anti-ulcer drugs.

Campylobacter pylori possesses powerful extracellular urease activity, which may be an important virulence or 'toxic' factor. Inhibition of this urease may lead to improvement in *C. pylori* gastritis or inhibit bacterial growth. Glupczynski, Labbe & Thibaumont have evaluated the inhibitory effect of acetohydroxamic acid, a potent urease inhibitor against *C. pylori*, alone and with antimicrobial agents (unpublished data). They found that *C. pylori* was inhibited by 400 mg/l of acetohydroxamic acid and that synergy occurred with antimicrobial agents against 10–45% of strains. This avenue of attack looks promising but the urease inhibitors available for clinical study are relatively toxic, and hydroxamates are mutagenic as judged by the Ames test [10].

It has been suggested that a mucolytic agent administered in combination with an antimicrobial agent may facilitate penetration of antimicrobials through the gastric mucus to reach *C. pylori*. Glupczynski *et al.* found that the mucolytic agent *N*-acetyl cysteine (5%) had no inherent activity against *C. pylori* or synergistic activity when combined with antimicrobial agents *in vitro* (unpublished data). It decreased the activity of minocycline and paromomycin. Although this approach seems promising other work suggests that it will not bear fruit — Slomiany *et al.* have found that *C. pylori* itself has mucolytic activity and suggest that this predisposes the host to peptic ulceration [11]. If this work is confirmed then mucolytic agents would be contraindicated in *C. pylori* infection.

Many of the antimicrobial agents with *in vitro* activity have now been used in clinical trials. Amoxycillin and nitrofurantoin give short-term clearance in over 80% of patients [5, 12]. However, erythromycin ethyl succinate [6], josamycin [13], cephalexin, ofloxacin [14] and ciprofloxacin [15, 16] are usually ineffective. The differences in clinical success may be related to reduced activity in the acid environs of the gastric mucosa or

Table 26.2 Susceptibility of *C. pylori* to anti-ulcer drugs

Agent (reference)	Isolates (*n*)	MIC (mg/l)	
		MIC_{90}	Range
Ranitidine [7]	20	1280	320–1280
Cimetidine [35]	27	3200	400–3200
Famotidine [36]	21	> 1024	128–> 1024
Omeprazole [7]	20	> 1000	> 1000
Sofalcone		50	25–50
Bismuth subcitrate [7]	20	16	2–16
Bismuth subsalicylate [35]	27	25	12.5–25
Sucralfate [35]	27	3200	400–> 3200

Table 26.3 Effect of pH on the activity of antimicrobial agents against *C. pylori*

Agent	Isolates (*n*)	MIC_{90} (mg/l)		
		pH 7.5	pH 6.0	pH 5.5
Penicillin	20	0.03	0.5	0.5
Ampicillin	20	0.06	0.25	0.5
Cephalexin	17	2	16	32
Erythromycin	20	0.06	2	8
Ciprofloxacin	20	0.12	0.5	2
Tetracycline	17	0.12	0.25	0.5
Nitrofurantoin	17	1	2	2
Metronidazole*	15	2	2	2
Bismuth subcitrate	17	16	8	–

*Sensitive isolates.

sub-inhibitory concentrations attained by the antimicrobials in the gastric crypts and mucus where *C. pylori* is found. To investigate this McNulty *et al.* have performed MICs at different pH values and have determined antimicrobial gastric mucosal concentrations [17].

In the pH study the activity of various antimicrobial agents was determined over the range of pH 5–8 at which *C. pylori* survives (Table 26.3). Penicillin was 10-fold less active at the most acidic pH than at neutral pH with an MIC_{90} of 0.5 mg/l at pH 5–6, and 0.03 mg/l at pH 7.5. There was a 10-fold change in the activity of ampicillin and a 16-fold change in the activity of cephalexin between pH 5.5 and 7.5 (ampicillin, 0.5–0.06 mg/l; cephalexin, 32–2 mg/l). Cephalexin was 50-fold less active than ampicillin. The erythromycin base was 100-fold less active at pH 5.5 (MIC_{90} = 8 mg/l) than at pH 7.5 (MIC_{90} = 0.12 mg/l). Ciprofloxacin was more than 10-fold less active at pH 5 (MIC_{90} = 2 mg/l) than at pH 7.5 (MIC_{90} = 0.12 mg/l). There was little change in the activity of tetracycline and nitrofurantoin; metronidazole and bismuth were unaffected by pH.

In the gastric mucosal concentration study 500 mg of amoxycillin (Bencard), pivampicillin (Leo), erythromycin ethylsuccinate suspension (Abbott), erythromycin stearate (Abbott) and ciprofloxacin (Bayer) were administered orally with 30 ml of water. During the endoscopic examination biopsy specimens were taken from the pyloric area for mucosal concentrations. All the antimicrobials attained gastric mucosal concentrations above their MIC_{90} values at neutral pH for *Campylobacter pylori* (Table 26.4) except in one patient who received erythromycin stearate. Excluding three high results (21.7, 36.7 and 40.9 mg/l) the two macrolide preparations attained consistent concentrations in the range 1.2–5.2 mg/kg; these concentrations are sub-inhibitory at pH 5.5. The range of concentrations with amoxycillin was great: from 15 mg/kg at 91 min to > 322 mg/kg at 100 min; the lowest concentration was still 30 times the MIC_{90} of ampicillin at pH 5.5 (0.5 mg/l). Pivampicillin attained ampicillin gastric mucosal concentrations of between 47.5 and 209 mg/kg; the lowest gastric mucosal concentration attained at 2 hours (47.5 mg/kg) was 100 times the MIC_{90} at pH 5.5. Ciprofloxacin attained the highest concentration (range < 1.5 to 1762 mg/kg); the concentrations were > 15 times the MIC_{90} at pH 5.5 (2 mg/ml) until 6 hours after the dose.

These two studies help to explain why some

Table 26.4 Range of antimicrobial concentrations in the gastric mucosa after a 500 mg oral dose in relation to time

Agent	Patients (*n*)	Concentration (mg/kg)	Time (min)
Erythromycin ethylsuccinate	8	1.8–40.9	85–53
Erythromycin stearate	10	< 0.7–36.7	60–42
Amoxycillin	7	15–> 322	91–100
Pivampicillin	5	47.5–209	120–49
Ciprofloxacin	14	35–1239	357–84

antimicrobial agents are ineffective *in vivo*. The pH of the stomach contents, and mucus and crypts where *C. pylori* colonizes, are probably the most important factors that affect drug activity. A significant pH gradient was observed by Quigley & Turnberg [18] from the gastric lumen (body pH = 1.82, antral pH = 3.52) through the gastric mucus to the mucosa (body pH = 5.5, antral pH = 5.42). *Campylobacter pylori* also produces urease that releases high concentrations of ammonium ions, which may raise the pH in the immediate environs of the bacterium that survives between pH 5 and 8 [19].

Many drugs including erythromycin [20] are unstable under acidic conditions and specific formulations have been marketed to protect them from gastric acid. Many of these formulations are designed to bypass the stomach and be absorbed lower down in the gastro-intestinal tract. The erythromycin ethylsuccinate ester is inactive but acid stable; it is absorbed in the intestine [21] and then hydrolysed to the active base [22]. Antimicrobials absorbed in the intestine must reach the gastric mucosa via the systemic circulation. The theoretical attainable concentration of drugs in an area such as the gastric mucus with a different pH to serum may be calculated from a modification of the Henderson–Hasselbalch equation. For a basic drug:

$$\frac{[\text{Tissue conc.}]}{[\text{Plasma conc.}]} = \frac{[1 + 10^{pK_a - \text{pH tissue}}]}{[1 + 10^{pK_a - \text{pH plasma}}]}$$

The un-ionized portion of basic drugs such as erythromycin and ciprofloxacin will diffuse into the acidic gastric secretions. At the lower pH of the gastric mucus 'trapping' in the ionized form occurs and concentrations in excess of those in the serum are achieved. Although active concentration may occur because of this increasing ionization, this will result in decreased activity — as confirmed by the pH studies.

Erythromycin activity was affected more than any of the other antimicrobials tested by the reduction in pH; the inhibitory concentrations required below pH 6 were not attained in the gastric mucosal study and would explain the lack of clinical efficacy. Cephalexin was also much less active with an MIC_{90} of 32 mg/l at pH 5.5. Although gastric mucosal concentrations of this drug have not been measured its lack of clinical efficacy is probably also due to sub-inhibitory concentrations in the gastric crypts.

Although the MIC of ciprofloxacin was much higher at pH 5 (2 mg/l) than at neutral pH, gastric mucosal concentrations were inhibitory up to 6 hours after the oral dose. Thus decreased activity at more acidic pH cannot completely explain the lack of clinical efficacy of the quinolones. Ofloxacin, another 4-quinolone, is ineffective at clearing *C. pylori*; and isolates from five patients who received 3 weeks ofloxacin showed a persistent increase in MIC_{90} from 0.5 to 32 mg/l [14]. These organisms also showed cross resistance to the other quinolones, but remained sensitive to amoxycillin. The acquired resistance of *C. pylori* to the quinolones may be partly due to sub-inhibitory concentrations of the active antimicrobial in the oesophagus or fundal areas of the stomach where concentrations were not measured.

Interestingly, when ofloxacin was given with an H_2-receptor antagonist to 32 patients, 26 were cleared of *C. pylori* [23]; raising the pH of the gastric lumen perhaps decreased ionization and thus increased the activity of this drug. In the six patients not cleared, resistance to ofloxacin had developed. The combination of this decreased activity at acid pH and acquired resistance make the quinolones unsuitable for the treatment of *C. pylori* infection.

The activity of metronidazole was unaffected by pH but acquired resistance has been reported with this and the other nitroimidazoles. Goodwin *et al.* [24] found that isolates from 19/27 (70%) patients after a course of cimetidine and tinidazole had become resistant to tinidazole. When colloidal bismuth subcitrate was given with tinidazole *C. pylori* was eliminated in a higher percentage (20/22) of patients but resistance to *C. pylori* still developed in the two patients not cleared. Resistance to the nitroimidazoles was also induced *in vitro* by Goodwin *et al.* [24] in 3/22 sensitive cultures plated on agar containing 8 mg/l of metronidazole. Nitroimidazoles are unsuitable single agents for the treatment of *C. pylori* infection and although combination with a bismuth salt reduces the development of resistance there are probably better alternative agents available.

Amoxycillin and ampicillin are probably more successful *in vivo* because although they achieved lower gastric mucus concentrations than ciprofloxacin, they are less ionized at low pH and retain good activity at acidic pH. They are both amphoteric drugs and at pH 3–6 are present as zwitterions which have a lower partition coefficient than the other ionic species. This may decrease tissue penetration, but may increase penetration into mucus, which is 90% water [25]. Acquired resistance of *C. pylori* to ampicillin or amoxycillin has so far not been described.

The concentrations and activity in the gastric mucosa, mucus and crypts will be affected by many other factors. Drug formulation is extremely important; the physical properties of the tablet or capsule determines dispersion in the stomach. Hey *et al.* [26] found that pivampicillin tablets dispersed over the gastric mucosa more effectively than capsules and that they achieved higher serum concentrations. Patchy distribution of a locally active antimicrobial over the gastric mucosa may cause a wide variation in concentration in different anatomical regions of the stomach. Inadequate concentrations in the fundus, duodenum and oesophagus may lead to relapse or development of resistance. Most of the treatment trials comparing bismuth salt formulations have been small and the results variable but there is some evidence that colloidal bismuth subcitrate is more effective when given as the liquid or chew tablet, than as the swallow tablet. The differences may be due to incomplete coating of the gastric mucosa by the swallow tablet which unlike the other formulations only disperses on reaching the stomach. Once a suitable therapeutic agent is found, trials with different formulations should be performed to establish which achieves maximum contact with the whole of the gastric mucosa, lower oesophagus and duodenum.

Gastric absorption is also determined by the physico-chemical properties of the molecule, presence of food or other drugs and the rate of gastric emptying. The physico-chemical properties of the penicillin esters such as pivampicillin and pivmecillinam may make these agents particularly suitable for the treatment of *C. pylori.* They are absorbed mainly in the gastric mucosa [27] and hydrolysed to the active component during diffusion through the mucus above pH 5 where *C. pylori* is found. The administration of an antimicrobial with food will delay gastric emptying and may increase gastric absorption or maintain inhibitory gastric mucosal concentrations for a longer period.

The presence of other drugs may alter the pH, or bind to the preparation. Fleming *et al.* [28] showed that ciprofloxacin binds to polyvalent cations, and therefore absorption is reduced by antacids containing Mg^{2+} and Al^{3+}. This could be problematical in dyspeptic patients who often take these medications. This drug interaction must also be taken into account for any other antimicrobial that forms complexes with divalent or trivalent cations (e.g. tetracycline). In treatment trials a combination of a bismuth salt (De-Nol, Brocades; or Pepto-Bismol, Procter & Gamble) and an antimicrobial such as amoxycillin, tetracycline or metronidazole leads to less relapse than treatment with a single agent [5, 24]. Triple therapy in turn is more effective than double therapy [29, 30]. The drug combination may be more effective for several reasons. The formation of bismuth–antimicrobial complexes in a similar way to the chelation by antacids probably decreases antimicrobial absorption and targets the antimicrobial to the area colonized by *C. pylori*. Alternatively, the combination of a locally active agent such as bismuth with a systemically active antimicrobial may attain inhibitory concentrations in more of the areas colonized by *C. pylori*. Work with *C. pylori in vitro* suggests that there is synergy between bismuth and some antimicrobial agents [31]. This may be due to the different mode of action of the antimicrobials. Borsch *et al.* used De-Nol chew tablets with amoxycillin and metronidazole to achieve 95% clearance [30]. Amoxycillin will lyse the rapidly dividing bacteria while the bismuth will bind and cause lysis by direct toxicity and may be equally effective against the slowly dividing, coccoid forms of *C. pylori*. These coccal forms appear in older cultures, but can also be induced by sub-inhibitory concentrations of antibiotics *in vitro*, or *in vivo* after unsuccessful treatment with antimicrobial agents. The coccal forms can be cultured only after prolonged incubation and are probably more resistant to antimicrobial treatment. Much more work is needed in this field, as triple therapy produces many side effects including nausea, diarrhoea and candidal infections.

Once an antimicrobial has been chosen, a dosage regimen must be decided upon. Bismuth has a purely local action. Gastric mucosal concentrations are sub-inhibitory 4 hours after a single oral dose of two De-Nol tablets (bismuth 215.4 mg) [32], suggesting that at least four times daily dosage will be needed. This has been confirmed in clinical trials by Lambert *et al.* [33] who found that a four times daily dosage with bismuth subcitrate chew tablets (70% eradication after 120 mg q.d.s. for 8 weeks) was more effective than a twice daily dosage regimen (30% eradication after 240 mg b.d. for 8 weeks). Local activity prior to absorption is probably also important with the other antimicrobials used so far. Therefore, again 6 hourly dosage would probably be appropriate. The duration of therapy required has not thus far been established in large studies. Borsch *et al.* found 1 or 2 weeks of triple therapy with bismuth subsalicylate, metronidazole and amoxycillin three times daily to be almost as effective as 4 weeks [30]. Thus the agent, dose, and formulation are probably more important determinants of success than the length of treatment.

It is possible that in the long term none of the currently marketed antimicrobial agents will be ideal for the treatment of *C. pylori* infection. Currently, the pharmaceutical companies are probably seeking a therapeutic agent with some or all of the following properties:

1 Formulation with rapid dissolution and good dispersion in the stomach.

2 Stability and activity over a wide range of pH but especially acidic pH.

3 Small molecular size to increase mucus penetration.

4 Absorption via the gastric mucosa.

5 Low susceptibility to acquired resistance.

6 Activity against the slowly dividing coccal forms of *C. pylori*.

7 Low incidence of gastro-intestinal side effects.

In conclusion, despite good *in vitro* activity, many antimicrobials are unsuitable for the treatment of *C. pylori* infection. The activity of the macrolides, aminoglycosides, quinolones and cephalexin are greatly affected by pH and use of the quinolones and imidazoles leads to acquired resistance. These agents are usually ineffective *in vivo*. The bismuth salts, penicillins, nitrofurantoin and tetracycline retain good activity at acid pH, and *in vivo* bismuth subcitrate [5], bismuth subsalicylate [6], pivampicillin (personal observations), amoxycillin [34] and nitrofurantoin [12] give good short-term results. Relapse is a major problem and combination treatment with drugs exhibiting different modes of action, especially three rather than two antimicrobial agents, achieves much better long-term results. Current triple therapy is probably an interim solution for patients with severe symptoms until the 'ideal' antimicrobial is found — however this may take some time.

Acknowledgements

Caroline Gaston is thanked for her help in preparing the manuscript.

References

1 Krakowka, S., Morgan, D.R., Kraft, W.G. & Leunk, R.D. Establishment of gastric *Campylobacter pylori* infection in the neonatal gnotobiotic piglet. *Infect. Immun.* 1987, **55**, 2789–2796.

2 Lambert, J.R., Borromeo, M., Turner, H., Smith, M., Chapman, C. & Pinkard, K.J. 1988 Colonisation of gnotobiotic piglets with *Campylobacter pyloridis*. In Kaijser, B. & Falsen, E. (eds.) *Campylobacter IV*. Kungälv, Gotterna. p. 342.

3 Marshall, B.J., Armstrong, J.A., McGechie, D.B. & Glancy, R. Attempt to fulfill Koch's postulates for pyloric *Campylobacter*. *Med. J. Australia* 1985, **142**, 436–439.

4 Morris, A. & Nicholson, G. Ingestion of *Campylobacter pyloridis* causes gastritis and raised fasting gastric pH. *Am. J. Gastroenterol.* 1987, **82**, 192–199.

5 Rauws, E.A.J., Langenberg, W., Houthoff, H.J., Zanen, H.C. & Tytgat, G.N.J. *Campylobacter pyloridis*-associated chronic active antral gastritis. A prospective study of its prevalence and the effects of antibacterial and antiulcer treatment. *Gastroenterology* 1988, **94**, 33–40.

6 McNulty, C.A.M., Gearty, J.C., Crump, B., Davis, M., Donovan, I.A., Melikian, V., Lister, D.M. & Wise, R. *Campylobacter pyloridis* and associated gastritis: investigator blind, placebo controlled trial of bismuth salicylate and erythromycin ethylsuccinate. *Br. Med. J.* 1986, **293**, 645–649.

7 Goodwin, C.S., Blake, P. & Blincow, E. The minimum inhibitory and bactericidal concentrations of antibiotics and anti-ulcer agents, against *Campylobacter pyloridis*. *J. Antimicrob. Chemother.* 1986, **17**, 309–314.

8 McNulty, C.A.M. & Dent, J.C. Susceptibility of *Campylobacter pylori* to twenty-one antimicrobial agents. *Eur. J. Clin. Microbiol. Infect. Dis.* 1988, **7**, 566–569.

9 McNulty, C.A.M., Dent, J.C. & Wise, R. Susceptibility of clinical isolates of *Campylobacter pyloridis* to 11 antimicrobial agents. *Antimicrob. Agents Chemother.* 1985, **28**, 837–838.

10 Wang, C. Mutagenicity of hydroxamic acids for *Salmonella typhimurium*. *Mutation Res.* 1977, **56**, 7–12.

11 Slomiany, B.L., Bilski, J., Sarosiek, J., Murty, V.L.N., Dworkin, B., Van Horn, K., Zielenski, J. & Slomiany, A. *Campylobacter pyloridis* degrades mucin and undermines gastric mucosal integrity. *Biochem. Biophys. Res. Commun.* 1987, **144**, 307–314.

12 Gilman, R., Leon-Barua, R., Ramirez-Ramos, A., Morgan, D., Recavarron, S., Spira, W., Watanabe, P., Kraft, W. & Pearson, A. Efficacy of nitrofurantoin in the treatment of antral gastritis associated with *Campylobacter pyloridis*. *Gastroenterology* 1986, **92**, 1405.

13 Lamouliatte, H., Megraud, F., De Mascarel, A. & Quinton, A. 1988. Placebo-controlled trial of josamycin in *Campylobacter pyloridis* associated gastritis. In Kaijser, B. & Falsen, E. (eds.), *Campylobacter IV*. Kungälv, Gotterna. pp. 388–389.

14 Glupczynski, Y., Labbe, M., Burette, A., Delmee, M., Avesani, V. & Bruck, C. Treatment failure of ofloxacin in *Campylobacter pylori* infection. *Lancet* 1987, **i**, 1096.

15 Hirschl, A.M., Stanek, G., Rotter, M.L., Hentschel, E., Schutze, K., Potzi, R., Gangi, A. & Weiss, W. 1988. *Campylobacter pyloridis*: frequency of occurrence, serology and susceptibility to antibiotics and ulcer-drugs. In Kaijser, B. & Falsen, E. (eds.) *Campylobacter IV*. Kungälv, Gotterna. pp. 363–364.

16 Stone, J.W., Wise, R., Donovan, I.A. & Gearty, J. Failure of ciprofloxacin to eradicate *Campylobacter pylori* from the stomach. *J. Antimicrobiol. Chemother.* 1988, **22**, 92–93.

17 McNulty, C.A.M., Dent, J.C., Ford, G.A. & Wilkinson, S.P. Inhibitory antimicrobial concentrations against *Campylobacter pylori* in gastric mucosa. *J. Antimicrob. Chemother.* 1988, **22**, 729–738.

18 Quigley, E.M.M. & Turnberg, L.A. pH of the microclimate lining human gastric and duodenal mucosa *in vivo*. *Gastroenterology* 1987, **92**, 1876–1884.

19 Hazell, S.L. & Lee, A. The adaption of motile strains of *Campylobacter pyloridis* in gastric mucus and their association with gastric epithelial intercellular spaces. *Proceedings of the Third International Workshop on Campylobacter Infections, Ottawa* 1985. Abstract No. 115, pp. 189–191.

20 Nelson, L.A. Physicochemical factors influencing the absorption of erythromycin and its esters. *Chem. Pharm. Bull. (Tokyo)* 1962, **10**, 1099–1101.

21 Lee, C.C., Anderson, R.C. & Chen, K.K. Site absorption of erythromycin in rats. *Antibiot. Chemother.* 1954, **4**, 926–930.

22 Tardrew, P.L., Mao, J.C.H. & Kenny, D. Antibacterial activity of 2′-esters of erythromycin. *Appl. Microbiol.* 1969, **18**, 159–165.

23 Bayerdorffer, E., Kasper, G., Sommer, A. & Ottenjann, R. Ofloxacin therapy of *Campylobacter pylori* positive ulcers. In Kaijser, B. & Falsen, E. (eds.) *Campylobacter IV*. Kungälv, Gotterna. pp. 372–373.

24 Goodwin, C.S., Marshall, B.J., Blincow, E.D., Wilson,

D.H., Blackbourn, S. & Phillips, M. Prevention of nitroimidazole resistance in *Campylobacter pylori* by coadministration of colloidal bismuth subcitrate: clinical and *in vitro* studies. *J. Clin Pathol.* 1988, **41**, 207–210.

25 Allen, A. 1981. Structure and function of gastrointestinal mucus. In Johnson, L.R. (ed.) *Physiology of the Gastrointestinal Tract.* Raven Press, New York. **21**, pp. 617–639.

26 Hey, H., Matzen, P., Thorup Andersen, J., Didriksen, E. & Nielsen, B. A gastroscopic and pharmacological study of the disintegration time and absorption of pivampicillin capsules and tablets. *Br. J. Clin. Pharmacol.* 1979, **8**, 237–242.

27 Swahn, A. Gastrointestinal absorption and metabolism of two ^{35}S-labelled ampicillin esters. *Eur. J. Clin. Pharmacol.* 1976, **9**, 299–306.

28 Fleming, L.W., Moreland, T.A., Stewart, W.K. & Scott, A.C. Ciprofloxacin and antacids. *Lancet* 1986, **ii**, 294.

29 Borody, T., Cole, P., Noonan, S., Morgan, A., Ossip, G., Maysey, J. & Brandl, S. Long-term *Campylobacter pylori* recurrence post-eradication. *Gastroenterology* 1988, **94** (5), A43.

30 Borsch, G., Mai, U. & Opferkuch, W. Oral triple therapy (OTT) may effectively eradicate *Campylobacter pylori* in man: a pilot study. *Gastroenterology* 1988, **94** (5), A44.

31 Caekenberghe, D.L. & Breyssens, V.J. *In vitro* synergistic activity between bismuth subcitrate and various antimicrobial agents against *Campylobacter pyloridis* (*C. pylori*). *Antimicrob. Agents Chemother.* 1987, **31**, 1429–1430.

32 Lambert, J.R., Way, D.J., King, R.G., Wan, A. & McLean, A.J. Bismuth pharmacokinetics in the human gastric mucosa. *Gastroenterology* 1988, **94** (5), A248.

33 Lambert, J.R., Borromeo, M., Eaves, E.R., Hansky, J. & Korman, M. Efficacy of different dosage regimes of bismuth in eradicating *Campylobacter pylori. Gastroenterology* 1988, **94** (5), A248.

34 Glupczynski, Y., Burette, A., Labbe, M., Deprez, C., De Reuck, M. & Deltenre, M. *Campylobacter pylori*-associated gastritis: A double blind placebo-controlled trial with amoxycillin. *Am. J. Gastroenterol.* 1988, **83**, 365–372.

35 Andreasen, J.J. & Andersen, L.P. *In vitro* susceptibility of *Campylobacter pyloridis* to cimetidine, sucralfate, bismuth and sixteen antibiotics. *Acta Pathol. Immunol. Scand.* 1987, **95**, 147–149.

36 Shungu, D.L., Nalin, D.R., Gilman, R.H., Gadesbusch, H.H., Cerami, A.T., Gill, C. & Weissberger, B. Comparative susceptibilities of *Campylobacter pylori* to norfloxacin and other agents. *Antimicrob. Agents Chemother.* 1987, **31**, 949–950.

27 Antibiotic treatment of Campylobacter pylori infection

A.M. HIRSCHL AND M. PLETSCHETTE

The discovery of *Campylobacter pylori* by Marshall & Warren in 1982 [1, 2] has not only brought new insights in the pathogenesis of gastritis and peptic ulcer disease but has also given an opportunity to critically review the treatment of these conditions. The supposition that both diseases could be of infectious origin has found some favour. Despite numerous innovations in the field of gastritis and ulcer disease therapy, one could not deny that none of the common therapeutic principles, such as neutralization or suppression of acid secretion and the so-called cytoprotection, could reach the final treatment goal, which is the permanent healing of the pathological lesion. The high rate of ulcer recurrence after H_2-receptor blocker therapy may stand as a suitable example of unfulfilled expectations of treatment [3, 4]. The fascinating possibility of a causative treatment for gastritis and peptic ulceration has given rise to numerous trials and studies which are now reviewed.

Influence of antimicrobial compounds on the detection of C. pylori

Table 27.1 summarizes data on the influence of different antimicrobial substances on the detection of *C. pylori*. Data listed here are derived from pilot studies as well as from placebo-controlled (double-blind) studies. In certain cases, additional treatment with substances such as H_2-receptor blockers is also listed as far as adequate evidence can be obtained from the literature.

In considering antimicrobial treatment of *C. pylori* infection, one has to differentiate between short-term effects (most likely not elimination, but only suppression) and long-term effects (true elimination). Rauws *et al.* [5] showed that failure to demonstrate the microorganism 4 weeks after cessation of treatment indicates true elimination. Results obtained prior to this can therefore be used only as evidence of failure but not for success of a treatment. The best rates of *C. pylori* elimination with antibiotic monotherapy are obtained with furazolidone (44%) [6], amoxycillin over 4 weeks (23%) [5] and nitrofurantoin (22%)[6]. All other substances have only a short-term effect or none at all. With the present state of knowledge and despite only a modest performance, oral ampicillin would qualify best for antibiotic monotherapy. The observation that treatment with ampicillin is not associated with the emergence of drug-resistant strains of *C. pylori* — as opposed to other antimicrobials — adds to the interest in this compound [7].

Long-term use of nitrofurans, which meets increasing disapproval from some authors [8] because of the not inconsiderable side-effects, would need to undergo an extensive cost-benefit study before being accepted into gastritis treatment.

Monotherapy with antimicrobial compounds results in bacterial elimination rates similar to those obtained with bismuth compounds (Table 27.2). Despite the diversity of their mechanisms of action [9], bismuth compounds are to be considered at least in part, if not principally, as antimicrobial compounds, the activity of which has been demonstrated *in vitro* as well as *in vivo* [10, 11].

With a combination of bismuth and/or antibiotics, elimination rates obtained are much higher (33–94%; Table 27.3). The best results were recently obtained with oral

Table 27.1 Effects of various antimicrobial agents on *C. pylori* colonization in patients with *C. pylori*-associated gastritis at < 1 month and ≥ 1 month after cessation of therapy

Therapy	Dosage	Additional treatment*	No. of patients *C. pylori* negative/total†		Reference
			at < 1 month	at ≥ 1 month	
Amoxycillin susp.	2 × 1000 mg/7days		20/22 (91%)	N.D. (0%)	Glupczynski *et al.* (1988)[16]
Amoxycillin susp.	2 × 1000 mg/14 days		13/18 (72%)	N.D. (0%)	Glupczynski *et al.* (1988)[16]
Amoxycillin	3 × 375 mg/28 days		15/22 (68%)	5/22 (23%)	Rauws *et al.* (1988)[5]
Bacampicillin	2 × 800 mg/21 days	H_2RA	3/15 (20%)	0/15 (0%)	Hirschl *et al.* (1987)[32]
Ciprofloxacin	2 × 500 mg/14 days	H_2RA	0/13 (0%)	–	Hirschl *et al.* (1987)[33]
Doxycycline	1 × 200 mg/10 days		0/3 (0%)	–	Schaub *et al.* (1987)[34]
Doxycycline	1 × 200 mg/1 day 1 × 100 mg/8 days	H_2RA/PA	2/7 (29%)	0/6 (0%)	Unge *et al.* (1988)[35]
Erythromycin susp.	4 × 500 mg/14 days		1/15 (7%)	N.D.	McNulty *et al.* (1986)[27]
Fusidic acid	3 × 500 mg/14 days	H_2RA	0/5 (0%)	–	Hirschl *et al.* (1988)[7]
Furazolidone susp.	4 × 100 mg/14 days		13/14 (93%)	4/9 (44%)	Gilman *et al.* (1987)[6]
Josamycin	4 × 500 mg/14 days		0/9 (0%)	–	Lamouilatte *et al.* (1987)[36]
Metronidazole	2 × 500 mg/21 days	H_2RA	0/5 (0%)	–	Hirschl *et al.* (1988)[7]
Nitrofurantoin susp.	4 × 100 mg/14 days		19/24 (79%)	4/18 (22%)	Gilman *et al.* (1987)[6]
Nitrofurantoin	3 × 100 mg/10 days	H_2RA	0/13 (0%)	–	Börsch *et al.* (1988)[37]
Ofloxacin	2 × 200 mg/14 days	H_2RA	0/5 (0%)	–	Hirschl *et al.* (1988)[7]
Ofloxacin	2 × 200 mg/21 days	H_2RA	1/5 (20%)	N.D.	Glupczynski *et al.* (1987)[29]
Penicillin V	3 × 600 mg/10 days	PI	0/6 (0%)	–	Hirschl *et al.* (1988)[7]
Phenylmercury borate	4 × 0.3 mg/30 days	H_2RA	0/6 (0%)	–	Hirschl *et al.* (1988)[7]
Pivmecillinam	3 × 400 mg/21 days	H_2RA	0/6 (0%)	–	Hirschl *et al.* (unpublished)
Rifampicin	1 × 600 mg/21 days	H_2RA	0/8 (0%)	–	Hirschl *et al.* (1988)[7]
Spiramycin	3 × 1000 mg/28 days		0/6 (0%)	–	Langenberg *et al.* (1985)[38]

*H_2RA = H_2-receptor-antagonists; PI = pirenzepine; PA = prostaglandin analogue.
†N.D. = data not available.

therapy using three compounds. Eradication rates at 4–8 weeks post-treatment have been reported as 90–94% [12, 13]. Borody *et al.* [12] have reported that 94% of patients initially cleared of *C. pylori* by triple therapy remain cleared at 18 months. They observed the following rates of side effects: nausea (32%), diarrhoea (7%), mouth/throat burning (4%), dizziness/drowsiness (4%), one case of *Clostridium difficile* diarrhoea and two cases of oral/vaginal candidiasis.

The addition of metronidazole seems to remarkably potentiate the action of bismuth + amoxycillin and this is in contrast to the only average performance with metronidazole and tinidazole as single drugs.

Table 27.2 Effect of bismuth preparations on *C. pylori* in patients with *C. pylori*-associated gastritis at < 1 month and ≥ 1 month after cessation of therapy

Therapy	Dosage	No. of patients *C. pylori* negative/total*		Reference
		< 1 month	≥ 1 month	
Bismuth subcitrate	4 × 120 mg/6 weeks	12/23 (52%)	N.D.	Coghlan *et al.* (1987)[23]
Bismuth subcitrate	4 × 120 mg/4 weeks	30/67 (45%)	10/67 (15%)	Rauws *et al.* (1988)[5]
Bismuth subcitrate	4 × 108 mg/8 weeks	20/45 (44%)	12/45 (27%)	Lambert *et al.* (1987)[22]
Bismuth subcitrate	4 × 120 mg/8 weeks	N.D.	7/21 (33%)	Goodwin *et al.* (1987)[21]
Bismuth subsalicylate	3 × 900 mg/4 weeks	8/17 (47%)	3/17 (18%)	Börsch *et al.* (1988)[37]
Bismuth subsalicylate	4 × 536 mg/3 weeks	14/18 (78%)	N.D.	McNulty *et al.* (1986)[27]

*N.D. = data not available.

Table 27.3 Effect of combined antimicrobial treatment on *C. pylori* colonization in patients with *C. pylori* associated gastritis at < 1 month and ⩾ 1 month after cessation of therapy

Therapy*	Dosage	No. of patients *C. pylori* negative/total†		Reference
		< 1 month	⩾ 1 month	
BSC + amoxycillin	4 × 120 mg/4 weeks 3 × 375 mg/4 weeks	18/20 (90%)	8/20 (40%)	Rauws *et al.* (1988)[5]
BSS + ofloxacin	3 × 600 mg/** 2 × 200 mg/4 weeks (max.)	11/14 (79%)	3/9 (33%)	Bayerdörffer *et al.* (1987)[24]
BSC + tinidazole	4 × 120 mg/8 weeks 2 × 500 mg/10 days	N.D.	21/28 (75%)	Goodwin *et al.* (1987)[21]
BSC + tetracycline*** + metronidazole	4 × 120 mg/4 weeks 4 × 500 mg/4 weeks 4 × 200 mg/10 days	N.D.	94/100 (94%)	Borody *et al.* (1988)[12]
BSS + metronidazole + amoxycillin susp.	3 × 600 mg/2 weeks 3 × 500 mg/2 weeks 3 × 500 mg/2 weeks	N.D.	9/10 (90%)	Börsch *et al.* (1988)[13]

†N.D. = Data not available.
*BSC = Bismuth subcitrate; BSS = bismuth subsalicylate.
**until 2 weeks beyond ulcer healing.
***amoxycillin (4 × 250 mg) was given in 8 patients instead of tetracycline.

However, the additional impact of metronidazole is not necessarily related to a true synergistic action with other antimicrobials but may be due to an action on the local conditions of the mucosa.

Monotherapy (antibiotics or bismuth) alone gives poor elimination when compared to combined treatment where elimination rates of 80–90% can be realistically expected. The ideal therapeutic regimen with regard to its real benefit, safety and tolerability is at present unclear.

Influence of antimicrobial compounds on C. pylori-associated antral gastritis

Studies dealing with this question clearly show that antibiotics by themselves do not have a significant influence on the degree of gastritis. This may also be approached from a different viewpoint, raising questions about the influence of suppression or elimination of *C. pylori* on the degree of the antral gastritis. From a clinical point of view, this question is even more interesting than the one dealing merely with the elimination of *C. pylori*, as the relation between detection of a bacterium and presence of a pathological process allows certain conclusions on the aetiological significance of this microorganism. Even though the exact clinical significance of antral gastritis is still unclear, one cannot doubt its true pathological nature since no one is born with gastritis. Nor may one consider it as being a part of a natural ageing process, despite its markedly increasing incidence with age [14, 15].

Rauws *et al.* [5] were able to impressively show how the degree of chronic active gastritis improved following true elimination of the microorganism. In comparison, no significant changes were observed in groups of patients in whom elimination of *C. pylori* had not been achieved. The degree of active gastritis considerably improved with the administration of amoxycillin alone. The strictly antibacterial mode of action of amoxycillin which differs from that of bismuth compounds strongly supports an association between *C. pylori* and chronic active gastritis. Elimination of bacteria and improvement of gastritis during bismuth therapy is also not likely to be primarily related to a cytoprotective effect of this substance and association with subsequent changes in the microenvironment deleterious to *C. pylori*; both are consequences of antibacterial activity. Glupczynski *et al.* [16] distinguish between the effects of transient disappearance of *C. pylori* on the histological signs of active and inactive gastritis. Failure to detect *C. pylori* is associated with a significant improvement in chronic active gastritis whereas the signs of chronic inactive gastritis remain unchanged. Recurrence of *C. pylori* leading to a renewed increase in activity of the mucosal inflammation further demonstrates the relation between active gastritis and *C. pylori*. These observations raise the question whether *C. pylori* only activates an already existing gastritis without being the prime cause. On the other hand, one may well believe that a longer bacteria-free period is necessary to improve chronic inactive gastritis. However, experimental infections produced in man and animals with a previously intact mucosa show that the pathogenic potential of *C. pylori* is not a function of an already damaged mucosa [17–20]. It has therefore been clearly shown that antibiotic treatment, by eliminating *C. pylori*, is able to improve if not even to heal chronic active gastritis. The influence of bacterial elimination on chronic inactive forms of Type-B gastritis has to be investigated further.

Influence of antimicrobial compounds on ulcer healing and ulcer recurrence rates

Campylobacter pylori-induced mucosal damage may be responsible for tilting the balance between protection (i.e. mucosal barrier) and aggressive (acid, pepsin) factors in the stomach towards the side of aggressive factors, thus favouring the evolution or the recurrence of a peptic ulcer. There is still an absence of long-term prospective studies indicating whether *C. pylori* colonization favours the development of ulcers. The few published studies are limited to the point where ulcer healing and recurrence rates with antimicrobial therapy are compared to classical ulcer treatment. The most extensive study of this kind is by Goodwin *et al.* [21] where 100 patients with duodenal ulcer and *C. pylori* in the stomach were entered into a double-blind placebo-controlled trial. One group was given cimetidine and placebo, the second cimetidine and tinidazole, the third tripotassium dicitratobismuthate (CBS) and placebo, and a fourth group CBS + tinidazole. The clearance of *C. pylori* in 28 patients was accompanied by ulcer healing in 27 (93%) but only 41 (57%) of the patients who remained infected with *C. pylori* showed ulcer healing. The relation between *C. pylori* elimination and ulcer recurrence rate is illustrated in Table 27.4 and shows a distinct link between *C. pylori* elimination and frequency of ulcer recurrence.

Lambert *et al.* [22] found an ulcer healing rate of 95% in 54 patients with duodenal ulcer after 8 weeks' treatment with CBS together with *C. pylori* elimination in 57% of these. However, no correlation between the elimination rates and primary ulcer healing was found. The ulcer recurrence rate after 12 months was 56% and all of these patients were positive for *C. pylori*. This study also demonstrated that persistence or recurrence of *C. pylori* constitutes a risk factor for recur-

Table 27.4 Elimination of *C. pylori* and ulcer relapse rate (12 months) in patients treated with cimetidine or colloidal bismuth subcitrate and tinidazole or placebo (Goodwin *et al.* 1987) [21]

Therapy	Elimination of *C. pylori*	Ulcer relapse rate (%)
Cimetidine	1/51 (both cimetidine groups)	100%
Cimetidine + tinidazole		57%
Bismuth subcitrate + placebo	7/21	29%
Bismuth subcitrate + tinidazole	21/28	15%

rence of the ulcer. A recurrence rate of 59% was found in 39 duodenal ulcer patients with healed ulcers by Coghlan *et al.* [23]. In this study, only the post-treatment *C. pylori* status was a significant predictor of endoscopically-proven recurrence; 79% of patients remaining *C. pylori* culture-positive had recurrence, compared with 27% of culture-negative patients. Recurrence was more likely (66%) in patients with a *de novo* detection of *C. pylori* after eradication was thought to have been achieved than in those who remained negative for *C. pylori* (10%).

Bayerdörffer *et al.* [24, 25] were able to show a faster healing of duodenal ulcers with ranitidine/ofloxacin or bismuth subsalicylate/ofloxacin than in patients treated with ranitidine alone. These authors also interpreted their observations as further evidence for a pathogenic role of *C. pylori* in duodenal ulcers.

Thus *C. pylori* elimination by treatment with antimicrobial agents may play a role in the prevention of ulcer recurrence.

Influence of antimicrobial compounds on gastro-intestinal symptoms

Unfortunately there is at present a lack of conclusive data on this clinically-important point. It is not established whether infection with *C. pylori* is responsible for specific symptomatology or has a particular clinical relevance independent of being a risk factor for ulcer occurrence. This particular aetiological uncertainty has not occurred only since the discovery of *C. pylori*; indeed the clinical significance of antral gastritis was under intensive discussion long before. At present, the question whether improvement of gastritis on the basis of an improved or normalized histological appearance (obtained with elimination of *C. pylori*) is followed by immediate benefits for the patients (i.e. relief of their complaints) remains open. A relation between *C. pylori* and non-ulcer dyspepsia (NUD) has been postulated. Rokkas & Sladen [26] compared the effects of bismuth subcitrate and placebo in patients with NUD in a double-blind study over 8 weeks. In relation to placebo the authors found a significant improvement in gastritis and symptomatology as well. These findings were corroborated by others [27, 28]. Nevertheless, these results cannot be considered as evidence that *C. pylori* elimination leads to symptom improvement, since bismuth for example has other effects on the gastric mucosa apart from its antimicrobial activity [9], and these could also be responsible for the relief of patients' complaints. This is consistent with the results of Glupczynski *et al.* [16] who could not find any difference in clinical symptomatology between patients temporarily cleared of *C. pylori* and those who remained infected — and this was despite a considerable improvement in the degree of active gastritis in the first group. Nevertheless, one has to consider that true elimination was not attained in this study. Hence there is a great need for controlled studies including recording and staging of symptoms.

The aim of a possible antimicrobial treatment for gastritis or NUD should also encompass improvement in patients' subjective symptoms in addition to elimination of *C. pylori* and improvement in the histological appearance of gastritis. If such an improvement of patients' symptoms could be shown, it will help antimicrobial treatment to achieve recognized status in gastritis therapy. Furthermore, antibiotic treatment of gastritis may be justified if it would constitute an effective way of preventing ulcer development.

Reason for poor response and features of an 'ideal' antimicrobial agent

The failure of many antimicrobial agents, and especially those used in monotherapy, to eliminate *C. pylori* despite good *in vitro* activity may occur for several reasons. For instance, many compounds lose activity at low pH values. Another reason for the failure of many compounds used in monotherapy is the emergence of resistant strains of *C. pylori* — for example under treatment with quinolones, metronidazole, nitrofurantoin and rifampicin [7, 29, 30]. Apart from those considerations, pharmacokinetic aspects may also be of great importance. The stomach mucus and the mucus–mucosa interface are compartments not extensively investigated and there may be areas where only sub-inhibitory concentrations of antimicrobial agents are achieved in which *C. pylori* may survive as persisting (perhaps coccoid) forms. This is supported by the findings of Langenberg *et al.* [31] showing that recurrence of infection after apparently successful elimination of *C. pylori* is nearly always due to the same strain of organism.

Many aspects of *C. pylori* infections are based on experience gained from other models of infectious processes. It is, however, a fact that all antimicrobial treatment is only effective in synergy with the immune system. Microbes causing chronic processes often possess special mechanisms of immune evasion or may induce tolerance by their host's immune system. Such an advantage would make *C. pylori* largely refractory to treatment by antibiotics alone. Enhancement of the immune response to *C. pylori* as well as considerations as to how one might utilize some metabolic peculiarity (for example urease activity) of *C. pylori* would be interesting as new approaches in the treatment of *C. pylori* infections.

As infection with *C. pylori* is mainly superficial, substances with topical activity should be adequate if complete spreading over the mucosa and retention in the stomach and duodenum could be achieved. As this is probably impossible to achieve, agents with both topical and systemic activity may be best. Such substances should have a narrow spectrum. Cross-resistance with other commonly used antimicrobials should not exist and resistance should not be mediated by transferable elements such as plasmids since these factors could induce emergence of resistant strains of different bacterial species. The combination of different agents may be a way to avoid the emergence of resistance [30]. Furthermore effectiveness in a wide range of pH values, a bactericidal mode of action, efficacy against quiescent coccoid forms, good penetration of mucus, concentration in the mucosal crypts and good tolerance are all important. Long shelf life and low pricing, particularly relevant in developing countries, are also important attributes of the 'ideal' compounds.

The best combinations of compounds (bismuth + amoxycillin/tetracycline + metronidazole) only partially fulfil these requirements. In addition, a proper rationale for the use of these combinations in treatment is lacking. Nevertheless, it is an undeniable fact that antibiotic treatment schedules exist, which lead, with a relatively high certainty, to elimination of *C. pylori*, thus providing treatment in at least certain well-selected cases. This is an acceptable interim solution until the ideal compound against *C. pylori* infection and therefore also for treatment of a number of diseases of the gastro-intestinal tract is found.

References

1 Marshall, B.J. Unidentified curved bacilli on gastric epithelium in active chronic gastritis. *Lancet* 1983, **i**, 1273–1275.

2 Warren, J.R. Unidentified curved bacilli on gastric epithelium in active chronic gastritis. *Lancet* 1983, **i**, 1273.

3 Bardhan, K., Cole, D.S., Hawkins, B.W. & Franks, C.R. Does treatment with cimetidine extended beyond initial healing of duodenal ulcer reduce the subsequent relapse rate? *Br. Med. J.* 1982, **284**, 621–623.

4 McLean, A.J., Harrison, P.M., Ionnides-Demos, L.L., Byrne, A.J., McCarthy, P. & Dudley, F.J. Microbes, peptic ulcer, and relapse rates with different drugs. *Lancet* 1984, **ii**, 525–526.

5 Rauws, E.A.J., Langenberg, W., Houthoff, H.J., Zanen, H.C. & Tytgat, G.N.J. *Campylobacter pylori*-associated chronic active antral gastritis. A prospective study of its prevalence and the effects of antibacterial and antiulcer treatment. *Gastroenterology* 1988, **94**, 33–40.

6 Gilman, R., Leon-Barva, R., Ramirez-Ramos, A., Morgan, D., Recavarron, S., Spira, W., Watanabe, P., Kraft, W. & Pearson, A. Efficacy of nitrofurans in the

treatment of antral gastritis with *Campylobacter pyloridis Gastroenterology.* 1987, **92**, 1405.

7 Hirschl, A.M., Hentschel, E., Schütze, K., Nemec, H., Pötzi, R., Gangl, A., Weiss, W., Pletschette, M., Stanek, G. & Rotter, M.L. The efficacy of antimicrobial treatment in *Campylobacter pylori* associated gastritis and duodenal ulcer. *Scand. J. Gastroenterol.* 1988, **23** (Suppl. 142), 76–81.

8 Simon, C. & Stille, W. 1985. *Antibiotika-Therapie in Klinik und Praxis.* Schattauer, Stuttgart, New York.

9 Konturek, S.J., Radecki, T., Piastucki, I. & Drozdowicz, D. Studies on the gastroprotective and ulcer healing effects of colloidal bismuth subcitrate. *Digestion* 1987, **37** (Suppl. 2), 8–15.

10 Armstrong, J.A., Wee, S.H., Goodwin, C.S. & Wilson, D.H. Response of *Campylobacter pyloridis* to antibiotics, bismuth and an acid reducing agent *in vitro* — an ultrastructural study. *J. Med. Microbiol.* 1987, **24**, 343–350.

11 Marshall, B.J., Armstrong, J.A., Francis, G.J., Nokes, N.T. & Wee, S.H. Antibacterial action of bismuth in relation to *Campylobacter pyloridis* colonization and gastritis. *Digestion* 1987, **37** (Suppl. 2), 16–30.

12 Borody, T., Cole, P., Noonan, S., Morgan, A. Ossip, G., Maysey, J. & Brandl, S. Long term *Campylobacter pylori* recurrence post-eradication. *Gastroenterology* 1988, **94**, A43.

13 Börsch, G., Mai, U. & Opferkuch, W. Oral triple therapy (OTT) may effectively eradicate *Campylobacter pylori* (CP) in man: a pilot study. *Gastroenterology* 1988, **94**, A44.

14 Siurala, M. & Salmi, H.J. Long term follow up of subjects with superficial gastritis or a normal gastric mucosa. *Scand. J. Gastroenterol.* 1971, **6**, 459–463.

15 Villako, M., Kekki, M., Tamm, A., Tammus, R., Savisaar, E., Viirsalu, V. & Sipponen, P. Epidemiology and dynamics of gastritis in a representative sample of an Estonian urban population. *Scand. J. Gastroenterol.* 1982, **17**, 601–607.

16 Glupczynski, Y., Burette, A., Labbe, M. Deprez, C., DeReuck, M. & Deltenre, M. *Campylobacter pylori*-associated gastritis: a double blind placebo controlled trial with amoxycillin. *Am. J. Gastroenterol.* 1988, **83**, 365–372.

17 Marshall, B.J., Armstrong, J.A., McGechie, D.B. & Glancy, R.J. Attempt to fulfil Koch's postulates for pyloric *Campylobacter. Med. J. Australia* 1985, **142**, 436–439.

18 Morris, A. & Nicholson, G. Ingestion of *Campylobacter pyloridis* causes gastritis and raised fasting gastric pH. *Am. J. Gastroenterol.* 1987, **82**, 192–199.

19 Lambert, J.R., Borromeo, M., Pinkard, K.J., Turner, H., Chapman, C.B. & Smith, M.L. Colonization of gnotobiotic piglets with *Campylobacter pyloridis* — an animal model. *J. Infect. Dis.* 1987, **155**, 1344.

20 Krakowka, S., Morgan, D.R., Kraft, W.G. & Leunk, R.D. Establishment of gastric *Campylobacter pylori* infection in the neonatal gnotobiotic piglet. *Infect. Immun.* 1987, **55**, 2789–2796.

21 Goodwin, C.S., Marshall, B.J., Warren, J.R., Blackbourn, S. & Blincow, E.D. 1987. Clearance of *Campylobacter pyloridis* and reduced duodenal ulcer relapse with bismuth and tinidazole compared to cimetidine. In Kaijser, B. & Falsen, E. (eds.) *Campylobacter IV.* University of Götebörg, Göteborg. pp. 368–369.

22 Lambert, J.R., Borromeo, M., Korman, M.G., Hansky, J. & Eaves, E.R. 1987. Effect of colloidal bismuth (De-Nol) on healing and relapse of duodenal ulcers — role of *Campylobacter pyloridis.* In Kaijser, B. & Falsen, E. (eds.) *Campylobacter IV.* University of Göteborg, Göteborg. p. 383.

23 Coghlan, J.G., Humphries, H., Dooley, C., Keane, C., Gilligan, D., McKenna, D., Sweeney, E. & O'Morain, C. *Campylobacter pylori* and recurrence of duodenal ulcers — a 12 months follow-up study. *Lancet* 1987, ii, 1109–1111.

24 Bayerdörffer, E., Simon, T.H., Bästlein, C.H., Ottenjann, R. & Kasper, G. Bismuth/ofloxacin combination for duodenal ulcer. *Lancet* 1987, ii, 1467–1468.

25 Bayerdörffer, E., Pirlet, T.H., Ottenjann, R. & Kasper, G. Peptisches Ulcus und *Campylobacter pyloridis. Dtsch. Med. Wochenschr.* 1986, **111**, 1459–1461.

26 Rokkas, T. & Sladen, G.E. Bismuth: effects on gastritis and peptic ulcer. *Scand. J. Gastroenterol.* 1988, **23** (Suppl. 142), 82–86.

27 McNulty, C.A.M., Gaerty, J.C., Crump, B., Davis, M., Donovan, I.A., Melikian, V., Lister, D.M. & Wise, R. *Campylobacter pyloridis* and associated gastritis: investigator blind, placebo controlled trial of bismuth salicylate and erythromycin ethylsuccinate. *Br. Med. J.* 1986, **293**, 645–649.

28 Lambert, J.R., Borromeo, M., Pinkard, K.J. & Turner, H. 1987. Role of *Campylobacter pyloridis* in essential non-ulcer dyspepsia — a randomized controlled trial. In Kaijser, B. & Falsen, E. (eds.) *Campylobacter IV.* University of Göteborg, Göteborg. pp. 367–368.

29 Glupczynski, Y., Labbe, M., Burette, A., Delmee, M., Avesani, V. & Bruck, C. Treatment failure of ofloxacin in *Camplyobacter pylori* infection. *Lancet* 1987, i, 1096.

30 Goodwin, C.S., Marshall, B.J., Blincow, E.D., Wilson, D.H., Blackbourn, S. & Phillips, M. Prevention of nitroimidazole resistance in *Campylobacter pylori* by coadministration of colloidal bismuth subcitrate: clinical and *in vitro* studies. *J. Clin. Pathol.* 1988, **41**, 207–210.

31 Langenberg, W., Rauws, E.A.J., Houthoff, H.J., Oudbier, J.H., Widjojokosumo, A., Tytgat, G.J.N. & Zanen, H.C. 1987. Follow up of *Campylobacter pyloridis* associated gastritis after treatment with amoxicillin and/or colloidal bismuth subcitrate. In Kaijser, B. & Falsen, E. (eds.) *Campylobacter IV.* University of Göteborg, Göteborg. pp. 366–367.

32 Hirschl, A.M., Stanek, G., Rotter, M., Pötzi, R., Gangl,

A., Hentschel, E., Schütze, K., Holzner, J.H. & Nemec, H. *Campylobacter pyloridis*, Gastritis und Ulcus pepticum. *Wien. Klin. Wochenschr.* 1987, **99**, 493–497.

33 Hirschl, A.M., Stanek, G., Rotter, M., Hentschel, E. & Schütze, K. Ulcus duodeni und Antibiotika Therapie. *Dtsch. Med. Wochenschr.* 1987, **112**, 781.

34 Schaub, N., Stalder, H., Vischer, W., Stalder, A., Affolter, H. & Wegmann, W. Versagen von Doxycyclin bei *Campylobacter pylori*-positiver Gastritis. *Dtsch. Med. Wochenschr.* 1987, **112**, 117–118.

35 Unge, P. & Gnarpe, H. Pharmakokinetic, bacteriological and clinical aspects on the use of doxycycline in patients with active duodenal ulcer associated with *Campylobacter pylori*. *Scand. J. Infect. Dis.* 1988 (Suppl. 53), 70–73.

36 Lamouliatte, H., Megraud, F., De Mascarel, A. & Quinton, A. 1987. Placebo controlled trial of josamycin in *Campylobacter pyloridis* associated gastritis. In Kaijser, B. & Falsen, E. (eds.) *Campylobacter IV*. University of Göteborg, Göteborg. pp. 388–389.

37 Börsch, G., Mai, U. & Müller, K.M. Monotherapy or polychemotherapy in the treatment of *Campylobacter pylori*-related gastroduodenal disease. *Scand. J. Gastroenterol.* 1988, **23** (Suppl. 142), 101–106.

38 Langenberg, M.L., Rauws, E.A.J., Schipper, M.E.I., Widjojokosumo, A., Tytgat, G.N.J., Rietra, P.J.G.M. & Zanen, H.C. 1985. The pathogenic role of *Campylobacter pyloridis* studied by attempts to eliminate this organism. In Pearson, A.D., Skirrow, M.B., Lior, H. & Rowe, B. (eds.) *Campylobacter III*, PHLS, London. pp. 162–163.

28 Campylobacter pylori: treatment of gastritis

E.A.J. RAUWS AND G.N. TYTGAT

The link between *Campylobacter pylori* and gastritis, as well as the association between gastritis and dyspepsia, led to therapeutic attempts intended to eradicate the microorganisms, to reduce gastric inflammation and to ameliorate dyspeptic symptoms [1–8]. If one succeeds in eradicating *C. pylori*, gastritis heals [9–11]. A definite link between histological gastritis and dyspeptic symptoms has not yet convincingly been established.

This chapter deals with the effect of ulcer healing agents, bismuth compounds and bismuth–antibiotic combinations in the treatment of *C. pylori* colonization, gastritis and their effects on symptoms.

Eradication of C. pylori

In vitro C. pylori is highly sensitive to the *β*-lactams, microlides, nitrofurans and the new quinolones [12–16]. The microorganism is also susceptible to bismuth compounds such as bismuth subcitrate (De-Nol, Gist Brocades, Delft, The Netherlands) and bismuth subsalicylate (Pepto-Bismol, Procter & Gamble, Cincinnati, USA). However, *in vitro* activity of an agent against *C. pylori* is no guarantee for *in vivo* success. This striking discordance between *in vitro* and *in vivo* efficacy has puzzled many investigators. It has been suggested that the clinical inefficacy of most drugs is caused by the inability to reach *C. pylori* within and below the gastric mucus layer in sufficient concentrations. At first sight several drugs and drug combinations seem capable of suppressing or eradicating the microorganism *in vivo*. However, during long-term follow-up, this 'eradication' turned out to be only temporary clearance or suppression of the bacteria [11].

Using restriction endonuclease analysis of the bacterial DNA before treatment and after relapse, we have shown that always the same strain or subtype is involved, suggesting recrudescence rather than reinfection when reappearance of the organisms occurs [17]. It has been suggested that some microorganisms survive presumably in the deeper layers within the gastric pits allowing recolonization to occur, when the effective drug is withdrawn.

The ultimate goal of successful therapy must be long-term eradication of the microorganism, that is *C. pylori* not detectable at least one month and preferably longer after the end of a treatment period.

Antacids

The effect of antacids on *C. pylori* gastritis has not yet been thoroughly studied. In a recent publication, Berstad *et al.* [18] conducted a prospective randomized, double-blind study to evaluate the effect of an aluminium magnesium antacid or pirenzepine (50 mg b.i.d.) versus placebo in non-ulcer dyspepsia patients with erosive pre-pyloric changes. *Campylobacter pylori* was found by light microscopy of silver-stained sections in 28% of 89 consecutive dyspeptic patients. After a 4-week period of treatment with the antacid there was a significant ($P < 0.001$) decreased density of *C. pylori* without improvement of the inflammatory changes. The number of patients with gastritis even tended to increase after the aluminium magnesium antacid as compared with placebo ($P < 0.10$).

Acid-inhibiting agents

Neither H_2-receptor-blockers [11, 19, 20] nor pirenzepine [18, 20] influence *C. pylori* colonization of the gastric mucosa. Humphreys *et al.* [21] found, however, seven patients who were initially *C. pylori* positive and became negative after 6 weeks' therapy with cimetidine 400 mg b.d. They also found five patients who were initially negative but became positive using this regimen. They explained these findings by the weak antibacterial effect of cimetidine or sampling error, while the five patients who became colonized during cimetidine treatment were explained by the more favourable conditions to colonize the gastric mucosa in a less acid environment. Several other studies report incidental disappearance of *C. pylori* while on H_2-receptor blocker therapy. However, the overall results of many studies demonstrate no change of *C. pylori* status in patients on H_2-receptor blockers.

Little information is available about the effect of powerful acid inhibition with omeprazole on *C. pylori* colonization. Whether acid-reducing drugs may even promote the bacterial colonization by allowing the intragastric pH to rise is still a matter of debate. We found, in two patients during maintenance therapy with omeprazole 20 mg/day, a higher urease activity in the stomach as measured by the ^{14}C-urea breath test after 6 months' treatment (unpublished observation). In the antral biopsies heavier colonization with *C. pylori* after treatment with omeprazole could not be detected compared to the pre-therapy colonization density. Unfortunately we took no biopsies from the corpus–fundic area where possibly more extensive colonization might have occurred.

Mucosal protective agents

Neither sucralfate [11, 12] nor prostaglandins [22] have any effect on *C. pylori* colonization in the stomach. In contrast, bismuth-containing preparations (colloidal bismuth subcitrate (CBS), bismuth subsalicylate (BSS)) are clearly effective in suppressing or clearing *C. pylori* from the stomach [9–11]. This may be due to its antimicrobial activity, its cytoprotective effects or both. *In vitro* colloidal bismuth subcitrate inhibits the growth of *C. pylori* at concentrations in the range of 4–32 mg/l [12, 23]. Marshall [9] was the first to report on *in vivo* bactericidal effect of CBS *C. pylori*. Clearance of *C. pylori* with CBS monotherapy has now been observed by many investigators, as has been demonstrated in a controlled study using bismuth subsalicylate (PeptoBismol)[10]. Clearance of *C. pylori* using bismuth preparations was hopeful, but eradication using bismuth monotherapy is rather disappointing (Table 28.1). The overall results of many studies demonstrate clearance of *C. pylori* while using bismuth monotherapy in about 45–100%, but long-term eradication is only achieved in 15% [11].

The mechanisms underlying the observed antibacterial effect of bismuth compounds have to be clarified. Administration of CBS leads to rapid onset of lytic changes and disappearance of the microorganisms within 24 hours, suggesting a direct bactericidal effect possibly by inactivating bacterial enzyme systems. In antral mucosal biopsies obtained a few hours after ingestion of CBS tablets, strong accumulation of bismuth particles along the bacterial membranes may readily be observed [24, 25].

It is unclear how the apparent preferential accumulation of bismuth is to be explained. The rather selective precipitation of bismuth complexes on the microorganisms may perhaps be explained by alteration of pH immediately adjacent to the bacteria. *Campylobacter pylori* produces large amounts of urease [26] and in the presence of urea can be expected to generate ammonium ions. High alkalinity would increase the solubility of colloidal bismuth subcitrate. Sudden reversion from an alkaline to acid milieu after *C. pylori* becomes metabolically inactivated by bismuth could

Table 28.1 Influence of treatment with bismuth compounds on *C. pylori* status immediately after treatment (= clearance) and during further follow-up (= eradication)

Investigator	*n*	Drug*	Clearance (%)	Eradication (%)**
Marshall *et al.* (1984)[9]	6	CBS	100	N.D.
McNulty *et al.* (1986)[10]	18	BSS	78	N.D.
Lambert *et al.* (1987)[35]	15	CBS	73	N.D.
Rokkas *et al.* (1987)[33]	11	CBS	82	N.D.
Rauws *et al.* (1988)[11]	67	CBS	45	15 (> 12 months)
Humphreys *et al.* (1988)[21]	10	CBS	100	N.D.

*CBS = colloidal bismuth subcitrate (tablet containing 120 mg as Bi_2O_3); BSS = bismuth subsalicylate (tablet containing 140 mg as Bi_2O_3).
**N.D. = not done.

explain the selective precipitation of the bismuth complex along the microorganisms. An alternative hypothesis could be selective absorption of bismuth particles along the lipopolysaccharide (LPS) layer of the bacterial outer membrane. A few hours after administration of CBS many bacteria show structural disintegration, especially the bacteria present at the luminal surface, but not those living deep in the foveolae.

This may suggest that the microorganisms deep in the gastric pits survive the bactericidal effects of CBS and recolonize the gastric mucosa after discontinuation of this drug in most patients. The clinical inefficacy might be caused by the inability of bismuth compounds to reach all microorganisms in sufficient concentration. This question has been addressed by Lambert [27] and Skoglund [28]. They measured bismuth concentrations in mucosal biopsies. Lambert found bismuth concentrations above MIC for 3 hours after ingestion of two tablets of CBS (= 240 mg Bi_2O_3). Skoglund measured bismuth concentrations above MIC_{90} in dogs for 3 hours of a single dose of bismuth subsalicylate liquid (30 ml containing 525 mg BSS) and for 6 hours with four times daily dosing. This might explain why continuous exposure (four times daily) revealed better clearance and eradication rates compared to twice daily dosing (Table 28.2). A prolonged treatment period and even a

Table 28.2 Efficacy of different dosage regimens of bismuth compounds in clearing and eradicating *C. pylori*

Investigator	Disease*	*n*	Drug†	Clearance (%)*	Eradication (%)*
Coghlan (unpublished)	D.U.	30	CBS 2 × 2 tablets	32	N.D.
		30	CBS 4 × 1 tablet	64	N.D.
Menge *et al.* [39]	N.U.D.	24	BSS 3 × 3 tablets (5 days)	83	N.D.
Lambert *et al.* [40]	N.U.D./D.U.	20	CBS 4 × 1 tablet (4 weeks)	65	N.D.
		20	CBS 4 × 1 tablet (8 weeks)	70	30 (6 months)
		20	CBS 2 × 2 tablets (8 weeks)	30	15 (6 months)
Rauws (unpublished)	N.U.D.	115	CBS 4 × 1 tablet (4 weeks)	31	10 (24 months)
		16	CBS 4 × 1 tablet (8 weeks)	31	N.D.
		17	CBS 4 × 2 tablets (4 weeks)	N.D.	6 (> 1 year)
		13	CBS 2 × 2 tablets (4 weeks)	15	0 (1 month)
		9	CBS 1 × 4 tablets (4 weeks)	11	0 (1 month)

*N.D. = not done; D.U. = duodenal ulcer; N.U.D. = non-ulcer dyspepsia.
†Daily divided doses. CBS = colloidal bismuth subcitrate; BSS = bismuth subcalicylate.

double dose (four times daily) are not superior to the standard dose regimen. No comparative studies are available, comparing CBS four times daily and BSS four times daily in eradicating *C. pylori.*

Bismuth–antibiotics combinations

Combination therapy of bismuth salts and various antibiotics has been tried to improve the long-term eradication of *C. pylori* [11, 29]. Triple therapy using two antibiotics has been used and the results obtained are superior to the eradication rates obtained with dual therapy alone [30, 31]. The success might be explained by the local bactericidal effect of the bismuth compound, in combination with the systemic effect of the antibiotic. An alternative hypothesis could be that the bismuth particles are bound to the lipopolysaccharide (LPS) layer of the bacterial outer membrane. An intact LPS-containing bacterial outer membrane is essential for bacterial survival. Thus CBS may damage the bacterial cytoplasmic membrane and make the microorganism sensitive (or more sensitive) to antibiotics such as metronidazole. This as monotherapy has hardly any effect but when combined with CBS and amoxycillin is very effective *in vivo* in eradicating *C. pylori.* One should be aware that combining many drugs and especially antibiotics can have serious side-effects, such as drug hypersensitivity and antibiotic-associated colitis. The best treatment available today seems to be combination therapy of CBS and amoxycillin or CBS and tinidazole; both are combinations with few side-effects. Triple therapy is more effective in eradicating *C. pylori*, but has significant side-effects.

Table 28.3 shows the eradication rates of *C. pylori* with combined bismuth and antibiotic therapy regimens.

Histological aspects of therapeutic attempts

The presence of *C. pylori* is strongly associated with chronic antral gastritis [1–8, 11] and *C. pylori* is absent from the normal antral mucosa. Many authors assess gastritis using the Whitehead classification [32] or a modification of it [11]. When eradication of *C. pylori* has been accomplished by drug treatment, histological improvement with disappearance of the active inflammation and much of the chronic inflammation was observed during further follow-up [9–11]. Recrudescence of the bacteria in the subsequent months [11] was associated with a return of the inflammatory response.

Symptoms of non-ulcer dyspepsia

Different authors use varying definitions for non-ulcer dyspepsia. There exists no generally accepted method of scoring for dyspepsia and a positive effect of about 30% whilst on placebo treatment makes interpretation of results difficult. Most studies published as abstracts

Table 28.3 Efficacy of bismuth compounds combined with antibiotics in clearing and eradicating *C. pylori*

Investigation	Disease*	*n*	Drug (combination)†	Clearance (%)	Eradication (%)*
Marshall *et al.* (1987)[29]	D.U.	28	CBS + tinidazole	75	N.D.
Rauws *et al.* (1988)[11]	D.U./N.U.D.	20	CBS + amoxycillin	90	35 (12 months)
Börsch *et al.* (1988)[30]	D.U.	23	CBS + amoxycillin + metronidazole	100	N.D.
Borody *et al.* (1988)[31]	D.U./N.U.D.	100	CBS + metronidazole + tetracyline or amoxycillin	94	94 (mean 18 months)

*N.D. = not done; D.U. = duodenal ulcer; N.U.D. = non-ulcer dyspepsia.
†CBS = colloidal bismuth subcitrate; BSS = bismuth subsalicylate.

have shown CBS to be superior to placebo in this situation and associated with *C. pylori* clearance. Rokkas *et al.* [33] found CBS to be superior in improving symptoms compared to placebo. Patients were allocated randomly to 8 weeks' treatment with either placebo or CBS after which they were re-evaluated. In patients receiving CBS, symptoms improved significantly compared with placebo. No significant correlation was reported between *C. pylori* eradication and disappearance of dyspeptic symptoms. It was concluded that CBS is effective in non-ulcer dyspepsia. Borody [34] found in a double-blind study a significant improvement in abdominal pain, burning, abdominal distension and nausea in non-ulcer dyspepsia patients treated for 4 weeks with CBS compared to placebo. After 4 weeks' treatment with either CBS or placebo, all patients were treated openly with CBS for a further 4 weeks. Patients receiving CBS following initial placebo treatment also experienced significant relief of abdominal pain and burning. Borody concluded that *C. pylori* colonization may be a common cause of non-ulcer dyspepsia and that patients with non-ulcer dyspepsia and *C. pylori* gastritis can expect a significant clinical improvement in abdominal pain, burning, distension and nausea when treated with CBS. Unfortunately, no data were given about the relationship between symptoms and *C. pylori* suppression or eradication rate.

Lambert *et al.* [35] treated non-ulcer dyspepsia patients with *C. pylori*-positive gastritis with either placebo or CBS for 4 weeks. Symptoms were scored before and after cessation of therapy. A significant improvement of day and night pain, nausea and vomiting was found in subjects cleared of *C. pylori.* McNulty and co-workers performed a controlled trial in which non-ulcer dyspepsia patients and *C. pylori*-associated gastritis were randomly assigned to treatment with antacid, BSS or erythromycin ethylsuccinate [10]. Treatment with bismuth subsalicylate was associated with clearance of *C. pylori* and improvement in symptomatology, although not significantly presumably due to the small number of patients studied. Most authors [33–35] report significant improvement in symptoms while using CBS compared to placebo, paralleled by improved gastritis scores. Only Lambert *et al.* [35] mention a significant correlation between symptom improvement and *C. pylori* clearance.

Justification for treatment

The discovery of *C. pylori* and its association with gastritis, peptic ulcer disease and probably non-ulcer dyspepsia has been well established. However, many gastroenterologists remain unconvinced of the pathogenic role of *C. pylori*, mainly because the microorganisms can also be found in many asymptomatic individuals together with histological evidence of gastritis. In addition, *C. pylori* is often absent and the gastric mucosa is normal in patients complaining of identical dyspeptic symptoms. A priority at the present is to identify those dyspeptic patients whose symptoms are truly caused by *C. pylori*-associated gastritis and perform properly conducted double-blind therapeutic studies to justify treatment with bismuth–antimicrobial combinations. Bismuth compounds have been used for more than a century for the treatment of various gastro-intestinal disorders, including dyspepsia. Nowadays, bismuth subnitrate (Roter) and bismuth subsalicylate (PeptoBismol) are still used as over-the-counter drugs for non-ulcer dyspepsia, apparently with some success although a significant placebo effect may be present [36]. H_2-receptor antagonists are often prescribed for patients with non-ulcer dyspepsia syndrome although the evidence to support this therapeutic approach is rather weak. Cimetidine, for example, has been found to be no better than placebo in relieving the symptoms in such patients [37].

As the placebo response is high [36, 38] and there is no clear correlation between the presence of gastritis and symptomatology, it is easy to understand why therapy for *C. pylori*-associated gastritis is rather controversial. Though many gastroenterologists are convinced from their own clinical experience of the benefits of antibacterial treatment, currently only triple therapy using colloidal bismuth subcitrate, amoxycillin or tetracycline, combined with metronidazole is adequate for eradication of *C. pylori* and for improvement of histological gastritis during long-term follow-up. This regimen is certainly not advisable for all non-ulcer dyspepsia patients colonized by *C. pylori*. Only patients with long-standing severe dyspeptic symptoms who respond poorly to conventional therapy should be offered triple therapy.

Conclusions

The very high correlation between the presence of *C. pylori* and active chronic gastritis has now been confirmed by many investigators. There is no clear correlation between the presence of gastritis and symptomatology. Currently it is almost impossible to identify those patients with non-ulcer dyspepsia whose symptoms are caused by *C. pylori*-associated gastritis.

Clearance of *C. pylori* using bismuth compounds often in combination with antibiotics is closely paralleled by normalization of the epithelial morphology. This probably explains the improvement of symptoms noted in some patients with non-ulcer dyspepsia and the protection against recurrence of peptic ulcer disease which has also been suggested.

References

1 Warren, J.R. & Marshall, B.J. Unidentified curved bacilli on gastric epithelium in active chronic gastritis. *Lancet* 1983, **i**, 1273–1275.

2 Marshall, B.J. & Warren, J.R. Unidentified bacilli in the stomach of patients with gastritis and peptic ulceration. *Lancet* 1984, **i**, 1311–1315.

3 Meyrick-Thomas, J., Poynter, D., Gooding, C. *et al.* Gastric spiral bacteria. *Lancet* 1984, **ii**, 100.

4 Rollason, T.P., Stone, J. & Rhodes, J.M. Spiral organisms in endoscopic biopsies of the human stomach. *J. Clin Pathol.* 1984, **37**, 23–26.,

5 Jones, D.M., Lessells, A.M. & Eldridge, J. Campylobacter like organisms on the gastric mucosa: culture, histological, and serological studies. *J. Clin. Pathol.* 1984, **37**, 1002–1006.

6 McNulty, C.A.M. & Watson, D.M. Spiral bacteria of the gastric antrum. *Lancet* 1984, **i**, 1068–1069.

7 Marshall, B.J., McGechie, O.B., Rogers, P.A. & Glancy, R.J. Pyloric *Campylobacter* infection and gastroduodenal disease. *Med. J. Australia* 1985, **142**, 439–444.

8 Buck, G.E., Gourley, W.K., Lee, W.K. *et al.* Relation of *Campylobacter pyloridis* to gastritis and peptic ulcer. *J. Infect. Dis.* 1986, **153**, 664–669.

9 Marshall, B.J., Hislop, I., Glancy, R. & Armstrong, J. Histological improvement of active gastritis in patients treated with De-Nol. *Austral. N.Z. J. Med.* 1984, **14** (Suppl. 4), 907.

10 McNulty, C.A.M., Gearty, J.C., Crump, B. *et al. Campylobacter pyloridis* and associated gastritis: investigator blind, placebo controlled trial of bismuth salicylate and erythromycin ethylsuccinate. *Br. Med. J.* 1986, **293**, 645–649.

11 Rauws, E.A.J., Langenberg, W., Houthoff, H.J., Zanen, H.C. & Tytgat, G.N.J. *Campylobacter pyloridis*-associated chronic active antral gastritis. *Gastroenterology* 1988, **94**, 33–40.

12 McNulty, C.A.M., Dent, J. & Wise, R. Susceptibility of clinical isolates of *Campylobacter pylori* to 11 antimicrobial agents. *Antimicrob. Agents Chemother.* 1985, **28**, 837–838.

13 Goodwin, C.S., Blake, P.K. & Blincow, E.D. The minimum inhibitory and bactericidal concentrations of antibiotics and anti-ulcer agents against *Campylobacter pyloridis. Chemotherapy* 1986, **17**, 309–314.

14 Lambert, J.R., Hansky, J., Davidson, A., Pinkerd, K. & Stockman, K. Campylobacter like organism (CLO) — *in vivo* and *in vitro* susceptibility to antimicrobial and antiulcer therapy. *Gastroenterology* 1985, **88**, 1462.

15 Morgan, D.R., Fitzpatrick, P.M., David, K.L. & Kraft, W.G. Susceptibility patterns of *Campylobacter pyloridis. FEMS Microbiol. Lett.* 1987, **42**, 245–248.

16 Shungu, D.L., Nalin, D.R., Gilman, R.H. *et al.* Comparative susceptibilities of *Campylobacter pylori* to norfloxacin and other agents. *Antimicrob. Agents Chemother.* 1987, **31**, 949–950.

17 Langenberg, W., Rauws, E.A.J., Widjojokusumo, A., Tytgat, G.N.J. & Zanen, H.C. Identification of *Campylobacter pyloridis* isolates by restriction endonuclease DNA analysis. *J. Clin. Microbiol.* 1986, 414–417.

18 Berstad, A., Alexander, B., Weberg, R. *et al.* Antacids reduce *Campylobacter pylori* colonization without healing the gastritis in patients with nonulcer dyspepsia and erosive prepyloric changes. *Gastroenterology* 1988, **95**, 619–624.

19 McIntyre, R.L.E., Piris, J. & Truelove, S.C. Effect of

cimetidine on chronic gastritis in gastric ulcer patients. *Austral. N.Z. J. Med.* 1982, **12**, 106. (Abstract)

20 Hirschl, A.M., Hentschl, E., Schütze, K. *et al.* The efficacy of antimicrobial treatment in *Campylobacter pylori* associated gastritis and duodenal ulcer. *Scand. J. Gastroenterol.* 1988, **23** (Suppl. 142), 76–81.

21 Humphreys, H., Bourke, S., Dooley, C. *et al.* Effect of treatment on *Campylobacter pylori* in peptic disease: a randomized prospective trial. *Gut* 1988, **29**, 279–283.

22 Ho, J., Lui, I., Hui, W.M., Ng, M.M.T. & Lam, S.K. A study on the correlation of duodenal ulcer healing with campylobacter-like organisms. *J. Gastroenterol. Hepatol.* 1986, **1**, 69–74.

23 Andreasen, J.J. & Andersen, L.P. *In vitro* susceptibility of *Campylobacter pyloridis* to cimetidine, sucralfate, bismuth and sixteen antibiotics. *Acta Pathol. Microbiol. Immunol. Scand. (Sect. B)* 1987, **95**, 147–149.

24 Marshall, B.J., Armstrong, J.A., Francis, G.J., Nokes, N.T. & Wee, S.H. Antibacterial action of bismuth in relation to *Campylobacter pyloridis* colonization and gastritis. *Digestion* 1987, **37** (Suppl. 2), 16–30.

25 Rauws, E.A.J. & Tytgat, G.N.J. Electronenmikroskopische Befunde während einer Behandlung Campylobacter-pylori-positiver Gastritiden mit Wismutsalzen. *Z. Gastroenterologie* 1987, **25** (Suppl. 4), 41–43.

26 Langenberg, M.L., Tytgat, G.N.J., Schipper, M.E.I., Rietra, P.J.G.M. & Zanen, H.C. Campylobacter-like organisms in the stomach of patients and healthy individuals. *Lancet* 1984, **i**, 1348.

27 Lambert, J.R., Way, D.J., King, R.S., Eaves, E.A. & Hansky, J. Bismuth pharmacokinetics in the human gastric mucosa. *Gastroenterology* 1988, **94**, A248.

28 Skoglund, M.L. & Watters, K. Bismuth concentration at the site of *Campylobacter pylori* colonization. *Gastroenterology* 1988, **94**, A430.

29 Marshall, B.J., Goodwin, C.S., Warren, J.R. *et al.* Long-term healing of gastritis and low duodenal ulcer relapse after eradication of *Campylobacter pyloridis*: A prospective double blind study. *Gastroenterology* 1987, **92**, 1518.

30 Börsch, G., Mai, U. & Opferkuch, W. Oral triple therapy may effectively eradicate *Campylobacter pylori* in man: a pilot study. *Gastroenterology* 1988, **94**, A44.

31 Borody, T., Cole, P., Noonan, S. *et al.* Long term *Campylobacter pylori* recurrence post-eradication. *Gastroenterology* 1988, **94**, A43.

32 Whitehead, R., Truelove, S.C. & Gear, M.W.L. The histological diagnosis of chronic gastritis in fibreoptic gastroscope biopsy specimens. *J. Clin. Pathol.* 1972, **25**, 1–11.

33 Rokkas, T., Pursey, C., Simmons, N.A., Filipe, M.I. & Sladen, G.E. Non-ulcer dyspepsia and colloidal bismuth subcitrate therapy: The role of *Campylobacter pyloridis*. *Gastroenterology* 1987, **92**, 1599.

34 Borody, T., Hennesey, W., Daskalopoulos, G., Carrick, J. & Hazell, S. Double-blind trial of De-Nol in non-ulcer dyspepsia associated with *Campylobacter pyloridis* gastritis. *Gastroenterology* 1987, **92**, 1324.

35 Lambert, J.R., Borromeo, M., Korman, M.G. & Hansky, J. Role of *Campylobacter pyloridis* in non-ulcer dyspepsia — A randomised controlled trial. *Gastroenterology* 1987, **91**, 1488.

36 Hailey, F.J. & Newson, J.H. Evaluation of bismuth subsalicylate in relieving symptoms of ingestion. *Arch. Intern. Med.* 1984, **144**, 269–272.

37 Nyren, O., Adami, H.O., Bates, S. *et al.* Absence of therapeutic benefit from antacids or cimetidine in non-ulcer dyspepsia. *N. Engl. J. Med.* 1986, **314**, 339–343.

38 Thomson, W.G. Non-ulcer dyspepsia. *Canad. Med. Assoc. J.* 1984, **130**, 565–569.

39 Menge, H., Hofmann, J. & Gregor, M. Dosis-Wirkungs-Studien mit Wismutsalzen zur Elimination von *Campylobacter pylori*. *Z. Gastroenterologie* 1987 (Suppl. 4), **25**, 44–46.

40 Lambert, J.R., Borromeo, M., Eaves, E.A., Hansky, J. & Korman, M. Efficacy of different dosage regimes of bismuth in eradicating *Campylobacter pylori*. *Gastroenterology* 1988, **94**, A248.

29 Campylobacter pylori and ulcer treatment

J.G. COGHLAN, A. TOBIN AND C. O'MORÁIN

Duodenal ulcer (DU) disease is characterized by remissions and relapses. Since the advent of H_2 antagonists in the 1960s, initial healing of DUs can usually be achieved medically, and resorting to surgical intervention is required less frequently. It has become clear, however, that H_2 antagonists do not alter the natural history of this disease, as up to 100% of patients relapse within two years. Thus, the long-term treatment of duodenal ulcer disease continues to be a major problem.

Long-term H_2 antagonist therapy has been proposed and is relatively effective in reducing the one-year relapse rate from 80–90% to 20–30% [1]. No evidence of tolerance to H_2 antagonist therapy is found after a year of continuous treatment [2]. Apart from the expense of such therapy, these drugs have only been available for just over 20 years and their long-term safety cannot be guaranteed. Chronic suppression of gastric acid production is associated with bacterial colonization of the stomach and increased local nitrite concentration which may promote the development of gastric cancer. Intermittent symptomatic therapy with H_2 antagonists is also being investigated and has for some time been practised by patients. This is economically preferable, but the frequency of asymptomatic relapse and incomplete healing may increase with this form of therapy. As one-third of ulcer deaths already occur at home [3], it is not inconceivable that intermittent therapy would increase the incidence of perforation and haemorrhage and delay hospital referral.

Surgical intervention is not necessarily 100% effective and only the more radical procedures, which themselves have a high complication rate, are associated with a low ulcer relapse rate [4]. The alternative approach is to remove the cause of DU disease. The pathogenesis of DU disease is, however, very poorly understood. The 'acid hypothesis' which provided the rationale for H_2 antagonist therapy is no longer adequate. Only 50% of DU patients have hypersecretion [5] and rebound acid secretion after discontinuation of H_2 antagonists is not clearly associated with a high early ulcer relapse rate [6]. The presence of gastric acid is, of course, a prerequisite for the development of DUs but is not in itself a sufficient cause alone.

The idea that peptic ulcer disease might have an infective cause is not new. In 1940 Freedberg & Barron suggested that the success of bismuth compounds in peptic ulcer disease might be due to the suppression of gastric spirochaetes [7]. Furthermore, antibiotics have been successfully used in the treatment of DU in several studies. Both furazolidone and metronidazole heal ulcers with an efficacy similar to cimetidine (Table 29.1). The discovery of *C. pylori*, and its strong association with DU disease, was certain to heighten interest in a possible infective cause for DU disease and lead to investigations to discover whether eradication of this organism would succeed where anti-acid therapy had failed.

Bismuth compounds

The association of bismuth compounds with the treatment of gastric disorders goes back over 200 years, and their use in France for DU disease was widespread until 1974. The advent of H_2 antagonists, which were simple

Table 29.1 Antibiotic versus cimetidine in the treatment of peptic ulcer disease

Reference	Agents	Number	% healing
Zheng *et al.* [8]	Furazolidone Cimetidine	70	73 55
Zhao *et al.* [9]	Furazolidone Cimetidine	270	69 68
Quintero & Sotto [10]	Metronidazole Cimetidine	55	76 77

and pleasant to take, were highly effective and whose mechanism of action seemed logical, rapidly displaced bismuth compounds in the treatment of ulcer disease. Reports from France of nearly 200 cases of bismuth encephalopathy during the 1970s further hastened the relegation of bismuth therapy [11].

Interest in bismuth therapy never disappeared completely and throughout the 1970s several trials demonstrated the superiority of colloidal bismuth subcitrate (CBS) over placebo (Table 29.2). Taking the results of these double-blind trials together, it is apparent that CBS is more effective than placebo in the treatment of DU, with a net therapeutic gain of approximately 45%. While direct comparisons are not entirely appropriate, this does compare favourably with trials of cimetidine versus placebo (net gain 34%) and ranitidine versus placebo (net gain 43%) [23].

Bismuth treatment versus H_2 treatment in duodenal ulcer disease

By the early 1980s, H_2 antagonists had become the 'Gold Standard' in the treatment of DU and if bismuth therapy was to be re-established, an efficacy comparable to H_2 antagonists had to be demonstrated. The results of unblinded and single-blind trials gave comparable results, and subsequent double-blind trials confirmed these findings (Table 29.3). Eight double-blind comparative trials, involving over 500 patients, have been published and no single trial shows a significant dif-

Table 29.2 Controlled trials of CBS versus placebo in DU healing

Reference	CBS dose formulation	No. studied (healing rate)	
		CBS	Placebo
Salmon *et al.* 1974 [12]	960 mg/day liq.	10 (70%)	10 (10%)†
Moshal 1975 [13]*	480 mg/day liq.	30 (67%)	30 (20%)‡
Shreeve 1975 [14]	480 mg/day liq.	19 (74%)	19 (21%)‡
Vissoulis *et al.* 1977 [15]	480 mg/day liq.	25 (72%)	22 (32%)‡
Coughlin *et al.* 1977 [16]	480 mg/day liq.	34 (82%)	23 (43%)‡
Connon 1977 [17]	480 mg/day liq.	19 (79%)	16 (38%)‡
Lee & Nicholson 1977 [18]	960 mg/day liq.	9 (89%)	10 (20%)‡
Poulantzas *et al.* 1978 [19]	480 mg/day liq.	24 (50%)	12 (8%)†
Dekker & Reisma 1979 [20]	480 mg/day liq.	27 (81%)	19 (42%)†
Bank *et al.* 1979 [21]*	480 mg/day tab.	7 (71%)	7 (29%)
Wilson & Alp 1982 [22]	480 mg/day tab.	22 (86%)	24 (25%)†

*6 weeks treatment, all others 4 weeks.
† $P < 0.05$.
‡ $P < 0.01$.

Table 29.3 Double-blind trials comparing the efficacy of CBS and H_2 antagonists in DU healing

Reference	Agent/dosage	Healing rate		
		4 weeks	6 weeks	8 weeks
Bianchi Porro *et al.*	CBS 480 mg/day*	21/25 (84%)		25/25 (100%)
1984 [24]	Cim 800 mg/day	18/25 (72%)		21/24 (88%)
Kang & Piper	CBS 480 mg/day		24/25 (96%)	
1982 [25]	Cim 1 g/day		25/26 (96%)	
Martin *et al.*	CBS 480 mg/day	25/38 (66%)		33/37 (89%)
1981 [26]	Cim 1 g/day	22/37 (60%)		29/34 (85%)
Moshal *et al.*	CBS 480 mg/day*		24/27 (89%)	
1981 [27]	Cim 1 g/day		22/24 (92%)	
Ward *et al.*	CBS 480 mg/day		14/19 (74%)	
1981 [28]	Cim 1 g/day		15/18 (83%)	
Bianchi Porro *et al.*	CBS 480 mg/day*	41/49 (84%)		47/49 (96%)
1984 [29]	Ran 300 mg/day	34/50 (68%)		46/50 (92%)
Ward *et al.*	CBS 480 mg/day*	25/33 (76%)		30/33 (91%)
1986 [30]	Ran 300 mg/day	28/32 (87%)		30/32 (94%)
Lee *et al.*	CBS 480 mg/day*	52/58 (90%)		56/58 (97%)
1985 [31]	Ran 300 mg/day	49/59 (81%)		57/59 (97%)

*CBS tablets.
Cim = cimetidine.
Ran = ranitidine.

ference between CBS and H_2 antagonists in respect of DU healing.

Further evidence of the short-term efficacy of CBS comes from two studies of the treatment of DUs resistant to standard doses of cimetidine. Lam *et al.* treated 25 patients with persistent ulcers with tripotassium dicitrato bismuthate (TDB), one tablet four times per day or cimetidine 1.6 g/day for 4 weeks [32]. In the first 4 weeks, 10/12 healed with TDB versus 5/13 in the cimetidine group. Those remaining unhealed then crossed over to the other treatment regime, and 7/8 receiving TDB and 1/2 receiving cimetidine subsequently healed [32]. In a similar study, Bianchi Porro *et al.* found 86% healing with CBS versus 46% with cimetidine in doses of up to 2 g/day in 43 patients with ulcers resistant to standard H_2 antagonist therapy [33].

In healing DUs, it has been noted that H_2 antagonists fail to improve the histological gastritis, which nearly always accompanies duodenal ulceration [34]. CBS, a known cytoprotective agent, has consistently been shown to improve this gastritis [34, 35]. The precise significance of this observation is as yet unclear, but it is possible that gastritis and duodenal ulceration are parts of the same disease process. Thus, it is speculated that more complete healing is achieved with mucosal protective agents and that relapse may be less common with these agents.

Duodenal ulcer relapse

Over the past decade, several studies have shown the relapse rate of DU to be lower in those initially treated with CBS than in those healed with H_2 antagonists. Of seven trials comparing bismuth subcitrate with an H_2 antagonist, where no maintenance therapy (including placebo) was given and at least 1-year endoscopic follow-up was available, the relapse rate is lower in the bismuth group in six, and in four statistical significance is reached (Table 29.4). It is of note that the relapse in the H_2 antagonist groups occurred throughout the period of follow-up, with only a slight excess of recurrences in the first 3–4 months of observation [26, 31, 37].

Table 29.4 Relapse rates of duodenal ulceration after CBS/H_2 antagonist treatment

Reference	Agent	No.	% Relapse		
			6 months	12 months	*P* value
Smith *et al.*	CBS	18	11%	35%*	
1988 [36]	Ran	18	55%	78%	<0.02
Martin *et al.*	CBS	30	13%	39%	
1981 [26]	Cim	27	70%	85%	<0.001
Hamilton *et al.*	CBS	33	30%	43%	
1986 [37]	Cim	29	43%	78%	<0.01
Lee *et al.*	CBS	53	41%†	62%	
1985 [31]	Ran	54	74%	89%	<0.02
Coghlan *et al.*	CBS	21		52%	
1987 [38]	Cim	18		63%	NS
Kang & Piper	CBS	17	47%	76%	
1982 [25]	Cim	18	44%	75%	NS
Shreeve *et al.*	CBS	15	33%	47%	
1983 [39]	Cim	10	50%	60%	NS

*Follow up at 18 months.
†Data refers to 4 months follow up.
Cim = cimetidine.
Ran = ranitidine.
NS = not significant.

Two studies by Bianchi Porro *et al.* have shown that cimetidine maintenance therapy is no more effective in preventing relapse than initial CBS therapy without maintenance [40, 41].

The above findings have recently been confirmed by Dobrilla *et al.* who, using pooled data from available trials, compared the relapse rates for DU patients whose ulcers had been healed with H_2 antagonists, to those whose ulcers were healed by any other treatment [42]. They showed that patients initially healed with bismuth subcitrate were less likely to relapse than their H_2-treated counterparts, by 26% at 6 months and 27% at 1 year. The data reviewed included patients treated with pirenzepine, sucralfate, carbenoxolone and antacids, and none of these other agents exhibited a significant advantage over H_2 antagonists. However, the total number of studies with each of these agents was too small to exclude such an effect. Nevertheless, it is interesting to note that only bismuth subcitrate, out of all the mucosal protective agents, has a significantly superior relapse profile to H_2 antagonists, and studies fail to demonstrate even a trend towards a reduction in ulcer recurrence with other cytoprotective agents. The reason why bismuth subcitrate should be associated with a lower relapse rate remains unclear. If H_2 antagonists induce relapse, then all other agents should demonstrate the same 'apparent' reduction in ulcer relapse. Furthermore, the excess relapse should occur within weeks of discontinuing H_2-antagonist therapy and relapse rates would be the same thereafter. Graphic representations of ulcer recurrence after bismuth and H_2 therapy, however, continue to show divergence at 1 year. It has been suggested that accumulated body stores of bismuth continue to exert a therapeutic effect long after dosage has ceased. As urinary excretion declines exponentially to normal levels within 2 months of stopping treatment [11], it is difficult to conceive that immeasurable drug levels continue to exert benefit, up to a year later.

Improvement of associated histological gastritis and more complete ulcer healing with normalization of the duodenal mucosa seems a reasonable explanation for the reduced recurrence rate. This would, however, apply equally to all cytoprotective agents and the available evidence does not support this concept.

Finally, one must consider the possibility that suppression of *C. pylori*, which colonizes the gastric antrum of over 90% of DU patients and which is strongly associated with histological gastritis, could in some way explain the lower relapse rate found with bismuth subcitrate.

Relationship between Campylobacter pylori and relapse

Campylobacter pylori is strongly associated with DU disease and histological gastritis. In one study from our unit, *C. pylori* has been found in 90% of DU patients, histological gastritis was present in 76% of DU patients and 95% of patients with histological gastritis were *C. pylori* positive [34]. These results are typical of those found world wide. Evidence of this association has been available from long before the organism was first cultured. In 1950, Fitzgerald & Murphy noted an excess of urease activity in the gastric mucosa of peptic ulcer patients [43]. The reason for this association remained obscure until the discovery of the urease-producing organism *C. pylori*, some 30 years later. Linking *C. pylori* pathogenically to DU disease has proved very difficult. As endoscopic studies of asymptomatic people are unusual, one cannot unequivocally prove that infection precedes ulceration. Studies of ulcer relapse therefore provide the logical method of resolving this problem, since if a causative relationship exists, then eradication of *C. pylori* should alter the natural history of DU disease.

Bismuth subcitrate, and to a lesser extent bismuth subsalicylate, have been shown to inhibit *C. pylori* both *in vitro* and *in vivo*. *In vitro* studies of over 100 strains of *C. pylori* have shown that the MIC 90 for CBS has a range of 2–32 μg/ml [44–46]. These levels are, however, achieved *in vivo*. Levels of 320 μg/g at 1 hour, 62 μg/g at 3 hours and 28 μg/g at 6 hours have been found in the gastric mucosa of dogs treated with bismuth subsalicylate 525 mg four times daily [47].

In a study from our unit [38], it was noted that the relapse rate of the CBS group was numerically, but not significantly, lower than in the cimetidine group. When patients were analysed according to post-treatment *C. pylori* status, it was seen that most relapse occurred in those remaining culture positive — 79%. Relapse was uncommon in those rendered culture negative. In addition, gastritis was only improved in those cleared of *C. pylori*, and recrudescence of *C. pylori* was associated with recurrence of histological gastritis and a high ulcer relapse rate — 66%. Smith *et al.* [36] obtained similar results when comparing relapse after ranitidine to bismuth subcitrate. At 18 months, 78% of ranitidine-treated patients had relapsed, versus 35% of TDB patients. Of 36 patients followed endoscopically for 18 months, *C. pylori* eradication was achieved in only seven, none of whom had ulcer relapse. Ulcer relapse, however, occurred in 20 of the 29 culture-positive patients. From these results it seems quite possible that eradication of *C. pylori* explains the superior relapse profile of bismuth subcitrate in DU disease. Five studies have now addressed the problem of *C. pylori* and DU relapse and the results have consistently shown that suppression or eradication of *C. pylori*, whether with CBS or antibiotics, is associated with a significantly lower ulcer recurrence rate (Table 29.5). Ulcers, when they do occur in those initially culture-negative post-treatment, are accompanied by recrudescence of the infection.

In respect of relapse, it is not yet entirely clear whether *C. pylori* or the ulcer reappears first. Langenberg *et al.* have shown that, despite apparent clearance of the *C. pylori*, recurrence up to 6 months later is probably due to recrudescence of the same organism, as identified by DNA pattern [51]. In this unit we

Table 29.5 Relapse (%) of DU according to post-treatment *C. pylori* status

Reference	No. studied	Relapse rate after 1 year	
		C. pylori positive	*C. pylori* negative
Coghlan *et al.* 1987 [38]	39	79%	27%
Smith *et al.* 1988 [36]	36	88%	25%*
Lambert *et al.* 1987 [48]	45	75%	40%
Borody *et al.* 1988 [49]	21	—	14%
Goodwin *et al.* 1988 [50]	68	74%	20%

*18 months.

have found that while CBS temporarily suppresses *C. pylori* in up to 70% of treated patients, probable eradication, i.e. culture negative at 1 month post-treatment, occurs in only 20–30%. Borody *et al.* have followed patients in whom eradication appeared to have been achieved 1 month after treatment, for a further 12–25 months. They found that reinfection was extremely uncommon, with only 3 of 50 having become *C. pylori* positive in this time [49].

This data indicates the potential for early recurrence of *C. pylori*, which is significantly earlier than ulcer relapse normally occurs. Since, however, reinfection is so uncommon, those who reconvert to culture positivity probably never cleared the organism. It is likely, therefore, that *C. pylori* infection precedes ulcer recurrence.

Further studies are necessary. This initial evidence, however, strongly suggests that agents which efficiently eradicate *C. pylori* infection will significantly reduce DU relapse, at least in the year after treatment, and may fundamentally alter the natural history of the disease.

Mechanism of action

The bactericidal action of bismuth on *C. pylori* is clearly an important factor in the healing of gastritis [34], ulcers [50], and in the prevention of ulcer relapse in humans. However, the correlation between post-treatment *C. pylori* status and ulcer healing or gastritis is far from absolute. Bismuth has multiple biochemical cytoprotective actions on the gastric and duodenal mucosae which are independent of *C. pylori.*

CBS binds to gastric mucin forming a glycoprotein bismuth complex which adheres to the ulcer crater and retards hydrogen ion diffusion [52]. It also increases local PGE2 [53] and bicarbonate concentrations [54], while reducing the concentrations of local aggressive factors such as leukotriene C4 [55] and pepsin [56].

CBS has been shown to protect against gastric damage due to ethanol and stress in rats, and to accelerate the healing of ulcers induced by acetic acid in rats [54]. Rats are not naturally colonized by *C. pylori.*

It is therefore important not to simply equate bismuth compound-induced ulcer healing with *C. pylori* eradication and consequent protection against ulcer relapse.

Formulation and dosage

CBS is available in liquid form, as chewable tablets and as coated 'swallow' tablets. There is no evidence that any of these particular formulations is superior in respect of ulcer healing efficacy. Three blind studies have compared liquid to chewable tablets and have shown equal efficacy in healing DUs [57–59]. A single blind study of 189 patients revealed no difference between chewable and coated tablets [60]. In respect of *C. pylori* suppression we have found coated tablets to be at least as effective as liquid, although no direct comparative study was undertaken. A reduction in culture positivity of 42% was achieved in 26 patients after 6 weeks' treatment with coated tablets, versus a 40% reduction in 32 patients treated with liquid CBS in sequential studies.

Frequency of administration is important in the treatment of *C. pylori*, but not in respect of ulcer healing. Three studies have shown indistinguishable ulcer healing rates with twice daily or four times daily CBS [61–63]. However, two blind studies have recently demonstrated that a 240 mg twice daily dosage is ineffective in suppressing the organism [61, 65]. This is believed to be due to the inability of 12-hourly doses to maintain cytotoxic concentrations of bismuth within the gastric mucin [64].

Duration of therapy may be more important in respect of ulcer healing than *C. pylori* suppression. From Table 29.3 it is apparent that 10–20% more ulcers are healed after 6 or 8 weeks of CBS therapy than after 4 weeks. However, Lambert *et al.* [65] found no major change in *C. pylori* status between 4 (65% *C. pylori* – ve) and 8 (70% *C. pylori* – ve) weeks of treatment.

Safety profile

While bismuth compounds are effectively topical agents, acting at a local level to heal ulcers and histological gastritis and also eradicate *C. pylori*, some of the drug is absorbed and may accumulate, particularly in the renal, hepatic and neural tissues. Our experience and that of other workers has been that bismuth is well tolerated with few side-effects which only cause a nuisance, darkening of the stool and blackening of teeth and tongue with liquid preparations and chewable tablets, other effects being rare. Possible drug toxicity has, however, been a major source of concern with regard to bismuth therapy, since nearly 200 cases of bismuth encephalopathy were recorded in France. These cases occurred as a result of prolonged high-dose treatment and were almost invariably associated with blood bismuth levels in excess of 100 μg/l. Current dosage schedules rarely cause levels of greater than 50 μg/l (two recorded cases) and have not been associated with serious CNS side-effects, although mild dizziness and headache have been noted on a few occasions [11].

The systemic effects of the recommended dosage (480 mg/day for a maximum of 2 months with at least 2 months between courses) have been thoroughly studied. No serious side-effects have been recorded in over 500 cases. Blood levels rarely exceed 50 μg/l and renal excretion returns to baseline level after 2 months off treatment [11]. However, as excretion is almost completely renal, caution is advised in patients with significant renal impairment. Finally, in the absence of proof that bismuth is not toxic to the fetus, these compounds should clearly be avoided in pregnancy.

Therapeutic indications

Despite limits to our knowledge, the benefits of CBS therapy in the long-term healing of histological gastritis and the reduction of DU recurrence are well documented. This therapy can be recommended for these reasons alone. For the purpose of eradicating *C. pylori*, bismuth compounds should not be used without adjuvant therapy, as their efficacy in this regard is inadequate.

Antibiotics

Two studies comparing furazolidone with cimetidine [8, 9] and one with metronidazole versus cimetidine [10] have shown similar healing rates for the antibiotics and the H_2 antagonists. As none of these agents has known anti-ulcer actions, these results are compatible with an infective pathogenesis for peptic ulcer disease, and suggest a possible role for antibiotic therapy in peptic ulceration.

Marshall *et al.* demonstrated that *C. pylori* is sensitive to a wide range of antibiotics [66]. Several studies of the *in vitro* sensitivities of this organism have since been completed. These have recently been reviewed [67] and the review shows that *C. pylori* is highly sensitive to all macrolides, tetracyclines, most beta lactams, aminoglycocides (neomycin not studied) and the older quinolone, ofloxacin. Moderate sensitivity is found with nitroimidazoles, nitrofurans and the newer quinolones. Resistance to sulphonamides, trimethoprim, vancomycin and amphotericin B is almost invariable. Given the above evidence it was clearly imperative to study clinically the effects of antibiotic therapy on *C. pylori* and DU.

Studies of antibiotic therapy *in vivo* have shown that many are indeed active against *C. pylori*. However, their efficacy used alone is not superior to CBS monotherapy and the organism readily develops resistance to most antibiotics. Synergy between CBS and many of the antibiotic groups has been demonstrated, significantly increasing the number of patients rendered *C. pylori* negative and reducing the likelihood of bacterial resistance. It is therefore suggested that in treating DUs, antibiotic therapy should only be used in combination with bismuth compounds.

Furthermore, studies of antibiotic adjuvant therapy (detailed below) show that, despite improved *C. pylori* eradication, the ulcer-healing efficacy of combination therapy is the same as CBS monotherapy. Thus, the sole aim of combination therapy is the eradication of *C. pylori*, consequently reducing the ulcer relapse rate.

Beta lactams

The beta lactam group of antibiotics has been extensively studied with respect to *in vivo* activity against *C. pylori* and ulcer healing. Penicillin, becampicillin and amoxycillin have each been investigated. Interestingly, neither penicillin nor becampicillin [68] appear to eradicate *C. pylori in vivo*. This is despite very low MIC 90s (0.03–0.225 μg/ml) and the failure of treated organisms to develop resistance during treatment. It has been postulated that this failure is partially due to the fact that penicillins only destroy replicating organisms [68]. This explanation is not the complete answer since amoxycillin is moderately active *in vivo* and benzylpenicillin rapidly destroys *C. pylori in vitro* [69]. It may be of relevance that Hollingsworth *et al.* have shown that ampicillin is not effectively secreted into the gastric pits [70]. Amoxycillin has not been specifically studied in this regard but is claimed to have better tissue penetration.

Amoxycillin monotherapy has been found to be as effective as CBS in respect of *C. pylori* suppression and eradication. Rauws *et al.* and Tytgat *et al.* respectively have found 68% and 64% culture negative immediately after treatment with amoxycillin 375 mg three times daily (the duration of therapy was not specified), and 1 month later 23% and 31% remained culture negative [71, 72].

Van Caekenberg & Breyssens have demonstrated *in vitro* synergy between ampicillin and CBS [45]. Therefore it might be expected that *in vivo* the combination of amoxycillin and CBS would produce dramatic results, both in terms of *C. pylori* eradication and ulcer healing. Indeed two studies have produced impressive initial suppression rates of 85% (in 23 patients) and 90% (in 20 patients) with combination therapy. Unfortunately, 1 month later only 35% and 40% respectively of patients remained *C. pylori* negative [71, 72], not dissimilar to CBS alone. In our comparative study (50 patients) of CBS alone versus CBS (for 6 weeks) and amoxycillin (250 mg three times daily for the first week), only 42% were rendered culture-negative by either treatment [73]. Follow-up is as yet incomplete, but already only 26% of the combined treatment group remain culture-negative. This study also revealed identical healing rates of DUs with CBS used alone or in combined therapy.

Duration of amoxycillin therapy would not appear to be critical either in respect of *C. pylori* suppression or ulcer healing. One open study found 87% ulcer healing and 47% *C. pylori* eradication in 15 patients receiving amoxycillin (500 mg four times daily) for 7 days plus CBS for 14 days. 68% ulcer healing with 44% *C. pylori* eradication was achieved when both agents were taken for 14 days [50].

Thus, the beta lactams studied to date are not superior to CBS as a monotherapy and confer little advantage over CBS alone when used in combination with CBS. However, the available data is limited, and the possible efficacy of the cephalosporins remains to be explored. Perhaps future studies will provide more encouraging results, but on present data beta lactams cannot be recommended as ideal adjuvant agents to CBS.

Nitroimidazoles

Theoretically, tinidazole and metronidazole should not contribute significantly to the eradication of *C. pylori* in the setting of DU. They are only moderately active against *C. pylori in vitro* with an MIC 90 of 8–16 μg/ml. Naturally occurring resistance has been documented [50] and, as a monotherapy, the number of resistant strains have been shown to increase dramatically with virtually no eradication [50, 68]. These agents are notorious for causing abdominal upset when used to treat other infections. However, metronidazole is known to be an effective ulcer healing agent [10], the mechanism of which is difficult to explain unless metronidazole is active against *C. pylori in vivo*.

Although synergy between CBS and metronidazole was found in only 4/12 strains tested [45], the nitroimidazoles have, to date, proven the most successful adjuvant to CBS in ulcer healing and *C. pylori* eradication. Comparing 8 weeks' CBS treatment alone with CBS plus 10 days of tinidazole, Goodwin *et al.* showed that the combination therapy did not improve the rate of DU healing but significantly increased the anti-*C. pylori* action. Eradication occurred in 7/22 in the CBS group versus 20/27 in the latter group. Furthermore, there was little change in the resistance profile, since virtually no strains remained to develop resistance. Not surprisingly the relapse rate in the combination treatment group was considerably less than in the CBS group [50].

Metronidazole 250 mg four times daily has been combined with bismuth subsalicylate with an initial suppression rate of 100% in five patients [74]. We found metronidazole 200 mg three times daily for 1 week plus CBS for 6 weeks numerically, but not significantly, superior to CBS alone in suppressing *C. pylori* in 25 DU patients [73].

In another study nearing completion, using metronidazole 400 mg three times daily for 1 weeks plus CBS for 6 weeks, 83% of 23 DU patients were culture-negative 4 weeks after the completion of therapy.

If further studies confirm these initial encouraging results, nitroimidazoles may yet prove to be the most appropriate adjuvant agent to CBS in the treatment of DU disease.

Other antibiotics

While numbers are as yet small, antibiotic monotherapy (with/without another inactive agent) has yielded disappointing results. With the exception of furazolidone (which has only moderate *in vitro* activity), none of the antibiotics achieved eradication rates comparable to CBS [67].

Both of the quinolones tested were ineffective, but ofloxacin and ciprofloxacin also readily induced the development of resistance in most strains tested [68].

Even fewer studies have, to date, addressed bismuth adjuvant chemotherapy with these antibiotics. However, encouraging results have come from one open study of erythromycin plus bismuth therapy with an eradication rate of 65% being achieved in 17 patients [50].

Nitrofurantoin combined with bismuth in one small study (6 patients) yielded an unimpressive eradication rate of 0/6 [75]. Bismuth plus ofloxacin in 14 patients was no more effective than could be expected of CBS alone [67]. Ofloxacin is relatively inactive, however, at a low pH and, unlike at least one of the newer quinolones (norfloxacin), exhibits no synergism with CBS *in vitro* [45].

Oral triple therapy

As treatment becomes more aggressive the risks increase. Thus, in treating just 100 patients Borody *et al.* found minor side-effects (e.g. nausea) to occur in over 30% and potentially lethal side-effects (*Clostridium difficile*-positive diarrhoea) in one [49]. However, the results with regard to *C. pylori* eradication and reduction of DU relapse in the two studies published to date have been impressive.

In the first study, 100 consecutive *C. pylori*-positive patients with DU or non-ulcer dyspepsia were treated with CBS 120 mg four times daily and tetracycline 500 mg four times daily (92 patients) or amoxycillin 250 mg four times daily (8 patients) for 4 weeks. Metronidazole 200 mg four times daily was also given for the first 2 weeks. Four weeks after the end of treatment, 94% of patients were culture-negative and a significant reduction of both acute and chronic gastritis was found. 94% of 50 reviewed over a year later remained *C. pylori*-negative. Of 21 previous DU patients reviewed, only three had ulcer relapse, all of whom were culture-positive at follow-up [49].

Borsch *et al.* reported their experience in *C. pylori*-positive patients treated with bismuth subsalicylate 600 mg three times daily, amoxycillin 500 mg three times daily and metronidazole 500 mg three times daily for 14 days. Review 1 month after therapy revealed that 9/10 were culture-negative [76].

The duration of highly toxic triple therapy in these studies was at least 2 weeks. Clearly, if such aggressive therapy proves necessary to consistently eradicate *C. pylori*, it will restrict *C. pylori* therapy to those with 'problem' ulcers. However, it may well be that shorter courses, e.g. 7 days' triple therapy, will be sufficient to eradicate the organism and only the bismuth need be continued to heal the ulcer. Studies of short duration triple therapy are now underway.

Topical antibiotic therapy

An interesting possibility, that topical antibiotics may succeed where larger doses given orally have failed, has been raised by Graham *et al.* [77]. They treated three *C. pylori*-positive patients with furazolidone suspension 7 mg seven times per day, having abolished acid secretion with famotidine. Two of these patients developed normal urea breath tests during therapy. Even if this initial attempt is unsuccessful, the concept is of interest since the high risk of side-effects from the currently most effective oral regimens (triple therapy) would be avoided.

Further studies will undoubtedly address the possibility of topical antibiotics as an adjuvant to CBS. Acid suppression is more likely to favour the continued growth of *C. pylori* since this organism is readily inactivated in an acid environment. It will also be important to study gastric pit antibiotic concentrations after topical doses to ensure that bactericidal concentrations are achieved in this important site.

Antacids

In a recent report Berstad *et al.* detailed their findings on the effects of a magnesium–aluminium antacid, pirenzepine, and placebo on *C. pylori* and gastric histology in patients with non-ulcer dyspepsia [78]. Of 88 patients studied, 47% had histological gastritis before treatment but only 20 were *C. pylori*-positive at that time. Antacid treatment, but not pirenzepine or placebo therapy, reduced *C. pylori* infection (*C. pylori* not detected in 5/10 post-treatment specimens) but increased histological gastritis. It is speculated that, far from being cytoprotective, this antacid is a chemical irritant.

This paper raises two important points: firstly, do antacids, often allowed liberally to DU patients during studies of *C. pylori* and healing agents, significantly bias results; and secondly, why was there no reduction in histological gastritis in parallel with the reduction of *C. pylori* infection.

The effect of antacids on *C. pylori* has not been the subject of detailed study to date. However, available data suggest that aluminium-based antacids are inactive against *C. pylori*. The MIC 90 of $Al(OH)_3$ in one *in vitro* study was >3200 μg/ml [79], and in one clinical study involving 10 patients aluminium phosphate was found to be ineffective against *C. pylori* [68]. 'Various' antacids have been found to have MIC 90s ranging between 40 and 1280 μg/ml [79], the lower levels of which are certainly achievable for a short period *in vivo*. Studies must now be undertaken to ascertain the MIC 90 for magnesium-based antacids and their *in vivo* activity. Until the result of these are available, magnesium-based antacids should be specifically avoided during studies of the effects of other agents on *C. pylori*.

Despite this, it is unlikely that incidental antacid consumption has undermined the results of previous studies on *C. pylori* eradication. CBS has been consistently shown to be superior to cimetidine in this regard and antacid consumption is always lower in the groups receiving CBS.

Conclusions

The treatment of DU has entered a new and exciting phase. While the presence of 'acid' remains a pre-requisite for ulcer development it is clear that a relative acid excess is not the sole aetiological agent in DU disease. Cytoprotective bismuth compounds provide an alternative mode of therapy known to be as effective as H_2 antagonists in initial healing but with reduced relapse rates. Finally, *C. pylori* may well be of aetiological significance in DU disease and would thus provide a rational basis for treatment with the possible elimination of DU relapse.

References

1 Boyd, E.J.S., Wilson, J.A. & Wormsley, K.G. 1984. Recurrent ulcer disease. In Misiewicz, J.J. & Woods, J.R. (eds.) *Ranitidine Therapeutic Advances*. Excerpta Medica, Amsterdam. pp. 14–42.

2 Feston, H.P., Lamers, C.B., Tangerman, A. & Van Tongeran, J.H.M. Effect of treatment with cimetidine for one year on gastrin cell and parietal cell function and sensitivity to cimetidine in patients with duodenal or gastric ulcers. *Postgrad. Med. J.* 1980, **56**, 698–701.

3 Armstrong, C.P. & Blower, A.L. Non-steroidal anti-inflammatory drugs and life threatening complications of peptic ulcer disease. *Gut* 1987, **28**(5), 527–32.

4 Welch, C.E., Rodkey, G.V. & von Ryll Grysku, P. A thousand operations for ulcer disease. *Ann. Surg.* 1986, **204**(4), 454–67.

5 Flemstrom, G. & Turnberg, L.A. Gastroduodenal defence mechanisms. *Clin. Gastroenterol.* 1984, 13(2), 327–54.

6 Ippoliti, A., Elashoff, J., Valenzuela, J., Cano, R., Frankl, H., Samloff, M. & Koretz, R. Recurrent ulcer after successful treatment with cimetidine or antacid. *Gastroenterol.* 1983, **85**, 875–80.

7 Freedberg, A.S. & Barron, L.E. The presence of spirochetes in human gastric mucosa. *Am. J. Dig. Dis.* 1940, 7, 433–45.

8 Zheng, Z.T., Wang, Z.Y., Chu, Y.X., Li, Y.N., Li, Q.F., Lin, S.R. & Xu, Z.M. Double blind short-term trial of furazolidone in peptic ulcer. *Lancet* 1985, **i**, 1048–9. (Letter)

9 Zhao, H.Y., Li, G., Guo, J., Yan, Z., Sun, S., Li, L., Duan, Y. & Yue, F. Furazolidone in peptic ulcer. *Lancet* 1985, **ii**, 276–7. (Letter)

10 Quintero Diaz, M. & Sotto Eschobar, A. Metronidazole versus cimetidine in the treatment of gastroduodenal disease. *Lancet* 1986, **i**, 907.

11 Bader, J.P. The safety profile of De-Nol. *Digestion*

1987, **37** (Suppl 2), 53–9.

12 Salmon, P.R., Brown, P., Williams, R. & Read, A.E. Evaluation of colloidal bismuth (De-Nol) in the treatment of duodenal ulcer employing endoscopic selection and follow-up. *Gut* 1974, **15**, 189–93.

13 Moshal, M.G. The treatment of duodenal ulcers with TDB: a duodenoscopic double blind cross-over investigation. *Postgrad. Med. J.* 1975, **51** (Suppl 5), 36–40.

14 Shreeve, D.R. A double blind study of tripotassium dicitrato bismuthate in duodenal ulcer. *Postgrad. Med. J.* 1975, **51**(Suppl 5), 33–6.

15 Vissoulis, C., Vissoulis, G., Golfis, G. & Psimenos, G. Endoscopic evaluation of TDB in the treatment of duodenal ulceration. *Iatrica Chronica* 1977, **17**, 157–64.

16 Coughlin, G.P., Kupa, A. & Alp, M.H. The effect of tripotassium dicitrato bismuthate (De-Nol) on the healing of chronic duodenal ulcers. *Med. J. Aust.* 1977, **1**, 294–8.

17 Connon, J.J. De-Nol, an effective drug in the therapy of duodenal ulceration. *J. Ir. Med. Assoc.* 1977, **70**(6), 206–7.

18 Lee, S.P. & Nicholson, G.I. Increased healing of gastric and duodenal ulcers in a controlled trial using tripotassium dicitrato bismuthate. *Med. J. Aust.* 1977, **1**, 808–12.

19 Poulantzas, J., Polymeropoulos, P.S. & Papasomatious, A. A double blind evaluation of the effect of tripotassium dicitrato bismuthate in peptic ulcer. *Br. J. Clin. Pract.* 1978, **32**, 147–8.

20 Dekker, W. & Reisma, K. Double blind controlled trial with colloidal bismuth subcitrate in the treatment of symptomatic duodenal ulcers with special reference to blood and urine bismuth levels. *Ann. Clin. Res.* 1979, **11**, 94–7.

21 Bank, S., Marks, I.N., Novis, B.H. & Barbezat, G.O. Colloidal bismuth tablets (tripotassium dicitrato bismuth) in the treatment of peptic ulceration. *Gastroenterol.* 1979, **76**, 1093.

22 Wilson, P. & Alp, M.H. Colloidal bismuth subcitrate tablets and placebo in chronic duodenal ulceration: a double blind randomised trial. *Med. J. Aust.* 1982, **1**, 222–3.

23 Tytgat, G.T.N. Colloidal bismuth subcitrate in peptic ulcer disease — a review. *Digestion* 1987, **37**(Suppl 2), 31–41.

24 Bianchi Porro, G., Petrillo, M., De Nichola, C. & Lazzaroni, M. A double blind endoscopic study with De-Nol tablets and cimetidine for duodenal ulcer. *Scand. J. Gastroenterol.* 1984, **19**, 905–8.

25 Kang, J.Y. & Piper, D.W. Cimetidine and colloidal bismuth subcitrate in the treatment of chronic duodenal ulcer, comparison of initial healing and recurrence after healing. *Digestion* 1982, **23**, 73–9.

26 Martin, D.F., Hollanders, D., May, S.J., Ravenscroft, M.M., Tweedle, D.E.F. & Miller, J.P. Difference in relapse rates of duodenal ulcer after healing with cimetidine or tripotassium dicitrato bismuthate. *Lancet* 1981, **i**, 7–10.

27 Moshal, M.G., Spitaels, J.M. & Khan, F. Tripotassium dicitrato bismuthate chewing tablets and cimetidine tablets in the treatment of duodenal ulcers — a double-blind double-dummy comparative study. *S.A. Med. J.* 1981, **60**, 420–3.

28 Ward, M., Pollard, E.J. & Cowen, A. A double-blind trial of cimetidine versus tripotassium dicitrato bismuthate in chronic duodenal ulceration. *Med. J. Aust.* 1981, **1**, 363–4.

29 Bianchi Porro, G., Barbara, L., Cheli, R., Dal Monte, P.R. & Mazzacca, G. Comparison of tripotassium dicitrato bismuthate (TDB) and ranitidine in healing and relapse of duodenal ulcers. *Gut* 1984, **25**, A565.

30 Ward, M., Halliday, C. & Cowen, A.E. A comparison of colloidal bismuth subcitrate tablets and ranitidine in the treatment of chronic duodenal ulcers. *Digestion* 1986, **34**, 173–7.

31 Lee, F.I., Samlof, I.M. & Hardman, M. Comparison of tripotassium dicitrato bismuthate tablets with ranitidine in healing and relapse of duodenal ulcers. *Lancet* 1985, **i**, 1299–1301.

32 Lam, S.K., Lee, N., Koo, J., Hui, W., Fok, K. & Ng, M. Randomised cross-over trial of tripotassium dicitrato bismuthate versus high dose cimetidine for duodenal ulcers resistant to standard doses of cimetidine. *Gut* 1984, **25**, 703–6.

33 Bianchi Porro, G., Parante, F., Lazzaroni, M. & Pace, F. Colloidal bismuth subcitrate and two different doses of cimetidine in the treatment of resistant duodenal ulcers. *Scand. J. Gastroenterol.* 1986, **21**(Suppl 122), 39–41.

34 Dooley, C.P., McKenna, D., Humphreys, H., Bourke, S., Keane, C.T., Sweeney, E. & O'Morain, C. Histological gastritis in duodenal ulcer: relationship to *Campylobacter pylori*, and effect of therapy. *Am. J. Gastroenterol.* 1988, **83**(3), 278–81.

35 Rokkas, T., Pursey, C., Simmons, N.A., Filipe, M.L. & Sladen, G.E. Non-ulcer dyspepsia and colloidal bismuth subcitrate: the role of *Campylobacter pyloridis*. *Gastroenterol.* 1987, **92** (No. 5, Part 2), 1599.

36 Smith, A.C., Price, A.B., Borriello, P. & Levi, A.J. A comparison of ranitidine and tripotassium dicitrato bismuthate (TDB) in relapse rates of duodenal ulcer. The role of *Campylobacter pylori* (CP). *Gastroenterol.* 1988, **94** (No. 5, Part 2), A431.

37 Hamilton, I., O'Connor, H.J., Wood, N.C., Bradbury, I. & Axon, A.T.R. Healing and recurrence of duodenal ulcer after treatment with tripotassium dicitrato bismuthate (TDB) tablets or cimetidine. *Gut* 1986, **27**, 106–10.

38 Coghlan, J.G., Gilligan, D., Humphreys, H., McKenna, D., Dooley, C., Sweeney, E., Keane, C. & O'Morain, C. *Campylobacter pylori* and recurrence of duodenal

ulcers — 12 months follow-up study. *Lancet* 1987, **ii**, 1109–11.

39 Shreeve, D.R., Klass, H.J. & Jones, P.E. Comparison of cimetidine and tripotassium dicitrato bismuthate in healing and relapse of duodenal ulcers. *Digestion* 1983, **28**, 96–101.

40 Bianchi Porro, G., Lazzaroni, M., Petrillo, M. & De Nicola, C. Relapse rates in duodenal ulcer patients formerly treated with bismuth subcitrate or maintained with cimetidine. *Lancet* 1984, **ii**, 698. (Letter)

41 Bianchi Porro, G. & Lazzaroni, M. *Campylobacter pylori* and ulcer recurrence. *Lancet* 1988, **i**, 593. (Letter)

42 Dobrilla, G., Vallaperta, P. & Amplatz, S. Influence of ulcer healing agents on ulcer relapse after discontinuation of acute treatment: a pooled estimate of controlled clinical trials. *Gut* 1988, **29**, 181–7.

43 Fitzgerald, O. & Murphy, P. Studies on the physiological chemistry and clinical significance of urease and urea with special reference to the stomach. *Ir. J. Med. Sci.* 1950, **292**, 99–153.

44 McNulty, C., Dent, J. & Wise, R. Susceptibility of clinical isolates of *Campylobacter pyloridis* to 11 antimicrobial agents. *Antimicrobial Agents & Chemotherapy* 1985, **28**(6), 837–8.

45 Van Ceakenberg, D. & Breyssens, J. *In vitro* synergistic activity between bismuth subcitrate and various antimicrobial agents against *Campylobacter pyloridis* (*C. pylori*). *Antimicrobial Agents & Chemotherapy* 1987, **31**(9), 1429–30.

46 Goodwin, C., Blake, P. & Glincow, E. The minimum inhibitory and bactericidal concentrations of antibiotics and anti-ulcer agents against *Campylobacter pyloridis*. *J. Antimicrobial Chemother.* 1986, **17**, 309–14.

47 Skoglund, M. & Watters, K. Bismuth concentration at site of *Campylobacter pylori* colonisation. *Gastroenterol.* 1988, **94**(No. 5, Part 2), A430.

48 Lambert, J.R., Borromeo, M., Korman, M.C., Hansky, J. & Eaves, E.R. Effect of colloidal bismuth (De-Nol) on healing and relapse of duodenal ulcers — role of *Campylobacter pyloridis*. *Gastroenterol.* 1987, **92**, 1489.

49 Borody, T., Cole, P., Noonan, S., Morgan, A., Ossip, G., Masey, J. & Brandl, S. Longterm *Campylobacter pylori* recurrence post eradication. *Gastroenterol.* 1988, **94**(No. 5, Part 2), A43.

50 Goodwin, C.S., Marshall, B.J., Blincow, E.D., Wilson, D.H., Blackbourn, S. & Phillips, M. Prevention of nitroimidazole resistance in *Campylobacter pyloridis* by co-administration of colloidal bismuth subcitrate: clinical and *in vitro* studies. *J. Clin. Pathol.* 1988, **41**, 207–10.

51 Langenberg, M.L., Rauws, E., Schipper, M.E., Widjojokosuo, A. & Tytgat, G.T.N. 1985. The pathogenic role of *Campylobacter pyloridis*, studied by attempts to eradicate these organisms. In Pearson, A.D., Skirrow, M.B., Lior, H. & Rowe, B. (eds.) *Campylobacter III.* Proceedings of the Third International Workshop on Campylobacter Infections. Ottawa 1985. PHLS, London. pp. 162–3.

52 Lee, S.P. A potential mechanism of action of colloidal bismuth subcitrate: diffusion barrier to hydrochloric acid. *Scand. J. Gastroenterol.* 1982, **17**(Suppl 80), 17–21.

53 Konturek, S.J., Radecki, R., Piastucki, I. & Drozdowicz, D. Studies on the gastroprotective and ulcer healing effects of colloidal bismuth subcitrate. *Digestion* 1987, **37**(Suppl 2), 8–15.

54 Konturek, S.J., Bilski, J., Kwiecien, N., Obtulowicz, W., Kopp, B. & Oleksy, J. De-Nol stimulates gastric and duodenal alkaline secretion through a prostaglandin dependent mechanism. *Gut* 1987, **28**, 1557–63.

55 Ahmed, A., Cairns, S.R., Varia, D., Hoult, J.R.S. & Salmon, P.R. Colloidal bismuth subcitrate (De-Nol) reduces leukotriene C4 formation in duodenal ulcers (DU). *Gastroenterol.* 1988, **94**(No. 5, Part 2), A.3.

56 Baron, J.H., Barr, J., Batten, J., Sidebotham, R. & Spencer, J. Acid, pepsin and mucin secretion in patients with gastric and duodenal ulcers before and after colloidal bismuth subcitrate (De-Nol). *Gut* 1986, **27**, 486–90.

57 Lane, M., Lee, S., Tasman-Jones, C. & Nicholson, G. Tripotassium dicitrato bismuthate tablets versus liquid in the treatment of duodenal ulcers. *NZ Med. J.* 1985, **98**, 191–2.

58 Hamilton, I. & Axon, A.T.R. Controlled trial comparing De-Nol tablets with De-Nol liquid in treatment of duodenal ulcer. *Br. Med. J.* 1981, **282**, 362.

59 Kellow, J.E., Barr, G.D., Middleton, W.R.J. & Piper, D.W. Comparison of colloidal bismuth subcitrate tablets and liquid in duodenal ulcer healing. *J. Clin. Gastroenterol.* 1983, **5**, 417–20.

60 Dekker, W., Dal Monte, P.R., Bianchi Porro, G., Van Bentem, N., Boekhorst, J.C., Crowe, J.P., Robinson, T.J., Thys, O. & Van Driel, A. An international multiclinic study comparing the therapeutic efficacy of colloidal bismuth subcitrate coated tablets with chewing tablets in the treatment of duodenal ulceration. *Scand. J. Gastroenterol.* 1986, **21**(Suppl 122), 46–50.

61 Coghlan, J.G., Gilligan, D., Humphreys, H., McKenna, D., Sweeney, E., Keane, C. & O'Morain, C. Efficacy of different dosage regimes in duodenal ulcer healing and eradication of *Campylobacter pylori*. *Gut* 1987, **28**, A1409.

62 Hollanders, D. Twice daily tripotassium dicitrato bismuthate in the treatment of duodenal ulceration. *Postgrad. Med. J.* 1986, **62**, 19–21.

63 Lazzaroni, M., Parante, F., Prada, A. & Bianchi Porro, G. Colloidal bismuth subcitrate as coated tablets, four

times versus twice daily dosage in duodenal ulcer. *Scand. J. Gastroenterol.* 1986, **21** (Suppl 122), 51–53.

64 Lambert, J.R., Way, D.J., King, R.G., Eaves, E.R. & Hansky, J. Bismuth pharmacokinetics in the human gastric mucosa. *Gastroenterol.* 1988, **94** (No. 5, part 2), A248.

65 Lambert, J.R., Borromeo, M., Eaves, E.R., Hansky, J. & Korman, M. Efficacy of different dosage regimes of bismuth in eradicating *Campylobacter pylori. Gastroenterol.* 1988, **94**(No. 5, Part 2), A248.

66 Marshall, B., McGechie, D., Rogers, P. & Glancy, R. Pyloric Campylobacter infection and gastroduodenal disease. *Med. J. Aust.* 1985, **142**, 439–44.

67 Bayerdorffer, E. & Ottenjann, R. The role of antibiotics in *Campylobacter pylori* associated peptic ulcer disease. *Scand. J. Gastroenterol.* 1988, **23**(Suppl 142), 93–100.

68 Hirshl, A.M., Hentschel, E., Schutze, K., Nemec, H., Potzi, R., Gangl, A., Weiss, W., Pletschette, M., Stanek, G. & Rotter, M.L. The efficacy of antimicrobial treatment in *Campylobacter pylori* associated gastritis and duodenal ulcer. *Scand. J. Gastroenterol.* 1988, **23**(Suppl 142), 76–81.

69 Armstrong, J.A., Wee, S.H., Goodwin, C.S. & Wilson, D.H. Response of *Campylobacter pyloridis* to antibiotics, bismuth and acid reducing agents *in vitro* — an ultrastructural study. *J. Med. Microbiol.* 1987, **24**, 343–50.

70 Hollingsworth, J.A., Goldie, J., Silletti Li, Y., Richardson, H. & Hunt, R.H. Gastric secretion of antibiotics used for *Campylobacter pyloridis. Gut* 1987, **28**, A1409.

71 Rauws, E.A., Langenberg, W., Houtoff, H., Zanen, H. & Tytgat, G.N.T. *Campylobacter pyloridis* associated chronic active antral gastritis. *Gastroenterol.* 1988, **94**, 33–41.

72 Tytgat, G.N.T., Rauws, E.A., Langenberg, W. & Houtoff, H. Long term follow-up of *Campylobacter pylori* associated gastritis after treatment with colloidal bismuth subcitrate and/or amoxycillin. *Gastroenterol.* 1988, **94** (No. 5, Part 2), A469.

73 O'Morain, C., Coghlan, G., Tobin, A., Casey, E., Hutchinson, L., Gilligan, D., McKenna, D., Sweeney, E. & Keane, C. Adjuvant antibiotic therapy and colloidal bismuth subcitrate in duodenal ulcer healing and the eradication of *Campylobacter pylori. Gastroenterol.* 1988, **94**(No. 5, Part 2), A334.

74 Murphy, R., Bahler, E.M., Appelman, H.D., Barnett, J.L. & Elta, G.H. Four year follow-up of gastroduodenitis and its association with *Campylobacter pylori. Gastroenterol.* 1988, **94**(No. 5, Part 2), A317.

75 Borsch, G., Mai, U. & Muller, K.M. Monotherapy or polychemotherapy in the treatment of Campylobacter related gastroduodenal disease. *Scand. J. Gastroenterol.* 1988, **23**(Supp 142), 101–6.

76 Borsch, G., Mai, U. & Opperkuch, W. Oral triple therapy (OTT) may effectively eradicate *Campylobacter pylori* (CP) in man: a pilot study. *Gastroenterol.* 1988, **94**(No. 5, Part 2), A44.

77 Graham, D.Y., Klein, P.D., Opekun, A.R., Boutton, Y.W., Evans, D.J., Alpert, L.C., Michaletz, P.A., Yoshimura, H.H. & Adam, E. Search for an effective and safe therapy for *Campylobacter pylori* infection. *Gastroenterol.* 1988, **94**(No. 5, Part 2), A152.

78 Berstad, A., Alexander, B., Weberg, R., Serck-Hanssen, A., Holland, S. & Hirschowitz, B.I. Antacids reduce *Campylobacter pylori* colonisation without healing the gastritis in patients with non-ulcer dyspepsia and erosive prepyloric changes. *Gastroenterol.* 1988, **95**, 619–24.

79 Andersen, L.P. Cytoprotective agents and *C. pylori* associated acid peptic disease. *Scand. J. Gastroenterol.* 1988, **23**(Suppl 142), 110–13.

30 Campylobacter pylori and CLO in animals: overview of mucus-colonizing organisms

A. LEE

To the microbial ecologist there is no more satisfying and challenging ecosystem than the gastro-intestinal tract. The contents of the human large bowel contain > 10^{11} bacteria/g with > 200 different species living in homeostatic balance. Over the period of evolution these organisms have adapted to the many ecological niches that are present in the gut [1]. The stomach has not featured strongly in discussions on intestinal ecology as it was always assumed to be mostly barren due to the antibacterial effects of gastric acidity. Some populations of gastric bacteria have been described, for example the acid tolerant lactobacilli and the yeast *Torulopsis* in the rodent stomach [2]. The isolation of *Campylobacter pylori* and the subsequent debate as to its role in gastro-duodenal disease has focused interest on the stomach as a bacterial habitat. Close examination of gastric biopsies [3] revealed that *C. pylori* was inhabiting a niche that was well protected from stomach acid, i.e. the organism colonizes the gastric mucus.

As the mucus lining the intestinal tract of all animals studied contains large numbers of highly specialized bacteria, it is not really surprising that stomach mucus provides a haven for similar bacteria. The difference with *C. pylori* is that the bacterium causes a histopathology in all persons colonized, albeit mild; this is not the case with the normal mucus-associated bacteria in other areas of the intestine. There are other pathogens belonging to this broad group of mucus-associated bacteria, for example *Campylobacter jejuni* colonizes intestinal mucus [4] as does the pig pathogen *Treponema hyodysenteriae* [5].

Spiral/helical-shaped bacteria in the stomachs of animals other than humans

Even though the focus of this book is *C. pylori*-associated disease in humans, it is appropriate to consider the inhabitants of the same ecological niche in other animal species. Spiral-shaped bacteria were seen in the stomach of animals late last century. Rappin [6] is credited with the first observation in 1881 followed by Bizzozero [7] in 1893 and Salomon [8] in 1896. A collage of the diagrams from some of these early papers of gastric bacteria in animals is shown in Fig. 30.1. It is clear that the organisms, even though they had a spiral morphology, were very different from *C. pylori*. They tended to be longer, consisted of a number of very tight spirals and were more helical than spiral. (Even though many of the bacteria discussed in this chapter are more helical in shape than spiral, the term 'spiral' has been used throughout to describe both spiral- and helically-shaped organisms as this is the description most commonly found in the literature.) Salomon's early study was the most extensive investigation of spiral-shaped bacteria in the stomach of animals; he saw large numbers in dogs, cats and the brown Norway rat but failed to find them in humans, apes, monkeys, cattle, pigs, mice, pigeons, crows and even owls! Dubosq & Lebailly [9] in 1912 found them in a fox. The most studied animals have been the cat and dog. Weber *et al.* [10, 11] examined the stomachs of 15 dogs, 12 cats and 4 kittens. Large numbers of these bacteria were seen in the fundic glands of the dogs and cats but none in the kittens. Organisms were located principally in the glandular lumina adjacent

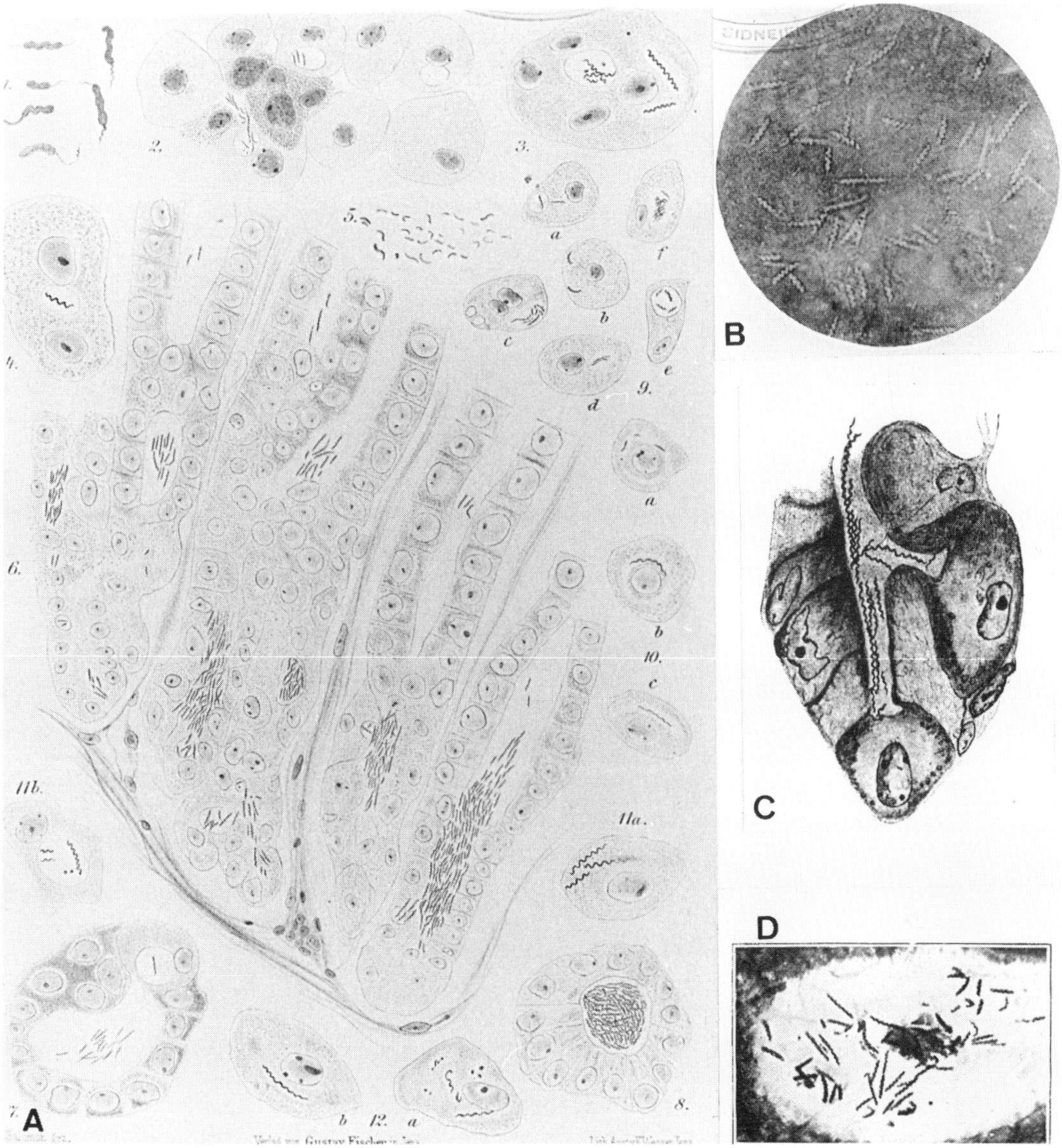

Fig. 30.1 A collage of early illustrations of non-*C. pylori* spiral-shaped bacteria in the stomachs of a number of animals. (A) Bacteria associated with the gastric mucosa and individual cells of the stomachs of mice which were colonized with the spiral organism from the dog stomach [8]; (B) a hanging-drop preparation of mucus from the stomach of a cat [8]; (C) bacteria in the stomach of the dog [34]; (D) section of the pylorus of a cat showing bacteria in the lumen of a duct [27].

to parietal cells, within the parietal cells, and large numbers in the gastric pits within and beneath the mucus. In an elegant electron microscopic study, Lockard & Boler [12] in 1970 noted three distinct morphological types in the gastric mucosa of the dog. Reproductions of their original pictures are shown in Fig. 30.2 as the characteristic mor-

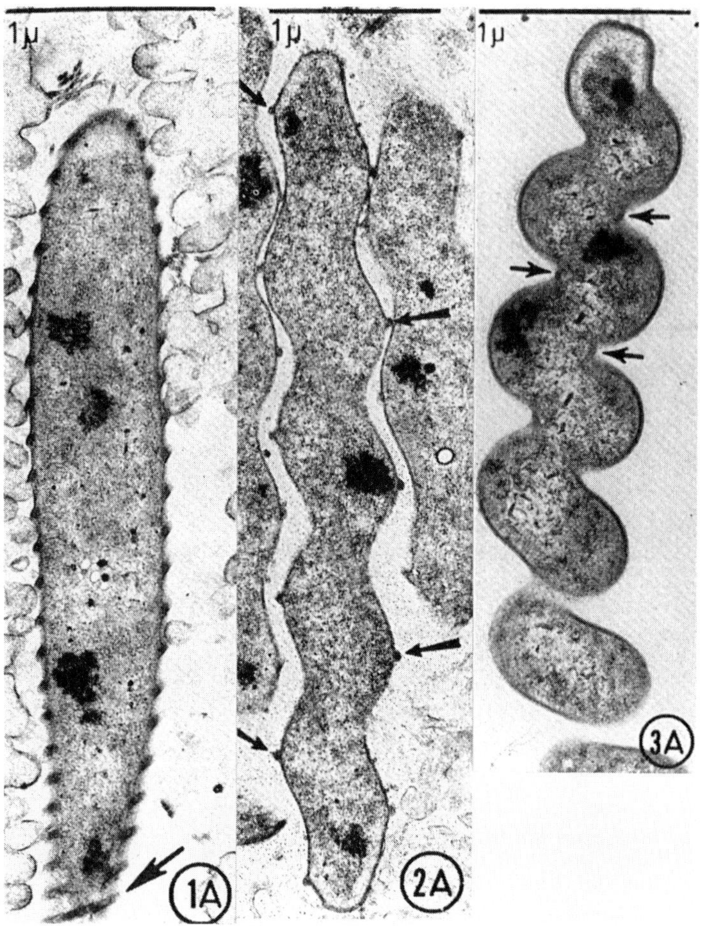

Fig. 30.2 A collage of figures taken from the work of Lockard & Boler [12] showing the three very different bacterial morphologies observed in the dog gastric mucosa. (1A) Straight form with a tightly-coiled periplasmic fibril wound around the body — cross-sections of the fibril appear as knobs along the periphery of the organism; (2A) a spiral form with the periplasmic fibril spread out along the body of the cell; (3A) a tightly coiled bacterium. The arrows in the original paper are supposed to indicate the fibril or filament as the authors called it 'in the valley of each spiral wave of the body'. (From Lockard, V.G. & Boler, R.K. Ultrastructure of a spiraled microorganism in the gastric mucosa of dogs, *American Journal of Veterinary Research*, 31, 1453–1462.)

phologies seen are relevant to later discussions. Salomon [8] in his early study had also reported three simultaneously occurring forms of spirilla as had Kasai & Kobayashi [13] in 1919. The main differences between these forms were in the tightness of the coils, the presence or absence of tapering ends and variations in body diameter. A study by Henry *et al.* [14] in 1987, showed these same three morphologies to be present in all of the 30 beagle dogs studied and it was concluded that these bacteria should be considered part of the natural gastric flora of dogs.

The stomach of the cat has a normal surface-associated microflora of spiral-shaped bacteria. Examples of the bacteria in cats and dogs are shown in Figs. 30.3 and 30.4. Gastric bacteria found in normal ferrets and monkeys have also been described. As further studies are done, other animals with a normal gastric flora of spiral organisms are likely to be found.

Pathology associated with non C. pylori spiral-shaped bacteria

All humans infected with *C. pylori* have an associated gastritis. To understand the basis of the pathogenicity of this organism it is important to consider whether the natural inhabitants of the stomachs of animals other

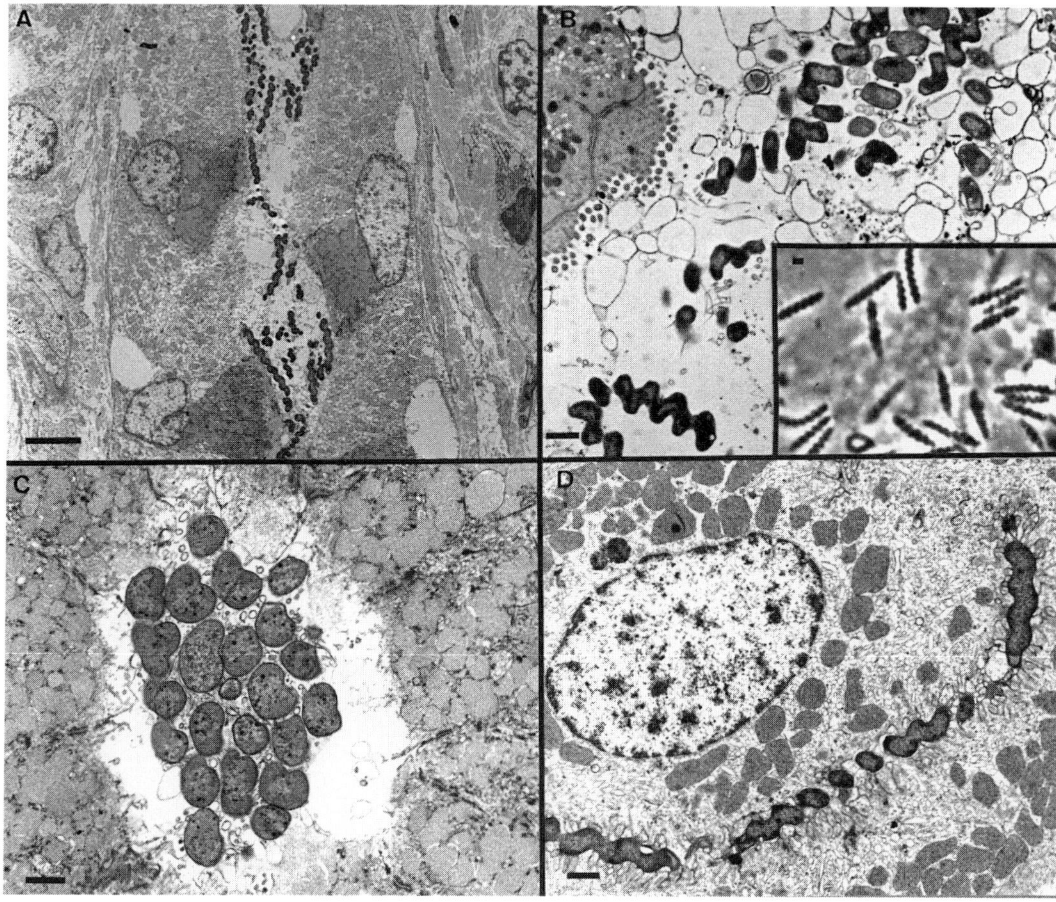

Fig. 30.3 Spiral-shaped bacteria in the gastric mucosa of normal cats. (A) Bar = 1 μm. (B) Bar = 1 μm. Inset is a scraping of cat gastric mucus viewed under phase-contrast microscopy. (C) Cross-section of a clump of cells (bar = 0.5 μm). (D) Four bacteria that have moved up into a caniculus of a parietal cell (bar = 1 μm).

than humans are associated with any obvious pathology. To date, there has been virtually no description of any disease linked to the presence of non-*C. pylori* spiral bacteria in animals other than humans. Hugo Salomon [8] reported no leukocytic response to be present, even in cases where there were large accumulations of spirilla. It is ironic that in the second edition of *Canine Medicine*, in pre-*C. pylori* days, it was reported that spirilla were present in the vomitus of dogs and cats with signs of gastritis [15]. The one report of associated pathology was a recent study on gastric spirillosis in beagles in which it was noted that bacteria were most numerous in the cardia and in the region of the fundic–pyloric junction, and it was in these areas of dense bacterial association that lymphoreticular hyperplasia was noted [14]. This may be equivalent to the chronic component of *C. pylori*-associated gastritis. Could it be that the stomach is an antigenically virgin site due to the protective effect of the gastric mucus? Bacteria which have evolved mechanisms to penetrate the mucus barrier stimulate a significant local immune response. In experiments with germ-free piglets administered *C. pylori* the only pathology noted was a very significant mononuclear response [16].

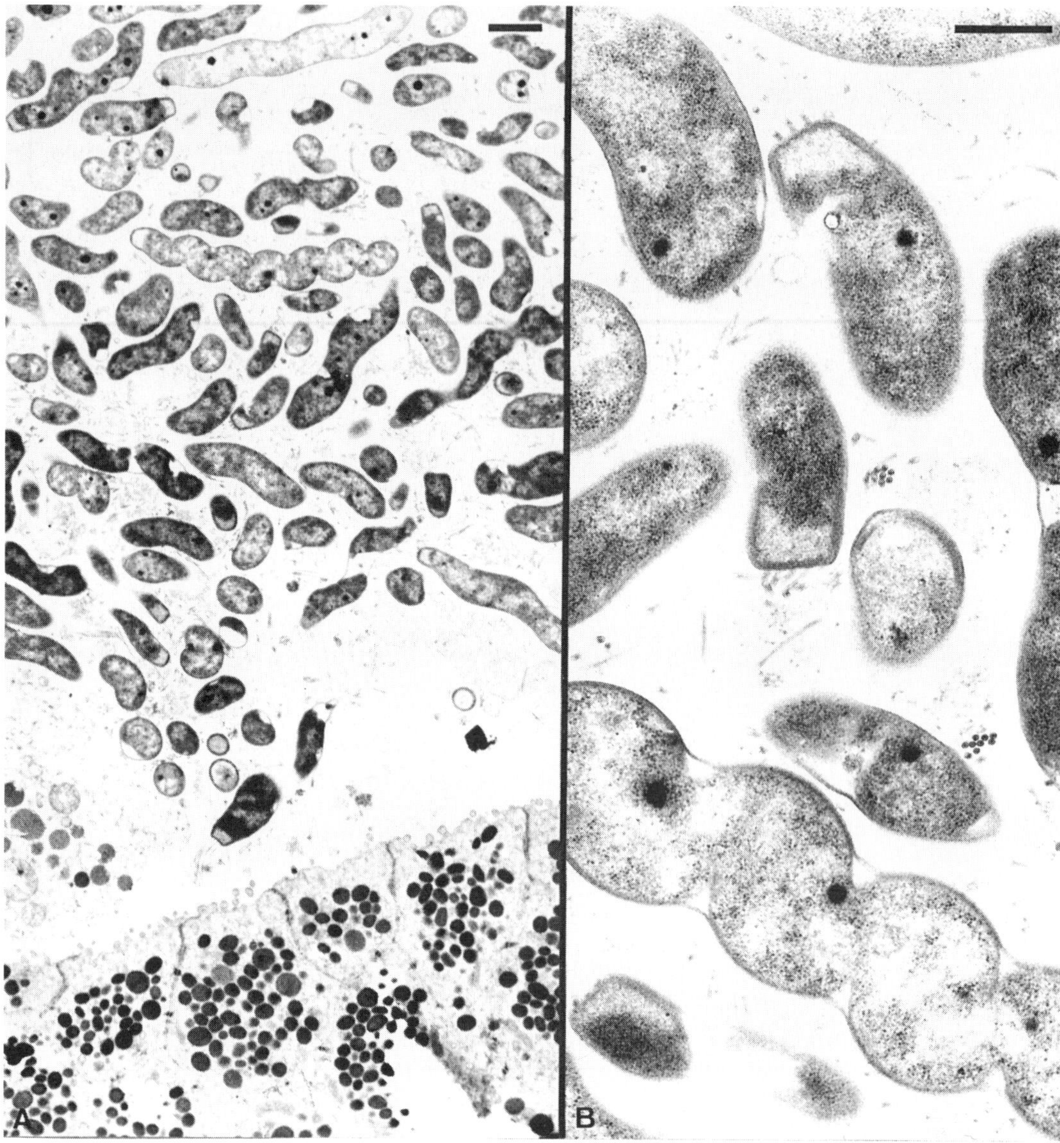

Fig. 30.4 Spiral-shaped bacteria in the gastric mucosa of a dog showing at least two morphological types. (A) Bar = 4 μm. (B) Bar = 1 μm.

Similarities and differences between bacteria colonizing the stomach

Under the light microscope the bacteria seen in the cat, dog and monkey stomach all look alike. The literature is confused as to how many different types there are either in one animal or in the different animal species. Thus, even though in any one animal up to three different morphologies are seen, the authors talk about 'the' organism in the cat or 'the' organism in the dog. Figure 30.2 shows the morphologies seen by Lockard & Boler [12] who hypothesize that the three forms are all of the same organism and represent stages

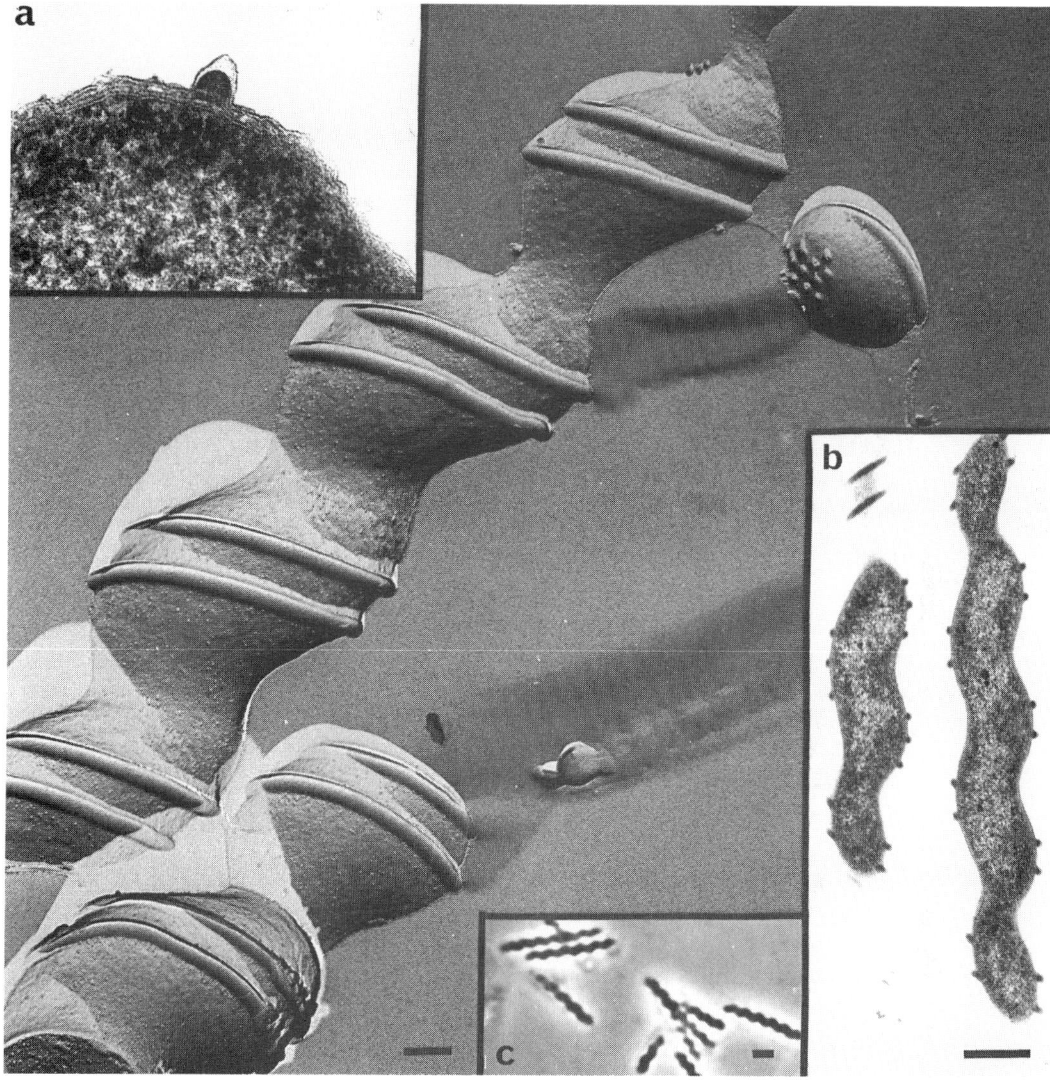

Fig. 30.5 Electron micrograph of a freeze fracture preparation of a pure culture of a spiral-shaped bacterium isolated from the stomach of a cat (bar = 0.1 μm). Inset: (a) Close-up of a section through a fibril showing the periplasmic location. (b) Transmission electron micrograph of two cells showing not only the spiral shape but also the variable location of the periplasmic fibrils (bar = 0.5 μm). (c) Pure culture of the cat isolate viewed under phase-contrast microscopy (bar = 1 μm).

in the mechanism of movement, which is accomplished by the organism changing the length of its body rather than by expansion and contraction of the filament. Recent success in our laboratory in growing animal spirals on laboratory media and in the mouse stomach has helped to clarify the situation [17]. The morphology of the spiral organism isolated from the cat stomach *in vitro* is shown in Fig. 30.5. It appears identical to Lockard & Boler's organism (Fig. 30.2, 2A) and to Henry's picture of an organism seen in the beagle dog [14]. When this isolate is fed to mice it establishes in the stomach and is seen in electron

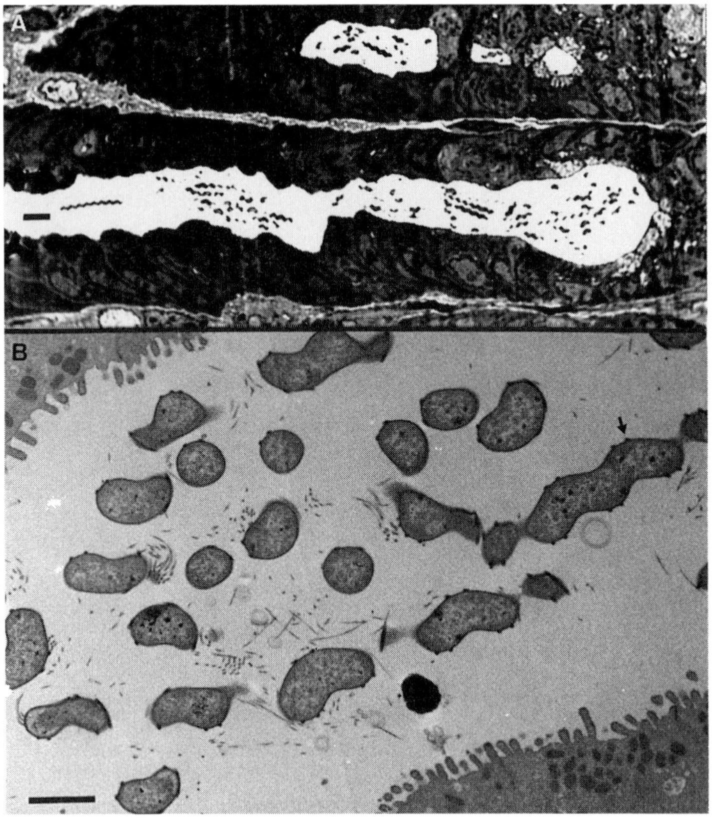

Fig. 30.6 (A) Light micrograph of a section of the glandular epithelium of a mouse stomach two weeks following oro-gastric administration of a pure culture of the spiral organism shown in Fig. 30.5 which was isolated from the antrum of a cat stomach. The spiral bacteria can be seen in the lumen of a gastric pit. Control animals showed no bacteria. (Bar = 5 μm.) (B) Transmission electron micrograph of the cat isolate in the stomach of a mouse showing the characteristic periplasmic fibrils in cross-section (arrow). (Bar = 0.5 μm.)

micrographs to have the identical morphology with the easily recognizable periplasmic fibrils (Fig. 30.6B). We have observed a second spiral organism in sections of the cat stomach viewed by electron microscopy and this is seen in Fig. 30.3. This appears to be identical to that of Lockard & Boler (Fig. 30.2, 3A). Attempts to culture the organism have been unsuccessful. However, we can maintain this organism in the laboratory by feeding cat mucus, in which it is dominant, to mice where it establishes in the stomach and is seen to have the same morphology. The conclusion from this work with mice is that the cat, dog and primate can be colonized with a variety of spirals, often more than one species being present at any one time. The species seen in the different animals may vary.

What are these bacteria? In the current literature based on morphologic studies they have been referred to as *Spirillum rappini* after Rappin [6] who apparently made the original discovery of spirilla in dogs in 1881 or *Spirillum stomachii* or *Spirillum canis*. The cultivation of one of these bacteria [17] will now allow some taxonomic studies. The distinguishing feature in this bacterium is the periplasmic fibrils that entwine the cell body (Fig. 30.5). This type of structure had been seen by us before in another spiral which we have also been successful in culturing [18]. Significantly, this organism is also a mucus colonizer whose normal habitat is the ileal crypts of the rodent. DNA analysis and 16S ribosomal RNA sequencing suggest that this rodent organism is closely related to *C. pylori* and is on the same branch of the phylogenetic tree. Recently, there have been reports of further isolations of bacteria with these distinctive periplasmic fibrils suggesting the existence of a whole new group of mucus-associated bacteria with a similar ancestory to *C. pylori*. There is the suggestion that some of these organisms may have a pathogenic potential. Romero *et al.* [19] isolated an organism with periplasmic fibrils from two humans with mild chronic gastroenteritis. Archer *et al.* [20] reported this organism to have similar biochemical reactions to *C. pylori*. The electron micrograph of the bacterium shown in their paper is exactly like that of Lockard & Boler (Fig. 30.2, 1A). Kirkbride *et al.* [21] have reported isolation of another bacterium with this same morphology from the liver and spleen of aborted sheep fetuses and Bryner [22] has given the sheep isolate to pregnant guinea-pigs where they induced abortion.

Determinants of colonization in bacteria adapted to mucus-secreting intestinal mucosa

The bowel is inhabited by a large variety of bacterial species but the number colonizing the mucosae is relatively few. These bacteria must share common properties that allow them to selectively colonize mucus-covered surfaces rather than live in the lumen. Some are likely to be general properties common to colonizers of all mucus environments while others are likely to determine the colonization of a particular location in the gut, e.g. stomach compared with small bowel. As factors influencing colonization are better understood, the more likely it becomes that rational intervention steps can be developed. As *C. pylori* colonizes only primates or gnotobiotic piglets [16, 23] the study of closely-related organisms that do grow in convenient laboratory animal models may elucidate factors influencing their colonization.

Colonizing factors shared by all organisms colonizing intestinal mucus

Spiral morphology

The one distinctive feature of all these bacteria, be they isolated from mouse colon, rat ileum, cat stomach or human stomach is their spiral morphology. We have proposed that this gives the organisms an increased ability to move in the viscous environment of mucus [24]. *In vitro* experiments with *C. jejuni* [25], *C. pylori* [3] and the mouse ileal isolate and cat stomach isolate show that they cope with viscous environments very well.

Colonizing factors of organisms preferentially colonizing stomach and small bowel mucus

Microaerophilism

The intestinal lumen is not oxygenated unlike the tissues that line it. Therefore at the interface, i.e. the intestinal mucus, one would predict a reduced oxygen tension. As organisms that have evolved to exist in a particular habitat are likely to have optimal growth requirements that reflect that environment, it is not surprising that the organisms that colonize the stomach and small bowel surfaces are microaerophilic. In the colon and rectum the spiral colonizers tend to be strictly anaerobic due to the highly reducing nature of the large bowel contents which mop up any oxygen diffusing into the mucus layer. Microaerophilism is a determinant of colonization that tends to be ignored even though it has significant survival value for these organisms. This is the probable reason why these bacteria remain on the surface and only penetrate stomach tissue to a limited degree, i.e. if the organism gets too far into the tissue the local environment is too oxygenated. Also, the host defence system of phagocytic cells is likely to be impaired in microaerophilic conditions, possibly explaining how these bacteria can survive so long in intimate contact with the host.

Urease activity

A striking feature of *C. pylori* is its remarkable urease activity which is so high that a urease test performed directly on tissue is an excellent predictor of active infection with the organism [3]. This property is not unique to *C. pylori*; the evidence to date would suggest that all the spiral organisms described above that colonize gastric mucus have high urease activity. This is certainly true for the isolate from the cat stomach shown in Fig. 30.5 and also for biopsies from adult cats, monkeys and dogs. Several theories have been proposed to explain this activity, in particular that it is a specialized adaptation to allow the bacteria to avoid the effects of stomach acidity and allow *C. pylori* some degree of acid tolerance. However as we have argued before [26], this is unlikely to be the main reason for high urease activity. The mucus colonizer *par excellence* is the helical-shaped organism we have isolated from the mouse small bowel [18]. This organism also has a very high urease activity and yet its natural habitat is the ileum which is non-acidic. There are many high urease producers that colonize the surface of the non-acidic rumen. A third explanation for the urease activity is that the stomach and small bowel cell surface are a rich source of urea which then acts as a nitrogen source for the organisms. The concept of urea as a preferred metabolite has already been proposed to explain the localization of *C. pylori* at intercellular junctions [3]. However, the bacteria colonizing other animal stomachs do not appear to have this association with cell junctions.

Colonizing factors unique to bacteria that colonize gastric mucosae

Specific attachment sites

Campylobacter pylori does demonstrate a high specificity for gastric type tissue as is shown by its association with gastric metaplasia. There have been no studies on metaplastic tissue in other animal species with respect to association with their natural gastric colonizers. The early reports on the cat and dog suggest some predilection for parietal cells and the bacteria certainly do appear to penetrate the gastric tissue more deeply including the caniculi of the parietal cells. This is illustrated dramatically in Fig. 30.3D, but no more is known as to tissue specificity. Salomon [8], Kasai & Kobayshi [13] and Weber *et al.* [11] were all successful in transmitting the gastric colonizers from one species to another, e.g. cat and dog to mouse and rabbit — although Salomon was unsuccessful in transmitting the organism when he fed a cat stomach to an owl! This lack of species specificity is in sharp contrast to *C. pylori* which appears to have a more restricted host range and only colonizes humans, primates and gnotobiotic pigs [16, 23].

Chemotaxis

Observation of wet scrapings of gastric mucosa from the dog or cat under phase-contrast microscopy is impressive because it allows one to see the remarkably active motility of the tightly spiralled normal inhabitants. As Lim observed [27] in 1920, they 'ram the cell and spiral or corkscrew furiously, occasionally giving large circular sweeps with their free ends. Suddenly they disengage themselves, turn a "somersault" and repeat the performance with the former free end'. This type of motility possibly explains how the bacteria get so deep into the tissue and into parietal cells although Lim never observed this. It is also suggestive of some type of chemotactic response. Kennedy *et al.* [5] recently demonstrated chemotaxis of *Treponema hyodysenteriae* to porcine mucus and suggested its importance in colonization of porcine mucosa.

Evidence for animal spiral-shaped bacteria colonizing the human stomach

The report by Doenges [28] in 1939 is a landmark study because it is the first paper that clearly shows that a large percentage of normal persons have many spiral bacteria in their stomachs. He looked at the stomachs of 242 persons who died in accidents and found that 103 of these were colonized. What is not often reported is that while the majority of the bacteria observed had a morphology consistent with *C. pylori*, two of the stomachs had spirals with 'sharp angulation'. He comments that a similar organism was frequently seen in monkeys (ref. [28], fig. 2). The morphology is the same as we saw in the cat and the dog (Figs. 30.3 and 30.4).

The possibility of one of the animal gastric bacteria infecting humans was not raised until a letter to the *Lancet* in July 1987 when Julie Dent, Cliodna McNulty and their colleagues [29] reported spiral organisms that were not campylobacters or spirochaetes on the gastric mucosa of 3/1300 endoscopy patients with upper gastro-intestinal symptoms. The organisms were seen in the necks of the pyloric glands on Gram-stained smears and on histological section of biopsy material of two patients. Chronic active gastritis was present in all three cases. Endoscopic examination had revealed severe oesophagitis in two cases although stomach and duodenum were normal in all three. The light and electron micrographs shown in the *Lancet* article showed the bacteria to be very similar to that seen in the cat and dog, shown in Figs. 30.2, 30.4 & 30.5. The authors were unable to culture the organism from any of these patients.

Having just succeeded in isolating the bacterium shown in Fig. 30.5 from the cat stomach, we had the opportunity to test the hypothesis that the organisms infecting the Dent patients were most likely acquired from an animal source. The sera were sent from England and we performed ELISAs on these three sera plus a known *C. pylori*-positive serum. The results from this study [30] showed that in the patients' sera there were high levels of specific antibody to the cat organism and a low level to *C. pylori*. The *C. pylori*-positive serum contained a high level of specific antibody to *C. pylori* with little specific antibody to the cat isolate. These studies allowed us to conclude that the three patients' organisms were more closely related to our cat isolate than *C. pylori*. It should be noted that the serological results for the three patients' sera were different, supporting the concept of more than one type of these bacteria being involved. Close inspection of the electron micrographs from one of the English patients showed it to have no periplasmic fibrils and to be like Lockard & Boler's organism (Fig. 30.2) and unlike the organism we used in the ELISA (Fig. 30.5). Our interpretation of these serological data is that this group of bacteria in the human patients is closely related but different to *C. pylori*, and originates from an animal source.

In our article reporting the serology results on these English patients we encouraged pathologists to examine their biopsy sections more closely and predicted that once people started looking for non-*C. pylori* infection in humans many more cases would be found. This has certainly proved to be the case. Within 2 weeks of presenting the data to gastroenterologists and histopathologists in Sydney (Australia), two new cases were reported to us. Subsequently, other workers have communicated with me about similar patients in the USA, New Zealand and Germany. While the details of the two Australian patients are to be published elsewhere, it is appropriate to review these two cases as examples of non-*C. pylori* infection.

Case 1

The first case was a 32-year-old male from Indonesia who had been living in Sydney for two years. He presented with a 5-year history of epigastric discomfort which was of moderate severity and either dull or cramping in nature; it was present during most of the day but increased post-prandially. Associated symptoms included anorexia, nausea with intermittent vomiting, flatulence, belching and abdominal bloating. He had lost 2 kg in

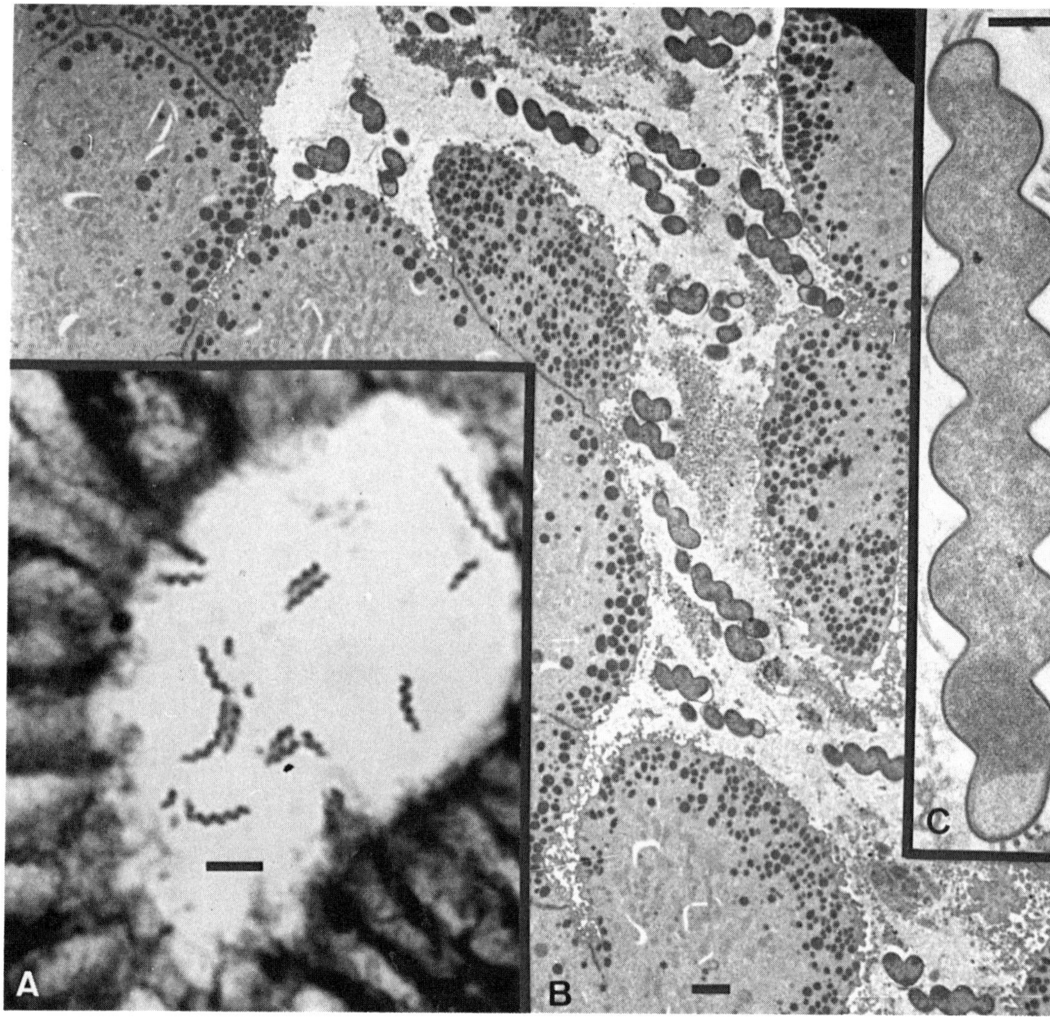

Fig. 30.7 Non-*C. pylori* spiral-shaped bacteria from the antrum of a human patient colonizing the stomach of a laboratory mouse. (A) Light micrograph of the spiral bacteria seen in a biopsy from a thirty-two-year-old Indonesian male (bar = 5 μm). (B) Transmission electron micrograph showing the same bacteria colonizing the glandular epithelium of a mouse stomach following orogastric administration of an homogenate of the human antral biopsy (bar = 1 μm). (C) High magnification view of a single organism originating from the human in the mouse stomach showing the insertion sites of the bipolar flagella and the regular spiral morphology (bar = 0.5 μm).

weight over the past 12 months. The bacterium in the patient is seen in Fig. 30.7(A). Large numbers of these bacteria with the now familiar morphology are seen. Culture of the bacterium using normal culture methods was unsuccessful. However, based on the papers of Kasai & Kobyashi [13] and Weber [11], we have started using the mouse as a vehicle for maintaining these organisms so that they can be studied further. An antral biopsy from the Indonesian patient was inoculated into mice. Within one week, the fundic stomach of the mouse was heavily colonized with the organism from the patient (Fig. 30.7, B, C).

Case 2

The second patient is especially interesting because she is only 2-years old. We were unable to culture the organism from the child

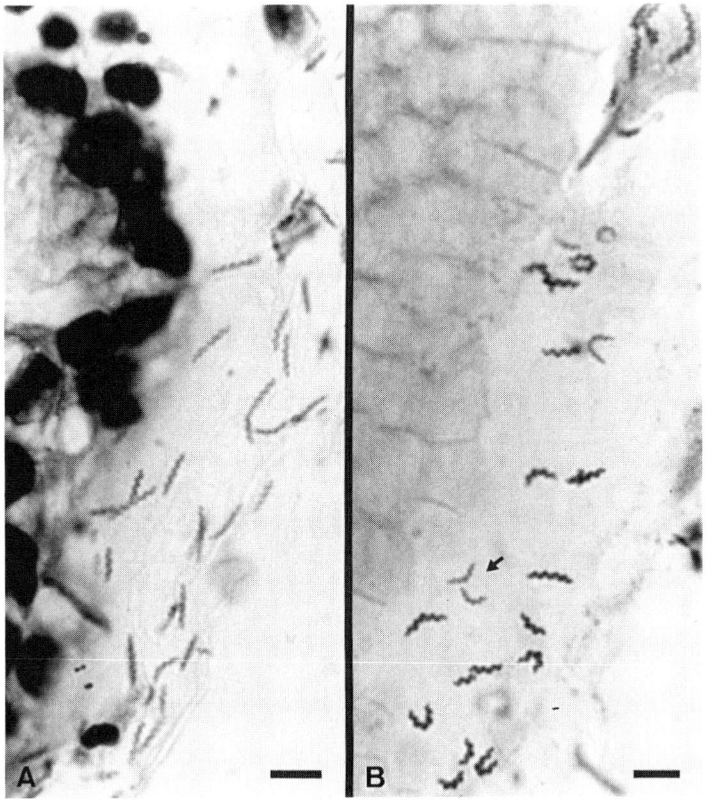

Fig. 30.8 A light micrograph of the surface mucus of the antral biosy of a 2-year-old girl showing numerous non-*C. pylori* spiral bacteria (bar = 5 μm). (B) Light micrograph of the surface mucus of the gastric mucus of the girl's cat showing spiral bacteria. Some of the organisms appear more darkly staining than the others suggesting two types. The organisms shown by the arrow have the same morphology as the spirals in the girl's stomach (bar = 5 μm).

and there was not enough material to inoculate a mouse. However, the family had two pet cats which recently were euthanased for independent reasons. Spiral organisms seen in gastric sections of these animals have not as yet been cultured but mice have been colonized with them. It is hoped that these stomachs will contain enough antigen for ELISA studies on the child's serum. Photographs of gastric mucosa from this patient and the organism in one of her cats are shown in Fig. 30.8.

ELISAs on the sera from the Indonesian man and the child showed the same results as two patients in the Dent study, i.e. the organisms infecting the patients were more closely related to our cat isolate than *C. pylori*, but were not the same organism.

Further evidence as to the source of these bacteria is likely to emerge as the incidence of infection amongst different groups is reported. There is a distinct possibility that the incidence of these organisms will be higher in groups that come in very close contact with domestic animals such as cats and dogs, i.e. children, or persons from cultures where family members live in very close contact with animals. The observation of Craige [15] that spirilla are present in the vomitus of cats and dogs with signs of gastritis is obviously relevant to the mode of transmission of these bacteria.

Culture of spiral-shaped bacteria from animal stomach mucosae

To date, very few of the non-*C. pylori* spiral bacteria from the stomachs of animals have been cultured. Fox *et al.* [31] have cultured an organism from the gastric mucosa of ferrets and we have grown one of the tight spiralled organisms from the cat and dog [17].

Culture in vitro

The secret to growing the mucus-associated spiral-shaped bacteria from the intestinal tract [18, 32] is patience and keen observation. The bacteria in this group do not normally form colonies but grow as a very fine spread which can easily be dismissed as a water mark on the plate. In the original isolation of the organism from the cat stomach, a *Campylobacter jejuni*-like spiral was also present and it took many sub-cultures from the edge of the spreading growth before the organism was obtained in pure culture. The organism was grown on 5% lysed horse blood (CSL, Parkville, Australia) in Blood Agar Base No. 2 (Oxoid, Basingstoke, UK) containing (per 500 ml) 2.5 mg amphotericin B (Fungizone, Squibb, Princeton, USA), 2.5 mg trimethoprim (Sigma, St Louis, USA), 1250 I.U. polymyxin B (Sigma, St Louis, USA), 5 mg vancomycin (Eli Lilly, West Ryde, Australia). They are very fickle bacteria and it is not unusual for them to suddenly not grow. It is essential once any growth is obtained to immediately put some of the cells in Brain Heart Infusion Broth (Oxoid, Basingstoke, UK containing 30% glycerol, v/v), then into liquid nitrogen in which they remain viable for years. Moisture is a critical feature; plates should be freshly prepared, not dried, and inoculated plates should be incubated with the lid uppermost. This latter point illustrates how critical the moisture level is. If one inoculates two plates with, for example, the rodent ileal isolate [18] and places one in the gas jar lid-uppermost and the other in the more conventional lid-down position, then the organism in some instances will only grow on the first plate.

We still have much to learn. The isolation of the non-fibril entwined organism has eluded us despite the use of very large numbers of media with many combinations of potential growth factors.

Culture in vivo

Given that we cannot grow all the bacteria seen in the stomachs of humans and animals we have resorted to a simple technique described by the early workers in the field to keep these bacteria alive [8, 11, 13]. The laboratory mouse which normally has no spiral-shaped bacteria in its stomach is used for *in vivo* culture. The mouse stomach is divided into two sections, viz. an area of stratified squamous epithelium that is heavily colonized with acid-tolerent lactobacilli, and a glandular fundic region with histology equivalent to the human stomach that is colonized on the outer surface by only an acid-tolerant yeast. This glandular epithelium becomes very readily colonized with spiral bacteria from a range of animals when inoculated oro-gastrically with either stomach scrapings or pure cultures. A pure culture of the cat isolate readily colonizes mice thus providing an interesting experimental model of colonization, and an organism from the dog stomach has been isolated in pure culture on the lysed blood agar via passage through the mouse. Small groups of mice have been colonized with the spiral bacterium from the human patient described above, and also with another spiral from a monkey which died in an accident at a local zoo, thus allowing investigation of bacteria that would have otherwise been lost.

Conclusion

A study of the spiral-shaped bacteria that inhabit the stomachs of animals can contribute important information to the understanding of *C. pylori*-associated disease. Firstly, drawing attention to these organisms highlights the fact that bacteria colonizing the stomach is not an unusual event. Many animal species have a natural stomach microbiota and thus it is not surprising that the human stomach should have a specialized adapted microflora. Given that in a developed country up to 50% of persons over the age of 36 years are actively infected with *C. pylori* and that in other cultures the incidence is much higher, it is reasonable to consider that *C. pylori* is a human-adapted bacterium; indeed it appears to be even more highly adapted than the other spirals described above because of the limited host range it exhibits. However, as has been proposed before [33], *C. pylori*, unlike the other gastric bacteria, is only 'almost normal flora' in the sense that it induces a histopathology in all colonized persons even if this pathology is in most cases not accompanied by any symptomatology.

Secondly, as *C. pylori* does not grow in a convenient animal model, it is not easy to study experimentally. As stated above, bacteria that colonize the stomach surface have many characteristics in common, e.g. spiral morphology, microaerophilism and urease activity. Therefore, study of the animal organisms is likely to provide information relevant to our understanding of *C. pylori*, particularly factors influencing colonization. The cat, dog or colonized mouse are likely to be very useful for investigation of, for example, antibacterial regimens. We have recently been using a paediatric endoscope to validate the cat as a useful model for treatment trials. Bismuth compounds and amoxycillin have been shown to be effective in removing the cat organisms from adult animals while erythromycin has proved ineffective. In all cases, cleared animals relapsed within one month. These results are similar to what is seen in the human with *C. pylori* and convince us of the value of this model.

Finally, as it has now been found that some of these bacteria can infect humans, much useful comparative information can be obtained relevant to an understanding of *C. pylori* pathogenesis, particularly as the animal bacteria are able to induce active chronic gastritis in the human while apparently not doing the same in their natural host. Can *C. pylori*-associated disease be explained in terms of the host response to bacteria in the human gastric mucosa rather than some special pathogenic determinant in *C. pylori*? Study of these non-*C. pylori* human isolates in animals is going to provide much interesting information relevant to this question.

Acknowledgements

The help of Jani O'Rourke, Sergei Kouprach, Elizabeth Dick and Nada Heneine in the production of the figures is grateful acknowledged as is the editorial editorial assistance of Nerissa Lee.

References

1 Lee, A. 1980. Normal flora of animal intestinal surfaces. In Bitton, G. & Marshall, K.C. (eds.) *Adsorption of Microorganisms to Surfaces*. Wiley, New York. pp. 145–173.

2 Savage, D.C. Microbial ecology of the gastrointestinal tract. *Ann. Rev. Microbiol.* 1977. **31**, 107–133.

3 Hazell, S.L., Lee, A., Brady, L. & Hennessy, W. *Campylobacter pyloridis* and gastritis: association with intercellular spaces and adaption to a mucus environment as important factors in colonisation of the gastric epithelium. *J. Infect. Dis.* 1986. **153**, 658–663.

4 Lee, A., O'Rourke, J.L., Barrington, P.J. & Trust, T.J. Mucus colonisation as a determinant of pathogenicity in intestinal infection by *Campylobacter jejuni*: A mouse cecal model. *Infect. Immun.* 1986. **51**, 536–546.

5 Kennedy, M.J., Rosnick, D.K., Ulrich, R.G. & Yancey, R.J. Association of *Treponema hyodysenteriae* with porcine intestinal mucosa. *J. Gen. Microbiol.* 1988. **134**, 1565–1576.

6 Rappin, J.P. 1881. Contr a l'étude de bacterium de la bouche a l'etat normal. Quoted by Breed, R.S., Murray, E.G.D. & Hitchens, A.P. in *Bergey's Manual of Determinative Bacteriology*, 6th edn. Williams & Wilkins, Baltimore. p. 68.

7 Bizzozero, G. Ueber die schlauchformigen Drüsen des Magendarmkanals und die bezeihungen ihres Epithels zu dem Oberflächenepithel der Schleimhaut. *Arch. f. Mikr. Anast.* 1893. **42**, 82.

8 Salomon, H. Ueber das Spirillum des Saugetiermagens und sein Verhalten zu den Belegzellen. *Centralbl. f. Bakt.* 1896. **19**, 422–441.

9 Dubosq, O. & Lebailly, C. *Spirilla canis* n.g., n.sp., spirille de l'estomac du chien. *Compt. Rend. Acad. Sci.* 1912,**154**, 835–837.

10 Weber, A.F., Hasa, O. & Sautter, J.H. Some observations concerning the presence of spirilla in the fundic glands of dogs and cats. *Am. J. Vet. Res.* 1958. **19**, 677–680.

11 Weber, A.F. & Schmittdiel, E.F. Electron microscopic and bacteriologic studies of spirilla isolated from the fundic stomach of cats and dogs. *Am. J. Vet. Res.* 1962. **23**, 422–427.

12 Lockard, V.G. & Boler, R.K. Ultrastructure of a spiralled microorganism in the gastric mucosa of dogs. *Am. J. Vet. Res.* 1970. **31**, 1453–1462.

13 Kasai, K. & Kobayashi, R. The stomach spirochaete occurring in mammals. *J. Parasitol.* 1919. **6**, 1–11.

14 Henry, G.A., Long, P.H., Burns, J.L. & Charbonneau, D.L. Gastric spirillosis in beagles. *Am. J. Vet. Res.* 1987. **48**, 831–836.

15 Craige, J. In Hoskins, H.P. Lacroix, J.V. & Mayer, K. (eds.) *Canine Medicine*, 2nd edn. 1959. American Veterinary Publications, Santa Barbara, California.

16 Krakowka, S., Morgan, D.M., Kraft, W.G. & Leunk, R.D. Establishment of gastric *Campylobacter pylori* infection in the neonatal gnotobiotic piglet. *Infect. Immun.* 1987. **55**, 2789–2796.

17 Lee, A. Hazell, S.L., O'Rourke, J. & Kouprach, S. Isolation of a spiral-shaped bacterium from the cat stomach. *Infect. Immun.* 1988. **56**, 2843–2850.

18 Phillips, M. W. & Lee, A. Isolation and characterization of a spiral bacterium from the crypts of the rodent gastrointestinal tract. *Appl. Envir. Microbiol.* 1983. **45**, 675–683.

19 Romero, S., Archer, J.R., Hamacher, M.E., Bologna, S.M. & Schell, R.F. Case report of an unclassified microaerophilic bacterium associated with gastroenteritis. *J. Clin. Microbiol.* 1988. **26**, 142–143.

20 Archer, J.R., Romero, S., Ritchie, A.L., Hamacher, M.E., Steiner, B.M., Bryner, J.H. & Schell, R.F. Characterisation of an unclassified microaerophilic bacterium associated with gastroenteritis. *J. Clin. Microbiol.* 1988. **26**, 101–105.

21 Kirkbride, C.A., Gates, C.E., Collins, J.E. & Ritchie, A.E. Ovine abortion associated with an anaerobic bacterium. *J. Am. Vet. Med. Assoc.* 1985. **186**, 789–791.

22 Bryner, J.H., Ritchie, A.E., Pollet, L. Kirbride, C.A & Collins, J.E. Experimental infection and abortion of pregnant guinea pigs with a unique spirillum-like bacterium isolated from aborted ovine fetuses. *Am. J. Vet. Res.* 1987. **48**, 91–95.

23 Lambert, J.R., Borromeo, M., Pinkard, K.J., Turner, H., Chapman, C.B. & Smith, M.L. Colonization of gnotobiotic piglets with *Campylobacter pyloridis*–An animal model? *J. Infect. Dis.* 1987. **155**, 1344.

24 Lee, A. 1984. Neglected niches. The microbial ecology of the gastrointestinal tract. In Marshall, K.C. (ed.) *Advances in Microbial Ecology*, Vol. 8. Plenum Press, New York. pp. 155–162.

25 Ferrero, R.L. & Lee, A. Motility of *Campylobacter jejuni* in a viscous environment: comparison with conventional rod shaped bacteria. *J. Gen. Microbiol*, 1988. **134**, 53–59.

26 Ferrero, R.L., Hazell, S.L. & Lee, A. The urease enzymes of *Campylobacter pylori* and a related bacterium. *J. Med. Microbiol.* 1988, **27**, 33–40.

27 Lim, R.K.S. A parasitic spiral organism in the stomach of the cat. *Parasitology*, 1920. **12**, 108–112.

28 Doenges, J.L. Spirochetes in the gastric glands of *Macacus rhesus* and of man without related disease. *Arch. Pathol.* 1939. **27**, 469–477.

29 Dent, J.C., McNulty, C.A.M., Uff, J.C., Wilkinson, S.P. & Gear, M.W.L. Spiral organisms in the gastric antrum. *Lancet* 1987. **ii**, 96. [Letter].

30 Lee, A., Dent, J. Hazell, S. & McNulty, C. Origin of spiral organisms in the gastric antrum. *Lancet*, 1988. **i**, 300–301.

31 Fox, J.G., Edrise, B.M., Cabot, E.B., Beaucage, C., Murphy, J.C. & Prostak, K.S. *Campylobacter*-like organisms isolated from gastric mucosa of ferrets. *Am. J. Vet. Res.* 1986. **47**, 236–239.

32 Lee, A. & Phillips, M. Isolation and cultivation of spirochaetes and other spiral shaped bacteria associated with the cecal mucosa of rats and mice. *Appl. Environ. Microbiol.* 1978. **35**, 610–613.

33 Lee, A. & Hazell, S.L. *Campylobacter pylori* in health and disease: an ecological perspective. *Microbiol. Ecol. Health Dis.* 1988. **1**, 1–16.

34 Kolmer, W. & Wagner, R.J. Über eine im Magenfundus des Hundes vorkommende saprophytische Spirochäte. *Centralbl. f. Bakt. (Abt. 1)* 1916. **78**, 382–383.

31 Ferret gastric campylobacter-like organisms

D.S. TOMPKINS

Campylobacter pylori has been isolated from the gastric tissues of primates, including human subjects, and laboratory-housed pigs but not from other animals [1]. Spiral organisms with a different morphology to *C. pylori* have been seen closely attached to gastric epithelium and in the gastric mucus of cats, dogs, monkeys and humans, and isolation of this type of organism from the stomach of a cat has been described [1]. A gastric campylobacter-like organism (GCLO) very similar in morphology and other characteristics to *C. pylori* has been isolated from the ferret; this raises the possibility of the ferret being a useful animal model for the study of chronic gastritis and *C. pylori*-related diseases.

Incidence

The incidence of GCLO colonization in ferrets is difficult to establish from the published reports. In the initial study in which ferret GCLO was isolated [2] 3/14 ferrets were colonized, and in a second study the organisms were isolated from the gastric tissues of 19/33 ferrets [3]. Mature ferrets were examined in two British studies; all of the 14 ferrets in the first investigation [4] and all of the 17 in the second [5] were found to be colonized by GCLO. As in humans colonized by *C. pylori*, the incidence of gastric colonization by GCLO of ferrets may increase with age. In a recent report, GCLO were isolated from only 2/33 ferrets aged < 173 days (the ferrets colonized being 24 and 31 days old) but were isolated from 25/27 ferrets > 1 year old [6].

Pathogenicity

The first isolation of a ferret GCLO was from a 9-month-old female which appeared clinically healthy and was examined as part of a study evaluating bile reflux and gastric carcinogenesis [2]. During the course of an operation, an ulcer was seen at the pyloric–duodenal junction and a GCLO was isolated from biopsy material taken from this area. Histological examination revealed an area of necrosis and ulceration with a mild lymphocytic infiltration of the lamina propria of an adjacent portion of intact mucosa. In this initial report, GCLO were also isolated from two healthy 10-month-old ferrets, both with normal gastric mucosa, but not isolated from another ferret with a small antral mucosal lesion. Sixteen healthy ferrets, including the two colonized by GCLO, all had lymphoid aggregates of various sizes present in the submucosa and lamina propria of the pyloric antral region. A more recent report by the same group [6] confirmed that these lymphoid aggregates were seen in many ferrets with a macroscopically-normal mucosa. In this series, ferrets with ulcers and severe gastritis had macrophages and lymphocytes infiltrating the lamina propria and sub-mucosa and GCLO demonstrated by Warthin–Starry stain on the gastric epithelial surface. On electron microscopy, GCLO were seen associated with disrupted microvilli and occasionally within the epithelial cells.

In one of the British studies [4], low-grade gastritis was seen in all 14 ferrets studied, deep in the antral mucosa and extending into the sub-mucosa in some cases; 1/14 ferrets had small antral erosions visualized macroscopically and GCLO were revealed on histology, close to these areas of deficient epithelium and vascular congestion. No evidence of gastritis was seen in any of the 17 ferrets investigated in the second British study [5]; there was no mononuclear cell infiltration of the lamina propria, no epithelial damage, no glandular atrophy, no epithelial mucin depletion, and no vascular congestion or oedema accompanying the surface bacteria. There were no lymphoid aggregates seen in the submucosa or lamina propria of these 17 ferrets which were all colonized by GCLO.

Pathologists who may be experienced in the interpretation of human pathology may differ in their interpretation of some histological sections. These differences may be more apparent when tissues from unfamiliar sources are examined. However, it is unlikely that the differences reported in the four studies are due to differences in the interpretation of the sections by the pathologists. It is possible that other factors may be involved in the production of gastric pathology in some colonies of laboratory animals. The animals examined in the second British study were obtained from a number of sources and not housed together for prolonged periods of time.

Gastric erosions with an associated minimal inflammatory response have been produced in one of five pre-weanling ferrets inoculated with a human strain of *C. pylori* [6]. Further study of the effects of *C. pylori*-colonization in the ferret may reveal why *C. pylori* colonization in humans is almost always associated with chronic gastritis.

Characteristics of Ferret GCLO

Campylobacter pylori isolates from baboons, pigs and monkeys closely resemble *C. pylori* isolates from humans morphologically, biochemically and serologically [7–9]. The ferret GCLO, although sharing many of the characteristics of *C. pylori*, differs in several respects. On electron microscopy, the ferret GCLO closely resembles *C. pylori* in size and morphology [2]. Like *C. pylori*, but unlike other campylobacters, the ferret GCLO has a smooth external cell wall with multiple (1–5) sheathed flagella which often have terminal bulbs [2, 10]. In *C. pylori*, the flagella usually arise from the one pole, with occasional bipolar, flagellated organisms seen at cell division [11]. Many of the ferret GCLO possess bipolar flagella, and also flagella arising from the sides of the bacterial cells [2].

The ferret GCLO are readily isolated from homogenized gastric tissue on blood agar media incubated at 37°C in a microaerobic atmosphere, suitable for the isolation of campylobacters [2, 4, 5]. Colonies are greyish in colour on blood agar, producing slight haemolysis, and tan-coloured on heated blood agar; they are slightly larger and more mucoid than *C. pylori* colonies, being about 1–2 mm in diameter after incubation for 72 hours. Ferret GCLO will not grow in air or anaerobically, but like *C. pylori* will grow on sub-culture in 8–10% CO_2 in a humidified CO_2 incubator. Most strains, but not all, will grow at 42°C but no strains are capable of growth at 27°C [2, 5]. Typically they are Gram-negative curved or U-shaped rods with few S-shaped forms. They are catalase and oxidase positive, do not hydrolyse hippurate and produce substantial urease activity, equivalent to that produced by *C. pylori* [2, 4, 5]. Growth is inhibited by concentrations of greater than 1.5% sodium chloride and 1% glycine [12]. Neither the ferret GCLO nor *C. pylori* are as tolerant of bile as other campylobacters and bacteria which normally inhabit the human lower intestinal tract [13]. Unlike human isolates of *C. pylori*, ferret GCLO reduce nitrate and are sensitive to nalidixic acid and resistant to cephalothin on sensitivity testing with 30 μg discs [2, 5, 7]. In one study [4], no differences between 11 isolates of *C. pylori* and 14 isolates of ferret GCLO were found in sensitivity to erythromycin and tripotassium dicitratobismuthate or in resistance to sulphathiazole and trimethoprim, but the ferret GCLO were more resistant than *C. pylori* to penicillins although neither organisms produced detectable β-lactamase. The ferret GCLO may be isolated on selective media incorporating Skirrow's selective supplement [4, 5].

Ferret GCLO and *C. pylori* both produce alkaline phosphatase, acid phosphatase, naphthol-AS-Bl-phosphohydralase and lesser amounts of esterase (C4) and esterase lipase (C8) [5]. Although all strains of *C. pylori* tested produce substantial leucine aminopeptidase activity, production of this enzyme is weak or absent in ferret GCLO [4, 5]. *Campylobacter pylori* and ferret GCLO also have different protein profiles demonstrated by polyacrylamide gel electrophoresis [5, 7] and differences in protein antigens by immuno-blotting techniques [7]. Isoelectric focusing of urease extracts yields identical isoelectric points (pI) of 5.9 for urease from human and pig isolates of *C. pylori*, a pI of 5.7 for baboon *C. pylori* urease and a pI of 5.4 for ferret GCLO urease [14]. Fatty acid methyl ester (FAME) profiles also reveal differences between ferret GCLO and *C. pylori* isolated from various species [5, 15]. The FAME profile of ferret GCLO resembles patterns of other campylobacters more closely than does the FAME profile of *C. pylori* [5].

DNA–DNA hybridization has confirmed the identity of *C. pylori* strains isolated from monkeys [8] and, with greater than 70% DNA relatedness between strains of *C. pylori* and ferret GCLO [12], suggests that ferret GCLO are of the same species as *C. pylori*. Partial 16S ribosomal RNA sequence analysis, which has indicated that *C. pylori* is more closely related to *Wolinella succinogenes* than it is to other campylobacter species [15], demonstrates a close phylogenetic relationship between these two organisms and the ferret GCLO [12]. However, FAME profiles, protein profiles, total polar lipid and phospholipid profiles all confirm a lack of similarities between *C. pylori* and *W. succinogenes* [16].

Conclusions

The ferret GCLO is closely related to *C. pylori* and shares many characteristics. These microaerophilic, urease positive, spiral bacilli occupy the same ecological niche beneath gastric mucus attached to the gastric epithelium [1]. Both organisms are sensitive to acid but unlike other, urease-negative campylobacters, have greater resistance to acid in the presence of physiological amounts of urea [13]. This characteristic may enable these organisms to survive when exposed for short periods of time to lethal gastric acid. Ferrets with no evidence of chronic gastritis may be colonized by GCLO; the association of colonization with the various pathologies observed in different studies needs clarification. Colonization of stomachs of Rhesus monkeys and pigtailed macaques by *C. pylori* is associated with chronic gastritis [8, 9], and the gnotobiotic piglet may be a useful animal model for the study of *C. pylori*-associated disease [1]. The ferret is a more practical laboratory animal model than either non-human primates or gnotobiotic piglets, but will be of value only if *C. pylori* established in the ferret gives rise to the effects seen in humans, or the association of GCLO colonization and chronic gastritis in the ferret can be clarified.

Two representative strains of ferret GCLO are deposited with the National Collection of Type Cultures, Central Public Health Laboratory, London, with the numbers NCTC 12031 and NCTC 12032. The ferret GCLO has been named as *Campylobacter pylori* subsp. *mustelae* and the type strain is ATCC 43772 [17].

References

1 Lee, A. & Hazell, S.L. *Campylobacter pylori* in health and disease: An ecological perspective. *Microbiol. Ecol. Health Dis.* 1988, **1**, 1–16.

2 Fox, J.G., Edrise, B.M., Cabot, E.B., Beaucage, C., Murphy, J.C. & Prostak, K.S. Campylobacter-like organisms isolated from gastric mucosa of ferrets. *Am. J. Vet. Res.* 1986, **47**, 236–239.

3 Cave, D.R., Taylor, N., Tuczynski, C. & Fox, J.G. Campylobacter-like organisms (CLO) from man and ferret: towards an animal model of CLO induced disease. *Gastroenterology*, 1986, **90**, 1368.

4 Jeffries, L., Buckley, D.E., Blower, P.R. & Plumb, J.N. Comparative sensitivities to antimicrobial agents of *Campylobacter pylori* and the gastric campylobacter-like organism from the ferret. *J. Clin. Pathol.* 1987, **40**, 1265–1267.

5 Tompkins, D.S., Wyatt, J.I., Rathbone, B.J. & West,

A.P. The characterization and pathological significance of gastric campylobacter-like organisms in the ferret: a model for chronic gastritis? *Epidem. Infect.* 1988, **101**, 269–278.

6 Fox, J.G., Hotaling, L., Connors, T., Goad, B., Taylor, N.S. & Murphy, J.C. 1987. Gastric mucosa in the ferret: bacteriologic and pathologic findings. In Kaijser, B. & Falsen, E. (eds.) *Campylobacter IV. Proceedings of the Fourth International Workshop on Campylobacter Infection, Göteborg.* University of Göteborg, Göteborg. p. 339.

7 Jones, D.M. & Eldridge, J. 1987. Gastric campylobacter-like organisms (GCLO) from man ('*C. pyloridis*') compared with GCLO strains from the pig, baboon and ferret. In Kaijser, B. & Falsen, E. (eds.) *Campylobacter IV. Proceedings of the Fourth International Workshop on Campylobacter Infection, Göteborg.* University of Göteborg, Göteborg. p. 44.

8 Bronsdon, M.A. & Schoenknecht, F.D. *Campylobacter pylori* isolated from the stomach of the monkey *Macaca nemestrina. J. Clin. Microbiol.* 1988, **26**, 1725–1728.

9 Newell, D.G., Hudson, M.J. & Baskerville, A. Naturally occurring gastritis associated with *Campylobacter pylori* infection in the Rhesus monkey. *Lancet* 1987, **ii**, 1338.

10 Jones, D.M., Curry, A. & Fox, A.J. An ultrastructural study of the gastric campylobacter-like organism '*Campylobacter pyloridis*', *J. Gen. Microbiol.* 1985, **131**, 2335–2341.

11 Goodwin, C.S., McCulloch, R.K., Armstrong, J.A. & Wee, S.H. Unusual cellular fatty acids and distinctive ultrastructure in a new spiral bacterium (*Campylobacter pyloridis*) from the human gastric mucosa. *J. Med. Microbiol.* 1985, **19**, 257–267.

12 Fox, J.G., Edmonds, P., Taylor, N.S., Paster, B. & Dewhirst, F. 1987. Comparison of the phylogenetic, biochemical, phenotypic and morphological characteristics of gastric campylobacters from humans and ferrets. In Kaijser, B. & Falsen, E. (eds.) *Campylobacter IV. Proceedings of the Fourth International Workshop on Campylobacter Infection, Göteborg.* University of Göteborg, Göteborg. pp. 50–53.

13 Tompkins, D.S. & West, A.P. *Campylobacter pylori*, acid and bile. *J. Clin. Pathol.* 1987, **40**, 1387.

14 Tompkins, D.S., Millar, M.R. & West, A.P. Isoelectric focusing of ureases from *Campylobacter pylori* and related organisms. *J. Clin. Microbiol.* 1988, **26**, 2678–2679.

15 Romaniuk, P.J., Zoltowska, B., Trust, T.J., Lane, D.J., Olsen, G.J., Pace, N.R. & Stahl, D.A. *Campylobacter pylori*, the spiral bacterium associated with human gastritis, is not a true *Campylobacter* sp. *J. Bacteriol.* 1987, **169**, 2137–2141.

16 Hudson, M.J., Bhavsar, P. & Wait, R. 1987. Chemotaxonomy of the campylobacters. In Kaijser, B. & Falsen E. (eds.) *Campylobacter IV. Proceedings of the Fourth International Workshop on Campylobacter Infection, Göteborg.* University of Göteborg, Göteborg. pp. 34–36.

17 Fox, J.G., Taylor, N.S., Edmonds, P. & Brenner, D.J. *Campylobacter pylori* subsp. *mustelae* subsp. nov. isolated from the gastric mucosa of ferrets (*Mustelae putorius furo*), and an emended description of *Campylobacter pylori. Int. J. Syst. Bacteriol.* 1988, **38**, 367–370.

32 Campylobacter pylori infections in non-human primates

D.G. NEWELL

Although the association of *Campylobacter pylori* infection of the human gastric antrum with histological gastritis and peptic ulceration is now widely accepted [1], the role of the organism in these diseases has yet to be established. Should *C. pylori* be shown to be pathogenic then the mechanisms of pathogenicity would need to be investigated and suitable therapeutic regimens developed.

In other infectious diseases such information has been derived from animal model studies. Unfortunately attempts to establish animal models of *C. pylori* infection have been largely unsuccessful. Although a gnotobiotic piglet model of *C. pylori* colonization has been described recently [2, 3], this model may not mimic the human disease very closely.

Several laboratories [4–6] have surveyed non-human primates, in particular the macaques, to determine whether there was any similar relationship between bacterial infections of the gastric mucosa and gastritis in these animals. Such natural infections would suggest that these animals may provide suitable models of the human disease.

Campylobacter-like organisms in the Rhesus monkey

Campylobacter-like organisms (CLO) have been cultured from the stomachs of the Rhesus monkey (*Macaca mulatta*), both at necropsy and from pinch biopsies obtained at endoscopy (Table 32.1).

Isolation and culture

Rhesus monkey CLO were cultured from biopsy material smeared onto blood agar plates containing 5% defibrinated blood or onto enriched chocolate agar. Like other CLO these organisms required a moist microaerophilic atmosphere and grew slowly at 37°C. Selective antibiotics [7] did not inhibit their growth and were essential in the isolation of organisms from necropsy material, which tends to be heavily contaminated with bacteria from the duodenum. The monkey CLO grew slowly on isolation then quickly adapted to laboratory conditions, after which they grew at a similar rate to reference strains of *C. pylori*.

Table 32.1 The incidence of CLO in macaques by culture and histology, of biopsies obtained at necropsy or endoscopy

Species	Cultured	Observed	N/E*	References
M. mulatta	6/10	8/13	N	Baskerville & Newell (1988) [4]
M. mulatta	–	4/7	N	Reed & Berridge (1988) [6]
M. mulatta	–	7/6	E	Dubois *et al.*(1988) [12]
M. mulatta	0/3	0/3	N	A. Euler[†]
M. nemestrina	6/24	–	E	Bronsdon & Schoenknech (1988) [5]

*N = necropsy; E = endoscopy.
[†] = Personal communication.

The rapid urease test [8] was not suitable for the presumptive detection of urease-positive CLO in monkey gastric biopsies, all of which tested positive. This was presumably because of the presence of large numbers of other ureolytic organisms such as urease positive staphylococci and enterobacteriaceae (E. Auler, personal communication).

Morphology

In Gram-stains of mucus smears, these CLO were frequently present in very large numbers and were short, S-shaped Gram-negative rods.

By transmission electron microscopy of negatively-stained grids, these organisms were approximately 2–3 μm long and 0.5 μm in diameter. Each bacterium had multiple (3–7) unipolar sheathed flagella with terminal paddles (Fig. 32.1).

Chemotaxonomy

The Rhesus monkey CLO had the same enzymic properties as *C. pylori*, including positive oxidase and catalase reactions. They very rapidly hydrolysed urea. They did not, however, hydrolyse hippurate, grow in the presence of 1% glycine or 1.5% sodium chloride or produce hydrogen sulphide.

The total protein profiles of these organisms by sodium dodecyl polyacrylamide gel electrophoresis (SDS-PAGE) were indistinguishable from human *C. pylori* strains and recent studies, using computer-assisted numerical methods, indicate that even the minor proteins of the Rhesus monkey strains have a

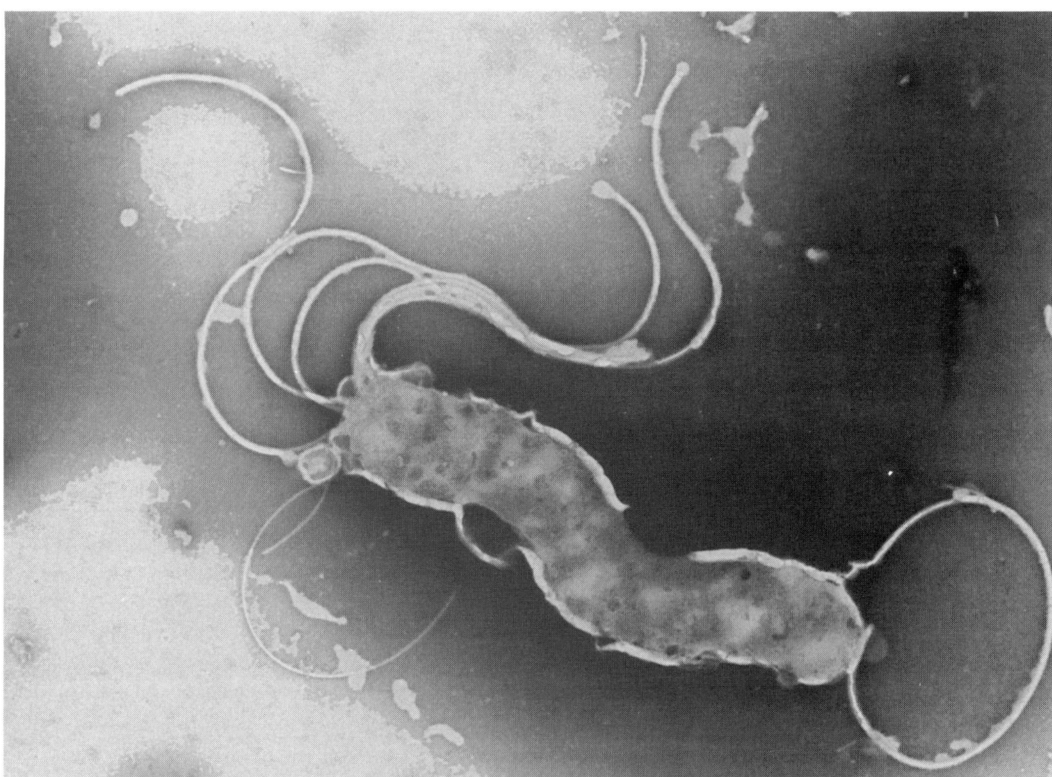

Fig. 32.1 Campylobacter-like organism isolated from the gastric antrum of the Rhesus monkey. Note the multiple unipolar sheathed flagella with terminal paddles. (Reproduced from Dent *et al.* [19].)

very high level of homology with the type strain of *C. pylori*. Moreover, the surface exposed proteins labelled by ^{125}I and separated by SDS-PAGE are also very similar to those of *C. pylori* (unpublished data).

The antigenic profile of the Rhesus monkey strains have been investigated by Western blotting techniques using polyclonal rabbit anti-*C. pylori* and rabbit anti-*C. jejuni* antisera and mouse anti-*C. pylori* monoclonal antibodies [9]. All the data suggested a marked similarity in the surface antigens of the Rhesus monkey strains and human *C. pylori* whilst both organisms are significantly different from *C. jejuni* isolates.

The cellular fatty acid profiles, of Rhesus monkey isolates, analysed by capillary gas–liquid chromatography (GLC) and GLC/mass spectrometry of their fatty acid methyl esters, are similar to those of *C. pylori* [9]. The % (G+C) was 35–38% for these Rhesus monkey isolates and there was about a 50% homology, by DNA hybridization, with *C. pylori* (S. Goodwin, personal communicaton).

By all criteria so far published, therefore, the Rhesus monkey CLO are identical to *C. pylori*.

CLO infection in the pig-tailed macaque

Six isolates of CLO have been recovered [5] from the pig-tailed monkey (*M. nemestrina*) (Table 32.1); infection seems to be at a lower prevalence than in the Rhesus monkey. These organisms appear to have a similar morphology, biochemistry and antibiotic susceptibility to *C. pylori*. Moreover, they reacted positively with genomic DNA probes specific for *C. pylori*. The only obvious differences observed [5] have been a smaller maximum colony diameter with a slower growth rate on isolation and the absence of colony colour or pigment. By cellular fatty acids, DNA hybridization and mol% (G+C) some of these organisms, the (type A) strains, were similar to *C. pylori*. However, at least one of the pig-tailed macaque isolates investigated, the type B strain, had a (G+C) of only 24%, a unique cellular fatty acid profile and only a 10% homology with *C. pylori*.

CLO in cynomolgus monkeys

A number of cynomolgus monkeys (*M. fascicularis*) have also been surveyed for the presence of CLO in the gastric antrum [4] (A. Euler, personal communication). CLO were not isolated from any of these monkeys. However, the presence of CLO cross-reactive antigen was suggested by immunohistological staining in at least one monkey [4] and recently cynomolgus monkey CLO have been isolated from animals. These isolates appear to have a slightly different cellular fatty acid profile from *C. pylori*.

CLO in other non-human primates

CLO have apparently been isolated from baboons [10] (S. Goodwin, personal communication); however, this does not appear to be a common infection as no CLO were isolated in a recent survey of 10 baboons at necropsy, though *C. jejuni* was occasionally found [4]. The baboon CLO was morphologically and serologically identical to *C. pylori* [10] with a similar cellular fatty acid profile, % (G+C) and DNA homology.

CLO infection and gastritis in non-human primates

Histopathology

In recent surveys a large proportion of the Rhesus and pig-tailed macaques investigated histologically had a chronic superficial gastritis associated with CLO colonization [4, 6, 11]. All of the affected monkeys appeared to be well and with good appetites. Few of these monkeys showed any gross pathological indicators of disease apart from mild erythema and none of the animals had peptic ulceration.

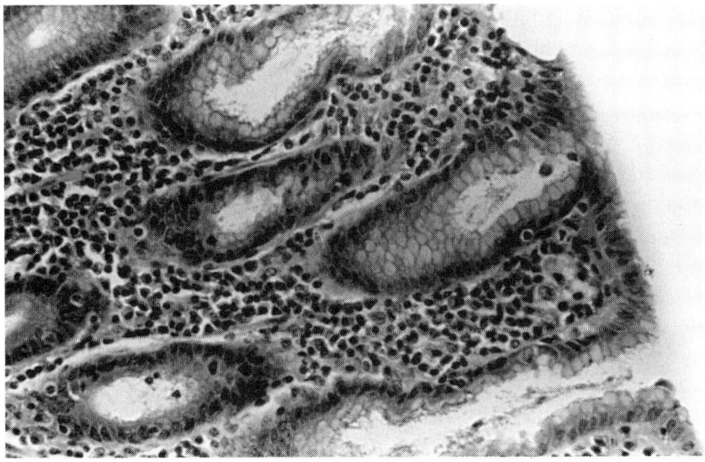

Fig. 32.2 Chronic gastritis, comprising an infiltration of plasma cells, histiocytes and lymphocytes, in the Rhesus monkey. Note the bacteria within the crypts. (Reproduced from Baskerville & Newell [4].)

The gastritis occurred throughout the cardia, body, pylorus, and duodenum but was most marked in the antrum. Atrophic gastritis with mucus cell depletion and loss of parietal and chief cells was not uncommon though intestinal metaplasia was rare. Special staining techniques showed a significant reduction in PAS-positive mucin content.

Large lymphoid follicles, some with prominent germinal centres occurred in most infected stomachs, frequently extending into the sub-mucosa. The inflammatory response comprised an intense infiltration of the lamina propria with plasma cells, lymphocytes and histiocytes (Fig. 32.2) but polymorphonuclear cells were rarely observed.

Bacteria were seen by haematoxylin & eosin, Warthin–Starry and modified Giemsa stains and by immunohistochemical staining techniques using monoclonal anti-*C. pylori* antibodies (Fig. 32.3). These bacteria were closely associated with the gastric epithelium beneath the mucus layer and in the gastric pits and glands (Fig. 32.2). Numerous bacteria were seen in the mucus by Gram stain. The numbers of organisms varied considerably but were sometimes related to the degree of gastritis [6].

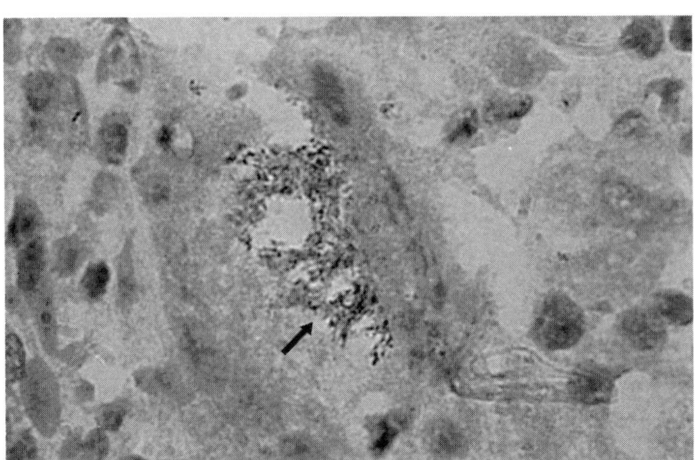

Fig. 32.3 *C. pylori*-like organisms (arrowed) in a crypt of the Rhesus monkey antrum immunohistologically stained with species-specific monoclonal anti-*C. pylori* antibody.

In all studies there was a direct relationship between those Rhesus monkeys having CLO and gastritis. Although some animals without detectable CLO had mild gastritis, all animals with CLO had moderate to severe antral gastritis.

Ultrastructural studies

By transmission electron microscopy small curved bacteria, approximately 2 μm long and 0.5 μm wide with several polar flagella [6], were found closely associated with the gastric epithelial cells and within the lower levels of the mucus layer (Figs 32.4 and 32.5). Many organisms formed intimate attachments to the plasma membrane, frequently giving the appearance of fusion between the bacterial wall and host cell membrane. Although no particular region of the bacteria was used for attachment, cup-like depressions of the cell surface formed at the site of bacterial adherence (Fig. 32.6). No phagocytosis of bacteria by inflammatory cells was observed and there was no obvious invasion of the epithelium or lamina propria [4, 6]. It seems likely that the organisms found by Dubois [12] within the intact parietal cell canaliculi and in injured parietal cells were not CLO but the spiral organism also frequently observed in non-human primates (A. Dubois, personal communication).

Immunological responses to CLO in the Rhesus monkey

Systemic immune responses

ELISA techniques developed for serodiagnosis of human *C. pylori* infections have been used

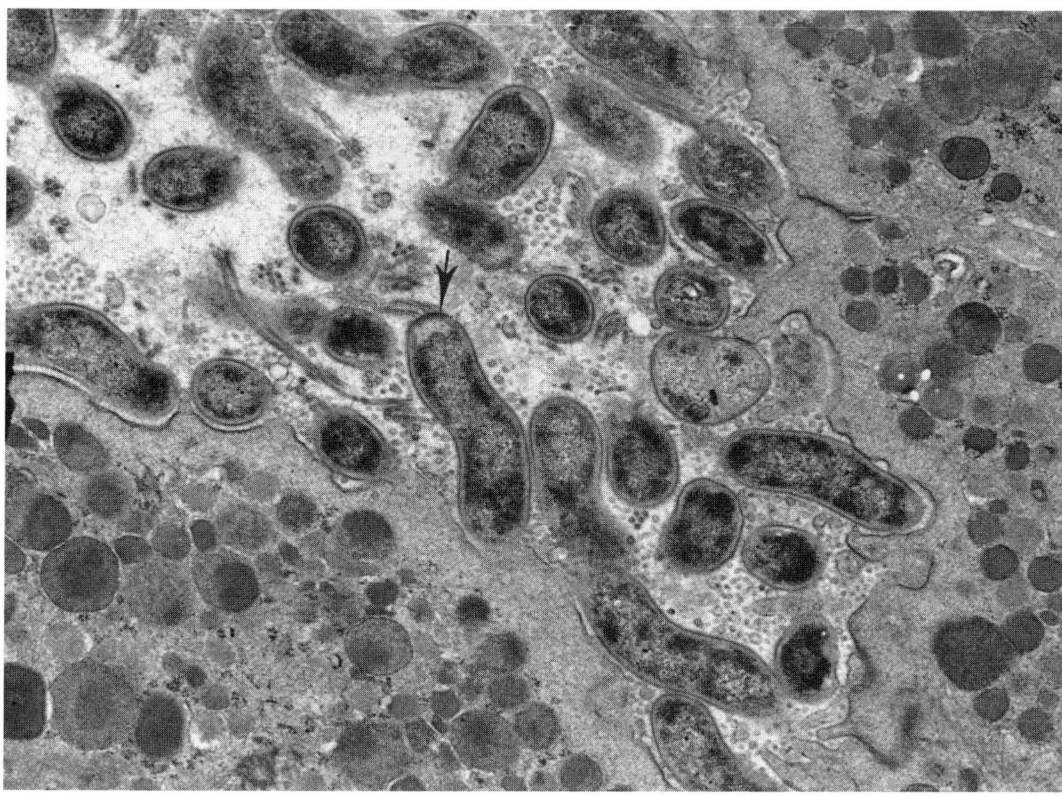

Fig. 32.4 S-shaped bacteria closely associated with the Rhesus monkey gastric epithelial cells. Note the terminal flagella (arrow). (Reproduced from Baskerville & Newell [4].)

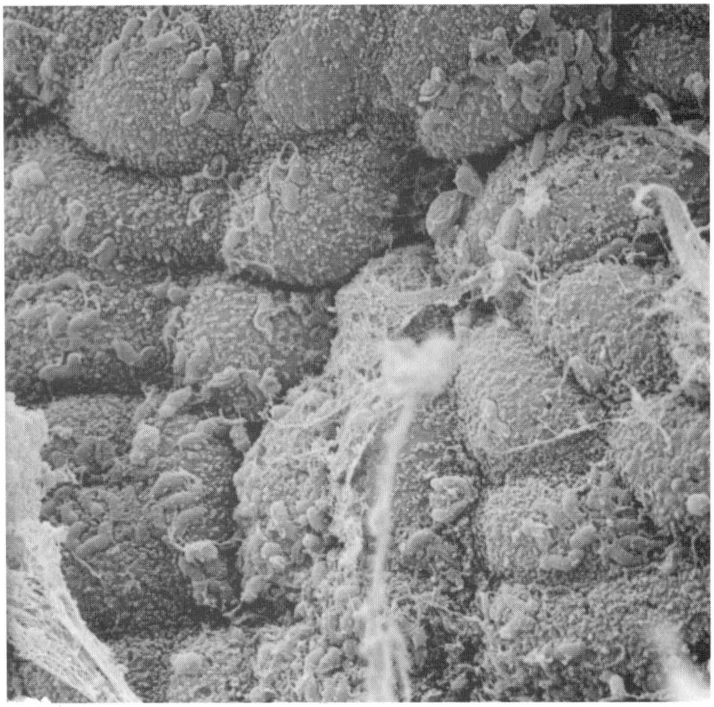

Fig. 32.5 Scanning electron micrograph of S-shaped organisms attached to the gastric epithelium of the Rhesus monkey.

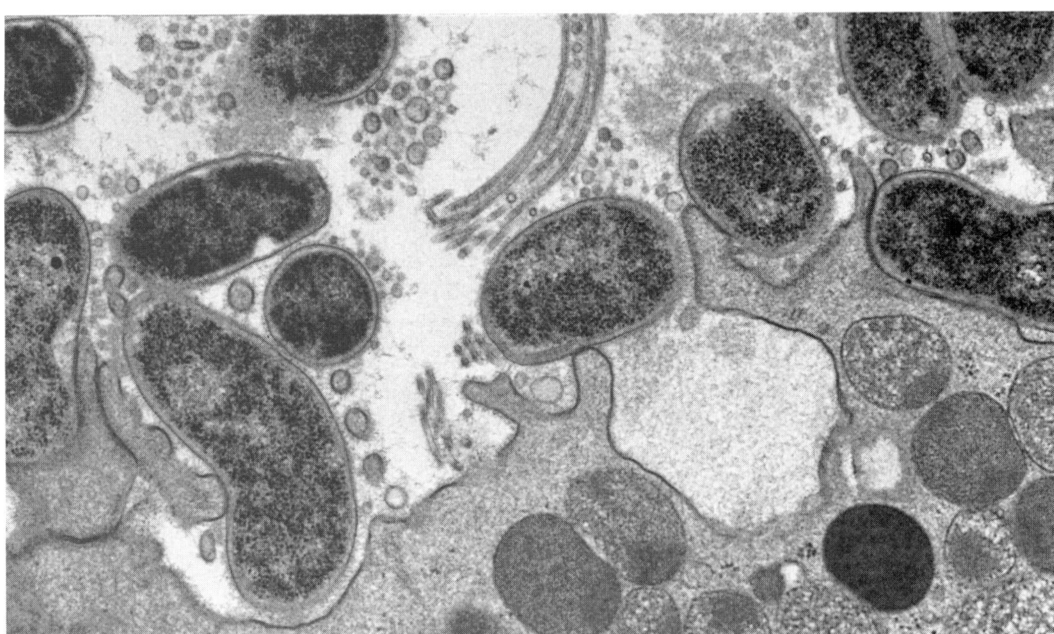

Fig. 32.6 Cup-like depressions and attachment pedestals formed at the site of bacterial adherence. (Reproduced from Baskerville & Newell [4].)

to determine the circulating antibody response of the Rhesus monkeys to their CLO. The results indicate that, in comparision with the infected humans, the specific antibody levels in the infected Rhesus monkeys were low. This is more likely to be a reflection of the efficiency of detection of the ELISA system than a poor antibody response. Anti-human immunoglobulin conjugates are routinely used to detect bound Rhesus monkey antibodies. Recent studies have shown that at least some anti-human conjugates have a very poor cross-reactivity with purified Rhesus monkey IgG (B. McBride, personal communication) so that these results should be interpreted carefully. Nevertheless, significantly higher antibody levels were observed in monkeys with CLO-associated gastritis than in those monkeys which had no evidence of CLO infection. In a survey of sera from 44 monkeys from the same breeding colony, 29 (66%) had convincing serological evidence of CLO infection (unpublished data). Because the majority of macaques in breeding colonies are infected with the thermophilic campylobacters [13] which have significant antigenic cross-reactivity with *C. pylori*, it was essential to establish the specificity of this antibody response. The antibody specificity was, therefore, confirmed by Western blotting and radioimmunoprecipitation techniques using both human *C. pylori* strains and Rhesus monkey CLO. In both cases the antigenic polypeptides detected by the CLO-positive Rhesus sera were variable but within the same spectrum of responses seen by the *C. pylori*-positive human sera.

Cellular immune responses

Immunohistological staining of cryostat sections of the gastric antral mucosa, using anti-Rhesus monkey T-cell monoclonal antibodies and anti-human plasma cell monoclonal antibodies, indicates that the inflammatory cells in the Rhesus monkey gastritis comprised mainly both mature B cells and T cells. The plasma cells and other immunoglobulin-producing B cells were generally located towards the surface of the mucosa while the T-cells tended to infiltrate from the basement membrane and accumulate in follicles. Those T-cells outside the follicles were mostly T-suppressor cells (CD8 positive).

Comparision of the CLO-associated gastritis in the non-human primate with C. pylori-associated gastritis in the human

Because the normal histopathology of the gastric mucosa of the macaques is remarkably similar to that of the human, it is not surprising that these animals have been investigated as potential models of *C. pylori*-associated gastritis. The evidence accumulated so far indicates that *M. mulatta*, *M. nemestrina* and, probably, *M. fascicularis* have a naturally-occurring infection with gastric CLO which is indistinguishable biochemically, morphologically and antigenically from *C. pylori*. Moreover, this infection is associated with a histologically-definable gastritis similar to that found in humans.

The absence of polymorphonuclear cells in the monkey CLO-associated gastritis is possibly a distinct difference from the human disease. It seems likely that in a chronic infection, present over several years, the presence of polymorphonuclear cells, usually part of the early inflammatory response to a mucosal infection, would be significantly reduced. However, there are no current data to suggest that these polymorphonuclear cells decrease with time in the human disease. In a long-term follow-up of a volunteer study, polymorphonuclear cells persisted throughout the infection (A. Morris, personal communication). Moreover, this active inflammatory response did not change in a series of patients followed over a period of up to 28 months [14]. Steer [15] found that the number of polymorphonuclear cells present in human infections related to the number of bacteria attached to the gastric mucosa. In the Rhesus monkey, polymorphonuclear cells, when present, were close to those pits or glands containing masses of bacteria. This may therefore be a major difference between the human and monkey local inflammatory responses to the infection.

The presence of some degree of gastric ulceration is not uncommon in non-human primates at necropsy [16]. Although this ulceration can be associated with a variety of infectious agents, there was no evidence of peptic ulcers in those monkeys with CLO infections. It therefore seems as though factors, such as psychological stress [16], in addition to the CLO infection are essential to the production of ulcers. This may also be true for humans.

Recently an association between atrophic gastritis and intestinal metaplasia, as precursors of gastric cancer, and *C. pylori* infection in the human has been proposed. Tumours of the stomach are the most frequently reported spontaneous tumours in this group of non-human primates [16] so the role of *C. pylori* in gastric cancers cannot be excluded on the basis of the monkey data available so far.

Experimental C. pylori infections of non-human primates

To date, only two brief reports of experimental infections have been made [11, 17]. Both suggest that human strains of *C. pylori* can colonize the antral mucosa of experimental primates, despite a pre-existing infection with the monkey CLO. Moreover, results indicate that monkeys infected with human *C. pylori* may produce an inflammatory response leading to a chronic gastritis which was more severe than that seen with the naturally occurring infection.

Other bacteria colonizing the gastric mucosa of the non-human primate

Campylobacter pylori (CLO) is not the only bacterium colonizing the gastric antrum of the macaques or baboons. Large Gram-positive rods and Gram-negative spiral organisms [10, 18] commonly occur in histological sections. The spiral organism, in particular, has generated considerable interest as it has similar morphological features to a spiral organism occasionally observed in humans with gastritis [19]. This spiral organism has not yet been cultured from either the human or non-human primate stomachs. Unlike the human infection the spiral organism is not associated with antral gastritis in the Rhesus monkey or in the baboon (unpublished data). However, this organism may invade the parietal cells causing cell damage (A. Dubois, personal communication).

The value of the macaques as an animal model of C. pylori-associated gastritis

The observations on the naturally-occurring CLO-associated gastritis in macaques indicates that these non-human primates closely mimic the human in terms of infection, histopathology and immune response. Indeed the few experimental infections performed so far confirm the usefulness of the animal as a model at least of gastritis. It would seem that as a model to investigate the role of *C. pylori* in peptic ulceration the Rhesus monkey will not be so useful.

Before experiments with these animals can be performed, however, it will either be necessary to eliminate the natural infection and cure the gastritis or to breed monkeys under specific pathogen-free conditions. The latter is evidently possible because some breeding colonies of Rhesus monkeys are apparently free of the CLO (A. Euler, personal communication). The source of infection for the monkeys, bred in captivity, is unknown. Although it has been suggested that the CLO infection in the Rhesus monkey is transmitted from human handlers [20], this seems extremely unlikely given the vicious nature of the animal. Moreover, the colony apparently free of CLO infection is housed in single cages, conditions under which close contact with handlers is much more likely to occur than in the open outdoor pens of other CLO-infected animals. Conversely, the suggestion that these animals provide an environmental source of the human infection is just as improbable.

In conclusion, the macaques, and possibly the baboons, should be considered as the most suitable *in vivo* model of the human diseases associated with *C. pylori* colonization of the stomach. Although endoscopic biopsies are easily obtained from the macaques using a paediatric endoscope so that necropsy investigations are unnecessary, it is assumed that cost, poor availability and the inconvenience of this animal will limit its usefulness as a laboratory model.

References

1 Blaser, M.J. Gastric campylobacter-like organism, gastritis and peptic ulcer disease. *Gastroenterology*. 1987, **93**, 371–383.

2 Lambert, J., Borromeo, M., Pinkard, K., Turner, H., Chapman, C. & Smith, M. Colonization of gnotobiotic piglets with *Campylobacter pylori* — An animal model? *J. Infect. Dis.* 1987, **155**, 1344.

3 Krakowa, S., Morgan, D., Krart, W. & Leunk, R. Establishment of gastritic *Campylobacter pylori* infection in the neonatal gnotobiotic piglet. *Infect. Immun.* 1987, **55**, 2789–2796.

4 Baskerville, A. & Newell, D.G. Naturally occurring chronic gastritis and *C. pylori* infection in the Rhesus monkey; a potential model for gastritis in man. *Gut.* 1988, **29**, 465–472.

5 Bronsdon, M.A. & Schoenknecht, F.D. *Campylobacter pylori* isolated from the stomach of the monkey, *Macaca nemestrina*. *J. Clin. Microbiol.* 1988, **26**, 1725–1728.

6 Reed, K.D. & Berridge, B.R. Campylobacter-like organisms in the gastric mucosa of Rhesus monkeys. *Lab. Animal Sci.* 1988, **38**, 329–331.

7 Skirrow, M.B. *Campylobacter enteritis*: a new disease. *Br. Med. J.* 1977, **ii**, 9–11.

8 McNulty, C.A.M. & Dent, J.C. Rapid identification of *Campylobacter pylori* by preformed enzymes. *J. Clin. Microbiol.* 1987, **25**, 1683–1686.

9 Newell, D.G., Hudson, M.J. & Baskerville, A. Isolation of a gastric campylobacter-like organism from the stomach of four Rhesus monkeys, and identification as *Campylobacter pylori*. *J. Med. Microbiol.* 1988, **27**, 41–44.

10 Curry, A., Jones, D. & Eldridge, J. Spiral organisms in the baboon stomach. *Lancet* 1987, **ii**, 634–635.

11 Bronsdon, M.A. & Schoenknecht, F.D. 1988. *C. pylori* isolated from non-human primate stomach: animal model studies. In Kaijser, B. & Falsen, E. (eds.) *Campylobacter IV*. University of Göteborg, Göteborg, pp. 340–341.

12 Dubois, A., Tarnawski, A., Fiala, N., Dabros, J., Stachura, J., Heman-Ackash, L.M., Montgomery, E., Lyons, F. & McQueen, C. Direct evidence for parietal cells invasion and injury by *Campylobacter pylori*-like (CPLO) organisms in Rhesus monkeys. An animal model for gastritis. *Gastroenterology*. 1988, **94**, A105.

13 Russell, R.G. & Sarmiento, J.I. 1988 Epidemiologic study of endemic *Campylobacter* in infant laboratory primates. In Kaijser, B. & Falsen, E. (eds.) *Campylobacter IV*. University of Göteborg, Göteborg, pp. 258–259.

14 Langenberg, W., Rauws, E.A., Houthoff, H.J., Oudbier, J.H., van Bohemen, C.G., Tytgat, G.N.J. & Rietra, P.J.G.M. Follow-up study of individuals with untreated *Campylobacter pylori*-associated gastritis and of uninfected persons with non-ulcer dyspepsia. *J. Infect. Dis.* 1988, **157**, 1245–1249.

15 Steer, H.W. The gastro-duodenal epithelium in peptic ulceration. *J. Pathol.* 1985, **146**, 355–362.

16 McClure, H.M., Chapman, W.L., Hooper, B.E., Smith, F.G. & Fletcher, O.J. 1978 The digestive system. In Benirschke, K., Garner, F.M. & Jones, T.C. (eds.), *Pathology of Laboratory Animals.* Springer-Verlag, New York, pp. 175–318.

17 Baskerville, A. & Newell, D.G. 1988 Gastritis associated with experimental *Campylobacter pylori* infection in the Rhesus monkey: A model for the human disease? In Kaijser, B. & Falsen, E. (eds.) *Campylobacter IV*. University of Göteborg, Göteborg, pp. 438–440.

18 Sato, T. & Takeuchi, A. Infection by spirilla in the stomach of the Rhesus monkey. *Vet. Pathol.* 1982, **19** (Suppl. 7), 17–25.

19 Dent, J., McNulty, C., Cliff, J., Wilkinson, S. & Gear, M. Spiral organisms in the gastric antrum. *Lancet*, 1987, **i**, 96.

20 Lee, A. & Hazell, S. *Campylobacter pylori* in health and disease: An ecological perspective. *Microbial. Ecology Health Dis.* 1988, **1**, 1–16.

33 Campylobacter pylori and the gnotobiotic pig

J.R. LAMBERT AND M. BORROMEO

Campylobacter pylori is found in the upper gastro-intestinal tract of humans and is associated with histological gastritis [1–3]. Whether this organism is merely a commensal on a previously altered mucosa or actually a pathogen has yet to be properly established, although evidence would support a pathogenic role.

Morris [4] and Marshall [5] have ingested the organism and reported their symptoms, signs, and histological findings in the stomach. The lack of effective and safe therapy as well as the pre-malignant potential of chronic gastritis precludes further human-challenge studies.

One of Koch's postulates to confirm the pathogenicity of a given micro-organism is that the micro-organism be cultivable outside the host and that it elicits, in other animal species, symptoms similar to those elicited in the originally infected host.

The development of a non-expensive and practical animal model may thus help to fulfil Koch's postulate and clarify the role of *C. pylori* infection in humans. A simple animal model is necessary also to study pathogenetic mechanisms as well as to assess the efficacy of therapeutic agents. Colonization of the gastro-intestinal tract of lower animals such as rodents, cats, opossums, rabbits, ferrets and monkeys with *C. pylori* has been difficult to establish [6–9]. The gastro-intestinal tract of these animals, with the exception of the monkey, is notably different from that of humans and may account for this difficulty.

Neonatal gnotobiotic piglets

We attempted to develop an animal model for *C. pylori* infection by inoculating piglets with *C. pylori*. The pig was chosen because it is a monogastric mammal with similar physiological and anatomical characteristics of the upper gastro-intestinal tract to humans [10]. Of particular relevance is the similarity between pig and human gastric mucus [11].

Three separate groups of piglets were assessed: weaned 6-week old, suckling newborn and gnotobiotic piglets. Attempts to infect the weaned 6-week old and suckling newborn piglets were unsuccessful. Other workers [12] have also been unsuccessful in infecting these groups of piglets. The gastro-intestinal tract of these piglets was heavily colonized by other bacteria. *Campylobacter pylori* may not have been able to compete with these organisms and hence was unable to colonize the gastric mucosa. Even if colonization of the gastric mucosa had occurred it was impossible to recover *C. pylori* from the cultured samples, because agar plates (both the selective and non-selective plates) were overgrown with normal flora. On Warthin–Starry silver-stained sections, the gastric mucosal surfaces and the necks and crypts of glands were covered with many different types of bacteria, making identification of *C. pylori* virtually impossible. Another possible explanation for the lack of colonization of the gastric mucosa by *C. pylori* may be the presence of mucosal immunity to organisms which were very similar to *C. pylori* [13].

Gnotobiotic piglets were successfully colonized with *C. pylori*. This model has also recently been used by Krakowka and colleagues [12]. The piglets were delivered by caesarian section, transferred and maintained in polyvinyl choride isolators throughout the experiment [14]. Inoculated and control piglets were housed in different areas to prevent contact between the two groups.

At 3 hours prior to inoculation, all piglets received 20 mg ranitidine (Glaxo, Australia). They were divided into two groups; one group was orally inoculated with 10^6 CFU suspended in alkaline peptone water (pH 8.0), while the other group received alkaline peptone water only and served as controls. Two of these piglets were re-inoculated with a further 10^6 CFU suspended in alkaline peptone water (pH 8.0) 3 days after the initial inoculation. These piglets were the last to be sacrificed. The organism used in the inoculum was isolated from the gastric mucosa of a patient diagnosed as having a duodenal ulcer at endoscopy. The cultures were used after enumeration by a standard plate-count method. Histology also showed the subject to have severe chronic active gastritis. The piglets were monitored daily for any signs of clinical illness.

The piglets were sacrificed at intervals over 24 days. All piglets were fasted overnight prior to being sacrificed. At post-mortem, tissues were taken for microbiological and histological examination from the distal oesophagus, fundus and antrum of the stomach, proximal duodenum, terminal ileum and colon. Gastric mucus, gastric juice and faeces were also collected.

Tissue samples for microbiological examination were collected in 5–10 ml sterile normal saline (0.85%), transported to the laboratory and processed within 3 hours. Tissues for histological examination were fixed in 10% phosphate-buffered neutral formalin. In the laboratory, the tissue was minced with two sterile scalpel blades and divided into three parts: one part was cultured on chocolate (horse) blood agar (Columbia agar, Oxoid CM331), the other part was transferred to brain heart infusion agar (Oxoid CM375) supplemented with 7% horse blood, 3 μg/ml vancomycin, 50 μg/ml nalidixic acid, 5 μg/ml trimethoprim and 2 μg/ml amphotericin. Plates were incubated in microaerophilic conditions for 3–6 days [15]. The final part of tissue was pressed between two slides; one slide was left to air dry, then fixed and stained with Gram stain, while the other slide was observed under phase-contrast microscopy [16].

All isolates and inoculum were tested in parallel and had the same growth characteristics, biochemical reactions and were sensitive to nalidixic acid and cephalothin (Table 33.1). Presence of constitutive enzymes was tested using an API Zym test. *Campylobacter pylori* recovered from the piglets and the inoculum had the enzymes present including acid and alkaline phosphatase, leucine arylamidase and phosphamidase (Table 33.1).

Distribution of C. pylori in the gastro-intestinal tract

Campylobacter pylori was recovered from all the inoculated piglets, but from none of the control non-inoculated piglets. Only the upper gastro-intestinal tract was colonized by *C. pylori*, with the organism recovered from the distal oesophagus, fundus and antrum of the stomach, gastric mucus and juice (the last present only on day 24) and from the first part of the duodenum. *Campylobacter pylori* was not recovered on any occasion from the terminal ileum, colon or faeces of any of the inoculated piglets (Table 33.2).

The colonization of these tissues was persistent as is seen with the human *C. pylori* infection. Bacteria were identified from at least one anatomical area of all inoculated piglets for each examination period over the 24 days of observation.

Warthin–Starry silver-stained sections from the fundus and antrum of the piglet sacrificed at day 4 showed a moderate number of curved organisms scattered within the mucus close to the apical surfaces of cells and in the necks and crypts of glands (Fig. 33.1). No bacterial invasion of submucosa and lamina propria was detected. *Campylobacter pylori*

Table 33.1 Phenotypic characteristics of *C. pylori* inoculum and isolates recovered from the upper gastro-intestinal tract of the gnotobiotic piglets and inoculum*

Test	Inoculum	Isolate recovered from gastric mucosa of piglets (n = 20)
Growth at:		
25°C	–	–
37°C	+	+
43°C	–	–
aerobic, 37°C	–	–
anaerobic, 37°C	–	–
10% CO_2 in air, 37°C	+	+
Kligler iron agar:		
aerobic	–	–
microaerophilic	–	–
Biochemical tests		
catalase	+	+
oxidase	+	+
urea hydrolysis (rapid)	+	+
require serum	+	+
nitrate reduction	–	–
acid from glucose	–	–
1% glycine	–	–
3.5% sodium chloride	–	–
1% bile	+	+
Sensitivity tests		
nalidixic acid (30 μg)	S	S
cephalothin (30 μg)	S	S
Constitutive enzymes		n = 3
acid phosphatase	+	+
alkaline phosphatase	+	+
leucine arylamidase	+	+
phosphamidase	+	+

*+ = growth/positive; – = no growth/negative; S = sensitive.

Table 33.2 Recovery of *C. pylori* from gnotobiotic piglets after oral inoculation[†]

Site	Time after oral inoculation with *C. pylori* (days)					
	3	4	7	9	14	24
Oesophagus	+	+	–	ND	–	–
Gastric fundus	+	+	+	+	+**	–
Gastric antrum	+	+*	+	+	+*	+*
Gastric mucus	ND	+	+	ND	+	+
Gastric juice	ND	ND	ND	ND	ND	+
Duodenum	–	–	+	+	–	–
Terminal ileum	–	–	–	–	–	–
Colon	–	–	–	–	–	–
Faeces	–	–	–	–	–	–

[†] + = *C. pylori* isolated; – = *C. pylori* not isolated; ND, not done.
Chronic inflammatory cells present: *focal, **moderate.

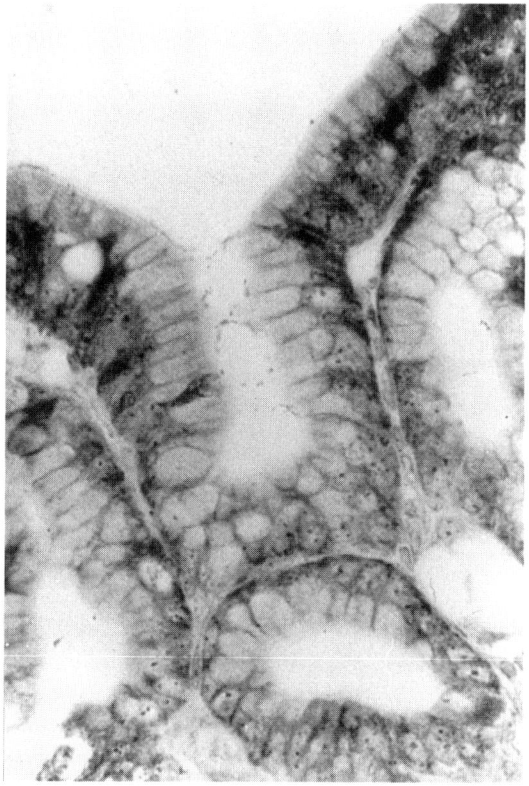

Fig. 33.1 Fundal mucosa from a piglet sacrificed 3 days after oral inoculation with a suspension of *C. pylori*, showing the curved bacteria within a gastric pit. (Warthin–Starry silver stain; × 200.)

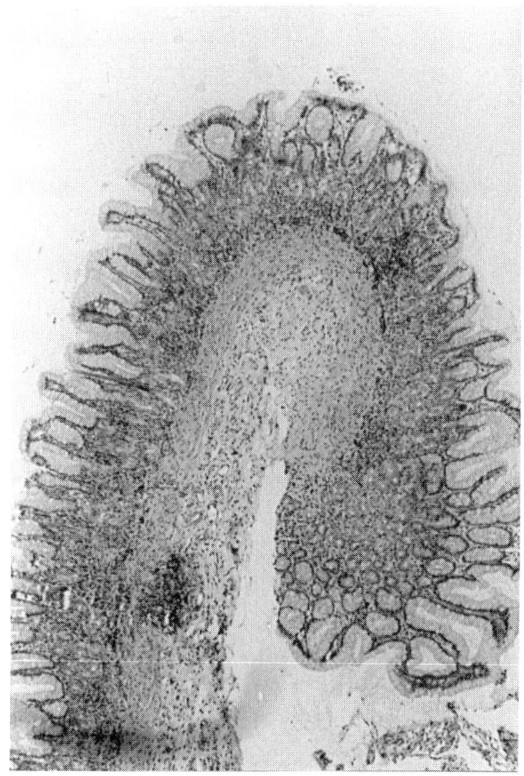

Fig. 33.2 Fundal mucosa of piglet sacrificed 14 days after oral inoculation of *C. pylori*. There is a moderate degree of inflammatory infiltrate within the lamina propria. (Haematoxylin & eosin; × 20.)

detected in non-gastric tissues, specifically the oesophagus and duodenum, did not exhibit the pathological changes characteristic of gastric epithelium colonized by the bacteria. It is likely that the *C. pylori* found in tissues adjacent to the stomach is due to colonization or contamination during the post-mortem period.

Clinical signs and macroscopic changes

Clinical disease in the piglets was not recognized except for one piglet that died 4 days after inoculation. At post-mortem, a gastric perforation was evident with no associated macroscopic change. No gastric mucosal erosions or ulcerations were seen in these animals nor were submucosal/mucosal follicles observed by day 24 in either control or non-inoculated animals. The latter have been noted in animals 4 weeks after challenge [12].

Patches of focal, chronic, inflammatory cellular infiltrate were detected in the gastric antrum of the piglets sacrificed 4, 14 and 24 days after oral inoculation with *C. pylori* (Table 33.2). Moderate, chronic gastritis was seen in the fundus of the piglet sacrificed at day 14 (Figs. 33.2 and 33.3). Occasional neutrophils were seen in the gastric mucosa of the piglet sacrificed at day 14. Cell aggregates and lymphoid follicles were present in the lamina propria and sub-mucosal regions. The severity of the chronic gastritis increased with time after inoculation.

Chronic inflammatory cell infiltrate was not seen in the sections from the terminal

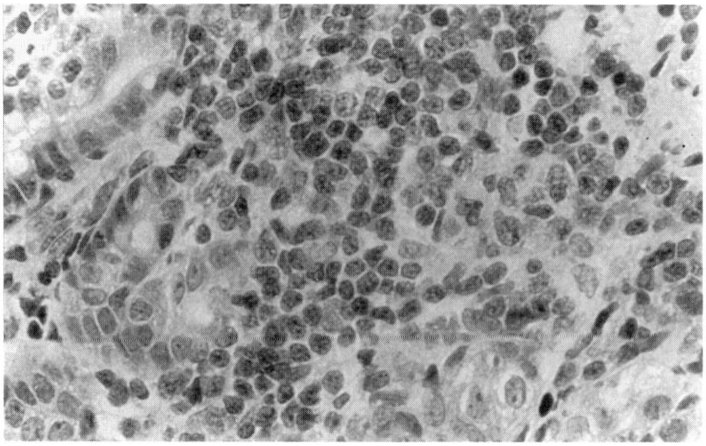

Fig. 33.3 Fundal mucosa of piglet sacrificed 14 days after oral inoculation of *C. pylori*. A chronic inflammatory infiltrate consisting of lymphocytes and plasma cells in the lamina propria is present. (Haematoxylin & eosin; × 154.)

ileum or colon of the inoculated piglets (Table 33.2). The non-inoculated piglets remained unchanged throughout the 24 days of the examination. Occasional mononuclear cells were observed along with a rare sub-mucosal lymphoid aggregate. No neutrophils were evident in gastric tissues in these piglets.

Comparison with human C. pylori infection

Colonization of the upper gastro-intestinal tract of the neonatal gnotobiotic piglets occurred after oral inoculation with the human strain of *C. pylori*. This has been confirmed by Krakowka and colleagues [12] who also successfully colonized gnotobiotic piglets with *C. pylori*. The site of *C. pylori* colonization was similar to that seen in humans [17, 18]. Bacteria colonized predominately gastric-type mucosa, on the luminal aspect of the gastric epithelium, and tissue invasion by *C. pylori* was not detected. *Campylobacter pylori* was recovered from the distal oesophagus and proximal duodenum. Human studies have shown that *C. pylori* colonizes gastric epithelium wherever it is located, including within the duodenum in areas of gastric metaplasia and, in Meckel's diverticulae, within the small intestine.

Development of a chronic infiltrate into the gastric mucosa of colonized piglets suggests an early pathogenic effect. Many subjects

Table 33.3 Summary of animal models used for assessing gastric *Campylobacter pylori*

Parameter*	Model			
	Ferret [7, 8]	Non-human primate [9, 13]	Swine [12, 13]	Gnotobiotic pig
Cost	inexpensive	expensive	expensive	expensive
Availability	readily available	not readily available	readily available	not readily available
Anatomy and physiology**	comparable	comparable	comparable	comparable
Ability to sequentially biopsy gastric mucosa	no	yes	no	no
Presence of native CLO in GIT	yes	yes	yes	no
Presence of other flora in GIT	no	no	yes	no

*CLO = Campylobacter-like organism; GIT = gastro-intestinal tract.
**Similarity to humans.

with *C. pylori* associated gastritis have active inflammation with polymorphonuclear leukocytes present [18]. Chronic lymphocytic infiltrate is commonly recognized and in one human volunteer a predominantly chronic inflammatory change was found 11 days after *C. pylori* ingestion [4]. In the early phase of infection, a transient active inflammation was seen with a predominant chronic inflammatory infiltrate. This active response dissipated and was replaced by a chronic inflammatory infiltrate. Aggregates of these latter cells increased in size and number as the infection persisted.

Mucus depletion was evident in infected animals compared to controls using PAS staining of tissue sections. This is similar to that reported by Krakowka *et al.* [12] in pigs and occurs in humans with *C. pylori* colonization of the gastric epithelium [18].

Support for a pathogenic role of *C. pylori* was also seen by the development of an antibody-specific response to the bacterium in the studies reported by Krakowka *et al.* [12]. The antibody was first detected in sera collected at 2 weeks after challenge and titres remained high through the remainder of the study period. No specific antibody response was detected in the sera of control piglets.

The use of this model may be limited due to the restricted duration that animals can be maintained in these germ-free isolators, which is usually a maximum period of 1 month. The gnotobiotic system is also labour intensive and thus expensive to develop and maintain. No clinical illness developed in the piglets nor was there any evidence of ulcers or erosions in the upper gastro-intestinal tract.

Table 33.3 briefly describes other animals which have been used as models of *C. pylori* infection. Many of these animals have organisms similar to *C. pylori* in their upper gastro-intestinal tract. Hence, *C. pylori* infection may be more difficult to establish due to the possible presence of gastric immunity to these organisms.

Despite the limitations of the neonatal gnotobiotic pigs, the model of *C. pylori*-associated gastritis was similar to that seen in humans. This model appears to be useful to characterize the pathogenicity of *C. pylori*, and to evaluate the usefulness of various antimicrobial therapies in bacterial eradication.

References

1 Marshall, B.J. & Warren, J.R. Unidentified curved bacilli in the stomach of patients with gastritis and peptic ulceration. *Lancet* 1984, **i**, 1311–1314.

2 Lambert, J.R., Dunn, K.L., Turner, H. & Korman, M.G. Effect on histological gastritis following eradication of *Campylobacter pyloridis*. *Gastroenterology* 1986, **90**, 1509. (Abstract)

3 Buck, G.E., Gourley, W.K., Lee, W.K., Subramanyam, K., Latimer, J.M. & DiNuzzo, A.R. Relation of *Campylobacter pyloridis* to gastritis and peptic ulcer. *J. Infect. Dis.* 1986, **153**, 664–669.

4 Morris, A. & Nicholson, G. Ingestion of *Campylobacter pyloridis* causes gastritis and raised fasting gastric pH. *Gastroenterology.* 1987, **82**, 192–199.

5 Marshall, B.J., Armstrong, J.A., McGechie, D.B. & Glancy, R.J. Attempt to fulfil Koch's postulates for pyloric *Campylobacter*. *Med. J. Australia* 1985, **142**, 436–439.

6 Warrelmann, M., Ehlers, S. & Hahn, H. 1987. In search of animal models for *Campylobacter pylori* induced diseases. In Menge, H., Gregor, M., Tytgat, G.N.T. & Marshall, B.J. (eds.) *Campylobacter pylori*. Springer-Verlag, Berlin. pp 40–45.

7 Fox, J.G., Edrise, B.M., Cabot, E.B., Beaucage, C., Murphy, J.C. & Prostak, K.S. Campylobacter-like organisms isolated from gastric mucosa of ferrets. *Am. J. Vet. Res.* 1986, **47**, 236–239.

8 Cave, D.R., Taylor, N., Tuczynski, C. & Fox, J.R. Campylobacter-like organisms (CLO) from man and ferret: Towards an animal model of CLO induced disease. *Gastroenterology* 1986, **90**, 1368.

9 Newell, D.C., Hudson, M.J. & Baskerville, A. Naturally occurring gastritis associated with *Campylobacter pylori* infection in the rhesus monkey. *Lancet* 1987, **ii**, 1338.

10 Jubb, K.V.F., Kennedy, P.C. & Palmer, N. 1985. *Pathology of Domestic Animals*, 3rd edn., Vol. 2, Academic Press, New York. pp. 24–49.

11 Bell, A.E., Sellers, L.A., Allen, A., Cunliffe, W.J., Morris, E.R. & Ross-Murphy, S.S. Properties of gastric and duodenal mucus: effect of proteolysis, disulfide reduction, bile, acid, ethanol, and hypertonicity on mucus gel structure. *Gastroenterology* 1985, **88**, 269–280.

12 Krakowka, S., Morgan, D.R. & Kraft, W.G. Establishment of gastric *Campylobacter pylori* infection in the neonatal gnotobiotic piglet. *Infect. Immun.* 1987, **55**, 2789–2796.
13 Jones, D.M. & Eldridge, J. 1987. Gastric Campylobacter-like organisms (GCLO) from man ('*C. pyloridis*') compared and GCLO strains from the pig, baboon and ferret. In Kaijeser, B. & Falsen, E. (eds.) *Campylobacter IV*. University of Göteborg, Göteborg. pp. 44–45.
14 Makin, T.J. & Tzipori, S. Inexpensive techniques for the production and maintenance of gnotobiotic piglets, calves and lambs. *Austral. Vet. J.* 1980, **56**, 353–358.
15 Borromeo, M., Lambert, J.R. & Pinkard, K.J. Evaluation of 'CLO-test' to detect *Campylobacter pyloridis* in gastric mucosa. *J. Clin. Pathol.* 1987, **40**, 462–463.
16 Pinkard, K.J., Harrison, B., Capstick, J.A., Medley, G. & Lambert, J.R. Detection of *Campylobacter pyloridis* in gastric mucosa by phase contrast microscopy. *J. Clin. Pathol.* 1986, **39**, 112–113.
17 Tricottet, V., Bruneval, P., Vice, O., Camilleri, J.P., Bloch, F., Bonte, N. & Roje, J. Campylobacter-like organisms and surface epithelium abnormalities in active, chronic gastritis in humans: an ultrastructural study. *Ultrastruct. Pathol.* 1986, **10**, 113–122.
18 Price, A.B., Leir, J., Dolby, J.M., Dunscombe, P.L., Smith, A., Clarke, J. & Stephenson, M.L. *Campylobacter pyloridis* in peptide ulcer disease: microbiology, pathology and scanning electron microscopy. *Gut* 1985, **26**, 1183–1188.

34 Perspective: current models of Campylobacter pylori infection and future research

M.J. BLASER

From total obscurity, *Campylobacter pylori* has emerged in a few short years as an object of considerable interest [1]. The evidence has been progressively accumulating that this essentially previously unrecognized inhabitant of the upper gastro-intestinal tract in fact may play an aetiological role in some of the most common and widespread of all human maladies. Though there have been rapid advances in our acquaintance with *C. pylori*, there nevertheless remain significant gaps in our knowledge.

That we are only at the preliminary stages of investigation is best exemplified by the dearth of information relating to the experimental pathology of *C. pylori* infection. Human experimentation is ultimately needed for medical progress to reach the consumer. However, experimental *C. pylori* infection of humans is fraught with hazard since, once acquired, the organisms persist for years, if not for life, and success at eradicating infection thus far has been partial, at best. The vast population of humans with chronic *C. pylori* infection represents one group that can be studied, but in such persons the crucial early stages of infection have already passed. Accidents of nature and *in vitro* studies offer avenues for better understanding of the pathogenesis and treatment of this disorder but they do not substitute for well-defined *in vivo* studies. Thus, it has become clear that appropriate animal models are necessary for the field of *C. pylori* research to continue its forward movement.

A number of observations have been made about gastric campylobacter-like organisms (GCLO) or *C. pylori* in animals, but each has its shortcomings in terms of applicability to humans. Despite these limitations, further work on animal models may be useful in a number of areas. The most important and amenable area is in the evaluation of antimicrobial therapy. Since most of the animals studied are chronically infected, they are analogous to infected humans, albeit with their own GCLO. It is likely that most of these organisms will resemble the predominant human GCLO, *C. pylori*, and the primate GCLO strain may in fact be identical [2]. Thus, it is likely that the pathophysiology of these infections has important concordances among several animal species. Careful serial observations of GCLO-infected animals may provide information on the relative importance of host versus environmental factors in the natural history of infection.

Nevertheless, in proper evaluation of microbial factors and host–microbial interactions, a model of experimental challenge is most useful since it can be readily controlled. However, the animal models described to date are far from optimal. The best current model of experimental *C. pylori* infection, in the gnotobiotic piglet [3, 4], is expensive, cumbersome, and limited to use in only a few facilities. As such, the development of an appropriate challenge model in a conventionally-raised animal should be a high priority for future research. One factor that should be considered is that GCLOs may have significant species adaptations.

Ideally it would be best to first challenge an animal with a homologous GCLO isolate and determine outcome before attempting challenge with a (heterologous) human isolate of *C. pylori*. Failure to produce disease in an animal with the heterologous strain does not necessarily imply that a model cannot be established in that species. This is an area that warrants re-exploration.

In summary, work on animal models of *C. pylori* infection of humans is just beginning. A high priority should be the establishment of a model of acute infection. Alternatively, study of chronically-infected animals, such as primates, may help answer questions related to natural history, sequelae, and treatment.

The future

Although much has been learned about *C. pylori* and its habitat since its discovery 5 years ago, substantial and ultimately crucial deficiencies in our knowledge of this interrelationship persist. The most important gaps may be considered as six separate questions, the answers to which will determine the significance of *C. pylori* as a human pathogen. These may be considered as follows.

1 *What is the natural history of C. pylori infection?*

While we do know that *C. pylori* may cause an acute gastritis, most infected persons have chronic gastritis. How are these two disorders related? Acute infection may be self-limited or the organisms may persist, but the determining factors have not been uncovered. Similarly, the natural history of the chronic phase of infection must be understood, since half, or more, of the adult populations in many countries are infected. Is chronic superficial *C. pylori*-associated gastritis a precursor for atrophic gastritis or gastric carcinoma [5, 6]? If progression of lesions occurs, their development is slow, evolving over decades rather than years [7, 8]. Again we need to know what are the critical determinants for this progression and whether it could be halted by courses of antimicrobial agents. Establishing simple ways of screening for *C. pylori* infection, such as serological assays with sufficient diagnostic accuracy [9], may allow for large-scale prospective or retrospective studies to answer some of these questions.

2 *What is the relation of C. pylori-associated gastritis to non-ulcer dyspepsia?*

Campylobacter pylori gastritis at present is a pathological entity searching for symptoms. Population-based studies indicate that most *C. pylori*-infected persons are asymptomatic. Among patients with non-ulcer dyspepsia, most are infected with *C. pylori* but some are not. Does *C. pylori* play a role in dyspeptic symptoms in the absence of ulcers? One partial solution to this problem may be to define terms very strictly. Non-ulcer dyspepsia represents a collection of symptoms and temporal sequences; it is a syndrome rather than a disease. Perhaps *C. pylori*-gastritis is associated with one specific symptom complex; careful categorization of clinical complaints may permit such a relationship to be seen. Along the same lines, many of the studies indicating high prevalence of *C. pylori* infection in the 'normal' population have been based on blood donors. Future investigations should note which of such 'normal' subjects have particular dyspeptic symptoms. As well as helping clarify the relationship of *C. pylori* infections to symptoms, such work also may help determine the prevalence of these symptoms in the normal population. Alternatively, evaluation of symptoms after temporary or permanent eradication of *C. pylori* using antimicrobial agents may permit definition of those that are related to the infection.

3 *What is the relation of C. pylori infection and duodenal ulceration?*

Campylobacter pylori-infection is clearly a marker of high risk for duodenal ulcer [10], and of likelihood of ulcer relapse after treatment [11]. Is this organism just a marker for a milieu favourable for ulcer development or does it contribute to pathogenesis as well? Certainly the hypothesis, first raised by Wyatt, that gastric metaplasia in the duodenum is the necessary link between *C. pylori* infection and duodenitis or ulcer is appealing [12]; however, this has not yet been confirmed. Alternatively, *C. pylori* infection of the stomach could lead to loss of cytoprotective factors elaborated by gastric tissue. In any case, the question of why most *C. pylori*-infected persons do not develop ulcers must be resolved. Clearly this question is of critical importance and requires rapid resolution.

4 *What is the relation of C. pylori infection and gastric ulceration?*

While gastric ulcer also is correlated with *C. pylori* infection, the association is not as strong as for duodenal ulcer. This dichotomy could suggest that duodenal ulcer disease commonly has only one cause while there are multiple aetiologies for gastric ulceration. Future analyses of gastric ulceration will need to carefully stratify the cases by whether salicylates or other barrier-breaking agents were used, by location of the ulcer, by acid production status, or by whether gastric carcinoma is present. *Campylobacter pylori* infection may be associated with one of the different types of gastric ulcer but not with others. Since high acid secretion is not associated with most gastric ulcers, any presumed pathogenic role of *C. pylori* must have a different basis than for duodenal ulcer. One hypothesis to examine is that gastric ulcer formation is dependent on extent or chronicity of gastritis such that only those persons with severe inflammation progress to ulcer formation. No doubt other hypotheses will be forthcoming.

5 *Are there environmental stimuli that affect C. pylori infection?*

Does diet affect the natural history of *C. pylori* infection? Do any drugs predispose to infection, such as acid-lowering agents, including histamine (H_2) blockers or marijuana? Does smoking remove cytoprotective agents necessary to keep inflammation in check? *Campylobacter pylori* infection is common among children in the developing world but rare among children in the developed countries. This difference could be due to dissimilar transmission patterns or could relate to predisposing environmental factors that permit the establishment of infection.

6 *How may C. pylori be eradicated from infected hosts?*

Ultimately, if *C. pylori* becomes well-established as a significant contributor to gastroduodenal inflammation, then antimicrobial treatment will become necessary under certain circumstances. Treatment might be limited to those with symptomatic infection, as in the case of recurrent ulcer, or severe non-ulcer dyspepsia, or may be indicated for the general population. Obviously the solution to this problem will be determined by the answers to the previous questions. If longstanding *C. pylori* infection is an important risk factor for the development of peptic ulcer disease, atrophic gastritis, or gastric carcinoma, then attempts at eradication of the organism becomes a necessary goal. If eradication cannot be achieved, intermittent suppression of the organism may be worthwhile.

The studies reported to date indicate that effective therapy will not be simple. Although the organism can be readily suppressed by single-drug therapy [13, 14], the recurrence rate approaches 100%. Combination treatment appears to offer longer periods of remission, but the long-term outlook is uncertain. One inherent difficulty in treatment is the privileged niche in which the organism lives. Neither invading tissue nor solely in the gastric lumen, *C. pylori* resides in the mucus gel. Luminal agents such as bismuth salts cannot adequately penetrate the gel or tissue to reach all organisms. Similarly, administration of systemic agents such as antibiotics cannot lead to sufficiently high levels in the gel for the organism to be removed. It seems likely that any effective therapy must be based on an agent that penetrates mucus in sufficiently high levels to permit bacteriolysis.

Combination therapy with a luminal agent and an agent that achieves satisfactory serum levels has theoretical appeal, but the benefits of such regimens must be weighed against the potential toxicity of a multi-drug regimen. Eradication of the organism should be the primary goal of therapy but alternatives exist. Long-term, low-dose, continuous or intermittent suppressive therapy may be worthwhile. At present, there are too many variables to consider before an optimal treatment trial can be started. As clinical investigators bring together fragments of information, a picture may emerge so that future treatment efforts are at least aimed in the right direction.

Research directed toward the key questions concerning *C. pylori* will require intensive study for many years. However, research is a collaborative process of building fact upon fact until a structure emerges. Progress in one area may have an unforeseen but salutary effect in an entirely different arena. The current flurry of activity in both clinical and basic investigation of *C. pylori* infections bodes well for answering the many questions before us.

References

1 Blaser, M.J. Gastric campylobacter-like organisms, gastritis, and peptic ulcer disease. *Gastroenterology*, 1987, **93**, 371–383.

2 Baskerville, A. & Newell, D.G. Naturally occurring chronic gastritis and *C. pylori* infection in the Rhesus monkey: a potential model for gastritis in man. *Gut*. 1988, **29**, 465–472.

3 Krakowka, S., Morgan, D.P., Kraft, W.G. *et al.* Establishment of gastric *Campylobacter pyloridis* infection in the neonatal gnotobiotic piglet. *Infect. Immun.* 1987, **55**, 2789–2796.

4 Lambert, J.R. Borromeo, M., Pinkard, K.J. *et al.* Colonization of gnotobiotic piglets with *Campylobacter pyloridis* — an animal model? *J. Infect. Dis.* 1987, **155**, 1344.

5 Siurala, M., Isokoski, M., Varis, K. *et al.* Prevalence of gastritis in a rural population: Bioptic study of subjects selected at random. *Scand. J. Gastroentrol.* 1968, **3**, 211–223.

6 Siurala, M. & Vuorinen, Y. Follow-up studies of patients with superficial gastritis and patients with a normal gastric mucosa. *Acta Med. Scand.* 1963, **173**, 45–52.

7 Siurala, M., Varis, K. & Wiljasalo, M. Studies of patients with atrophic gastritis: A 10–15 year follow-up. *Scand. J. Gastroenterol.* 1966, **1**, 40–48.

8 Kekki, M., Villako, K., Tamm, A. *et al.* Dynamics of antral and fundal gastritis in an Estonian rural population sample. *Scand. J. Gastroenterol.* 1977, **12**, 321–324.

9 Perez-Perez, G.I., Dworkin, B.M., Chodos, J.E. & Blaser, M.J. *Campylobacter pylori* antibodies in humans. *Ann. Intern. Med.* 1988, **109**, 11–17.

10 Johnston, B.J. Reed, P.I. & Ali, M.H. *Campylobacter*-like organisms in duodenal and antral endoscopic biopsies: Relationship to inflammation. *Gut* 1986, **27**, 1132–1137.

11 Coghlan, J.G., Gilligan, D., Humphrey, H. *et al.* *Campylobacter pylori* and recurrence of duodenal ulcers — a 12 month follow-up study. *Lancet* 1987, **ii**, 1109–1111.

12 Wyatt, J.I., Rathbone, B.J. Dixon, M.F. *et al.* *Campylobacter pyloridis* and acid-induced gastric metaplasia in the pathogenesis of duodenitis. *J. Clin. Pathol.* 1987, **40**, 841–848.

13 Rauws, E.A.J., Langenberg, W.L., Houthoff, H.J. *et al.* *Campylobacter pyloridis*-associated chronic active antral gastritis. *Gastroenterology* 1988, **94**, 33–40.

14 Gilman, R., Leon-Barua, R., Ramirez-Ramos, A. *et al.* Efficacy of nitrofurans in the treatment of antral gastritis associated with *Campylobacter pyloridis*. *Gastroenterology* 1987, **92**, 1405. (Abstract).

Index

Page references in *italics* refer to figures or tables